PROJECT MANAGEMENT

PROJECT MANAGEMENT

Strategic Design and Implementation

David I. Cleland
Lewis R. Ireland

Fourth Edition

McGRAW-HILL
New York Chicago San Francisco Lisbon London Madrid
Mexico City Milan New Delhi San Juan Seoul
Singapore Sydney Toronto

Library of Congress Cataloging-in-Publication Data

Cleland, David I.
Project management : strategic design and implementation / David I. Cleland, Lewis R. Ireland—4th ed.
p. cm.
Includes bibliographical references and index.
ISBN 0-07-139310-2
1. Project management. I. Ireland, Lewis R. II. Title.

HD69.P75 C526 2002
658.4′04—dc21 2002070294

McGraw-Hill

A Division of The ***McGraw-Hill*** *Companies*

2 3 4 5 6 7 8 9 0 DOC/DOC 0 8 7 6 5 4 3

ISBN 0-07-139310-2

The sponsoring editor for this book was Larry Hager, the editing supervisor was Caroline Levine, and the production supervisor was Sherri Souffrance. It was set in the HB1 design in Times Roman by Wayne Palmer of the McGraw-Hill Professional's Hightstown, New Jersey composition unit.

Printed and bound by R. R. Donnelley & Sons Company.

This book was printed on recycled, acid-free paper containing a minimum of 50% recycled, de-inked fiber.

CONTENTS

Part 2 The Strategic Context of Projects

Chapter 3. When to Use Project Management 63

PM 513

Chapter 4. The Strategic Context of Projects 105

PM 513

Chapter 5. The Board of Directors and Major Projects 133

PM 613

Part 5 Interpersonal Dynamics in the Management of Projects

Part 6 The Cultural Elements

Part 7 New Prospects

PREFACE

In today's environment, certainty of change is without precedent. Although the practice of project management has been with us for centuries, only in the past few decades has an expression in the literature of a philosophy and process of managing projects taken place. In recent years, there has been a growing interest in the use of projects as building blocks in the strategic management of the enterprise. This book's fourth edition continues to hold the commanding lead, taken by previous editions, in showing how to use projects for the management of product, service, and organizational process change to prepare the organization for its competitive future.

Today, project management has reached a maturity level in which it is applied to many uses. It is the principal means by which operational and strategic changes are managed in contemporary organizations, in both for-profit and not-for-profit enterprises. This growing maturity of project management has fostered the use of "nontraditional" project teams in the management of change. Benchmarking teams, concurrent engineering teams, reengineering teams, and self-managed production teams are a few of these nontraditional teams that are gaining popularity in strategies used by today's managers.

Formal project management emerged in an unobtrusive manner in the late 1950s and began taking on the characteristics of a distinct discipline. No one can claim to have invented project management. Its beginnings can be found in the creation and construction of many different historical architectural artifacts, such as in cathedrals, canals, highways, and in voyages of discovery and military campaigns, to name a few. In more recent times, project management has gained special attention in the military weapons and systems development businesses. The modern-day origins of project management concepts and techniques can be found in such large-scale ad hoc endeavors as the Manhattan Project and the Polaris submarine program, in large construction initiatives, and in the use of naval task forces.

The emergence of professional societies has helped stimulate the development and dissemination of project management knowledge and skills. There are many such professional societies in existence—with a commanding lead being taken by the Project Management Institute (PMI®). The growth of PMI in recent years in many ways reflects the increasing interest in the use of project management. PMI has over 90,000 members drawn from a wide variety of different industries and organizations.

Many books and articles that have been published about project management treat the subject as if it were a nearly separate entity in the management of organizations. Little is found in this literature that puts project management in its proper place in the strategic management of organizations. This book tries to do just that. Our study of how contemporary organizations deal with change reinforced our belief that, in spite of an abundance of books and periodical literature, there was a serious lack of theoretical and practical literature that placed project management in the context of the design and execution of organizational strategies.

We found that too many leaders and managers, particularly at the upper and senior levels, were inclined to view project management as a special case of management—a minor departure from the proper or expected ways of managing the organization. Too often these managers failed to appreciate the strategic role that projects can play in the management of their organizations. Up until the last few years, many managers tended to tolerate rather than fully accept project management as the way to enhance organizational effectiveness. This caused project managers, functional managers, and project professionals to see themselves in ambiguous roles in supporting project initiatives. However, once upper and senior managers recognized project management for what it is—a philosophy and process for managing change—they embraced the use of project management in the enterprise.

In this fourth edition of *Project Management: Strategic Design and Implementation*, special care has been taken to update the material in each chapter. New material has been added that has emerged as part of the growing literature supporting project management, such as earned value, project management maturity, nontraditional teams, project partnering, and the outsourcing of project management, to name a few. In some cases, the growing literature in project management is adequately described in the text. In some cases, an area is only mentioned with guidance on where the reader can find expert references on the subjects. To give a detailed description of all of the emerging areas of thought that relate to project management would greatly lengthen the book to an unwieldy size.

Updated examples of the use of project management in many different contexts have been added. End-of-chapter material has been strengthened through the use of detailed chapter summaries, additional sources of information, discussion questions, user checklists, project management principles, a project management situation, and a student/reader assignment for further investigation of project management areas. Sufficient end-of-chapter material exists to support the use of the text in undergraduate and graduate programs as well as in short training courses. The book is valuable as well for the professional practitioners, who want to increase their knowledge and skills in the practice of project management. Upper-level and senior managers will find an abundance of information that can be used to enhance their use of project initiatives in the management of the enterprise.

We believe that this book is both "student" and "user" friendly!

David I. Cleland, Ph.D.
Lewis R. Ireland, Ph.D.

ACKNOWLEDGMENTS

The development and publication of this fourth edition required the cooperative effort of many people, who served as advocates, consultants, and facilitators. Our deep appreciation goes to all of these "stakeholders" who contributed in many important ways to this book. Much appreciation is due our students, clients, and project-community associates who allowed us to talk with them about the theory and practice of project management.

Our thanks to Dr. Bopaya Bidanda, chairperson of the Industrial Engineering Department and Dr. Gerald D. Holder, dean of the School of Engineering at the University of Pittsburgh, who provided an intellectual and supportive environment where projects such as this book can be pursued.

We thank Claire Zubritzky of the Industrial Engineering Department, who provided outstanding administrative support in the development of this book. We also thank Lisa Bopp of the Industrial Engineering Department, who provided supporting administrative assistance when needed on this project. Ouida Ireland is recognized for her support in reviewing drafts to identify unclear or confusing sentences, all the while encouraging the authors in writing this book.

We recognize and thank Dr. Hans Thamhain of Bentley College, who prepared Chapter 18 of this book. Hans is one of the notables of the project management community who has made outstanding contributions to project management literature. We are pleased to have him on our team.

We acknowledge and recognize the support of our wives and families who provided logistic and emotional support, as well as tolerated our absences from the family activities during the preparation of the manuscript for this book.

INTRODUCTION

This fourth edition of *Project Management: Strategic Design and Implementation* has been put together with the objective of further contributing to the project management knowledge of professionals at all levels of responsibility and to the student aspiring to be a part of a project team.

Managers and professionals engaged in project management, who desire to be more effective contributors in their organization's success, will find this book useful. The format of the book is adaptable to many different uses. Readers may read sections and topics in whatever order best suits their interests. The 7 parts and 22 chapters provide an easy division of information for readers. In Figure I.1 these parts and chapters of the fourth edition are portrayed in the context of strategic management and project management and are briefly described in the following text.

Part 1, Introduction, consists of two chapters that introduce project management and describe the management processes that are involved in the practice of this discipline. Here, a historical perspective shows early efforts that could only be called projects in our modern interpretation.

Part 2, The Strategic Context of Projects, shows how projects are used in both the strategic and operational management of the enterprise. The theme emphasized in this section is how projects are building blocks in preparing the enterprise for its uncertain future. When to use project management, the strategic context of projects, strategic issues, and the role of stakeholders and boards of directors are examined in this part.

Part 3, Organizational Design for Project Management, looks at how to organize human resources, project authority, and project management maturity. Management of people and gain in project maturity capability are important to the achievement of enterprise goals.

Part 4, Project Operations, reviews project planning, information systems, project control, project termination, and earned value systems. Foundation areas of project operational framework provide views of what is needed to successfully manage a project.

Part 5, Interpersonal Dynamics in the Management of Projects, presents information on project leadership, communications, and effective project teamwork. Because only people can make things happen, various ways an individual or a team is motivated comes into play.

Part 6, The Cultural Elements, reviews a strategy of continuous improvement through projects, and cultural considerations in project management. A new aware-

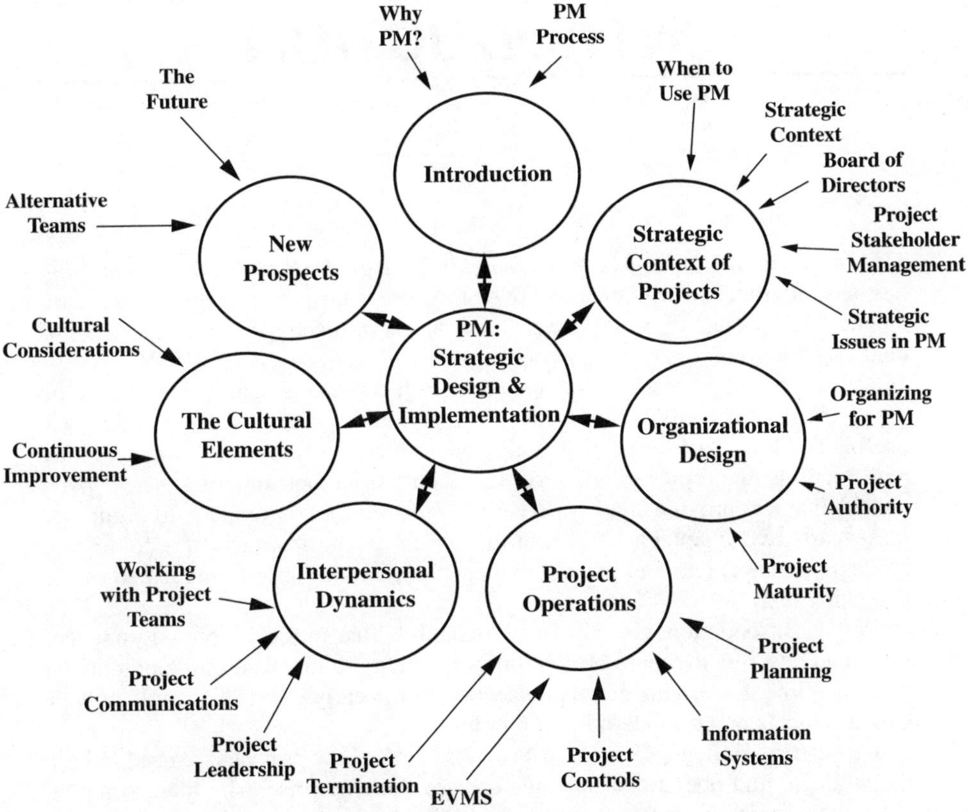

FIGURE I.1 Strategic management context of project management.

ness is raised in project management that cultural aspects are perhaps as important as the "traditional" cost, schedule, and technical performance issues.

Part 7, New Prospects, considers the alternative uses to which project teams can be put. The final chapter of the book speculates on what the future of project management might be. The challenges of the future give rise to new and unique applications for project management.

The reader can do further reading about project management from the additional sources of information noted at the end of each chapter. Then, too, each reference cited in the text of the book can be a valuable source of additional information.

We wish readers much success in their project management work!

PROJECT MANAGEMENT

P • A • R • T • 1

INTRODUCTION

CHAPTER 1
WHY PROJECT MANAGEMENT?

"There is nothing permanent except change."

HERACLITUS OF GREECE, 513 B.C.
Rogers' Student's History of Philosophy

1.1 INTRODUCTION

It has been said that there is nothing as powerful as an idea whose time has come. Project management, an evolving field of theory and practice, has emerged slowly as the field of management has come forth. Since the 1950s there has been acceleration in the development of the theory, literature, and practice of project management. Today there is a sufficient body of knowledge about project management so that this discipline has taken an important position in the lexicon of management and in the practices of modern organizations.

In this first chapter of the book an historical perspective of projects will be given, augmented with examples of the early literature. The evolution of project management from organizational liaison devices will be explored. In addition a preliminary philosophy of the discipline will be offered. As a mark of the growing maturity of project management a description will be provided of the Project Management Institute, the commanding professional society in the field. The narrative of the chapter starts with a definition of just what a project is, followed by the way in which projects reflect how past societies have coped, in part, with changes in their environment.

This is a book about project management, a "field of study" and practice that has evolved over decades and now promises to take its rightful place in the lexicon of management and in contemporary organizations. In this chapter, the overall concept of a project will be presented along with some examples of historical and contemporary projects.

Just what is a *project?* Two early definitions are helpful. For example, it is "any undertaking that has definite, final objectives representing specified values to be used in the satisfaction of some need or desire."[1]

[1]Ralph Currier Davis, *The Fundamentals of Top Management* (New York: Harper, 1951), p. 268.

Newman, Warren, and McGill defined a project and described its value as

> simply a cluster of activities that is relatively separate and clear-cut. Building a plant, designing a new package, soliciting gifts of $500,000 for a men's dormitory are examples. A project typically has a distinct mission and a clear termination point.
>
> The task of management is eased when work can be set up in projects. The assignment of duties is sharpened, control is simplified, and the men who do the work can sense their accomplishment.
>
> [A project might be part of a broader program, yet the] chief virtue of a project lies in identifying a nice, neat work package within a bewildering array of objectives, alternatives, and activities.[2]

The authors define a project as a combination of organizational resources pulled together to create something that did not previously exist and that will provide a performance capability in the design and execution of organizational strategies. Projects have a distinct life cycle, starting with an idea and progressing through design, engineering, and manufacturing or construction, through use by a project owner.

Four key considerations always are involved in a project:

- What will it cost?
- What time is required?
- What technical performance capability will it provide?
- How will the project results fit into the design and execution of organizational strategies?

The questions noted above must be answered on an ongoing basis for each project in the enterprise that is being considered, or for projects on which organizational resources are being used. The answers to these questions must also be evaluated in the context of the project's fit into the organization's operational (short-term) or strategic (long-term) strategies. Figure 1.1 portrays these considerations.

Project management and strategic management are highly interdependent. In the material that follows, this interdependence will be presented.

1.2 STRATEGIC MANAGEMENT—THE PROJECT LINKAGES

One of the areas of management this book addresses is the strategic context in which projects are found in contemporary enterprises. In the material that follows, *choice elements* are found in the theory and practice of strategic management—*the management of the enterprise as if its future mattered.* A key choice element

[2]William H. Newman, E. Kirby Warren, and Andrew R. McGill, *The Process of Management: Strategy, Action, Results,* 6th ed. (Englewood Cliffs, N.J.: Prentice Hall, 1987), p. 140. Reprinted by permission.

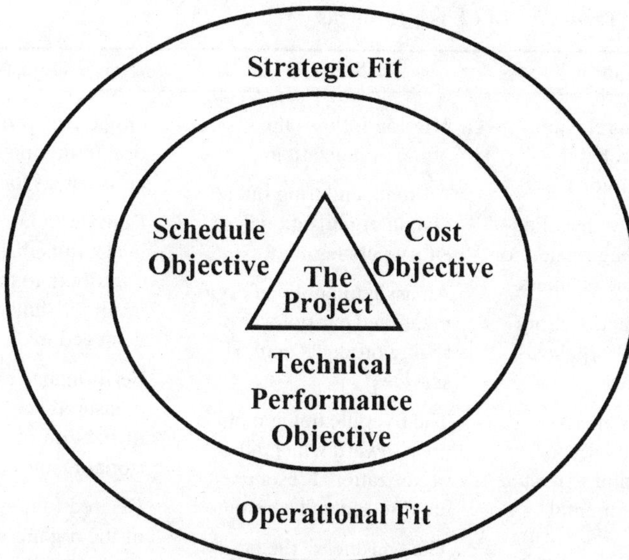

FIGURE 1.1 Interrelationships of project objectives and organizational fit.

of strategic management is the emerging projects that are building blocks in the design and execution of strategies for the enterprise's future.[3]

In the management of an enterprise as if its future mattered, nine key choice elements are involved. These choice elements for the enterprise are:

- Vision
- Mission
- Objectives
- Goals
- Strategies
- Programs
- Projects
- Operational plans
- Organizational design

These choice elements provide for the major performance standards by which enterprise resources will be identified, selected, committed, and reviewed in the enterprise for survival and growth in its future products, services, and organizational processes. These choice elements are defined below and portrayed in Table 1.1 and Fig. 1.2.

---

[3]The term "choice elements" was previously used in William R. King and David I. Cleland, *Strategic Planning and Policy* (New York: Van Nostrand Reinhold, 1978), chap. 5.

TABLE 1.1 Taxonomy of Choice Elements

Vision	Mission	Objectives
Intelligent and relevant foresight of probable future opportunities A mental image that anticipates something related to the future of the business Discernment and sharing of values to all organizational members Reflects assessment of enterprise strengths or weaknesses, and expected future environmental conditions *Example:* Customers must have a quality experience and must be pleased, not just satisfied. (COMPUTER COMPANY)	Mission follows the vision of the organization A broad, enduring intent that an organizational entity pursues An assignment to an organizational entity for providing products and/or services The overall strategic purpose toward which all organizational resources are directed and committed The "business" the organization is in What the organizational entity is and what it intends to become The symbol around which all organizational effort is focused Supported directly by objectives *Example:* "...to be the number one aerospace company in the world, and among the premier industrial firms, as measured by quality, profitability, and growth." (AIRCRAFT MANUFACTURER)	Objectives provide direction to the mission and define areas for pursuance Long-term target and critical results that directly contribute to mission accomplishment to be achieved in an enterprise Performance criteria to be measured and achieved in utilization of organizational resources Desired future destination of the organizational entity stated in qualitative and/or quantitative terms Performance results (financial, productivity, market share, etc.) and qualitative results (image, personnel development, research) are included *Example:* Lead in state-of-the art technology in our product lines. (COMPUTER COMPANY)

TABLE 1.1 Taxonomy of Choice Elements (*Continued*)

Goals	Strategies	Programs	Projects
Specific, time-sense milestones to be accomplished in using organizational resources Attainment of a goal signifies that progress has been made toward attaining organizational objectives in support of the mission Include quantitative performance goals (e.g., 15 percent ROI by a specific date) and qualitative goals (product development, project/program completion) Basic component for measuring progress toward an organization's desired end results Successful completion of a project means that one or more goals have been achieved for the organization *Example:* "We will initiate our basic research program strategy by January 1, 2002. (R&D ORGANIZATION)	Design of means through use of resources to accomplish organizational purposes Development of action plans for setting coordinated use of resources Designation and implementation of programs, projects, policies, procedures, and other protocols for use of resources Designation of organizational design initiatives to facilitate use of resources Selection among alternatives of the "best" method for implementing processes to achieve goals *Example:* Commit 5 percent of corporate earnings to product applied research programs. (INDUSTRIAL EQUIPMENT MANUFACTURER)	Resource-consuming element of organizational capabilities that have a common purpose An output, which serves the broad purpose of similar objectives of the enterprise Defined in the context of what the enterprise is trying to achieve (i.e., vision, mission, objectives, goals, strategies) Usually cast in context of extended enterprise horizons: say, 5 to 10 years Usually implemented through projects that define the specific work to be accomplished *Example:* Productivity improvement program. (ELECTRICAL EQUIPMENT MANUFACTURER)	Ad hoc resource-consuming initiatives having cost schedule and technical performance results that support organizational purposes Basic building blocks of organizational strategies to accomplish work and meet goals Any ad hoc undertaking that has definitive, final objectives to deal with product, service, or organizational process change Provide a philosophy and strategy for the management of change in the enterprise The primary means of planning and executing work within the organization *Example:* Highway construction project. (CONSTRUCTION PROJECT COMPANY)

Facilitative Services

Policies Procedures Protocols Systems

FIGURE 1.2 Choice elements of strategic management.

Vision A vision is a mental image of what could be anticipated for the enterprise's future—such as becoming a world-class competitor. One company defined its vision to be a "world-class competitor—and to keep it that way....We have programs in place to do just that such as a total quality management process whereby we *live* quality." Another company included in its vision statement: "We will enhance our competitiveness by being first in the development of advanced technology that supports our world-class products and services."

A telecommunications company conceived its vision in the following fashion: "As we enter the new millennium, AT&T is successfully transforming itself from a domestic long distance company to an any-distance, any-service global company. We've made the right strategic decisions, invested in the right assets, and have the right people to get the job done."[4]

Mission The mission of an enterprise answers the basic question: What business are we in? One project-driven firm defined its mission in the following way: "We are in the business of designing, developing, and installing energy management systems and services for the domestic industrial market." The Boeing Company, which uses project management widely, describes its mission in the following fashion: "to be the number one aerospace company in the world, and among the premier industrial firms, as measured by quality, profitability, and growth." Boeing uses projects as building blocks in the design and execution of strategies to fulfill its mission.

[4]"Delivering," Letter to AT&T Shareholders, Annual Report, 1999.

Objective An objective is a statement of the ongoing purposes in the enterprise that must be carried out to support the organizational mission. A computer company defines one of its objectives as "leading the state-of-the-art in its products and services." Another company defines its objectives as achieving a compounded earnings growth rate of 15 percent and a 20 percent return on capital. The major part of this strategy is to be the leader in providing scientists and educators worldwide with laboratory product and service systems created through technology, integrity, and a commitment to excellence. Objectives directly support the enterprise's mission. Thus a failure to maintain an organizational objective can put the accomplishment of the enterprise's mission in jeopardy.

Goal A goal is a specific achievement in the satisfaction of enterprise objectives. As a performance measurement for progress in the use of resources to support corporate purposes, a goal has a specific time element. One company defined its goals as the realization of a certain percentage of return on invested assets by a specific date. Another company stated one of its principal goals as follows: "We intend, by the end of 1999, to complete the construction of a new manufacturing facility, which will complete the transition begun in 1997 from a predominately R&D services company to an industrial manufacturer."

Further distinction between an objective and a goal is offered. An objective is an aspiration to be working toward on a continuous basis. A goal is an achievement to be realized in future times. Further differentiation between these two terms can be distinguished using a few measures:

- *Time frame* An objective is timeless and unending; a goal is time-based and intended to be overrun by subsequent goals.
- *Specificity* Objectives are usually stated in general terms, dealing with the attainment of desirable conditions in the future. A goal is much more specific, stated in terms of a particular result to be expected at a specific time point. Objectives are open-ended, and are sought on a continuous basis, regardless of the time element. Goals are milestones.
- *Focus* Objectives are usually stated in terms of some ongoing achievement in a relevant external environment, whereas goals are internally focused, whose achievement can be measured by a selected date. Objectives are often stated in the context of achieving leadership or recognition in certain desirable conditions for the enterprise. A goal implies a specific resource commitment to be used by a certain date.
- *Measurement* Both objectives and goals can be stated and measured in quantitative terms or qualitative terms. A company that states one of its objectives in terms of "achieving a compound rate of growth in earnings per share, placing its performance in the top 10 percent of all corporations" may attain that benchmark in 1 year, but it is timeless and means that the ensuing years must reflect the same performance unless changed. A goal that is quantified is expressed in absolute terms—a president of a company could state that the enterprise would "achieve half of its sales revenue from a particular industry by 2002." The achievement of that goal can be specifically measured. Once attained, the goal would be restated for the ensuing year.

- *Organizational goals and projects are inextricably interwoven* The successful completion of a project means that an organizational goal has been achieved, which in turn means that progress has been made toward the realization of the enterprise's objectives and mission. When a project is behind schedule, or overrunning costs, or unlikely to attain its performance objective, the enterprise's objectives and mission could be impaired.

Strategy An organizational strategy is the design of the means, through the use of resources, to accomplish end purposes. Strategies also include action plans for establishing the direction for the coordinated acquisition and use of resources through organizational design choices. Strategies also provide for the means to obtain resources for the enterprise, and how to use such resources effectively and efficiently in the fulfillment of organizational purposes.

Programs Programs are resource-consuming combinations of organizational resources, which have a common purpose in supporting the enterprise's purposes. For example, a productivity improvement program could be composed of projects such as the following:

- The use of self-managed production teams on the assembly line
- Plan and equipment modernization initiatives
- Use of computer-aided design and manufacturing
- Changeover of a production facility from conventional manufacturing to manufacturing cells

A capital investment program would consist of a number of new projects such as improved equipment development, new facilities, acquisition of equity or debt funding initiatives, and training of personnel.

Projects Projects are ad hoc, resource-consuming activities used to implement organizational strategies, achieve enterprise goals and objectives, and contribute to the realization of the enterprise's mission. An early definition of a project stated that it was "an undertaking that has definite final objectives representing specified values to be used in the satisfaction of some need or desire."[5]

Operational Plans Operational plans are those documents developed to guide the organization in a consistent fashion toward meeting its mission, objectives, and goals through designated strategies. These plans form the overarching polices, procedures, and practices for when and how program and project work will be accomplished.

Organizational Design Organizational design is the organizational structure that facilitates performing the work. Organizational design considers the business that is being conducted, the manner in which work will be conducted, the practices

---

[5]Ralph Currier Davis, *The Fundamentals of Top Management* (New York: Harper, 1951), p. 268.

for managing the work effort, and strategies for work accomplishment. An optimal organization design supports the enterprise in getting its work accomplished in the most competitive way.

1.3 A HISTORICAL PERSPECTIVE OF PROJECTS

Projects have played a key role in some instances and have initiated changes in the societies of antiquity that are still being felt today. A few of these projects are cited and portrayed in Fig. 1.3 and in the material that follows:

1. In Europe there was a great advance in building projects for places of worship. Between 1050 and 1350, in France alone, over 500 large churches were built, as well as 1000 parish churches, so that there was a church or chapel for every 200 people. The Cologne Cathedral, considered by some to be the most perfect specimen of Gothic architecture, undoubtedly took the longest to build. The foundation stone was laid in 1248. By 1437, one of the towers was finished to one-third its present height, but at the time of the Reformation its roof was still covered with boards. Finally, the cathedral was completed in 1880, over 650 years after construction first began!
2. In the United States, the second half of the 1860s witnessed the presence of a project to join the continent of the United States by railroad. The two biggest corporations in American, the Central Pacific and Union Pacific, had armies of men at work building separate railroad lines. This immense project was an epic of logistics, organization, and endurance—as well as an opportunity for the railroad companies to get very rich. The federal government issued land grants along the right-of-way and low-interest bonds underwriting construction costs of up to $48,000 per mile in the mountainous regions. Although there were many things about this project that were marvelous, one record set by the Central Pacific workforce remains unequaled today: 10 miles and 56 feet of track were constructed in 1 day. When the Golden Spike went in the last tie to connect the last rail, it brought together the lines from east and west, thus initiating a transportation system that held the East and West together for the first time. As Stephen E. Ambrose noted, "Things happened as they happened. It is possible to imagine all kinds of different routes across the continent, or a better way to have the government build a railroad and own it. But those things didn't happen, and what did take place is grand. So we admire those who did it—even if it was far from perfect—for what they were and what they accomplished and how much each one of us owes them."[6] The reader can only ponder with fascination the "strategic effects" that this project set in motion.
3. In the early years of the fifteenth century, Prince Henry the Navigator developed and operated what could be called today a primitive research and

[6]Paraphrased and quoted from Stephen E. Ambrose, "The Big Road," *American Heritage,* October 2000, pp. 55–66.

- Egyptian pyramids
- European castles
- European cathedrals
- Great Wall of China
- Voyages of discovery
- Canals
- Military campaigns

FIGURE 1.3 Project management was born in antiquity.

development laboratory, located in Sagres, Portugal. During these early years, Prince Henry initiated, organized, and directed expeditions on the frontier of discovery. The voyages of the discovery that he set forth could be described as "projects." These projects of discovery made important contributions to the evolving body of knowledge in cartography, navigation, and shipbuilding. Prince Henry required his mariners, who also functioned as "project managers," to keep accurate logbooks and charts, and to make a record of everything they saw during their exploration of the waters. The knowledge base contributed by his discoveries helped add to the latest navigating instruments and newest navigating techniques. At Sagres and at the nearby port of Lagos, experiments in shipbuilding produced a new type of ship—the caravel—without which Prince Henry's exploring projects would not have been possible. This light sailing vessel was designed for explorers' needs, combining some cargo-carrying features and enhanced maneuverability of previous ships. The caravel had enough capacity to return from its voyages of commerce and discovery. Its shallow draft qualified it to explore inshore waterways, as well as made it easier to beach the vessel for repair.

Although Prince Henry did not actually build a research laboratory, he did through his strategies, collect the books, charts, sea captains, pilots, map makers, instrument makers, the shipbuilders and other craftspersons to plan voyages, learn from each voyage, assess the findings, and add to the growing knowledge base about waterways, new ships, and new lands. Indeed, Prince Henry's strategies might be called the first organized project-driven enterprise for continuous discovery.[7]

4. Another early explorer, Amerigo Vespucci, was in project work and indeed could be called a "project manager." In 1501, commanding three caravels, he arrived at a new land, which he called a "new continent." What he did was follow the South American coast for about eight hundred leagues, which took him well down into Patagonia, near the present San Lulian, only some four hundred miles north of the southern tip of Tierra del Fuego. The new continent that Vespucci discovered was not named by himself. Rather the name America came from the efforts of Martin Waldseemiller (1410?–1518), an obscure clergyman, who had studied at the University of Freiburg. In one of Waldseemiller's books, *Cosmoqraphiae Introductio,* which summarized the traditional principles of cosmography, he observed that "Inasmuch as both Europe and Asia received their names from women, I see no reason why any one should justly object to calling this new land Amerigo (from Greek "ge" meaning "land of," the land of Amerigo, or America, after Amerigo, its discoverer, a man of great ability."[8]

5. The "project" to discover the cause, and the cure of yellow fever was surely one of the health challenges of the nineteenth and twentieth centuries. Yellow fever had killed thousands of victims in epidemics that raged in tropical and coastal cities, especially in the Caribbean. Walter Reed (1851–1912) was an American army surgeon, who went to Cuba in 1900 to investigate an outbreak among U.S. soldiers. By intentionally subjecting volunteers to bites, he proved that, like malaria, yellow fever was carried by mosquitoes, not people. The success of his research efforts on this project is best described in a letter Walter Reed wrote to his wife:

> Columbia Barracks, Quesmados, Cuba, December 9, 1900 It is with a great deal of pleasure that I hasten to tell you that we have succeeded in producing a case of unmistakable yellow fever by the bite of the mosquito. Our first case in the experimental camp developed at 11:30 last night, commencing with a sudden chill followed by fever. He had been bitten at 11:30 December 5th, and hence his attack followed just three and a half days after the bite. As he had been in our camp 15 days before being inoculated and had had no other possible exposure, the case is as clear as the sun at noon-day, and sustains brilliantly and conclusively our conclusions. Thus, just 18 days from the time we began our experimental work we have succeeded in demonstrating this mode of propagation of the disease, so that the most doubtful and skeptical must

[7]Material paraphrased and embellished from Daniel Boorstin, *The Discoverers* (New York: Vintage Books, a division of Random House, 1983).

[8]Ibid.

yield. Rejoice with me, sweetheart, as aside from the antitoxin of diphtheria and Koch's discovery of the tubercle bacillus, it will be regarded as the most important piece of work, scientifically, during the 19th Century. I do not exaggerate, and I could shout for very joy that heaven has permitted me to establish this wonderful way of propagating yellow fever.…Major Kean says that the discovery is worth more than the cost of the Spanish War, including the lives lost and money expended.[9]

6. The creation of the Panama Canal was far more than a vast, unprecedented feat of engineering. It was a profoundly important historic event and a sweeping human drama not unlike that of war. Apart from wars, it represents the largest, most costly single effort ever before mounted anywhere on earth. It held the world's attention over a span of 40 years. It affected the lives of tens of thousands of people at every level of society and virtually every race and nationality. Great reputations were made and destroyed. For a large number of men and women it was the adventure of a lifetime.

Because of it one nation, France, was rocked to its foundations. Another, Colombia, lost its most prized possession, the Isthmus of Panama. Nicaragua, on the verge of becoming a world crossroads, was left to wait for some future chance. The Republic of Panama was born. The United States embarked on a role of global involvement.

In the history of financial capitalism and in the history of medicine, it was an event of single consequence. It marked a score of advances in engineering, government planning, labor relations—the first grandiose and assertive show of American power at the dawn of the new century.[10]

7. The Manhattan Project for the development and delivery of the atomic bomb was put under the charge of General Leslie R. Groves for the period September 17, 1942, through December 31, 1946. There was, according to General Groves, a "cohesive entity" that was the Manhattan Project, a factor in its success. The memorandum of appointment for General Groves (Colonel Groves) is shown in Fig. 1.4. The organization chart that identifies the position of General Groves in May 1945 is shown in Fig. 1.5.

Much has been discussed about the importance of the use of the atomic bomb as a key strategy in the U.S. pursuit of World War II. Perhaps there is no better example of how a research and development project led to a major building block in the design and execution of a nation's war strategy. The individual who wishes to pursue additional reading on the Manhattan Project could start with the book *Now It Can Be Told,* by Leslie R. Groves (New York: Harper, 1962).

8. On a rainy day in May 1804, Meriwether Lewis and William Clark started up the Missouri River. Their expedition's (project's) objective was to explore and find an easy water route across the continent, the fabled Northwest Passage that geographers believed lay somewhere to the west. It was the first

[9] "Walter Reed to His Wife," *The Wall Street Journal,* October 22, 1999, p. B1.

[10] Drawn from David G. McCullough, *The Path Between the Seas—The Creation of the Panama Canal, 1870–1913* (New York: Simon and Schuster, 1976), pp. 11–12.

September 17, 1942
MEMORANDUM FOR THE CHIEF OF ENGINEERS

SUBJECT: Release of Colonel L. R. Groves, C.E., for Special Assignment

1. It is directed that Colonel L. R. Groves be relieved from his present assignment in the Office of Engineers for special duty in connection with the DSM[*] project. You should, therefore, make the necessary arrangements in the Construction Division of your office so that Colonel Groves may be released for full time duty on this special work. He will report to the Commanding General, Services of Supply, for necessary instructions, but will operate in close conjunction with the Construction Division of your office and other facilities of the Corps of Engineers.

2. Colonel Groves' duty will be to take complete charge of the entire DSM project as outlined to Colonel Groves this morning by General Styer.

a. He will take steps immediately to arrange for the necessary priorities.

b. Arrange for a working committee on the application of the product.

c. Arrange for the immediate procurement of the site of the TVA and the transfer of activities to that area.

d. Initiate the preparation of bills of materials needed for construction and their *earmarking* for use when required.

e. Draw up the plans for the organization, construction, operation and security of the project, and after approval, take the necessary steps to put it into effect.

BREHOM SOMERVELL
Lieutenant General
Commanding

[*]The then code name for the atomic energy project.

FIGURE 1.4 Appointment memorandum for Colonel Groves to the Manhattan Project.

Some of these instructions were never carried out because, as the work progressed, they no longer seemed appropriate. No working committee was ever established and it proved impracticable to transfer all activities to the Tennessee site.

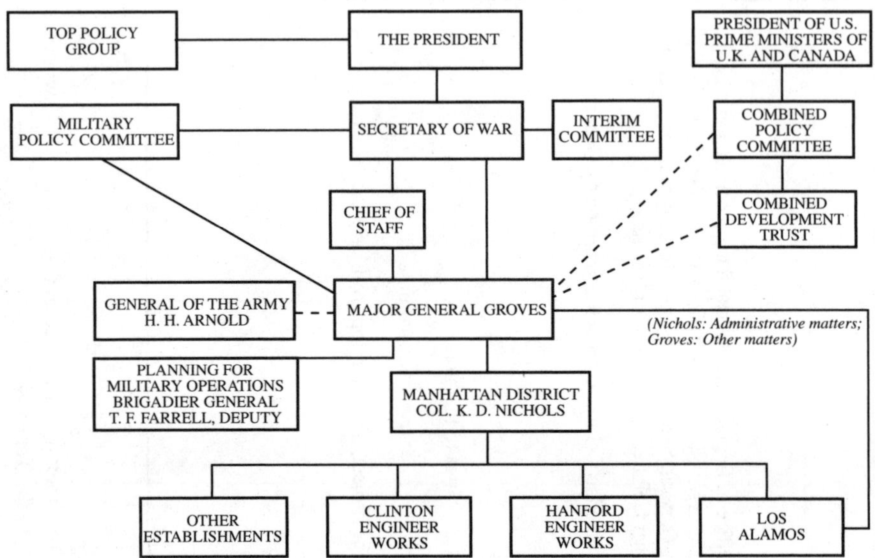

FIGURE 1.5 Organization chart of the Manhattan Project, May 1945.

American expedition to cross the continent and reach the Pacific Ocean by land. Today they are still America's best-known explorers. The project they launched had an original budget of $2,500 and was to involve only a dozen men. But the project team that left the St. Louis area in 1804 numbered 45 men, and the final cost to the American taxpayers reached $38,722.25, more than 10 times the original estimate. In retrospect this project cost was a bargain—but it also was one of the nation's earliest examples of a government cost overrun. The project might be considered a failure, because the explorers did not find that water route. But the success of the project was in proving that no such Northwest Passage existed—and the initial disappointment was soon forgotten when it was realized that Lewis and Clark showed what was out there beyond the sunset, and created a desire for a westering nation of explorers and settlers.[11]

9. A military initiative, or project, that was a major turning point for the United States during World War II was the battle of Midway in June of 1942. In the initial phase of this battle, three squadrons of U.S. torpedo bombers attacked the Japanese aircraft carriers *Agai, Kago, Soryo,* and *Hiryu,* an attempt to draw first blood. A total of 41 U.S. torpedo planes left the carriers *Enterprise, Yorktown,* and *Hornet.* Traditional strategy called for such planes to have a fighter escort to protect them from air attacks as they made their torpedo runs. The U.S. planes attacked in three successive waves—from the start they were doomed. Japanese fighters and antiaircraft batteries on the Japanese fleet destroyed every plane in the first wave, and the next two waves were almost

[11]Paraphrased from *The Old Farmer's Almanac* (Dublin, N.H., 1994), pp. 98–101.

completely destroyed, with a loss of over 80 percent of the pilots. Only a few torpedoes were launched and none hit their targets. A tragic failure?

Historians have noted that the attack by the U.S. torpedo planes was ineffective by itself. But in the larger context of the battle of Midway, the Japanese fleet maneuvered to avoid the torpedo attack and was unable to sail into the wind to launch their planes against the U.S. fleet. When the U.S. SDB-3 *Dauntless* dive-bombers came in at 15,000 feet, there were no Japanese fighter planes to stop them. The Japanese carriers had heavily armed aircraft on their decks. During the next several minutes the Japanese fleet suffered a decisive blow to its carriers, from which it never recovered—and the naval war in the Pacific shifted in favor of the United States.[12]

Thus, what was a tragic failure in a part of the overall strategy for the Midway "project" actually provided an opportunity, or window, for the project to be successful, as the final project results contributed to the evolving Allied military strategy in the South Pacific.

10. In the United States the building of the Pennsylvania Turnpike provided the opportunity to use an early process of project management. The Pennsylvania Turnpike was America's first superhighway. It paved the way for the superhighway system in the United States. Although the existing interstate superhighway system is tax-supported, the Pennsylvania Turnpike is financed from tolls and revenue bonds. The design engineers for the turnpike decided on several key design standards: a right-of-way width of 200 feet and a four-lane divided configuration with 12-foot-wide concrete traffic lanes with a 10-foot-wide median strip. Other design requirements included:

- A maximum grade of 3 percent
- A maximum curvature of 6 degrees
- Substantial banking on curves
- Limited access with 1200-foot-long entrance and exit ramps
- A minimum 600-foot sight distance from motorist to traffic ahead
- No cross streets, driveways, traffic signals, crosswalks, or railroad crossings

By July 1939, all 160 miles of highway, seven tunnels, and more than 300 structures were under contract, and by August 1939, every part of the project was under construction. Contractors worked usually two shifts a day, and often three. The project called for construction of 160 miles of highway, 7 two-lane tunnels, 11 interchanges, and 10 service plazas. The turnpike opened October 1, 1940. By the end of 1940, the turnpike had carried 514,231 cars, 48,170 trucks, and 2409 buses, and had collected total revenues of $562,464.[13]

The Pennsylvania Turnpike project was completed on time and within budget, and attained the expected technical performance objective—an innovative means for improving the efficiency of motor vehicle travel. Modern-day project managers should remember that most of today's project management

[12]This battle has been written about extensively. The account that is given here has been paraphrased from Owen Edwards, "Tragic Lost Cause?" *Forbes,* June 2, 1997, pp. 71–72.

[13]Dan Cupper, *The Pennsylvania Turnpike* (Lebanon, Pa.: Applied Arts Publishers, 1990), pp. 10–22.

processes and techniques did not exist: no cost and schedule software existed, no PERT scheduling techniques existed, the "matrix" organizational design had not been described, little literature existed on how to manage projects, and computers and modern communication means did not exist. Why was the Pennsylvania Turnpike project so successful? First, the project had a high priority in the U.S. defense initiative. Second, unemployment was high, and the opportunity to work on the project was a godsend for many people and families. Motivation of the work force was not a problem! Finally, the opportunity to be part of an innovative project, such as the turnpike provided, stimulated extraordinary support from the project's stakeholders.

1.4 OTHER EXAMPLES

Indicated below are a few examples of other projects that have provided the project "owner" or "sponsor" with an enhanced capability:

- Theatrical production involves the use of project teams, in part to evaluate the significant risks inherent in any theatrical production. Such productions involve a high financial investment with limited forecasting of probable success, use of highly skilled and expensive personnel, and great dependence on the producer's professional experiences. Improvisation while production itself is still in progress is a particular challenge. One author has examined the use of project management techniques in the performing arts and has concluded that the use of such techniques could help improve the planning, cost control, and schedule control of such productions.[14]
- The H. J. Heinz Company has used alternative project teams in the management of that enterprise. An old factory in Pittsburgh, Pennsylvania, was shut down and a new one was designed and built at the same time—resulting in not a day of lost production. A massive investment in training was initiated to enhance the skills of its seasoned work force. The company provided workers with an array of training tools—evaluation of basic skills, educational counseling, literacy education, classroom instruction, and training on the factory floor. The process of shutting the old factory down and starting up the new one provided the opportunity to bring about an unprecedented degree of employee involvement in day-to-day management. Teams of workers were provided the opportunity to solve problems and take on responsibility. Team-based quality and safety efforts slashed workers' compensation costs by 60 percent and helped make Heinz the quality leader of the pet food industry. In England, at Heinz's Harlesden and Kitt Green factories, worker-led project process evaluation teams helped streamline factory operations and improve quality, in some cases reducing overhead by as much as 40 percent. The teams developed their own plans, presented them to coworkers, and worked to implement the changes with scarcely any direct intervention by management.[15]

[14]Itzak Wirth, "Project Management in the Performing Arts," *Proceedings,* Project Management Institute, 25th Annual Seminar/Symposium, Vancouver, British Columbia, October 17–19, 1994, pp. 636–637.

[15]"Working Smarter," H. J. Heinz Company Annual Report, pp. 16–17.

- When the Chrysler Corporation sets out to create a new automobile, it forms a project team of about 700 people from the different principal disciplines of the company to work on the project. A corporate vice president acts as a "godfather" to the team, but the team and its leaders plan and direct the work. After a suitable contract has been worked out with management, the team is turned loose to design and develop the product. There are no committees, no hierarchy outside of the team. By using these concurrent engineering teams, the time to design and develop a new vehicle model has been reduced nearly 50 percent.[16]
- An important and exciting project that captured the attention and emotions of millions of people around the world occurred in the form of Desert Storm, a military project that changed the world and the use of military forces to bring about political change. Max Wideman, a Fellow of the Project Management Institute, described this extraordinary project as follows:

Project Desert Storm (July 2, 1991)
Projects come in all sizes and shapes, so they tell us. Whether differentiated by duration, complexity or area of application makes no difference. A project is a project. So a [military] project like Desert Storm lasting just 100 hours should be nothing out of the ordinary. Of course, that covered only the duration of the execution [battle] and completion [victory] phases. The prior phases and stages of concept, planning, design, and procurement of a complex set of commitments by a large diverse group of culturally different participants [the United Nations], plus preparation for execution [the prior air campaign] which preceded the project accomplishment phases, added considerably to the real overall project duration.

Nevertheless, the project was a managerial triumph of successful project management [resounding military victory] even though one of the potential deliverables [the opposing Commander in Chief] was not included. The project had some unique features. The location [miles of empty desert] was hardly one of the choicest. Project success would rely heavily on teamwork [joint military command], a decisive logistical achievement [assembling, supplying, and transporting over unheard-of distances the most fearsome strike force in history] and innovation [military surprise]. The project manager [General Schwarzkopf] did well to give recognition to his logistics manager for a job well done [battlefield promotion].

The project was full of risk, and was opposed by a large number of stakeholders [both at home and overseas]. Once committed, success depended on utterly logical and overwhelmingly powerful and determined courses of action.

For example, Project Desert Storm was superbly equipped. Firstly, its technology was unsurpassed [world-beating] and state-of-the-art [high-technology], the product of intensive and highly successful R&D. Secondly, the human resources [troops] working with the equipment and materials [weaponry and firepower] were rigorously trained to the highest standards—that showed not only in their effectiveness but also in their morale. Thirdly, an able team of leaders, highly trained, experienced, qualified and selected, had only reached the top [through military promotion] after much education and training both technical and general.

Another equally important aspect was the organizational structure [army command] within which the team operated. The project manager had full delegated power to

[16]Marshall Loeb, "Empowerment That Pays Off," *Fortune,* March 20, 1995, pp. 145–146.

run the operation his way. His instructions from above [the President's office] were absolutely clear, and his immediate sponsors [such as General Powell] gave their total support. Authority and responsibility were passed down the organization structure in the same way. Once allotted their role by the project manager, work package managers [field commanders] made their own plans and executed them decisively.

It is true that the project team was not very keen when the project manager first proposed his outlandish and risky strategy [a mighty encircling sweep behind the Republican guard] to save man-hours [lives], time and, ultimately, cost [subsequent prolonged war effort]. His team [tactical commanders] gave the classical response "It can't be done" but that only made the project manager more determined. After all, had he not found a very tempting market niche ignored by others? [Iraq's generals no doubt also thought that no army could drive their tanks over all that desert and that far without breaking down and going to ruin.] "They could never make it," they said.

That's what made the project so exciting. The project manager performed a crucial role of any project leader—he converted a tremendous risk into a tremendous opportunity by insisting that his team had to achieve the impossible. However, the team members only agreed to the course of action after the project manager had ordered his logistics manager to pledge in writing that everything would be in place by the scheduled deadline [February 21, 1991]. But that too is an elementary lesson of project management—give people the tools if you want them to do the job.

There's even more to it than that, of course. Project Desert Storm had an enormous supply of two vital elements: quality information and planning. They are necessary in that order because nothing can be effectively planned without solid information. But the planning and information experts were not an isolated function to satisfy some latest management theory whether in the design office [Washington] or in the field [war zone]. They were an integral part of the project process [war effort]. Line and staff relations were not an issue; the focus was on getting the job done [winning].

One advantage the project did enjoy, and that was it was not constrained by budget. Yet by the very acts of determination, precision, and quality organization [to say nothing of superb timing and decision making by the Chief Executive in the White House for launching the project] the project has proved to be highly cost-effective, compared to most similar ventures. It may even be a significant revenue generator when indirect benefits are taken into account. All of this in a most unappetizing market [the Middle East], where competition at the outset appeared to be overwhelming [the fourth largest army in the world]. Indeed, the project represents a powerful argument in support of establishing highly selective strategic alliances in order to achieve project goals.

Surely this must be one of the best object lessons for any project manager today?[17]

One of the outcomes of Desert Storm was that it demonstrated the need for further project management.

The Iraqi invasion of Kuwait in August of 1990, and the subsequent Gulf War, caused almost all of the country's oil production facilities to suffer extensive damage from fires and other causes. The reconstruction of the oil field infrastructure was planned in November 1990 before the war ended in February 1991. At the time the amount of work that would be required to bring the oil fields up to production was

[17]With apologies to *Management Today,* British Institute of Management, May 1991, p. 34. Appreciation is extended to R. Max Wideman, P. Eng., MCSCE, FEIC, FICE, Fellow PMI, for contributing the information on project Desert Storm.

not known—but some of the project work packages could be identified and front-end planning was undertaken for these.[18]

1.5 EARLY LITERATURE

One of the first comprehensive articles that caught the attention of the project management community was published by Paul O. Gaddis in the *Harvard Business Review* (May–June 1959). This article, titled "The Project Manager," describes the role of that individual in an advanced technology industry, the prerequisites for performing the project management job, and the type of training recommended to prepare an individual to manage projects. Several basic notions put forth by Gaddis contributed to a conceptual framework for the management of projects that holds true today. These basic notions were:

- A project is an organizational unit dedicated to delivering a development project on time, within budget, and within predetermined technical performance specifications.
- The project team consists of specialists representing the disciplines needed to bring the project to a successful conclusion.
- Projects are organized by tasks that require integration across the traditional functional structure of the organization.
- The project manager manages a high proportion of professionals organized on a team basis.
- The superior/subordinate relationship is modified, resulting in a unique set of authority, responsibility, and accountability relationships.
- The project is finite in duration.
- A clear delineation of authority and responsibility is essential.
- The project manager is a person of action, a person of thought, and a front person.
- Project planning is vital to project success.
- The project manager is the person between management and the technologist.
- The subject of communication deserves a great deal of attention in project management.
- Project teams will begin to break up when the members sense the project has started to end.
- The integrative function of the project manager should be emphasized.
- Status reporting is appropriate and valuable to management of the project.
- The role played by project management in the years ahead will be challenging, exciting, and crucial.

[18] For a fuller description of the management of this project see Mehdi Adib, "Managing Kuwait Oil Fields Reconstruction Projects," *Proceedings,* Project Management Institute, 25th Annual Seminar/Symposium, Vancouver, British Columbia, October 17–19, 1994.

Project management, as an important and growing *philosophy of management,* came into its present conceptual framework as a culmination of experimentation with a variety of organizational liaison devices. In a sense project management is the formalization of organizational liaison devices.

1.6 ORGANIZATIONAL LIAISON DEVICES

Project management evolved from a series of liaison devices that have been developed in contemporary organizations. These liaison devices, both formal and informal, have encouraged experimentation in integrating activities across organizational structures. Jay R. Galbraith is one of several researchers who have studied these liaison devices.[19] His research provides in part the basis for a description of the following types:

- Individual liaison
- Standing committees
- Product managers
- Managerial liaison
- Task forces
- Project engineer
- Liaison position

Figure 1.6 depicts the interrelationships of these liaison devices in the context of the emergence of project management. A brief discussion of these types follows.

FIGURE 1.6 Organizational liaison activities.

[19]Jay R. Galbraith, *Organization Design* (Reading, Mass.: Addison-Wesley, 1977).

Individual Liaison

The simplest and perhaps best form of liaison is that brought about by people who sense the need to work together and go about maintaining contact with others in the organization who have a vested interest in an activity under way. This liaison is usually self-motivated.

Standing Committees

Standing committees are used extensively to integrate organizational activities. These committees are found at all levels in the organization. At the top level such committees are called plural executives. They bring about synergy in the making and execution of key operational and strategic decisions for the organization.

Product Managers

Product managers usually are appointed to act as a focal point for the marketing and sales promotion of a product. Originating in the personal products area, the first product manager appeared before 1930. Persons occupying these positions usually were provided a small administrative staff and might have had profit/loss responsibility. They usually were not backed up by a specific team but rather worked closely in a coordinating role with other key individuals.

Managerial Liaison

When a more formal linkage is needed, a manager or supervisor is appointed who is in charge of several people through setting direction for the organizational unit and providing supervisory jurisdiction over the people. This form is widely used in modern organizations. As the organization increases in size and the work becomes more complex, additional managers are added, resulting in the creation of a chain of command eventually leading to large management structures. Other liaison roles as described in this chapter deal with the organizational complexity and bureaucracy to encourage contacts between individuals and organizational units.

Task Forces

Task forces often are used to bring a focus to organizational activities, usually those that are short-term. Members are appointed to the task force to work on an ad hoc problem or opportunity. During the time they are on the task force, members have a reporting relationship to their regular organizational unit and to the task force chairperson as well. When the purpose for which the task force was created is accomplished, the task force is dissolved.

Project Engineer

Sometimes a liaison position evolves through practice. Such is the case of the project engineer, who is responsible for directing and integrating the technical aspects of the design/development process. These positions have evolved in contemporary organizations to the point where the project engineer manages a product through all its engineering steps, from initial design to manufacture or construction.

Liaison Position

When a significant amount of contact is required to coordinate the activities of two or more organizational entities, a liaison position is formally established to bring about synergy and communication between the units. Usually this position has no direct formal authority over the organizational units but is expected to communicate, coordinate, pull together, and informally integrate work among the organizational units. Examples of liaison roles are an engineering or construction liaison person and a production coordinator who mediates between the production control, product engineering, and manufacturing. A purchase engineer who sits between purchasing and engineering is another example.

Other liaison positions may join line and staff groups. In the military establishment, the position of military aide-de-camp, a military officer acting as a secretary and confidential assistant to the superior officer of general or flag rank, is a liaison role. Originating in the French army, this position originally served as a camp assistant. The aide-de-camp carried out a coordinating and liaison role for his or her commanding officer; because the aide was close to the senior officer, there was a good deal of implied authority attached to the aide's role.

Modern teams are used for a variety of purposes.

1.7 TEAMS

In modern organizations, project teams are used to complement an existing organizational design. An overriding feature of the team design is a departure from the traditional form of management in favor of a team form in which there are multiple authority, responsibility, and accountability relationships, resulting in shared decisions, results, and rewards. These teams include some of the teams already mentioned—project teams, project engineering teams, and task forces—but also include production teams, quality circles, product design teams, and crisis management teams. The importance of the use of teams in contemporary organizations cannot be underestimated. Peters and Austin found that small-scale team organization and decentralized units are vital components of top performance.[20]

[20]Tom Peters and Nancy Austin, "A Passion for Excellence," *Fortune,* May 13, 1985, pp. 20–32.

Some examples of project teams follow:

- At one end of a large IBM plant in Charlotte, North Carolina, 40 workers toil at building 12 products at once—hand-held bar-code scanners, portable medical computers, fiber-optic connectors for mainframes, satellite communications devices for truck drivers—a typical half day's output on a line designed to make simultaneously as many as 27 different products. Each worker has a computer screen hooked into the factory network. Called a digital factory for its dependence on information technology, it is sometimes called soft manufacturing, in which software and computer networks have emerged along with people, which will set the tone for years, perhaps decades, in manufacturing. Soft manufacturing blurs the boundaries of the traditional factory by integrating production closer to suppliers and customers. Often an order is complete within 80 minutes, and depending on where the customer lives, he or she can have the product the same day or the day after. Workers on the factory floor are organized into teams that manage themselves and have real decision-making power. Teams of knowledge workers are the force that makes the digital factory go—teams of people are tied together along with equipment through the medium of information.[21]
- The Advanced Development Company, or "Skunk Works," of the Lockheed Corporation continues its super-secret research and development. The results produced by the Skunk Works are unparalleled. In 1943, the XP 80 *Shooting Star,* the first operational U.S. fighter jet, was put together in 4 short months. During the Cold War, the Skunk Works came up with the famous U-2 and the SR-71 *Blackbird* spy planes that soared at the edge of space taking pictures of Russian and Chinese military installations. In 1991, a product developed at the Skunk Works, the F-117A *Stealth* fighter, slipped through Iraqi radar undetected to deliver "smart" bombs with pinpoint accuracy. The Skunk Works has a unique culture, a small-team approach to projects managed by project leaders, who have extraordinary latitude in recruiting in-house specialists. The project team is isolated from Lockheed's sprawling bureaucracy and is able to have direct contact with the military project team without going through "channels" at Lockheed. As the design teams developed product and process designs, they were able to alter their designs without the approval of a hierarchy of executives at corporate headquarters.[22]
- At the riot-ravaged Taco Bell restaurant in Compton, California, a joint effort of the city of Compton and Taco Bell management launched a 48-hour rebuilding plan. Fluor Daniel, Inc., an international engineering, construction, maintenance, and technical services subsidiary of Fluor Corporation, was called in to plan for the new restaurant less than 3 weeks before the building was to begin. Taco Bell wanted everything done on-site with nothing prefabricated and wanted to have the restaurant open and ready for business in 48 hours. Fluor Daniel used Primavera's Finest Hour software, which allows its users to schedule

[21]Gene Bylinsky, "The Digital Factory," *Fortune,* November 14, 1994, pp. 93–110.
[22]Eric Schine, "Out at the Skunk Works, the Sweet Smell of Success," *Business Week,* April 26, 1993.

by 1-hour increments to handle the complexities of a short-term, intensive project. Fluor Daniel organized a dry run of everything 3 days before construction started to ensure that all team members knew the plan. Progress meetings were held every 3 hours to distribute the earned-value report, just one of the software's several productivity reports, comparing plans versus actuals in terms of both budget and work accomplished. At hour 46 the health inspector gave the okay, and at hour 47 the certificate of occupancy was signed. At hour 48 the first new tacos were served.[23]

- At the Pennsylvania Electric Company's Generation Division (Penelec) project management is used as a "way of life." A centralized division planning group has been set up to link the different functional units required for a project and to integrate these units into a single project control system. Project teams are responsible for satisfying project objectives in the areas of life extension, maintenance, plant improvement, and the environment. On one project created to study turbine outage, the company estimated that the computerized project management system saved the company $300,000. Project management is recognized by senior division management as having credit to successfully allocate organizational resources to satisfy company objectives.[24]
- At Johnson Controls Automotive Systems Group, product development activities are done by using a company-institutionalized project management system. Use of project management by the company has prompted an increase in the training of employees and the creation of a standard approach for project management. The use of a common approach in project management has facilitated the development of organizational strategies, policies, procedures, and other ways of working on projects. Employees are educated in the company's project management process; the improvement of the culture for project management—and the development of a common approach for the management of projects—has enabled the company to complete project development efforts in a timely and efficient manner.[25]
- At MBI, Inc., a company that produces collectibles, such as porcelain birds and plates, program (project) managers are used, one for each series of collectibles. Two key objectives guide these program managers: Get new customers at the lowest possible cost and retain them for follow-on purchases. The company "piggybacks" on the Franklin Mint's product development costs and market research. Accordingly, industry observers believe that MBI, Inc.'s costs are much lower than its competitors'.[26]
- A central blood bank in a major U.S. city sees project management as a key management strategy in its world-class center for general support of the area hospitals' blood banks. Today over 400 staff members collect and test the units of over 152,000 donations, distributing more than 400,000 blood products per

[23]"Riot-Ravaged Taco Bell Rebuilt in 48 Hours Using Project Planning Software," *Industrial Engineering,* September 1992, p. 18.

[24]Anthony J. Catanese, "At Penelec Project Management Is a Way of Life," *Project Management Journal,* Project Management Institute, December 1990, p. 7.

[25]W. D. Keith and D. B. Kandt, "Project Management at a Major Automotive Seating Supplier," *Project Management Journal,* Project Management Institute, September 1991, p. 28.

[26]Phyllis Berman with E. Lee Sullivan, "Getting Even," *Forbes,* August 31, 1992, pp. 54–55.

year for 22 member hospitals. In addition, the bank provides pretransfusion testing services to three hospitals, reference testing to all areas involving clotting and bleeding disorders, and outpatient transfusion services for patients not requiring hospitalization.

- Nicolas G. Hayek and his colleagues at the Swiss Corporation for Microelectronics and Watchmaking have brought about one of the most spectacular industrial comebacks in the world—the revitalization of the Swiss watch industry. According to Hayek, "we are big believers in project teams." He describes the use of project teams in the context of finding your best people, letting them take on a problem, disbanding them, and then moving on to the next problem. According to Hayek, the whole process of using projects works only if the whole management team focuses on developing products and improving operations—not fighting with each other.[27]
- Project management is used in the U.S. Justice Department. In the early 1980s the Reagan administration came close to merging the Drug Enforcement Agency (DEA) and the Federal Bureau of Investigation (FBI)—a merger that was evaluated through the use of a project team. In 1984, over 56 separate projects were under way to integrate various functions of the two agencies. Only 9 of the 56 projects were completed; the others dropped into a state of bureaucratic limbo. An area that was successfully merged was training. Now the DEA has received $11 million to build its own training center in Quantico, Virginia—an effort that used project management to design, construct, and start up this new facility.[28]
- A large electronics company uses product design teams in simultaneous engineering to ensure the right timing and integration required during product and process development. These teams provide a focus for bringing together the people on a product development activity to coordinate and integrate an effort to support the product and process synergy. A product design team might include design engineers, technical writers, customer support people, marketing representatives, regulator and legal experts, purchasing agents, human factors analysts, and representatives from manufacturing and quality. These team members, acting in concert, provide both a focus and the necessary cross-fertilization of information and strategies to reduce the time required to get the product developed, manufactured, and in the customer's hands.
- A large agricultural and industrial equipment manufacturer that does material, manufacturing, and product-applied research at the business-unit level uses concurrent engineering to accelerate product and process development cycles. Product development research is not usually considered high-risk because it is primarily applied research. Few new product development efforts are carried out; rather, the research is aimed at incremental product and process improvements. Product improvement includes the enhancement of the product's performance as well as cost reduction and improvement of product quality. The research follows product lines and is evolutionary.

---

[27] Reported in William Taylor, "Message and Muscle," *Harvard Business Review,* March–April 1993, pp. 99–110.
[28] Janet Novack, "How About a Little Restructuring?" *Forbes,* March 15, 1993, pp. 91–96.

Companies today, facing unprecedented global competition, are finding it advantageous to cooperate with partners around the world to share resources, risks, and rewards. These partnerships take the form of strategic alliances and are used for many purposes, such as sharing the design of products and processes, sharing manufacturing and marketing facilities, and sharing in the financial risk and rewards. Technology is changing so fast today that companies are finding it impossible to assemble the resources to keep ahead of the competition. Indeed, a form of "cooperative competition" is becoming the standard for success in the unforgiving global marketplace.

Once the opportunity for a strategic alliance has been established, a joint project team is often appointed to begin the analysis and work on the alliance. This project team establishes the rationale for the alliance, makes recommendations for the selection of the partner(s), and initiates the analysis required for the development of a suitable working agreement among the partners. Key matters considered by these teams include the mission, objectives, goals, and strategies for the alliance. The team develops the alliance performance standards and builds a recommended strategy for how the joint arrangement will be integrated and managed. A key responsibility of the joint project team is to prepare a strategy for and participate in the execution of the negotiations required to bring about a meeting of the minds on the partnership alliance.

Once the alliance is consummated, the project team that managed the alliance during its development can be disbanded. Then the alliance will start the process of becoming "institutionalized"—merged into the ongoing businesses of the partners. Something that did not previously exist has been created through the use of project management technologies. Sometimes the partners will continue the project team's existence to oversee the alliance in its early period and until the alliance can be integrated into the ongoing operations of the partners.

Many projects are becoming global, in some cases coming forth out of strategic alliances that global partners have negotiated. IBM alone has joined hundreds of strategic alliances with various companies in the United States and abroad, reflecting the fact that alliances have become a part of strategic thinking.

The challenges in a strategic alliance lie in the comparative management of the business and in the personal relationships between managers from different organizational cultures. Perhaps the biggest stumbling block to making an alliance work is the lack of trust among the partners.[29]

Project teams can be used for a wide variety of projects:

- Design, engineering, and construction of a civil engineering projects such as a highway, bridge, building, dam, or canal
- Design and production of a military project such as a submarine, fighter aircraft, tank, or military communications system
- Building of a nuclear power generating plant
- Research and development of a new machine tool

[29]Paraphrased from Ricardo Sookedeo, "Are Strategic Alliances Working?" *Fortune,* September 21, 1992, pp. 77–78.

- Development of a new product or manufacturing process
- Reorganization of a corporation
- Landing an astronaut on the moon and returning her or him safely to earth

Project work in the engineering, architecture, construction, defense, and manufacturing environments is easy to recognize. A new plant, bridge, building, aircraft, or product is something tangible; however, the project model applies to many fields, even to our personal lives.

These projects are the leading edge of change, in both our professional and our private lives. Change encourages—or may force—us to do something different, at some cost, and on some time or schedule basis. These changes often take the form of projects, such as:

- Writing a book or article
- Painting a picture
- Having a cocktail party and dinner
- Restoring an antique piece of furniture or an automobile
- Getting married or divorced
- Having children
- Adopting a child
- Designing and teaching a course
- Organizing and developing a sports team
- Building a house or modifying an existing house

Students are feeling the impact of project management. In May 1985, the National Academy of Engineering held a symposium on U.S. industrial competitiveness. The symposium brought together some of the nation's leading industrial and academic technological leaders to discuss the industrial competitiveness challenge and how the National Academy of Engineering might formulate its programs to improve U.S. competitiveness. During the symposium's discussion of engineering education, it was recommended that the education of engineers for a future technological age require that the students develop the skills of leadership "for projects and programs...as well as technical leadership in their respective discipline."[30]

1.8 THE PROJECT MANAGEMENT PROFESSIONAL SOCIETIES

Emerging professional associations are dedicated to project management. The largest in number is the Project Management Institute with more than 85,000 members. In Europe, the International Project Management Association has more

[30]*The Bridge,* Summer 1985, pp. 22–25.

than 30,000 members and represents national project management associations throughout Europe. The Australian Institute for Project Management and the Japanese Project Management Forum have a few thousand members. There are a wealth of small professional societies that are either directly promoting project management principles, practices, and processes or have formed to exchange project management information within a particular segment of industry. It is estimated that there are perhaps more than a million individuals who could benefit from membership in a professional society.

Professional societies typically provide collectively through members what one member or organization cannot provide. Out of this concept, there have emerged over the past 20 years several project management bodies of knowledge—some compatible except for the cultural aspects of a particular region and others more comprehensive as the knowledge areas are defined.

These bodies of knowledge are used for certification of individuals as to their qualifications in the project management field. Two types of certifications have emerged over the past decade—certification based on a person's knowledge of the profession and certification based on a person's competency in the profession. Each certification has its merits and challenges.

A sample of the areas of knowledge is given in Table 1.2. This table is not representative of any particular society, but given to promote thought on the full range of knowledge and skills that a project management practitioner might need to be successful.

One of the distinguishing characteristics between different bodies of knowledge is the scope. Some bodies of knowledge are limited to the project's life cycle, that is, that of a single project from start to finish, whereas others take a larger view and address the aspects of projects within an enterprise or even within a global context. Either body of knowledge is valid and the value of it is dependent upon the application.

The scope of the body of knowledge, of course, defines any certification program and whether it can be a "knowledge-based" or a "competency-based" certification. The number of areas included in the body of knowledge will show the range of knowledge needed to master the profession.

1.9 A PHILOSOPHY

A philosophy is a synthesis of all the knowledge, skills, and attitudes that one has about a field of learning and practice, the critique and analysis of fundamental beliefs about a discipline. A philosophy is also the system of motivating concepts and principles surrounding a field of study and practice. A field of thought or, to put it into more pragmatic terms, a "way of thinking" about a field of learning and practices, is what a philosophy is all about. Anyone who has been exposed to the field of management as either a manager or the objective of management has a philosophy or way of thinking about management. The study of the management discipline—

TABLE 1.2 Sample Project Management Knowledge and Skill Areas

No.	Knowledge area
1.0	*Primary knowledge and skill areas (samples)*
1.1	Scope management
1.2	Technical performance management
1.3	Schedule management
1.4	Cost management
1.5	Configuration management
1.6	Planning
1.7	Resource management
2.0	*Supporting knowledge and skill areas (samples)*
2.1	Risk management
2.2	Communication management
2.3	Contract administration
2.4	Negotiation
2.5	Leadership
2.6	Decision making
2.7	Marketing
2.8	Customer relationship
2.9	Personnel conflicts

and of project management in particular—enables one to broaden and sharpen the way one thinks about project management concepts and processes. Remember: To a large degree we participate on a project team, either as the team's leader or as a member of team, based on the way we think about the project management discipline. Although we may not recognize it, the philosophy that we hold about project management influences the decisions we make and implement in the project management way of doing things.

1.10 BREAKING DOWN HIERARCHIES

Project management has caused many changes in how contemporary organizations operate. One major change deals with organizational hierarchies. Paramount companies today are tearing down traditional hierarchies. In today's fast-changing, information-driven, and computer-facilitated competitive economy, new paradigms on how to manage are coming forth. Some of the more important paradigms are described below:

- Project management and strategic management are highly interdependent.
- Work is organized around processes carried out by teams of employees working in an ever-changing organizational design.
- Temporary teams are drawn from a range of functional expertise and are formed around specific organizational projects.
- Few remaining vestiges of the traditional organization such as rigid hierarchy, command and control management styles, and bureaucratic policies and procedures remain.
- In the team-driven organization the organizational design is more like a web of teams and projects rather than a clearly defined vertical hierarchy with clear discipline boundaries.
- Managers constantly move people from projects and teams that are phasing down and seek out promising teamwork and projects positioned for the future.
- E-mail, Internet, and other forms of electronic communication which enable people at all levels of the enterprise to keep abreast of what is happening are used.
- The work force is constantly trained and retrained.
- More cooperation with suppliers, customers, and even competitors should be developed.
- Egalitarian cultures should be fostered, but not to excess.
- A general sense of urgency and importance to speed up product, service, and process development is required.

Alternative teams that provide for broad cross-organizational cutting such as new product development, new facilities, benchmarking, and reengineering initiatives have become the new organizational design replacing narrowly focused departments and functions. By organizing the resources so that focus can be brought to the management of organizational processes—such as order fulfillment and new-product development—a synergy is possible that could not be realized through using the traditional organizational design based on functional specialization. Under functional specialization each organizational unit became a fiefdom—a collection of talent and resources working in silos, usually independent of others, and developing and implementing strategies on its own.

The growth of project management is reflected to some degree by the recognition that was given to this discipline by contemporary literature. An excellent and timely article that appeared in *Fortune* magazine in mid-1995 doubtless helped to accelerate the growth of project management. According to this article:

- Midlevel management positions are being cut.
- Project managers are a new class of managers to fill the niche formerly held by middle managers.

- Project management is the wave of the future.
- Project management is spreading out of its traditional uses.
- Managing projects is managing change.
- Expertise in project management is a source of power for middle managers.
- Job security is elusive in project management—because each project has a beginning and an end.
- Project leadership is what project managers do.[31]

Some of the unique characteristics of project management today include the following:

- Projects are ad hoc endeavors, which have a defined life cycle.
- Projects are building blocks in the design and execution of enterprise strategies.
- Projects are the leading edge of new and improved organizational products, services, and enterprise processes.
- Projects provide a philosophy and strategy for the management of change in the enterprise.
- The management of projects entails the crossing of functional and organizational boundaries.
- The management of a project requires that an interfunctional and interorganizational focal point be established in the enterprise.
- The traditional management functions of planning, organizing, motivation, directing, and control are carried out in the management of a project.
- Both leadership and managerial capabilities are required for the successful completion of a project.
- The principal outcomes of a project are the accomplishment of technical performance, cost, and schedule objectives.
- Projects are terminated upon successful completion of the cost, schedule, and technical performance objectives—or earlier in their life cycle when the project results no longer promise or have a strategic fit in the enterprise's future.

1.11 TO SUMMARIZE

The major points that have been expressed in this chapter include:

- Strategic management and project management are interrelated in the management of an organization.
- The origins of project management are rooted in antiquity. The practice of project management has been carried out for centuries—if only in an unsophisticated

[31]Thomas A. Stewart, "The Corporate Jungle Spawns a New Species: The Project Manager," *Fortune,* July 10, 1995, pp. 179–180.

manner as compared to today's practices. Nevertheless, the results of ancient project management are found in many places in the world.

- Project management is an idea whose time has come, in terms of the continued design and development of project-driven management strategies for industrial, military, educational, ecclesiastical, and social entities. Project management processes and techniques can be used for the management of personal resources such as getting married or divorced, building a house, having a cocktail party, or pursuing a hobby such as forming and managing a sports team.
- The basic considerations in any project center around the cost, schedule, and technical performance parameters—and how well the project results fit into the operational or strategic purposes of the enterprise.
- The results of project management usually take the form of a new or improved organizational product, service, and process.
- Many examples of the use of project management were provided in the chapter to include representation from many different organizations.
- A project tends to be ad hoc in nature, and the project results can be considered to be building blocks in the design and execution of operational and strategic initiatives for the enterprise.
- No one can claim to have invented project management—rather the concept and process evolved over a long period of time.
- The Project Management Institute (PMI®) is the leading professional association in the discipline. Other professional associations also exist whose purpose is to facilitate the spread of the theory and practice of project management.
- The results produced by projects have had an impact on history—including a major influence on the infrastructure of many institutions and societies.
- Sometimes projects fail to produce the results that were planned because of such factors as technology, economics, and political and social imperatives.
- Project management began to become conceptualized and documented in the 1950s in the sense of a philosophy and process for dealing with ad hoc opportunities.
- Prior to the emergence of project management, various organizational devices evolved to provide the means for an integration of activities across organizational structures.
- Project management has laid down the strategic pathway for the emergence of alternative teams in the modern organization to deal with such change initiatives as reengineering, benchmarking, simultaneous engineering, and self-managed production teams.
- In the early days of project management it was considered to be a "special case" of management. Today it has taken its rightful place in the theory and practice of management.
- When projects are managed, there tend to be a breakdown and an alteration of the traditional organizational hierarchies in favor of a horizontal form of organizational design.

- Project managers are emerging as a new class of managers to fill the niche formerly held by middle managers.
- Projects provide a philosophy, strategy, and process for the management of change in the enterprise.
- The management of a project usually requires the crossing of functional boundaries of the enterprise.
- The traditional management functions of planning, organizing, motivation, direction, and control are carried out in dealing with a project.

1.12 ADDITIONAL SOURCES OF INFORMATION

The following additional sources of project management information may be used to complement this chapter's topic material. This material complements and expands on various concepts, practices, and theory of project management as it relates to areas covered here.

- Curtis R. Cook and Carl Pritchard, "Why Project Management?" chap. 3 in David I. Cleland (ed.), *Field Guide to Project Management* (New York: Van Nostrand Reinhold, 1997).
- Mark Maremont, "Kodak's New Focus," in David I. Cleland, Karen M. Bursic, Richard J. Puerzer, and Alberto Y. Vlasak, *Project Management Casebook,* Project Management Institute (PMI). (First published in *Business Week,* January 30, 1995, pp. 62–68.)
- Henry Fayol, *General and Industrial Management* (London: Sir Isaac Pitman & Sons, 1949). This book is a scientific exposition of the general principles of management—written by one of the greatest pioneers of the field. The book is truly a "classic" of management, an indispensable work of reference for those engaged in the practice or teaching of management concepts and processes.
- Francis M. Webster, Jr., *PM 101, According to the Old Curmudgeon,* Project Management Institute, Newtown Square, PA 19073, 2000. The author offers this book as a basic introduction to the fundamental concepts and processes of modern project management, and he delivers just that in this readable and enjoyable book. It is about the principles of modern project management—how many principles can be applied in modern organizations. Basic and sufficient information and explanations on how to manage projects are the themes of this book. It is designed to appeal to the professional who has been assigned in some capacity to the management of projects in the enterprise.
- Paul O. Gaddis, "The Project Manager," *Harvard Business Review,* May–June 1959. This is one of the earliest articles on project management that appeared in the

business literature. "The Project Manager" describes the role of that individual in an advanced technology industry, and the type of training recommended to prepare an individual to manage projects.

- Peter F. Drucker, "The Coming of the New Organization," *Harvard Business Review,* January–February 1988, pp. 45–53. In this article, Peter F. Drucker opines that organizations of the future will be "information-based" and have reduced numbers of hierarchical levels, with much of the work being done in task-focused teams. He further believes that these teams will work on new product and process development from the conceptual state of the product until it is established in the market.
- Dr. Alaa A. Zeitoun and Dr. Andy W. Helmy, "The Pyramids and Implementing Project Management Processes," *Proceedings,* Project Management Institute, 28th Annual Seminar/Symposium, Chicago, 1997, pp. 593–596. This paper addresses the building of the pyramids and the concepts behind this enormous project. The authors build a case that many of the concepts and processes of modern project management were likely applied in the building of this antique.

1.13 DISCUSSION QUESTIONS

1. How do projects fit into the overall design of enterprise purposes, in particular with the choice elements of an enterprise?
2. Describe and discuss situations in your work or personal experience that fit the definition of a project. How effectively were/are these managed?
3. In what ways do the concepts of project management appear to violate traditional, established ways of managing?
4. How do the three parameters of a project—cost, time, and technical performance—interact?
5. What are the various roles that need to be accounted for on a project team?
6. How do the leader's and the project manager's styles affect how these roles are played?
7. List and discuss the various liaison devices described in the chapter.
8. What are some of the advantages of the use of teams in organizations?
9. Why is it important for project managers to adapt "synergistic thinking"?
10. Discuss the steps involved in the management of change. What additional steps can be taken?
11. How can a young professional's experience in working on small projects benefit his or her professional development?
12. Describe what is meant by team management.

1.14 USER CHECKLIST

1. Are there clear and appropriate "choice elements" identified in your organization?
2. Does the management of your organization recognize projects and understand the concepts of project management?
3. How well does your organization use project management in dealing with change?
4. Are clear lines of authority, responsibility, and accountability defined for project team members?
5. How well are the liaison devices described in the chapter used to integrate activities across organizational lines?
6. Are cost, time, and technical performance objectives defined for each project? Are they properly managed? Do existing projects have a probable and suitable operation or strategic fit?
7. Does your organization use teams to its advantage? In what ways?
8. Is your organization prepared for change? Is change being managed effectively?
9. Are young professionals being properly trained in the concepts of project management so that they are prepared to take on the responsibilities of a project or team manager?
10. Does top management provide support and opportunities for functional and project managers to plan, organize, motivate, direct, and control those project activities for which they are responsible?
11. Does the organization use contemporaneous, state-of-the-art project management techniques in the management of projects?
12. How is project management integrated into the strategic management philosophies of the organization?

1.15 PRINCIPLES OF PROJECT MANAGEMENT

1. Project management has earned its rightful place in the evolution of the management discipline.
2. Strategic management and project management are interdependent in the management of an enterprise.
3. Projects are a key "choice element" in the management of an organization.
4. Projects are the building blocks of change in organizations.
5. The evolution of project management has influenced the continued evolution of general management theory and practice.

1.16 PROJECT MANAGEMENT SITUATION—EXTERNAL AND INTERNAL PROJECTS

Projects, as building blocks in the design and execution of enterprise strategies, can be either external or internal in nature. An *external* project is one undertaken for, or on behalf of, stakeholders who are not part of the enterprise structure, such as design and construction of a bridge, highway, or new product design. In an external project, the customer is located outside the enterprise, such as another company, government, or military organization. An *internal* project is one to be carried out primarily for the improvement of organizational processes, such as productivity improvements, training initiatives, organizational restructuring, or reengineering. Internal projects usually have an internal customer, such as a manufacturing manager who wishes to update the company's manufacturing equipment, build a new plant, or develop enhanced information systems capability.

Companies that are in economic difficulties often undergo downsizing or restructuring. Improvements in organizational processes can be gained from reengineering projects. The development of new award systems, flexible work practices, improvement of quality, or the flow of work on the production line, can be accomplished by using project teams. Although many of these projects are modest, compared to large projects that are being developed for an outside customer, for the members of the enterprise the internal projects usually indicate that a change in the operating policies is forthcoming.

Organizations are basically systems of people using resources to accomplish enterprise mission and purposes. Changing how an organization works is thus fundamentally about changing how people work and relate to each other. Organizational development projects are meant to be a planned process of change for the people in an organization's culture, often using management principles and behavioral processes. A major reason why some of these changes fail is because the action that is planned and undertaken is not treated as a project and—not managed as a project.

1.17 STUDENT/READER ASSIGNMENT

The reader/student should select an organization with which he or she is familiar and accomplish the following:

1. Identify the *internal* and *external* projects that are underway in the organization.
2. Determine the strategic or operational changes that each of the projects will likely impact.
3. Assess the effectiveness with which each project is being managed. Are there differences in how such projects are being managed?
4. Identify some probable forthcoming changes likely to impact the organization for which project management concepts and process can be applied.
5. Give thought to what project management principles might be applied in the management of these projects.

CHAPTER 2
THE PROJECT MANAGEMENT PROCESS

"The distance is nothing; it is only the first step which counts."
MADAME DUDEFFARD, 1697–1784

2.1 INTRODUCTION

Project management is a series of activities embodied in a process of getting things done on a project by working with project team members and other stakeholders to attain project schedule, cost, and technical performance objectives. The project management process is adapted from the general management process.

In this chapter the project management process will be explained along with an exploration of the project life cycle. How to manage this life cycle will be examined, along with an explanation of how project life cycles and uncertainty are linked. An early linkage of a *project* life cycle and a *product* life cycle will be presented.

Project management is a series of activities embodied in a process of getting things done on a project by working with members of the project team and with other people in order to reach the project schedule, cost, and technical performance objectives. This description helps identify project management, but it does not tell too much about how a project manager reaches project goals and objectives. This chapter will describe the project management process along with the idea of the life cycle. First we describe the management process.

2.2 THE GENERAL MANAGEMENT PROCESS

A process is defined as a system of operations in the design, development, and production of something, such as a project. Inherent in such a process is a series of actions, changes, or operations that bring about an end result, in the case of a project attainment of its cost, schedule, and technical performance objectives. Another meaning of a process is that it is a course or passage of time in which something is created—an ongoing movement or progression.

As a manner and means of progressing, a project management process sets the tone for the conceptualization of project management; the planning and execution of concepts, methods, and policies; and the commitment of resources to the project endeavors. Taken in its entirety a project management process provides a paradigm for how the management functions of planning, organizing, motivation, directing, and control will be carried out in the commitment of resources on the project.

Figure 2.1 provides a simple model of the management functions portrayed in the larger context of the management process. Each of these functions can stand alone—yet in their design and execution they are interdependent in the overall management process of an organization or a project.

The management discipline is usually described as a *process* consisting of distinct yet overlapping major activities or functions. A brief review of the early conceptualization of the management discipline in terms of the major activities or functions involved follows.

The management discipline that received recognition early in the twentieth century reflected to some degree the practices of the time. Although there were a few singular writings in historical times, there was no attempt to organize and portray an overall philosophy and concept of management. But in the early writings of Frederick W. Taylor and Henri Fayol the first integrated ideas about management started to take form. Taylor's book, *The Principles of Scientific Management* (1911), centered around the improvement of capabilities of people on the production line. Fayol, on the other hand, wrote his classic *General and Industrial Management* from the perspective of the overall management of the enterprise. Fayol's definition of management as consisting of forecasting and planning to organize, to command, and to coordinate and to control, set the stage for the differentiation of managerial activities from the technical activities of the enterprise. To quote Fayol: "To plan is to foresee and provide a means of examining the future and drawing up the plan of action. To organize means building up the dual structure, material and human, of the undertaking. To command means maintaining activity among the personnel. To co-ordinate means binding together, unifying, and harmonizing all activity and effort. To control means seeing that everything occurs

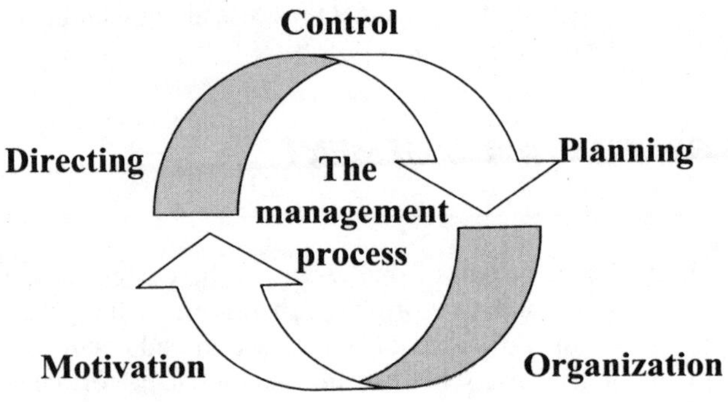

FIGURE 2.1 The management process.

in conformity with established rules and expressed command."[1] Fayol goes on to note that management is an activity spread between the head and members of the body corporate. His early description of managerial functions being carried out from the highest to the lowest levels of an enterprise established the traditional hierarchical state of management—a state of perception and thought which characterized both the theory and practice of management for many decades. Any perception of the horizontal nature of management was largely confined to the idea of coordinating the technical activities of the enterprise. In the traditional paradigm of management authority, responsibility and accountability were primarily considered to be vertical forces, extending from the senior level of the enterprise down to the worker level.

The idea of using teams as an alternative organizational design was described very little in the literature, but there were singular examples where the use of teams to integrate functional activities in the enterprise was recommended. An early advocate of teams was Mary Parker Follett.

In the 1920s she extolled the benefits of teams and participative management, and said that leadership comes from ability rather than hierarchy. She advocated empowerment and tapping the knowledge of workers, and supported the notion of cross-functioning in which a horizontal rather than a vertical authority would foster a freer exchange of knowledge within the organization. She fervently believed that knowledge and experience determine who should lead.[2]

A simple yet important way of further describing the management process through its major functions is indicated below:

- *Planning* What are we aiming for and why? In the execution of this function, the organization's mission, objectives, goals, and strategies are determined.
- *Organizing* What's involved and why? In carrying out the organizing function, a determination is made of the need for human and nonhuman resources—and how those resources will be aligned and used to accomplish the organization's mission. Authority, responsibility, and accountability are the "glue" that holds an organization together.
- *Motivation* What brings out the best performance of people in supporting the organization's purposes?
- *Directing* Who decides what and when? In the discharge of this management function, the manager provides the face-to-face leadership of the organizational members.
- *Controlling* Who judges results and by what standards? In this function the manager monitors, evaluates, and controls the effectiveness and efficiency in the utilization of organizational resources.

Figure 2.2 portrays the relationship of project management resources and the core functions of project management. There is much literature describing these management functions. Thousands of articles and hundreds of books are published every year about the management discipline.

---

[1]Henri Fayol, *General and Industrial Management* (London: Sir Isaac Pitman & Sons, 1949), pp. 5–6.
[2]Dana Wechsler Linden, "The Mother of Them All," *Forbes,* January 16, 1995, pp. 75–76.

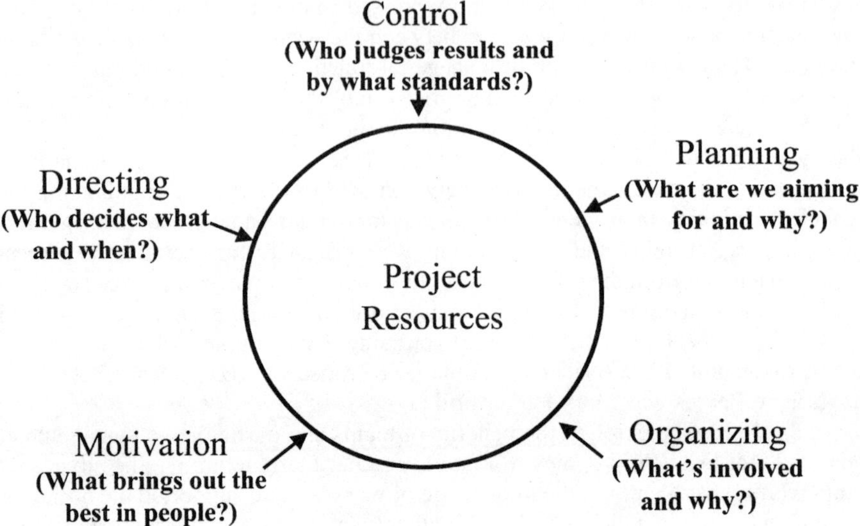

FIGURE 2.2 The core functions of project management.

TABLE 2.1 Principal Responsibilities: Project Management and General Management

Project management	General management
• Specific cost, schedule, and technical performance objectives	• Strategic management of the enterprise
• Matrix organizational design	• Vertical organizational design
• Ad hoc in nature	• Concerned with enterprise mission, objectives, and goals
• Focal point for functional and enterprise interfaces	• Ongoing enterprise
• Concerned with product, service, and enterprise process design and development	• Concerned with enterprise stakeholders
• Supports organizational strategies	• Seeks enterprise efficiency and effectiveness
• Concerned with project stakeholders	• Integrates functional and project activities

Table 2.1 shows the principal responsibilities between project management and general management. Both draw on the theory and practice reflected in the management discipline. There are some subtle differences, however, reflected in the main management considerations that are involved in the management of the project or the enterprise, as the case may be. Both general management and project management have the same basic philosophies, even though the application of the management process may differ depending on the applications in each area. Both make and implement decisions, allocate resources, manage organizational interfaces, and provide leadership of the people who are involved in the enterprise and the project. The differences and similarities are subtle yet important for both the managers and the professionals that are involved.

2.3 THE PROJECT MANAGEMENT PROCESS

In Table 2.2, the project management process is portrayed in terms of its major functions. The activities noted under each of these functions are only representative. Effective project management requires many more activities or "work packages" under each of these functions. More descriptions of these functions are found elsewhere in this book.

TABLE 2.2 Representative Functions/Processes of Project Management

Planning: What are we aiming for and why?
Develop project objectives, goals, and strategies.
Develop project work breakdown structure.
Develop precedence diagrams to establish logical relationship of project activities and milestones.
Develop time-based schedule for the project, based on the precedence diagram.
Plan for the resource support of the project.
Organizing: What's involved and why?
Establish organizational design for the team.
Identify and assign project roles to members of the project team.
Define project management policies, procedures, and techniques.
Prepare project management charter and other delegation instruments.
Establish standards for the authority, responsibility, and accountability of the project team.
Motivation: What motivates people to do their best work?
Define project team member needs.
Assess factors that motivate people to do their best work.
Provide appropriate counseling and mentoring as required.
Conduct initial study of impact of motivation on productivity.
Directing: Who decides what and when?
Establish "limits" of authority for decision making for the allocation of project resources.
Develop leadership style.
Enhance interpersonal skills
Prepare plan for increasing participative management techniques in managing the project team.
Develop consensus décision-making techniques for the project team.
Control: Who judges results and by what standards?
Establish cost, schedule, and technical performance standards for the project.
Prepare plans for the means to evaluate project progress.
Establish a project management information system for the project.
Evaluate project progress.

The management functions used in the management of a project are the principal elements in making and implementing decisions about the use of resources applied to project purposes. A checklist to review how well these functions are carried out can be useful. A representative general checklist is shown in Table 2.3.

2.4 THE PROJECT LIFE CYCLE

Project management is a continuing process. New demands always are put on the project team and have to be coordinated by the project manager through a process of planning, organizing, motivating, directing, and controlling. As new needs come up before the project, someone has to satisfy these needs, solve the problems, and exploit the opportunities. The project originates as an idea in someone's mind, takes a conceptual form, and eventually has enough substance so that key decision makers in the organization select the project as a means of executing elements of strategy in the organization. In practice, the project manager must learn to deal with a wide range of problems and opportunities, each in a different stage of evolution and each having different relationships with the evolving project. This continuing flow of problems and opportunities, in a continuous life-cycle mode, underscores the need to comprehend a project management process which, if effectively and efficiently planned for and executed, results in the creation of project results that complement the organizational strategy.

Managing a large project is so complex that it is difficult to comprehend all the actions that have to be taken to successfully plan and execute the project. We need to divide the project into parts in order to grasp the full significance of each part and just where that part fits in the scheme of the project. We have to look at the project parts, its "work packages," its logical flow of activities, and the phases that the project goes through in its evolution, growth, and decline.

The management of a project is like the management of any activity. Two fundamental steps are involved in such management, namely, the making and implementation of decisions. There is a substantial body of knowledge regarding how decisions can be made—in particular how to consider the evaluation of risk and uncertainty in the potential use of resources committed through the decision process. Decision analysis in projects is an important responsibility of the project team, facilitated by the project manager. A 12-part series that reviews and summarizes the concepts and processes behind good project decisions was published in *PM Network.*[3]

Using a model of the project's life cycle is useful in identifying and understanding the total breadth and longevity of the project and as a means to identify the management functions involved in the project life cycle. A project's life cycle contains a series of major steps in the process of conceptualizing, designing, developing, and putting in operation the project's technical performance "deliverables." These major steps are the key work elements around which the project is managed. The context of a project life cycle—and how the conceptualization and

[3]See John R. Schuler, "Decision Analysis in Projects: Summary and Recommendations," *PM Network,* October 1995, pp. 23–27.

TABLE 2.3 Decisions of the Team Management Functions

Team planning
What is the mission or "business" of the team?
What are the team's principal objectives?
What team goals must be attained in order to reach team objectives?
What is the strategy used by the team to accomplish its purposes?
What resources are available for the team's use in accomplishing its mission?
Team organization
What is the basic organizational design of the team?
What are the individual and collective roles of the team that must be identified, defined, and negotiated?
Will the team members understand and accept the authority, responsibility, and accountability that is assigned to them as individuals and as a team?
Do the team members understand their authority and responsibility to make decisions?
How can the team effort be coordinated so that the members will work in harmony, not against one another?
Team motivation
What motivates the team to do their best work?
Does the team manager provide the leadership style acceptable to the members of the team?
Is the team "productive"? If not, why not?
What can be done to increase the satisfaction and productivity of the team members?
Are the team meetings conducted in such a manner that people attending are encouraged? Discouraged?
Team direction
Is the team leader qualified to lead the team?
Is the team leader's style acceptable to the members of the team?
Do individual members of the team assume leadership in the areas where they are expected to lead?
Is there anything that the team leader can do to increase the satisfaction of the team members?
Does the team leader inspire confidence, trust, loyalty, and commitment among the team members?
Team control
Have performance standards been established for the team? For the individual members?
What feedback on the team's performance does the manager have who appointed the team?
How often does the team get together to formally review its progress?
Has the team attained its objectives and goals in an effective and efficient manner?
Do the team members understand the nature of control in the operation of the team?

Source: Adapted from M. H. Mescon, et al., *Management,* (New York: Harper & Row, 1991), p. 167.

development of that life cycle provide a useful model for project management—will be described in the material that follows.

But first, an example of a project's life cycle. All projects go through a series of phases in their life cycle as they progress to completion, transforming the project resources to a product, service, or organizational process. As the project results are transformed into a product, service, or organizational process, they create value for the enterprise. Modifications and improvements are typically added to the project results as they are provided to customers in the marketplace by way of new models, modified configuration, reduced price, and so forth, and as the results of the projects compete in the marketplace for which they were designed and developed. Project results, like most other things in the world, are always undergoing change in order to remain competitive in their marketplaces. For example, a new car, or a car that has been modified from the original configuration, may eventually be discontinued because of a key decision made by the car manufacturer. For example, in December 2000 the General Motors Company announced their intention to kill the 103-year-old Oldsmobile name, cut 15,000 jobs, and reduce 15 percent of GM's factory capacity in Europe. These were the first steps in a sweeping overhaul of the biggest U.S. automaker since the dark days of the early 1990s. GM is also setting up special project teams to work with suppliers on how to save money for parts and supplies. The management of GM hopes that the elimination of the Oldsmobile car and cost savings with suppliers will provide funds for the development of "hot new cars and trucks" that the number 1 automaker is counting on to reverse the decades of declining market share.[4]

The reader can note the many opportunities in the GM situation presented above for the use of project management:

- Development of the original Oldsmobile auto—although not recognized as such since the project management discipline did not exist at that time. What did exist was the need to bring together many disciplines and organizational functions to develop the original car and all subsequent models.
- Canceling the Oldsmobile auto required the use of project management, as well as using the discipline in downsizing the manufacturer's plants and other facilities.
- Setting up project teams to work with suppliers to reduce vendor costs.
- Need for project teams to develop new "hot cars" and trucks to reverse GM's declining market share.

Thus project management was used to create change in the strategic and operational purposes of the company, and to deal with the change coming from the marketplace that was impacting the company.

The phases of a project life cycle—and what happens to the project during its life cycle—depend on the distinctive nature of the project. The phases that are

[4]Gregory L. White, "Killing Off Oldsmobile Was Just the Beginning," *The Wall Street Journal,* December 18, 2000, pp. A1, A13.

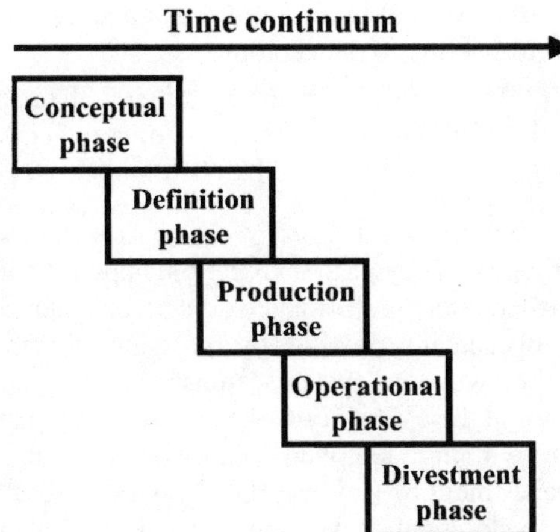

FIGURE 2.3 Generic model of project life cycle.

described in this chapter are generic and representative. Figure 2.3 is intended to provide a broad notion of the life cycle of a generic project to show how the project starts off with a conceptual model, goes through definition of its cost, schedule, and technical performance objectives, becomes operational, and will finally go into a divestment phase and is likely to be replaced by a new or improved project. Many different yet similar phases have been described in the life of a project. These phases typically include:

- *Idea* The generation of the notion or concept for a new product, service, or process that provides the basis for the creation of something new for the enterprise, which did not previously exist.
- *Research* The patient, systematic search and inquiry and examination into a field of knowledge. Such an inquiry is taken to establish facts or principles, and when successful, should convert an idea into a practical plan for further work. If the applied research does not result in anything of value, the project will be redirected or terminated as appropriate.
- *Design* The means for the conversion of the idea into a plan for a product, service, or organizational process.
- *Development* Usually means taking a design specification and converting it into an actual product, service, or process. This is done through added features of appearance and configuration change, and through the stages of experimental models, breadboard models, experimental prototypes, and production prototypes. The resulting outcome of the development phase is a product or service ready for production.
- *Marketing* Involves determining the need for the product or service, and the development of a sales and marketing plan to deliver the results to customers. The marketing effort is usually under way prior to or during the design and development efforts.

- *Production* The conversion of human and nonhuman resources into a product or service that provides value to the customers.
- *After-sales services* Means to provide the customer with maintenance, technical documentation, and logistics support for the product or service during the time that the product or service is being used by customers.

Projects, like organizations, are always in motion as each proceeds along its life cycle. Projects go through a life cycle to completion, hopefully on time, within budget, and satisfying the technical performance objective. When completed, the project joins an inventory of capability provided by the organization that owns the project.

All projects—be they weapons systems, transportation systems, or new products—begin as a gleam in the eye of someone and undergo many different phases of development before being deployed, made operational, or marketed. For instance, the U.S. Department of Defense (DOD) uses a life-cycle concept in the management of the development of weapons systems and other defense systems. An early U.S. Air Force version of this life cycle identifies a number of phases, each with specific content and management approaches. Between the various phases are decision points, at which an explicit decision is made concerning whether the next phase should be undertaken, its timing, and so on. Generically, these phases are as follows:

1. *The conceptual phase* During this phase, the technical, military, and economic bases are established, and the management approach is formulated.
2. *The validation phase* During this phase, major program characteristics are validated and refined, and program risks and costs are assessed, resolved, or minimized. An affirmative decision concerning further work is sought when the success and cost realism become sufficient to warrant progression to the next phase.
3. *The full-scale development phase* In the third phase, design, fabrication, and testing are completed. Costs are assessed to ensure that the program is ready for the production phase.
4. *The production phase* In this period, the system is produced and delivered as an effective, economical, and supportable weapons system. When this phase begins, the weapons system has reached its operational ready state and is turned over to the using command. During this period, responsibility for program management is transferred as an Air Force logistics supporting capability within the Air Force.
5. *The deployment phase* In this phase, the weapons system is actually deployed as an integral organizational combat or support unit somewhere within the Air Force.

The management of technology can be viewed in a life cycle context. Cleland and Bursic have, in a research project, studied the management of technology within a major corporation. One of their conclusions is that technology can be managed from the context of a life cycle. Figure 2.4 illustrates this life cycle.

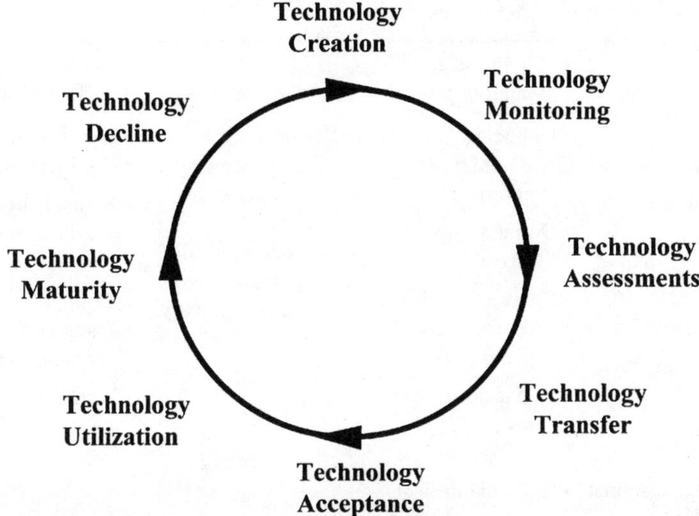

FIGURE 2.4 The life cycle of technology. (*Source: David I. Cleland and Karen M. Bursic,* Strategic Technology Management: Systems for Products and Processes, *AMACOM American Management Association, 1992, p. 23.*)

The results of this research project set a new standard in the corporation for the use of project management as a focus for the development of this new product.[5]

2.5 PRODUCT DEVELOPMENT

The National Society of Professional Engineers (NSPE) has developed a valuable and comprehensive document describing the new-product stages of development. These stages are defined, the objectives of the stage are presented, the engineering activities are described, and the information needed to communicate the actions and activities in each stage is provided.[6]

2.6 MANAGING THE LIFE CYCLE

One of the first undertakings in planning for a project is to develop a rough first estimate of the major tasks or work packages to be done in each phase.

There are many ways of looking at a project life cycle. Adams and Brandt suggest two ways of looking at the managerial actions by project phase and the tasks accomplished by project phase. See Table 2.4 and Fig. 2.5.

---

[5]David I. Cleland and Karen M. Bursic, *Strategic Technology Management: Systems for Products and Processes,* AMACOM American Management Association, 1992, p. 23.

[6]See, for example, *Engineering Stages of New Product Development,* Publication 1018, pp. 16–23, National Society of Professional Engineers, 1420 King Street, Alexandria, VA 22314.

TABLE 2.4 Managerial Actions by Project Phase

Phase 1 Conceptual phase	Phase 2 Planning phase	Phase 3 Execution phase	Phase 4 Termination
• Determine that a project is needed. • Establish goals. • Estimate the resources that the organization is willing to commit. • "Sell" the organization on the need for a project approach. • Make key personnel appointments.	• Define the project organization approach. • Define project targets. • Prepare the schedule for execution phase. • Define and allocate tasks and resources. • Build the project team.	• Perform the work of the project (i.e., design, construction, production, site activation, testing, delivery, etc.).	• Assist in transfer of project product. • Transfer human and nonhuman resources to other organizations. • Transfer or complete commitments. • Terminate project. • Reward personnel.

Source: John R. Adams and Stephen Brandt, "Behavioral Implications of the Project Life Cycle," in David I. Cleland and William R. King (eds.), *Project Management Handbook* (New York: Van Nostrand Reinhold, 1983), p. 227.

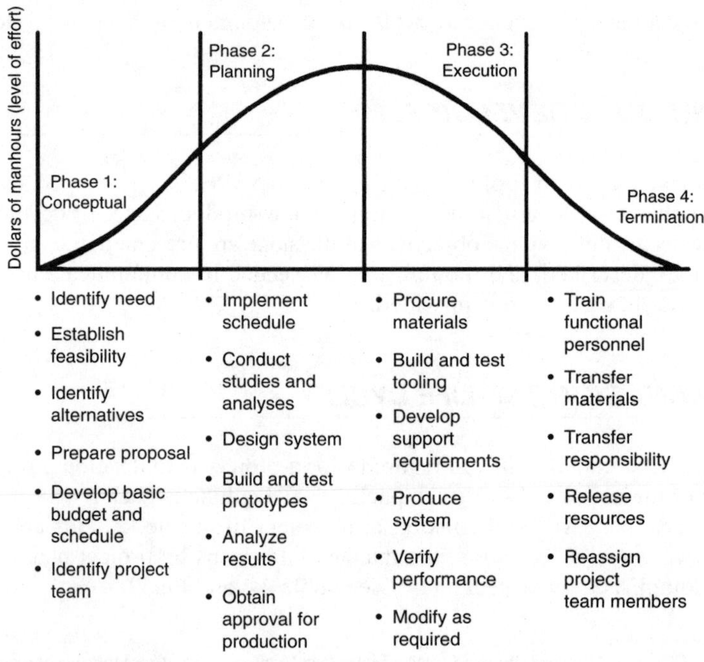

FIGURE 2.5 Tasks accomplished by project phase. [*Source: John R. Adams and Stephen E. Brandt, "Behavioral Implications of the Project Life Cycle," in David I. Cleland and William R. King (eds.),* Project Management Handbook *(New York: Van Nostrand Reinhold, 1983), p. 227.*]

Once established, the life-cycle model should be updated as more is learned about the project. As the project progresses through its life cycle, the project exhibits ever-changing levels of cost, time, and performance. The project manager must make correspondingly dynamic responses by changing the mix of resources assigned to the project as a whole and to its various work packages. Thus budgets will fluctuate substantially in total and in terms of the allocation to the various project work packages. The need for resources and various kinds of expertise will similarly fluctuate, as will virtually everything else. This is portrayed in Fig. 2.6, which shows changing levels of budget and of engineering and marketing personnel for various stages of the life cycle.

This constantly changing picture of the life cycle is an underlying structural rationale for project management. The traditional hierarchical organization is not fully designed to cope with managing such an always-changing mix of resources. Rather, it is designed to control and monitor a much more static entity that, day to day, involves stable levels of expenditures, numbers of persons, and so forth.

As has been stated earlier, project management is used by many different organizations. Banks, such as the Security Pacific National Bank in Los Angeles, California, use project management. At this bank project management was used in the automation of its loan collection system. Security Pacific had decided to centralize all of its collections, scattered throughout some 600 offices and collection centers. The plan was to devise six regional adjustment centers and a charge card center for all collection operations. Using project management, the development project was completed on time and within the budget that was allocated for it. There was a 100 percent increase in collector productivity in the first 6 months of

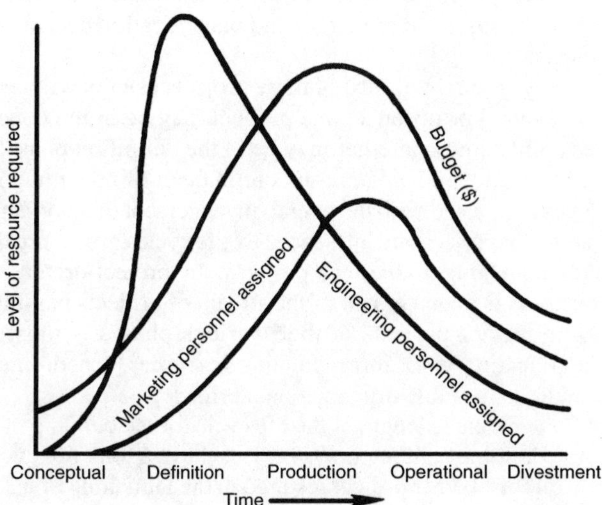

FIGURE 2.6 Changing resource requirements over the life cycle. [*Source: John R. Adams and Stephen E. Brandt, "Behavioral Implications of the Project Life Cycle," in David I. Cleland and William R. King (eds.),* Project Management Handbook *(New York: Van Nostrand Reinhold, 1983), p. 227.*]

operations under the new collection system. There was also a 95 percent decrease in paperwork generated by the collectors. Also, there was a significant reduction in loan delinquency and charge-offs. It was estimated that Security Pacific would save about $4 million by the end of the year because of the implementation of the new collection system.[7]

2.7 PROJECT LIFE CYCLES AND UNCERTAINTY

As the project life cycle progresses, the cost, time, and performance parameters must be "managed." This involves continuous replanning of the as yet undone phases in the light of emerging data on what has actually been accomplished.

The project team must rethink much during the project life cycle to modify and fine-tune the work packages for each phase. Archibald notes, "The area of uncertainty is reduced with each succeeding phase until the actual point of completion is reached."[8]

Many organizations can be characterized at any time by a "stream of projects" that place demands on its resources. The combined effect of all the projects facing an organization at any given time determines the overall product, service, and process status of the organization at that time and gives insight into the organization's future.

The projects facing a given organization at a given time typically are diverse—some products are in various stages of their life cycles and embody different technologies; other products are in various stages of development. Management subsystems are undergoing development. Organizational units are in transition. Major decision problems, such as merger and plant location decisions, are usually studied as projects.

Moreover, at any given time, each of these projects usually will be in a different phase of its life cycle. For instance, one product may be in the conceptual phase undergoing feasibility study; another may be in the definition phase. Some might be in production. Others are being phased out in favor of upcoming models.

The challenges associated with the overall management of an organization that is involved in a stream of projects are influenced by life cycles, just as are the challenges associated with managing individual projects. In project-driven organizations whose main business is management of the stream of projects passing through the organization, the mix of projects in their various phases is most challenging, particularly in allocating work force, funding resources, scheduling work loads, and so on, to maintain a stable organizational effort.

In Chapter 1, the probable length of the life cycle for the building of the historical artifacts was noted. Some contemporary projects have a long life cycle as well. At Motorola, the Iridium 11-year project resulted in the launching of a 250-ton rocket carrying the first 3 of 66 planned satellites into orbit, 420 nautical miles above the

---

[7]H. B. Einstein, "Project Management: A Banking Case Study," *The Magazine of Bank Administration,* vol. 58, issue 43, 1982, p. 36.

[8]Russell D. Archibald, *Managing High-Technology Programs and Projects* (New York: Wiley, 1976), p. 23.

earth. By the time the entire constellation of satellites was up in September 1998, Motorola and its 16 coinvestors expected to spend close to $5 billion, making the project one of the largest privately funded infrastructure projects ever. The technical, regulatory, and political complexity of the project is numbing. More than 25,000 complex design elements have come forth. The project team scoured the globe, seeking partners and money to build the project, which began a new era of humanity-helping global interconnectedness. At the end of 1996, 2000 people were working on the project, up from 20 at the start. Marketing of the Iridium project was challenging. A sophisticated global marketing campaign is under way to sell its phones—aiming at vastly different markets around the world. Even if the project has less than hoped for success, it has yielded valuable indirect benefits, such as enhanced technology, greater attention to satellite technology, and a modern production facility that can be used for other satellite systems.[9]

Drug research projects take an average of 12 years and cost an average of about $359 million. During the life cycle of these projects the disciplines critical to the project change. Projects often start with a biologist, and then a chemist and other disciplines become involved. Once developed, the new drugs have to be tested on animals and finally on human beings; then, on to manufacturing. Development projects require enormous amounts of expertise with a willingness to promote the free flow of information across disciplines and organizational boundaries.

The point to be remembered is that the management process has direct application to the management of the project resources and should so be approached in a life-cycle context.

2.8 TO SUMMARIZE

The major points that have been covered in this chapter include:

- The project management process was described as a guide for the management of those major activities involved in having an idea for a project and carrying that idea through to attainment of the project objectives.
- A process was described as a system of operations in the design, development, and production of something—such as a project.
- Henri Fayol, the noted French author, was the first individual to conceptualize and define the management functions. His definition set the stage for the subsequent examination of these functions in the management of an enterprise.
- Planning deals with how to determine the likely future forces facing an enterprise—and how to prepare the enterprise for its future.
- Organizing deals with how best to provide an orderly alignment of the people and the resources used by the enterprise for the accomplishment of its purposes.
- Motivation establishes the philosophy, attitude, and means for bringing out the best in people.

[9] Quentin Hardy, "Higher Calling: How a Wife's Question Led Motorola to Chase Global Cell-Phone Plan," *The Wall Street Journal,* December 16, 1996, p. 1.

- Directing is the face-to-face situation in which the leader provides a vision and the means of accomplishing that vision for the project.
- Monitoring, evaluation, and control provide the means for determining how well the project and organizational strategies are being used in meeting objectives and goals through the employment of predetermined strategy.
- A checklist to assist the project manager in determining how effectively decisions are being made on the project was provided in the chapter.
- Everything living in the world today goes through a life cycle. Projects are no different.
- Several paradigms of a project life cycle were provided to assist the reader in understanding how such paradigms can guide the management of a project.
- A project life cycle can be portrayed using the work packages of the project appropriately placed in the phases in that life cycle.
- Some projects have a long life cycle, such as that presumed about the building of the Great Pyramids. However, contemporary projects can have a long life cycle as well.
- Any new or improved product, service, or process goes through a life cycle as conceptual models are built, and design, development, production, and after-sales support initiatives are carried out in such a manner that the project results provide value to the customer.
- The management system for a project is built around the design and execution of the managerial functions that have been described in this chapter.

2.9 ADDITIONAL SOURCES OF INFORMATION

The following additional sources of project management information may be used to complement this chapter's topic material. This material complements and expands on various concepts, practices, and the theory of project management as it relates to areas covered here.

- John R. Adams and Miguel E. Caldentey, "A Project Management Model," and J. Davidson Frame, "Tools to Achieve On-Time Performance," chaps. 5 and 8 in David I. Cleland (ed.), *Field Guide to Project Management* (New York: Van Nostrand Reinhold, 1997).
- David I. Cleland, "Prudent and Reasonable Project Management," in David I. Cleland, Karen M. Bursic, Richard J. Puerzer, and Alberto Y. Vlasak, *Project Management Casebook,* Project Management Institute (PMI). (First published in *Project Management Journal,* December 1985, pp. 90–97.)
- Harold Koontz and Cyril O'Donnell, *Principles of Management,* 2d ed. (New York: McGraw-Hill Book Company, 1959). This book is considered to be one of the early "classics" in presenting a management theory that embodies a

principle and a process perspective. The authors believe that management is one of the more important activities through its task of getting things done through people.

- Charles B. Randall, *The Folklore of Management* (New York: Wiley, 1997). The first edition of this book was written more than 30 years ago. A basic *commonsense* treatment of business management, Randall reveals the elements of success as well as failure in the corporate world. The simplicity and humor of this book, plus its insight into the management discipline, makes it a "management classic" in the literature. Any manager, whether at the senior or junior level of the enterprise—or a project manager—will find excellent insight into the management challenge in modern times by a reading of this book.
- Thomas C. Belanger, "Choosing a Project Life Cycle," in David I. Cleland (ed.), *Field Guide to Project Management* (New York: Van Nostrand Reinhold, 1997), pp. 61–73. The author believes that the standardization of project management practices in an organization requires a flexible life-cycle model. Customizing a life-cycle model consists of deleting nonapplicable activities and tasks, and adding unique activities and tasks. By examining these ideas, a flexible life-cycle model can be developed for an organization that will increase the proportion of successful projects.
- William R. Duncan, "The Process of Project Management," *Project Management Journal,* vol. 24, no. 3, September 1, 1993. In this article the process of project management is integrative, principally because a change in one department usually affects other departments. Trade-offs among project objectives are often required; successful project management can only be realized through the optimum handling of such interactions.
- Stephen E. Brandt, Julius C. Larsen, et al., "Organizational Climate Change in the Project Life Cycle," *Research Management,* vol. 20, no. 5, 1977. This article from a leading research journal puts forth the idea that organizational climate differs among project phases, but not always in a logical pattern. The results of the authors' research indicate that it may be important to think in terms of changing or differing organizational climates, not just the climate of the project organization.

2.10 DISCUSSION QUESTIONS

1. Identify and define the management functions discussed in the chapter.
2. Within the management function of planning, project managers and team members should settle on project objectives, goals, and strategies. How can this task help define the "strategic fit" of the project to the organization?
3. Discuss the difference between organizational policies and procedures.
4. Project review meetings often are used as a tool for controlling projects. In general, what kinds of questions should be addressed in these meetings?
5. List and briefly define the phases of a generic project life cycle.

6. What management functions are most important in the conceptual phase of the project life cycle?
7. During the definition phase of a project, it is important for management to design and develop a management system to support the project. Discuss the importance of each system.
8. Discuss what is meant by the project "losing its identity" and being "assimilated into the ongoing business of the user."
9. Discuss the importance of the divestment phase as a tool for avoiding technological obsolescence.
10. What is meant by the phased approach for project development?
11. How can a project manager, by understanding the project life cycle, use the concepts of this chapter to help direct and control projects?
12. Discuss the challenge of managing a project-driven organization with ongoing projects at various stages in their life cycles.

2.11 USER CHECKLIST

1. Do the managers of your organization understand and use the management functions in the management of projects?
2. Do the managers of your organization understand the project management process?
3. In the early stages of projects within your organization, are objectives, goals, and strategies clearly defined?
4. Are project management roles assigned? Are standards for responsibility and accountability established?
5. Does management consider the needs of individual team members in order to motivate people to do their best work?
6. How well does your organization use participative management techniques and consensus decision making in your project management work?
7. What techniques do project managers, within your organization, use to control project problems?
8. Do your organization's managers truly understand the implications of the project life cycle?
9. Does your organization use a systematic approach for managing projects?
10. During the conceptual phase of a project, does the organization attempt to determine the potential strategic fit of the project?
11. Does the organization recognize the need for a management system for managing projects?
12. Are projects purposely put through the divestment phase in order to avoid technical obsolescence? Why, or why not?

2.12 PROJECT MANAGEMENT PRINCIPLES

1. Project management is a series of activities comprising the process of applying management principles to project activities.
2. The management process consists of the organic functions of management.
3. All projects go through a series of phases in their life cycle as they progress to completion, transforming the project resources to a product, service, or organizational process to support organizational strategies.
4. The area of uncertainty in a project is reduced with each succeeding phase until the actual point of completion is reached.
5. A project life cycle can be portrayed using the work packages of the project appropriately placed in the phases of that life cycle.

2.13 PROJECT MANAGEMENT SITUATION—STRATEGIC MONITORING AND CONTROL

In Chapter 1 the interdependent relationship between strategic management and project management was presented. In Chapter 2 the project management process was described within the organic functions of management: planning, organizing, motivation, direction, and control. In the material that follows, the *control* function is presented from the perspective of strategic management.

Strategic Management Monitoring, Evaluation, and Control

Strategic management, like any other management activity, needs to be continually monitored, evaluated, and controlled to ascertain how the actual results compared with the results that were planned.

Steps in the Control Cycle There are several distinct steps in the project-oriented strategic management control system. In the material that follows a brief insight into the nature of these steps is provided. We will consider performance standards first.

Performance Standards Performance standards are based on the "choice elements" that have been established during the strategic process for the enterprise. Of these choice elements, projects and goals are particularly important. Goals represent milestones in the progress of the enterprise in accomplishing its objectives. Many of the goals of the enterprise are made up of projects under development and completion. For example, a goal for the enterprise to improve its manufacturing operations would have several key projects such as:

- Acquisition of state-of-the-art machine tools
- Construction of a new "green field" manufacturing facility

- Design and implementation of new training programs
- Design and development of automated factory production capabilities
- Redesign of the organizational structure to facilitate the operation of "self-managed" production teams
- Benchmark manufacturing capabilities of "best in the industry" firms as well as enterprise competitors

Following good project management practices, each of these projects would have appropriate objectives, schedules, and cost estimates. By reviewing the status of these projects, valuable insight into the progress that is being made toward the realization of company goals would be gained.

Comparing Planned and Actual Performance

An explicit review of the actual progress vis-à-vis planned progress provides strategic managers the intelligence to make an informed judgment of how well the strategic goals of the enterprise are being developed and implemented through the use of projects. An explicit review helps give answers to the following key questions:

- Is the project's progress consistent with the elements needed to support the strategic purposes of the enterprise?
- If there are deviations from the planned progress, how significant are these deviations?
- Will any changes in the resources directed to the project be required to more fully support the strategic purposes of the enterprise?
- Will the project's progress, or lack thereof, adversely impact the chances of the choice elements being adequately executed in the enterprise?

Corrective Action

Corrective action on the projects can take many forms of reprogramming, reallocation of assets, cancellation of the project, or reformulation of the goals of the enterprise, which the project was destined to support. Corrective actions directed to a particular project may have the result of impacting other projects—or even other goals of the enterprise.

An effective policy and process for reviewing the progress being made on those projects that support enterprise choice elements will enable the senior managers to tune their use of resources in preparing the entity for its future as if that future mattered.

Strategic management should be carried out at every level in the enterprise. Accordingly, the choice elements described in Chapter 1 have applications at each level; of course, the time dimension surrounding these choice elements is different. As these choice elements are developed and are used for each level in the enterprise, opportunities exist for the coordination and assessment of how well the overall enterprise accomplishes its strategic purposes.

As a result of the use of a strategic control system, certain initiatives of the enterprise will be changed. A few examples of such changes are listed in Table 2.5.

2.14 STUDENT/READER ASSIGNMENT

The reader should do a "self-test" by seeking the answers to the following questions—as well as following the instructions indicated for a fuller appreciation of the processes involved in tracking, monitoring, and controlling the design and execution of strategic management done in the context of projects.

1. Why does it make sense to review the status of the projects in an enterprise when evaluating the performance of strategic management within the enterprise?
2. Select an organization with which the reader is familiar and identify the "stream of projects" in that organization that should be reviewed to ascertain how well the organization is being prepared for its future.
3. What does a "failing" project in an enterprise do to the development and execution of future strategies for the organization? Identify some "failing" projects in an organization with which you are familiar. Relate these "failures" to difficulties on the organization's part in the design and execution of future changes expected in the organization's environments.
4. Based on your current understanding of project management, what alternative strategies might be available to use in managing change in the enterprise?
5. What information should the project team have to make a strategic management control system operable?

TABLE 2.5 Examples of Changes Coming Out of a Strategic Control System

Changes in choice elements
Development of new/modified products, services, processes
Changes in market strategy
Modification of R&D programs
Cancellation of enterprise projects
Downsizing
Restructuring of organizational design
Changes in resource requirements

P • A • R • T • 2

THE STRATEGIC CONTEXT OF PROJECTS

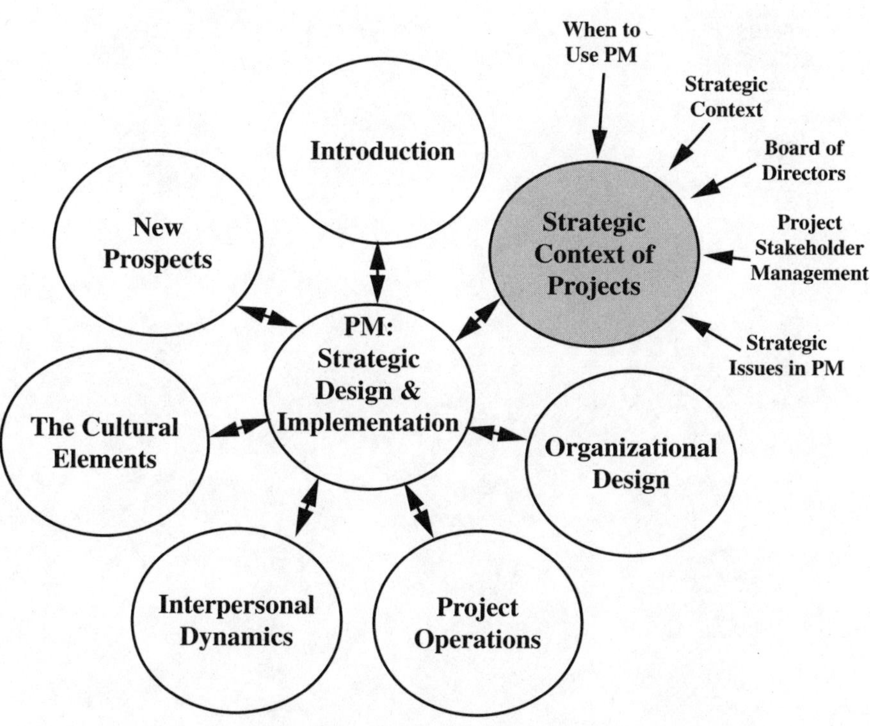

CHAPTER 3
WHEN TO USE PROJECT MANAGEMENT[1]

"I keep six honest serving-men (they taught me all I knew);
their names are What and Why and When and How and Where and Who."
RUDYARD KIPLING, 1865–1936

3.1 INTRODUCTION

The principal reason to use project management is to provide an organizational focus and a philosophy on how to deal with the inevitable changes facing contemporary organizations. There is a close relationship between project management and the organizational process changes that are required to cope with future opportunities. Projects are inexorably related to the design and implementation of strategic and operational initiatives.

In this chapter specific guidance will be presented on when projects are needed in an enterprise. Examples of project management applications will be given. The responsibilities of senior managers in the organization with regard to project management will be provided. Also, how to manage "small projects" will be presented—because many changes needed by organizations may be at minor levels of effort in that organization.

The primary reason for using project management is to provide an organizational design and a strategy to bring an organizational focus to those ad hoc activities needed to effect change in the organization. Modification of organizational products, services, and processes is required to accommodate the inevitable environmental changes that affect all enterprises today. Reaction to these changes usually requires an organized and focused use of resources to design new strategies in order to commit current organizational resources to prepare the enterprise for its future. An organization today that wants to remain competitive in providing its customers with continually improving products, services, and organizational processes has no choice but to use modern project management concepts and

[1]Some material in this chapter has been paraphrased from D. I. Cleland and W. R. King, *Systems Analysis and Project Management,* 3d ed. (New York: McGraw-Hill, 1983).

processes. The current use of reengineering initiatives has sharpened our assessment of organizational processes.

3.2 BUSINESS PROCESS CHANGES

If there is any basic belief that enterprise managers have today, it is that the cost of change in products, services, and organizational processes can be significant. In the early 1980s new products and services tended to last for 5 to 10 years—today some of them may last for only 2 years or less. Change cuts across the functions and locations of the enterprise as well as across the various professional groups of the enterprise.

Business process changes brought to our attention through the vehicle of reengineering projects have changed the way in which we deal with change in the enterprise. A business process reengineering project team provides for an excellent organizational design initiative to bring about the desired change in the management of the enterprise. Taken in its most basic form, a business process change provides for an outcome that reflects the need to change the way in which people work together. The business process change identifies a series of changes for the conversion of inputs to outputs, which represent an integrated assessment of the way in which value is created in the enterprise. Policies, procedures, rules, standards, and the manner in which resources are used are changed.

A business process is composed of logical steps that cross organizational functions and extend outward and cross stakeholder organizations such as vendors, customers, regulators, and even unions if labor issues are involved in the process change. All the aspects of the change coming about in creating a business process strategy must be managed as a whole. It is through project management that these changes are brought about. The business process change can be described in terms of performance objectives and goals that must be defined and must be capable of being measured. The major components of a process change brought about by a reengineering project could include:

- Designation of a business process work flow
- Job design
- Organizational redesign
- Redesignation of individual and collective roles of people
- Utilization of new or improved technology
- Modified management system to include changes in objectives, goals, policies, procedures, and rules for the use of resources
- Cultural changes to include modified behavior patterns by the people involved
- Finally, improved effectiveness and efficiency in the design and execution of strategies leading to improved products, services, and organizational processes in support of organizational goals, objectives, and mission

Business process strategies provide for the simultaneous design and execution of the manner in which resources are used to produce products and services. When properly designed and executed, such processes result in:

- An integrated strategy for the creation of value in the enterprise, at lower cost and of higher quality, that meets the timing required for market acceptance and meeting and beating the competition
- Greater customer satisfaction, enhanced enterprise performance, and state-of-the-art use of technology
- New policies, procedures, and the manner in which individual and collective roles are carried out in the enterprise
- Improvements in morale and motivation of people
- New skills, improved utilization of resources, and streamlined operations
- Improved productivity, quality, and organizational capabilities
- A new management philosophy with supporting policies, procedures, and strategies
- New knowledge, skills, and attitudes of people
- Improved competitive advantage
- Greater shareholder and stakeholder satisfaction

In the development of new products, services, and processes in organizations there is the need to provide an organizational focal point through which resources can be directed to keep abreast of—and even more, beyond—changing technologies. A "champion," such as a project manager, is needed to provide the leadership and management skills to bring about the needed changes—some of which may be global in nature. For example, strategic alliances—a form of long-term partnership being seen more and more in global competitive markets—are being conceptualized, developed, and managed through the use of project management.

In today's fiercely competitive world, sole dependence on traditional boundaries will ensure obsolescence. You have to reach out to the larger systems context and build new initiatives with and around key stakeholders. For example, the creative act of Wal-Mart and its unprecedented cooperation with its suppliers, through a sharing of information systems to improve manufacturing and distribution efficiencies, set a new competitive standard in the world of retailing.

3.3 SPECIFIC USES

Project management concepts and processes can be used to support an organization's crisis management strategies. Crises can arise from such mishaps as plane crashes, toxic chemical spills, hostage taking, product liability lawsuits, poisoned products, natural disasters, storms, and earthquakes.

For example, in the early morning of January 17, 1994, a 6.8 magnitude earthquake on the Richter scale struck the Los Angeles area centered in the San Fernando Valley. After the full extent of the highway damage was determined, coordinated planning started to respond. Construction crews immediately started demolition of failed structures and design engineers were alerted to start the redesign effort. Detours were worked out with participating local agencies.

A governor's task force was organized consisting of affected agencies, which met regularly to coordinate and evaluate progress. Project management played a key role in restoring the freeway system. Progress was tracked daily, and innovative contract procedures were developed. An extensive use of bonus/penalty clauses, invitation-only bids, and awards based on partially completed plans was carried out. Within 6 months four major freeways were restored.[2] Sawle recommends the use of a crisis control model consisting of the logical steps in preparing for and resolving a major crisis. He believes that project management and crisis management skills can be combined to deal more effectively with crisis situations.[3]

Project management is being increasingly used to support a company's factory operations. The entire field of manufacturing systems technology is changing rapidly. Just-in-time (JIT) inventory management, material requirements planning (MRP), total quality management, computer-integrated manufacturing, computer-aided design, and flexible manufacturing systems are some of the primary new technologies that have been developed to support manufacturing operations.

No doubt the pace of technology in manufacturing systems will continue to advance, resulting in "systems" changes impacting manufacturing as well as the supporting functions in the enterprise such as R&D, marketing, finance, and after-sales maintenance and support. What all these changes have done is create an environment in the modern company that is too multidisciplinary to be organized solely along traditional functional entities. Contemporary factory managers need a management philosophy that allows them to bring an organizational focus to the management of resources in the factory that are dedicated to change—to the creation of something that does not currently exist, but that is needed to remain competitive in the global manufacturing environment.

The factory manager has a wide range of options to consider in using resources to manufacture products and provide supporting resources. The "traditional" factory retains most of the basic functional characteristics tied together by hierarchical relationships where clearly established lines of command are exercised through authority and responsibility relationships. The functional subunits of the factory are headed by a department manager, along with a person designated to be in charge of the production workers—this individual is traditionally called a *first-level supervisor, foreman, production boss,* or some such title. The role of this first-level supervisor has changed, and will continue to change, significantly.

Peter Drucker noted that no job is going to change more in the future than that of the first-level supervisor. Production teams, concurrent engineering teams,

---

[2]Jerry B. Baxter, "Northridge Earthquake Response," *Proceedings,* Project Management Institute, 25th Annual Seminar/Symposium, Vancouver, British Columbia, October 17–19, 1994, p. 102.

[3]See W. Stephen Sawle, "Crisis Project Management," *PM Network,* January 1991, pp. 25–29.

quality teams, task forces, and such organizational designs dedicated to managing change in today's factories all draw heavily on the concepts and processes of project management to pull together resources across traditional factory operations.

The use of robots to perform simple repetitive manufacturing tasks reliably at relatively low cost is another application of manufacturing systems technology facilitated by the use of project management concepts and processes. However, the use of robots has not become widespread in manufacturing operations; only a modest percentage of the industrial enterprises that could benefit from robots have any. Part of the problem has been that in most companies the design process is not adequately integrated with manufacturing. The typical robot system takes up to 12 months or more to go through a life cycle of concept, design, fabrication, installation, debugging, and start-up. The use of a project manager, such as an industrial engineer, to define robot tasks establishes operating parameters, and the designing and interfacing of the human/material/robot system provide for a means to integrate different disciplines to support a common objective. A project team led by an industrial engineer, including representatives from engineering, maintenance, production control, manufacturing, management, safety, personnel, labor reporting, and accounting, can effectively address the issues involved in setting up robots for use in the factory. Other responsibilities of this project team include addressing the issues of what will happen to the employees who are displaced by robots and how the supporting functional elements of the factory will be realigned to support the use of robots in the changed factory operation.

An industrial engineer whose education and experience are in forecasting techniques, economic order quantity calculations, material requirements planning, flow process charts, human-machine charts, time and motion study, time balancing Gantt charts, from-to charts, queuing networks, computer simulation, and of course project management, is well suited to perform the key role of a project manager in the factory environment.

More and more literature is coming forth that describes how project management can be applied to the factory. A typical contribution to the literature has been offered by Professor Hans Thamhain, "Project Management in the Factory," chap. 5 in David I. Cleland and Bopaya Bidanda (eds.), *The Automated Factory Handbook* (New York: McGraw-Hill, 1990).

3.4 PROJECTS AND STRATEGIC PLANNING

Strategic planning establishes the mission, objectives, goals, and strategies for where the organization wants to go in the future. Project planning is discussed more fully in Chap. 11.

Projects play an important role in the enterprise. The importance of projects is related to how they provide the transformation process from enterprise resources to strategic initiatives. This transformation process is depicted in Fig. 3.1.

Strategic design and *implementation* are concerned with how the organization is going to get there through the planned use of resources. *Strategies* include things

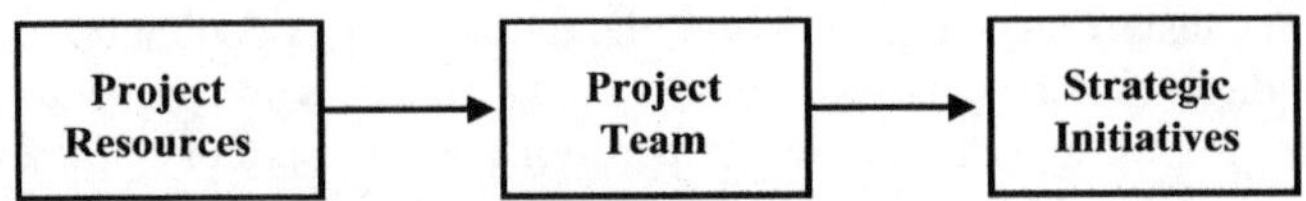

FIGURE 3.1 Project transformation process.

such as short-term action plans, policies, procedures, resource allocation directions, programs, and projects. Of these, *programs and projects* are of special interest. *Programs* are resource-consuming sets of organizational resources, which have a common purpose. For example, a productivity improvement program could be composed of the following projects:

- A participative management-style training project
- Realignment of manufacturing assembly processes
- Professional management development projects
- A project for development and use of autonomous production teams
- An integrated design-manufacturing information system

A principal reason to use project management is to facilitate the implementation of organizational strategy, although project management can be used effectively in other organizational contexts. Another reason for electing to manage things on a project basis is the fragmentation of functions and skills throughout the organizational structure. When an activity that is too large for any one functional department to manage is introduced into the organization, a single focal point must integrate the functional efforts through a matrix organizational design. The matrix organizational design is discussed in Chapter 8.

Davis and Lawrence insist that one should turn to a project only when the following conditions exist simultaneously:

- When outside pressures require that intensive attention be focused on two or more different kinds of organizational tasks simultaneously, e.g., functional groupings around technical specialties and project groupings around unique customer needs
- When tasks become so uncertain, complex, and interdependent that the information-processing load threatens to overwhelm competent managers
- When the organization must achieve economies of scale and high performance through the shared and flexible use of scarce human resources[4]

Projects are resource-consuming activities used to implement organizational strategies, achieve goals, and contribute to the accomplishment of the organizational mission. All of these suggest that when an enterprise considers the use of project management in its strategy, it should first determine if the proposed projects could be associated with the following:

- The core "product line" business being pursued in the organization's market strategy

[4]Stanley M. Davis and Paul R. Lawrence, *Matrix* (Reading, Mass.: Addison-Wesley, 1977).

- A proposed new or improved product, service, or process design and development effort
- The development of resources to support the enterprise's product lines, such as facilities construction, productivity improvement programs, quality assurance programs, and employee participation projects

Whether to use project management raises the fundamental question of how to organize to implement our organizational strategy, because it is the organizational strategy that sets the organizational design that follows. In some situations, the decision to use project management techniques is made by the customer. Companies that bid on government contracts will probably find that they are expected to establish a project management system as a prerequisite to winning a contract.

3.5 WHEN IS A PROJECT NEEDED?

In general, project management may be applied effectively to any ad hoc undertaking. If such an undertaking is unique or unfamiliar, the need for project management is intensified. In some cases, such as that of an undertaking whose successful accomplishment involves complex and interdependent activities, a project manager can pull everything together to accomplish an organizational purpose. Basic to successful project management is recognizing when the project is needed—in other words, when to form a project as opposed to when to use another form of organizational design to do the job.

At what time do the forces in the organization and its environment add up to project management? The senior executives must have a basis for identifying undertakings, which the regular departments cannot manage. There are no simple rules to follow, but several general criteria can be applied in considering the use of project management. The justification for project management arises from the need for new or improved products, services, or organizational processes. Within this context Fig. 3.2 shows the principal criteria that can be applied in considering the use of project management. These criteria are discussed in the sections that follow.

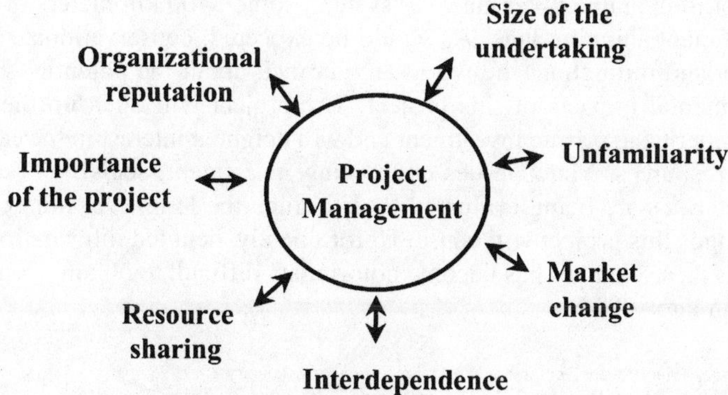

FIGURE 3.2 The need for project management.

Size of the Undertaking

The question of size is difficult to pin down because size is a relative matter. When an undertaking requires substantially more resources (people, money, and equipment) than are normally employed in the business, project techniques may be indicated. Even though the functional elements for the end product are discernible in the organization, the diversity and complexity of the task easily can overwhelm a department. In these cases, project management provides a logical approach to the organizational relationships and problems encountered in the integration of the work. For example, let us consider the move of a company from an eastern city to one in a southern state. This may appear to be a simple operation, but the complex development and correlation of plans, the coordination required in constructing the new site, and the task of answering numerous inquiries about the new site easily can swamp the existing organizational structure. These difficulties are compounded by the fact that the company must continue its normal operations during the period of the move. In such a situation, managing the move along traditional lines would be difficult, if not impossible.

In the future, the use of project management to manage changes in major infrastructures of societies can be expected. Such changes have already started. Two major projects give examples of the types of change that must be considered.

- In China, work has been inaugurated on the world's largest hydroelectric dam project, the Three Gorges effort on the Yangtze River. The dam will be nearly 1 kilometer long and some 100 meters high, and will consume enough material to build 44 Great Pyramids. The reservoir will stretch 600 kilometers upstream. Between 15 and 20 years will be required to build the dam—consequently nobody is reasonably sure how much the project will cost. Present estimates center on $12 billion at 1990 prices. Major environmental considerations are involved, to include landslides, potential military attacks, and earthquake issues, increasing the risk that the dam could fail. Continued planning to evaluate these and other risks will need to be carried out even though the dam is already under construction.[5]
- The proposed Hidrovia Project in Brazil is a proposal to reengineer the natural infrastructure of a continent. The Hidrovia project is planned to run the entire length of the Paraguay-Paraná river system, some 3400 kilometers of it, into a superefficient shipping lane. As would be expected, conservationists in South America and throughout the world are alarmed about the potential social and environmental impacts of this project. The project will open up the heart of South America to private investment and will heighten interest in the continent's natural resources. Major issues concerning investment, social, and economic considerations are being evaluated to determine the long-term implications of the changes this project will cause. Unfortunately, detailed information needed to assess these changes has become notoriously difficult to obtain. Construction started in late 1997.[6]

---

[5]Fred Pearce, "The Biggest Dam in the World," *New Scientist,* January 28, 1995.

[6]Raphael Heath, "Hell's Highway," *New Scientist,* June 3, 1995, pp. 22–25.

The minimum magnitude of project-oriented effort depends on the basic strategy of the organization. A company engaged in routine manufacturing probably does not require much project management. However, if the company were to go through a major redesign of its product line, which dictated significant special tooling and facility changes, project management could be set up to manage the change.

In the development and production of a weapons system or a nuclear power plant, an awesome inventory of human and nonhuman resources has to be synchronized and integrated into an operable system. The use of project management in these situations is clearly needed.

Indeed, the need for formal project management was heightened during the early days of the large DOD-NASA awards for the design, development, and production of major systems. In managing these projects, companies recognized that two or more functional elements of their organizations had to be pulled together in order to develop these systems.

The U.S. Air Force provided a strong impetus for the use of project management in the early 1950s. At that time it became clear to the USAF and the aerospace industry that a key point in the selection of system contractors would be whether the contractor had a project-driven organizational structure where work on a project could be centralized. By late 1958 and early 1959, aerospace companies had begun to establish project-driven organizations that cut across functional lines in order to accomplish project objectives. Factors that emerged in this period to account for the trend to centralize the management of projects within these organizations were as follows:

- A rapid technical advance in U.S. Air Force weaponry that led to the demand for minimum time in developing an operational system
- A change in technology that fostered new doctrines for the employment of costly weapons systems, along with the urgent need to produce project results at a minimum cost
- In the development of long-range ballistic missile systems, the motivation that arose from the management of the technology as well as the technology itself
- The extremely tight schedules, limited funding, state-of-the-art pressures, and increasingly complex procurement regulations that provided an unparalleled challenge to the aerospace industry

The current realignment of the DOD budget and the conversion to nondefense products by defense contractors will also provide ample opportunity for the use of project management.

Unfamiliarity

An *ad hoc undertaking* is a project out of the ordinary, something different from a normal, routine affair in the organization. But the degree of unfamiliarity also must be considered. For example, the redesign of a major product would require project management. An engineering change to an existing product, however, could be conducted without setting up a project, although there might be a loss of overall efficiency in accomplishing the objective. In the first instance, the changes in cost, schedule, and technology would require a central management point to bring together the functional activities required and relate them to compatibility. In the second case, each of the functional managers could draw on experience to accomplish the work.

Unique opportunities or problems are generally project-oriented. Work on these opportunities is usually scattered in the organization, yet it is all interrelated, because various functional groups have to provide different disciplines to support the undertaking. Project management handles such opportunities well.

At NCR a corporate project manager was made responsible for overseeing all activities relating to a postmerger integration of AT&T's computer systems with those of NCR. The use of project management techniques helped reduce the time-consuming and complicated process of integrating AT&T's worldwide computer services division with NCR, without interrupting business as usual.[7]

There are other important and unfamiliar activities in an organization that require a management focus so that resources can be marshaled and closely controlled. Some of the particular problems or opportunities that fit into the category of the "unfamiliar" are:

- A major reorganization
- A takeover threat by an unfriendly suitor
- A crisis, such as a serious product failure, legal action, or other nonroutine occurrence, that seriously threatens the integrity of the enterprise
- Any unfamiliar undertaking that is of critical importance to the enterprise, such as new product or market development, a new business venture, or acquisition of another company

In these ad hoc situations, management may not know how to integrate many different profit centers in the corporate structure. To meet this problem, one large corporation created a projects division as a profit center to enhance its capabilities in competing for projects business in its industry products business unit. The mission of this profit center was to act as the project management arm of the corporation for those jobs that required teamwork among the corporate divisions that produced the products, the various sales organizations, and the supporting corporate staff. This projects division was chartered to handle projects:

- Of $10 million or more in the industrial products province
- With products from two or more divisions
- Managed under one contract
- Either domestic or international in scope
- Without any other appropriate lead division

[7]Eva Hofstadter, "The Science of the Deal: Project Management Needs Wall Street," *PM Network,* November 1992, pp. 11–19.

This division acts as a "strike force" to develop project markets in a market segment for the overall benefit of the corporation. This division also organizes, coordinates, delivers, and installs the electrical equipment package for large industrial projects. Here, project management is used for the advantage of single-source responsibility for all stages of a project, including:

- Up-front studies and analyses
- A single, coordinated proposal
- A single contract covering all products and services
- An interface with customers, or with other contractors as required by the contract
- Large and complex projects packaged from smaller pieces
- Integrated equipment design and installation
- Stringent control of scheduling, shipping, and installation
- A single point of contact for problem resolution
- Centralized invoicing

In this corporation, project management is carried out by a team composed of representatives from the participating profit centers. The team is formed during proposal development and has responsibility through the warranty period. In addition to the design, development, and production phases of a project, the team manages installation, erection, and equipment start-ups (all integrated into the overall project) as well as support activities such as personnel training and start-up engineering. Some projects include field service and contract maintenance and repairs.

It took many years for this "interdivisional project management" to evolve, principally because of the territorial restrictions of the divisional charters that existed within the product group structure of the corporation. As the market for industrial equipment systems emerged, it became clear that no single division had the familiarity with the "systems" capability to serve the industry/construction project needs. What was needed was an organizational design to facilitate strategy across the many corporate profit centers that produced product components.

Market Change

Many firms and organizations operate in a turbulent market that is characterized by continually changing products, rapid technological innovations, and rapid changes in the values and behavior of customers and competitors. Such conditions place a premium on innovation, creativity, rapid response, and flexibility. Heterogeneous, changing markets require a management system that can flourish in the ambiguity of changing objectives and goals with the life cycle of many projects placing varying demands on managerial and professional support. These rapid market changes require an organizational approach that permits flexibility in the use of resources.

One company whose products are well known throughout America's kitchens uses project management as part of its competitive strategy. Wooster, Ohio-based Rubbermaid Company's products exhibit such high quality that they rarely need replacing. So this company must depend on new products and new markets for growth. The company's CEO plans to add a new market segment every 12 to 18 months. In the design of new products, the company depends heavily on a new generation of computer-aided design (CAD) workstations—so advanced that they reduce new-product design time from months to days. These workstations enable Rubbermaid to go directly from rough sketches to finished products in weeks rather than months. The company is moving from sequential to simultaneous design through the use of project teams—and in so doing is able to reduce cycle time, duplication of effort, and errors. In their markets, the reduction of cycle time is critical to enable the company to have a market as long as possible and before cheaper versions appear. The company, to survive, has to "reinvent itself" continuously. Simultaneous engineering facilitated by CAD technology is critical to this reinvention.[8]

A senior executive in a project-driven organization comments on the flexibility that project management provides:

- In the short term, the project team provides for flexible use of key technical resources, both people and facilities. (Functional "fiefdoms" don't have to be reorganized to move the talent from program to program to meet fluctuating demands.)
- For the long term, the team expands the avenues for business benefit from broadly applicable strategic investment. (A pattern of shared resources and shared responsibilities obviates the traditional "technology transfer" issue altogether.)
- Basically, through the program management dimension, the team establishes minigeneral managers (project managers) who are extensions of the general manager for a subset of the business, but without imposing the inflexibility and communication isolation of the functional resources that are characteristic of self-contained business segment departments.[9]

Interdependence

Another decision criterion for establishing a project is the degree of interdependence between the departments of the organization. If the effort calls for many functionally separated activities to be pulled together and if these activities are so closely related that moving one affects the others, project techniques are needed.

Consider the development and introduction of a new product. The early planning would require sales forecasts to be completed before plans for manufacturing processes, industrial facilities, special tooling, and marketing could be developed. Sales promotions cannot be completed before plans for manufacturing processes, industrial facilities, special tooling, and marketing. Sales promotions also cannot

---

[8]Seth Lubove, "Okay, Call Me a Predator," *Forbes,* February 15, 1993, pp. 150–153.

[9]John W. Stuntz, "A General Manager Talks about Matrix Management," in David I. Cleland (ed.), *Matrix Management Handbook* (New York: Van Nostrand Reinhold, 1984), p. 211. All rights reserved.

be completed until the marketing research points the direction for the promotions. Performance and technical specifications, as well as the many interdependencies among the production, marketing, finance, advertising, and administration groups, must be resolved. Provincialism cannot be tolerated. If no one agency can pull all the separate parts together, if the functional groups fail to make credible estimates, or if the plans submitted by the different departments cannot be reconciled, then the activity needs the singleness of purpose of project management.

Sometimes project management comes about through a reorganization. In Norwalk, Connecticut, Perkin-Elmer Corporation reorganized into a functional alignment from its former geographic organization—one designed to compete in a global market. Project teams composed of people from engineering, manufacturing, sales, and services have the responsibility of developing new products.[10]

Product and service development projects are the lifeblood for success in the future. Manufacturing and marketing process development provide the basis for determining the resources needed to support the new product or service and the customer-related processes that will be used to get the product or service to the customer. In such situations, the interrelatedness factor in justifying a project is an important consideration. The timing of the development of projects is important as well.

Sometimes the risks and costs of developing new technology dictate the use of project management. For example, a project team with IBM, Siemens, and Toshiba participation was formed to design the first 256-megabit memory chip and its fabrication process at an estimated cost of $1 billion. Factories to produce the chips in volume will run another $1 billion each. These huge costs are the reason why such huge international alliances are likely to become the norm in the future. The new project-process team is centered at IBM's new Advanced Semiconductor Technology Center in East Fishkill, New York, where a trilateral team of some 200 engineers will report to a Toshiba manager.[11]

World-class manufacturers are skillful at both product and process development and become expert in the design and manufacture of production equipment, either doing the work themselves or subcontracting the work to outside suppliers. Product development and process development are closely intertwined, usually through the use of a project team, often called a *product-process design team,* or *simultaneous engineering.* A brief continuing discussion of the subject of simultaneous engineering is presented here to demonstrate the project relevance of such engineering. This intertwining provides for the continual improvement of all the "systems" that support future organizational strategies. This subject is also briefly discussed in Chap. 1 and is further discussed in Chap. 21.

Using project-process design teams in simultaneous engineering increases the probability of close interaction among engineering, marketing, and manufacturing groups. The teamwork across disciplines helps ensure that everything that can influence the success of the new product or service in the marketplace is considered. With suitable senior management involvement and surveillance, a final core value is added to the development of new businesses.

---

[10]Reed Abelson, "Getting Its Act Together," *Forbes,* August 31, 1992, pp. 44–45.

[11]Otis Port, "Talk about Your Dream Team," *Business Week,* July 27, 1992, pp. 59–60.

The importance of interdependence is clearly demonstrated in the use of product-process design teams. Not only does the use of such teams provide the opportunity to get the product or service to the market sooner, but also it ensures more "systems" considerations in the strategic management of the enterprise. Other benefits include these:

- Organizational resources are used more effectively and efficiently.
- People working on the teams sense a higher degree of ownership with the product or process being developed.
- The composition of the team, with people having different backgrounds and coming from different disciplines in the organization, provides an enhanced "checks and balances" in the design and execution of strategies for the product, service, or process being developed.
- Time is saved. Time represents money—and profit—when the product or service is introduced earlier into the market.

When a large research laboratory must pull together many different specialties, that is another example of the importance of interrelatedness. The laboratory must establish criteria for the use of formal project management when research and/or development projects require assembling diverse technologies and when larger projects require an engineering/design output as well as advanced technology. Laboratories typically use project management when a research project exhibits the following characteristics:

- A potential and significant long-range impact on the corporation
- The need to pinpoint corporate responsibility for the project
- The need for fast response
- The need for integrating widely varying disciplines, technical skills, backgrounds, and facilities
- The need for close coordination with corporate profit centers on product design, development, manufacturing equipment, and processes
- Significant size and duration
- Corporate technical and financial reporting requirements

Each year research projects are funded in a laboratory. A principal investigator or project manager provides the technical direction and integration of the project as well as accomplishing tasks on schedule and within budget. An informal project management system is used for small projects where only a few disciplines are required. For larger projects meeting the criteria just given, a formal project management system is used.

Resource Sharing

Projects are characterized by strong lateral working relationships requiring continuing coordination and decisions by many individuals, both within the parent organization and in outside companies. During the development of a major product, there is close collaboration between the process and design engineers, and even closer collaboration between the individuals of a single department. These horizontal relationships do not function to the exclusion of the vertical relationships. Resource sharing becomes common practice.

Successful project management was executed in a resource-sharing partnership between General Electric and a group of Mexican appliance manufacturers to build a gas range manufacturing plant in San Luis Potosi in Mexico. GE management realized that the project's complexity was increased by working with a foreign partner in a foreign land. All members of the project team were given extensive Spanish language instructions; Mexican members of the team were put through an extensive U.S. language and culture training program. Care was taken during the project to identify cultural, linguistic, and other related issues that were liable to create barriers to communication.[12]

Project management makes sense when increasing professional specialization and its attendant higher cost lead to the need to share professional resources throughout the organization. It also makes sense when there are critical or scarce resources; when, in the ebb and flow of the life cycles of modern business products and services, it is difficult to keep a professional work force fully and effectively employed; or when certain types of professional skills are in short supply. Project management techniques can be utilized to share resources, potentially reducing both direct and indirect costs and delivering needed results.

Resource sharing in a project management context can lead to the best use of resources to promote the objectives and goals of the organization. Project management places priorities on the work efforts to allow resource assignment to those most critical aspects of business. Using resources from either internal or external sources on a shared basis leads to the most efficient and effective support of the organization's purpose.

Importance of the Project

Another reason for using project management techniques is the importance of the project to the enterprise. Managers might not want to place it in the "bureaucracy" of the organization, where it might become lost in the daily operational workings. When an ad hoc activity has high risks and uncertainty factors, then the use of project management techniques may be required. If an emerging problem or project is viewed as a potential building block in the design and implementation of future strategies for the enterprise, then project management techniques are required.

A new product line requires financing, design, development, and production—clearly an opportunity for project management, particularly if the emerging opportunity constitutes an effort that is too large to manage in a "business as usual" approach or if the product is very important to the company's future business. If such an emerging product carries high risk and has an apparent direct relationship to the company's objectives, then project management is usually required.

[12]Robert J. Butler, "A Project Milestone Bonus Plan: Bringing a Plant Startup On-Line, On-Time, On-Cost," *National Productivity Series,* Winter 1991/1992, pp. 31–39.

An important part of an organization's policy should be a statement of the conditions under which project management will be used. Senior managers will develop these criteria when they realize the important role that projects can play in the management of the enterprise.

The important thing to remember is that a project, as an ad hoc activity, cannot stand on its own; it is interrelated to the strategic mission of the organization. A project contributes something to the ability of an organization to change to meet its future. A project is the opportunity for an enterprise to complete a goal that leads to accomplishing its objectives, and ultimately its mission. Thus, the basic purpose for starting a project is to accomplish some goals that are held by the larger organizational unit—the department, the division, or the corporate entity. The reason for using a project is to provide a focus for organizational resources to be applied against the organizational problem or opportunity so that an enterprise goal can be attained.

Organizational Reputation

The overall organizational stake in the undertaking is another crucial determinant in the decision of whether to use the project techniques. For instance, if a failure to complete a contract on time and within cost and performance goals would seriously damage the company's image and result in customer and stockholder dissatisfaction, then the case for using project management is strong. A company's financial position can be seriously damaged if its performance on a contract fails to meet standards. In the case of government contracting, the company faces a single, knowledgeable customer, and failure to perform successfully can be catastrophic in terms of obtaining future contracts with the government.

Project management is no panacea, but it does provide a means for effective use of resources in ad hoc efforts. Project managers who see their role as that of integrator-generalists, responsible for meeting time, cost, and performance objectives, can do much to lessen the dangers inherent in an ad hoc undertaking. Project management concentrates into one person the attention demanded by a complex and unique undertaking, which will affect the enterprise's reputation.

Before a decision is made about whether to use project techniques, the effects of the company's environment on the project must be weighed and evaluated. The objective of the undertaking must be considered. Methodological improvements that might take some time to implement require considerable thought. The size and complexity of the project must be considered, because too much sophistication is also an ever-present danger. Other factors that merit consideration are the number of current projects in the company, the number in prospect, and the time remaining to complete the project. For example, establishing project management would be more appropriate at the start of an undertaking or at least early in its life, before large expenditures of work hours and resources are made. Each situation is unique, and the decision of whether to manage by a project or another approach should be made on the basis of specific problems expected as well as the concepts of organization presently used in the enterprise.

No company takes a purely project-oriented or a purely functional management approach. All companies combine the two, although one form may dominate, to focus the efforts and meet organizational objectives.

3.6 PROMOTING PARTICIPATIVE MANAGEMENT

In some cases, project management is used to provide an opportunity for an individual to take an idea and see that idea through to a successful product or service. Some companies have recognized the value of individual initiative and have organized their corporate structure and management philosophy to accommodate the entrepreneurial abilities of individuals.

At Honda, a fixed percentage (approximately 10 percent) of the R&D budget is set aside to fund new-concept development. Anyone can propose a new concept. It is reviewed by a peer group. If it is accepted, Honda organizes a small project team and provides funds to develop the concept to the point of a formal evaluation by senior management.

Texas Instruments' IDEA Program has a small pool of funds, distributed by senior technical people, to pay for concept development. Anyone who can get a concept development idea funded can manage a small project. Some ideas have led to full-scale product development projects and even commercial products.

In 3M Company, anyone who invents a new product, or promotes an idea when others lose faith, or figures out how to mass-produce a product economically has a chance to manage that product as though it were her or his own business, with a minimum of constraints from higher management. Called the process of *divide and grow,* the practice is aimed at keeping 3M a company of entrepreneurs. 3M's culture and its organizational structure are all directed to encouraging its people to take an idea and run with it. The new-product enterprises are broken out into self-sustaining units, each with considerable responsibility for its future. These ideas, developed into small projects managed by a team of professionals, may grow into departments and then into divisions within the corporate organizational structure. The depth of 3M's faith in allowing people to manage their ideas into projects was summed up by one manager at 3M, who stated:

> If you put fences around people, you get sheep. If you want the best from people, give them all the room they need to grow, and all the responsibility they can handle.[13]

3M's growth has been compared to cell mitosis because of the company's history of allowing small projects to grow and then dividing them. As described in a letter to the author:

> ...a product idea may emerge from a laboratory and link with a small amount of test-marketing assistance from a parent organization.

[13]"Getting to Know Us," 3M pamphlet.

This product moves into the marketplace and reaches a level where a project is created as a profit center having responsibility for creating additional business and products.

The project succeeds and becomes a department. The department succeeds and becomes a division, which is the basic business unit at 3M. Divisions, in turn, are organized into product groups which themselves form business sectors.

The "mitosis" usually occurs at the division level when a department achieves certain goals for profitability and sales. It is spun off from the parent division to create another, new division. Simplistically but accurately, a former board chairman once explained, "Split a $100 million division and you get two $60 million divisions."

This process allows the product champions who have built a business to be rewarded without their having to wait for their bosses to retire or advance.

For example:

1. In the 1950s, Lewis W. Lehr was working as a tape engineer when he had some contacts with physicians with an idea for a surgical tape. Lehr developed such a product, which languished. 3M wanted to drop the idea, but Lehr asked to buy the line, and 3M regained its interest. The product succeeded eventually.
2. Autoclave tapes and other medical products followed.
3. In 1960, Lehr was named manager (not even general manager) of a Medical Products Division.
4. Through new technologies and acquisitions, the Medical Products Division grew to become a group, with medical, surgical, orthopedic and dental products divisions. This group, Health Care Products and Services, is a significant portion of our Life Sciences Sector, one of four major business areas of the company.

As for Lehr, he moved with [the] expanding business—becoming division general manager, division vice president and group vice president. He then became president of U.S. Operations and board chairman and CEO, retiring March 1, 1986.[14]

Truly, 3M's organizational design and operating philosophy emphasize project management, which in turn supports their corporate mission: "We Are in the Business of Building Businesses."[15]

3.7 SENIOR MANAGEMENT RESPONSIBILITY

Although it does not always happen, project management should be used only when senior management fully understands its own role in the strategic management of the enterprise and is fully committed to making it happen. The responsibilities of the senior managers of an organization and their willingness to provide an environment for the growth and propagation of project management depend on

[14]Letter from H. G. Owen, 3M Center, St. Paul, Minn., to D. I. Cleland, March 20, 1986.

[15]*3M 1985 Annual Report,* title page.

how well they establish an organizational culture for project management by doing the following:

- Maintaining the balance of power between the project office and the functional elements of the organization
- Providing facilitating services such as budget, finance, accounting, general administrative accommodation, and so on, to the project
- Developing and promulgating a philosophy of how resource priorities will be determined in the organization's matrix and how conflict over these resources will be resolved
- Providing performance standards for both project success and adequacy of functional support
- Establishing criteria for performance evaluation and wage and salary classification schemes in the organization's matrix
- Acquainting key individuals with the theory of matrix organization and presenting a process model of how the organization is intended to operate
- Providing models of organizational interfaces—developing authority, responsibility, and accountability relationships
- Defining decision parameters within the matrix organization
- Providing the project manager and the functional manager with strategic direction

If an executive wants to implement project management, the first requirement is for that executive to believe in project management and what it can do for the enterprise. With such belief can come commitment—and the demonstration of that commitment to the people who will be working on the new organizational design and project management strategies. To make project management work, strong leadership by the management cadre concerned with the use of project management is required. If senior management becomes too busy to take a strong leadership role in moving the organization to a project management strategy, the development effort will likely fail. What are some of the reasons for the failure of project management start-up strategies? The failures can usually be traced to one or more of the following:

- Design and implementation of a well-designed, well-developed strategy for project management in the enterprise are not done.
- Project management is viewed as the "fad of the month" and is not taken seriously, particularly by the managers and professionals in the product and service development part of the enterprise.
- Project management is viewed as a separate entity in the organization, and not as a building block in the management of change in the enterprise.

For those of us who believe in the project management process as a means for dealing with major change in the enterprise, it is difficult to believe that there are some who find fault with that process. One study of project management groups in *Fortune* 500 companies conducted at the *Fortune* 500 Benchmarking Forum in Anaheim, California, found that communicating the financial benefits of project management to senior management and other corporate areas is the major project-related challenge. Even in dealing with peer groups within a functional area, project management can be seen as a threat to existing work groups and functional hierarchies.[16]

3.8 SELLING PROJECT MANAGEMENT TO SENIOR MANAGERS

It is often a challenge to obtain the approval of senior managers to change to something new when it is perceived that the present system is working. The question is typically, Why change something that is working? Responses from senior managers could also be the time-tested cliché, "If it ain't broken, why fix it?" Getting acceptance from decision makers for a change that has dramatic impacts on the organization, both in terms of what it takes to implement a new management approach and the time that it takes to adapt the culture, can be a challenge to the person offering a new way of doing business.

Project Management Application

Project management is used extensively in some form within many organizations. Regardless of industry or product being produced, project management has application for improved productivity. Surgeons can use these techniques to plan, implement, and follow through on surgical procedures. Grocery markets can use these techniques to change the layout in the stores. Book writers can use the techniques to plan, write, and review manuscripts. There has been no identified profession or industry where project management practices will not work.

Complete adoption of project management by an enterprise requires the acceptance by senior managers to champion and support its formalization and implementation. Senior managers may recognize the benefits of project management, but frequently do not become involved in the process. Perhaps, senior managers focus on more familiar processes for management and decision making. These senior managers are, therefore, not committed to the project management process that can make a significant difference in the organization's future.

This unrecognized potential of project management overlooks its contribution to an organization's capability to efficiently develop and deliver products and services as well as effectively implement organizational change. Project management's latent potential is not tapped to its fullest and organizations do not realize their optimum potential. Senior managers can "discover" this potential and cause changes in the policies, procedures, and practices for conducting the organization's business.

[16] Frank Toney, PMP, "Good Results Yield Resistance?" *PM Network*, October 1996, pp. 35–38.

Awareness of the Capability by Senior Managers

There is a vital need to raise the level of awareness of the potential for using project management in senior managers, those individuals charged with the responsibility to guide the organization in the most efficient and effective manner. Raising the level of awareness, or "selling project management to senior managers," is vital to maintaining the organizations' competitive edge. Without a competitive edge, organizations will shrink in the marketplace and their capability to develop new products and services to meet emerging needs will decline.

Project management, as an essential element of business solutions, is often discovered accidentally.[17] Only a few companies train their executives to understand project management and its effective use as a competitive-edge approach to developing products and services for customers. Also, the use of project management techniques is not fully appreciated as one of the most efficient and effective means of implementing organizational change.

Benefits Derived from Using Project Management

What are the benefits of project management that would demonstrate the need for full and complete implementation of a formal system? In a 1995 paper, Bradford Price reported the results of an assessment of implementing project management in the U.S. Army Corps of Engineers. Major benefits realized by the U.S. Army Corps of Engineers were described as follows:

- Helped reduce project duration from 26 years to approximately 10 years
- Increased cost estimating accuracy that resulted in 30 percent fewer projects experiencing cost increases
- Reduced schedule slippage by 23 percent with more reliable project schedule completion dates
- Enhanced the ability of local sponsors (customers) to influence the final design, cost, schedule, and plan
- Identified problems for resolution at all organizational levels
- Increased staff productivity
- Reduced total project management costs through shorter project phases

Further, Price reports that in a follow-on assessment on five projects, managed under the new project management system, the average project development cycle was reduced to less than 7 years from the previous 10-year cycle. Reduced project development cycle time resulted in significant savings for the organization.[18]

---

[17]Jeffrey K. Pinto and O.P. Kharbanda, "Lessons for an Accidental Profession (Project Management)," *Business Horizons,* vol. 38, issue 2, March 1, 1995, pp. 41–45.

[18]Bradford S. Price, P.E., "Implementation of Project Management in the U.S. Army Corps of Engineers," Project Management Institute 26th Annual Seminar/Symposium, New Orleans, La., October 16–18, 1995.

Change to an organization's culture and its way of doing business takes time and concerted efforts on the part of many people. Senior managers direct the change and midlevel and project managers implement the change. Senior managers must have a vision for the future state to be achieved, sufficient goals to define the change, and a commitment to continue in the face of obstacles. Commitment to continue includes providing resources for the change and giving people the authority and responsibility for the tasks to be accomplished. Senior managers committed to adopting project management as the system of choice is the key to changing to a new project management culture.

The U.S. General Services Administration (GSA), a governmental agency, estimates that the change to a project-driven culture and full implementation took nearly 12 years. GSA's charter requires that it purchase supplies, information technology, and real estate for use by governmental activities. Each purchase could be viewed as a project and each delivery to a governmental activity as another project. GSA found that the change to a project-driven culture was slow and incremental. The change process needed to be stabilized at each step in the evolution process.[19]

AT&T's complex voice/data telecommunications/information systems projects benefited significantly through project management practices taught in project start-up workshops. These workshops established the requirements for the projects and set the baseline for project initiation. The workshops materially contributed to successful planning and successes for medium and large projects.[20]

Convincing Senior Managers to Use Project Management

"Selling project management to senior managers" is a task of developing a business case that clearly demonstrates the added value of changing to procedures, practices, and processes embraced by project management. Some questions that may be asked by senior managers are:

- What is the cost benefit of changing to a project management system?
- What impact does changing to project management have on customers?
- How is this going to improve business?
- What is the cost of converting to project management?
- How do our competitors use project management?
- How long will it take to convert our business to a project management culture?
- What are the immediate benefits?

Table 3.1 provides a comparison of some examples that would use project management. These examples are derived from experience and demonstrate the type of comparison that senior managers may find compelling cases for change to project

[19] Al Delucia, "The Evolution of Project Management at GSA," *PM Network,* September 1999, pp. 57–61.

[20] Dan Ono and Russell D. Archibald, "Project Startup Workshops: Gateway to Project Success," Workshop at Project Management Institute Seminar/Symposium, San Francisco, Calif., September 17, 1988. Also Russell D. Archibald, *Managing High-Technology Programs and Projects,* 2d ed. (New York: Wiley, 1992), chap. 11.

TABLE 3.1 Comparison for Project Management and Ongoing Functions

Item (examples)	Current method	Project management method	Differences (benefits and costs)
New product development	Team focuses on a technical solution without the requisite business solution (cost and time).	Focused planning and implementation against a schedule that incorporates the cost of development.	Reduced time to develop and market a new product.
Competitive analysis assessment	Typically, not performed. If done, uses an ad hoc team.	Dedicated team plans, implements, and reports assessment of competition vs. own business.	Rapid identification of strengths and weaknesses in relation to others. Provides basis for change to meet future business needs.
Organizational change	Ad hoc team to assess and plan change.	Dedicated team to assess and plan change.	Sharp focus on the change and improved planning for change. Should result in more effective organizational outcome.

management. The primary advantage envisioned is a sharp focus on the work to be accomplished and a dedicated effort to complete that work for the benefit of the organization.

Assessing the Opportunity to Implement Project Management

Many senior managers may view project management as a fad or something that is already in place within the organization. How many times does one hear, "Project management, yes, I do that all the time." Past experience will prevail unless there is compelling evidence that a change or new method will materially contribute to advancing the organization's goals. The old saying, "Change is inevitable, progress is optional," applies to those who deny the need for change.

To sell project management, one must first identify the stakeholder to champion change. There are those stakeholders who have the responsibility and authority to optimize the organization for the "best infrastructure" to meet strategic and business goals. There are those individuals who may desire "stability" in the workplace and oppose change. Then, there are those who are not key decision makers, but desire to improve the way work is accomplished. Table 3.2 lists some of the stakeholders, their responsibilities, and their potential for accepting or rejecting change.

There is a different time domain that each stakeholder views and each may have different interest. It does not appear that the stakeholders share common interests or common times for results. This shows the challenge in getting agreement on a change to use project management as the primary methodology for business solutions.

TABLE 3.2 Stakeholders for Change in an Organization

Stakeholder	Responsibilities	Accept/reject change
Corporate stockholders	Focus on the profitability of the organization and return on investment.	Probably would support a change to project management if the cost of change is low, the return on investment is high, and the risk is low or moderate.
Board of directors	Focuses on the long-term business and guides the organization in short-term decisions. This typically means setting goals 3 to 5 years for the future and conducting oversight on major ongoing activities.	Probably will accept change to project management if a compelling case is made. May also institute a system that links strategic goals with ongoing projects.
CEO, COO, president, vice presidents	Focus on the current business goals (current year and 1 year in future). May also serve on the board of directors.	May be reluctant to accept project management as a major change to the method of doing business. Any change to business procedures may negatively affect current plans, i.e., would cause a change to the plans.
Functional managers (e.g., finance, engineering, marketing, manufacturing)	Focus on repetitive type of work with emphasis on maintaining the status quo.	May reject the concept of using project management. May see a threat to their positions and status. May not understand project management concepts and would argue that project management does not fit in this organization.
Project directors, program directors	Focus on ongoing projects and projects being planned or initiated.	May accept changes to the project management system in the future, but would be reluctant to change procedures on current projects.
Project managers, project team members	Focus on one or more projects in the planning, implementation, or closeout phases.	Probably would support converting the organization to a "project-driven enterprise." These stakeholders are in a position to observe the positive and negative aspects of projects.

Table 3.3 identifies some of the differences in a typical organization's stakeholders. The approximations give some idea as to the barriers to successfully "selling" project management to senior managers. Take, for example, that stockholders are looking for either immediate return on investment and the board of directors is charged with developing a strategic position that may take 3 years. Expenditures in

TABLE 3.3 Stakeholder Focus and Interests

Stakeholder	Window focus	Principal interest
Corporate stockholders	3 to 12 months for dividends 3 to 5 years for capital growth	Return on investment
Board of directors	1 to 5 years for planning 1 to 12 months for implementation of plans	Strategic direction and future business opportunities
CEO, COO, president, vice presidents	0 to 3 months for operations 3 to 12 months for changes to operations	Meeting quarterly and annual objectives
Functional managers (e.g., finance, engineering, marketing, manufacturing)	0 to 12 months to meet organization's objectives	Maintain the status quo with some improvements in operations
Project directors, program directors	0 to 12 months to complete projects ongoing and new projects	Maintaining the momentum on project execution and new project initiation
Project managers, project team members	1 to 12 months to complete current projects	Complete projects and start new ones

the current year will affect the dividends paid to stockholders and defer their return on investment. On the other hand, if the investment is not made the business will suffer and perhaps not be competitive in the future.

The previous discussion gives mostly an internal sell of project management. This internal selling needs someone within the current system to propose a change. Most changes are generated as a result of external threats or opportunities. For example, if a company views its competition becoming more productive through the use of project management practices, that company may change to meet or beat the competition. If that same company was in the leading position, would it make changes to move farther ahead of the competition? Probably it would not.

3.9 EXTERNAL PROJECT MANAGEMENT SELLING

With the predictions for the information technology industry that more than 50 percent of project management will be outsourced in less than 5 years, this presents a different approach to "selling project management to senior managers." A scenario that places the project management provider outside the organization and in a competitive situation with similar service providers dictates that the seller be able to describe his/her product and service in such a manner that senior managers decide to buy.

A decision to sell project management services by an external consultant requires defining the services as well as being able to deliver in a manner that provides satisfaction with the customer.[21] Selling project management or being a service provider is typically defining the services being provided and convincing senior managers that the services can be delivered to meet their needs. Project management services from external sources may take on different characteristics, based on the needs of the customer. A range of services may be similar to the following:

- Provide interim project planning services to assist an organization in building its capability while delivering a product that is immediately put to use.
- Provide project planning and control services to support project managers. These services often entail only the schedule and cost functions. Specialists develop resource-loaded schedules and track progress as well as expenditure of labor hours.
- Provide a project manager, a project control team, and technical specialists to plan, execute, and close out the project. The work may be conducted on the customer's site or at the provider's location. The customer's expectation is for an end product that meets specifications and periodic reports to confirm the project is progressing in a satisfactory manner.

3.10 WHAT IT TAKES TO SELL PROJECT MANAGEMENT

A compelling story for senior managers and decision makers must be developed on the basis of facts to obtain their commitment to change to project management approaches for developing and delivering products and services. This commitment can only be obtained if there is value to the proposed change and the change replaces current methods that are less effective. One or more of the senior managers must champion the change.

The sequence for selling project management to an organization follows a series of steps. These steps may be sequential or in parallel, depending upon the situation:

- Define the organization that should change to project management. Identify the areas that should change or would change.
- Assess the positive and negative impacts of change to project management.
- Identify the stakeholders and their probable support or rejection of the proposal to change functions to project management.
- Identify success stories for conversion to project management and quantify the benefits.

[21] David I. Cleland and Lewis R. Ireland, *Project Manager's Portable Handbook* (New York: McGraw-Hill, 2000), pp. 7.52–7.53.

- Prepare a proposal with an implementation plan. This document must make the compelling case for change and the implementation plan should use project management techniques to show the process of change.
- Generate support for the change with those individuals who will directly benefit from the change.
- Reduce resistance for the change with those individuals who may oppose the change regardless of benefits to the organization.
- Schedule an informal meeting with the champion and determine the approach to other organization decision makers.
- Make a formal presentation to the organization's decision makers and present examples of where immediate improvements may be made and the resultant benefits derived by the organization.
- Obtain approval of the implementation plan to include a schedule of milestones and a budget.
- Implement the plan.

3.11 TWO VIEWS OF SELLING PROJECT MANAGEMENT

Selling project management takes on two views. One is the view of the internal person attempting to change the way the organization creates business solutions. The second view is the external consultant or project management provider attempting to convince senior management that his/her project management services can add value to the organization.

The internal selling is difficult because of the number of stakeholders with different interests to serve and the time in which each would like to see a solution. A crisis will energize senior managers to act and use project management when it is viewed as the solution to the problem. A crisis usually dictates some action and the implementation of project management may be identified as the solution. A crisis also has the challenge of immediate action for which the organization may not be prepared.

The external selling is primarily a transfer of responsibility for work and the work being accomplished through project management. There may be some instances of augmenting an organization's project management team to either perform the work or to gain experience from the person providing the services. Outsourcing of project management work is an organization's option if they do not want to build it as a core competency. As a core competency, it would not be outsourced.

Whether the "selling of project management" is internal or external, there must be perceived value by the senior manager. There must be some compelling reason for the senior manager to become the champion and support the change. The value must be to the organization and it must outweigh considerations such as economic, political, and stakeholder objections, and personal interests. The change must also advance the organization on the competitive scale to improve its position with relation to competitors.

Selling project management is no different than selling a product or service to an organization. One must show the benefits of project management to the organization and the differences between what is currently done compared to the future. Benefits might include faster delivery of products to customers, improved productivity that results in lower costs, better products to meet customer needs, and greater confidence in the organization's ability to plan, implement, and close out work areas.

Senior managers are looking for value-added solutions in project management and not the features, characteristics, or process. Anyone selling project management or project management services must focus on the value to the organization. This value must be quantified and show the advantages or benefits of changing to project management practices.

In this chapter, the matter of when to use project management has been emphasized. The final answer in this regard is that project management provides an important means of how to manage change. Figure 3.3 poses a provocative question in the matter of dealing with change facing the enterprise.

3.12 TYPES OF PROJECTS

When assessing whether to use project management and how to use it effectively to meet the objectives of the organization, one should have an understanding of the types of projects that will be applied. Typically, all projects have not been defined by an industry or organization, but are simply viewed as the work to be accomplished. Project types can be useful in determining how to approach the work and what degree of planning is needed, for example.

Understanding the types of projects that an organization will or will not perform and the types that an organization has the capability to perform can assist in developing the organization's competencies. Developing the organizational competencies through understanding what is needed to use project management should give a useful insight on individual and team competencies needed to pursue the organization's mission.

If we could — by a deliberate and diabolic stroke — eliminate all projects and project management literature, how would we (could we) deal with change in the world?

FIGURE 3.3 The management of change.

Some typical project classifications are:

- *Product projects.* Projects that design, develop, and deliver a product as a result of the work effort. The customer for this project is typically an external stakeholder, who has contracted for a product to meet a specific need.
- *Service projects.* Projects that plan, design, and deliver services to external customers. The customer has contracted for specific expertise that is not available in-house.
- *Continuous improvement projects.* Projects that are internal to the organization and result in process change. The customer is an internal manager, or group of managers, who wants to change the manner in which current operations occur.

This simple approach to "types of projects" leads to an expanded listing of attributes or characteristics of projects or those that describe projects. This list may be expanded to include other features of projects, both from the physical features and the need features, to create a type of project specific to an industry or organization.

The 12 feature areas in Table 3.4 are a start to describing or classifying projects by type. These features are not necessarily a complete list and other features that may be more meaningful for an industry or organization could be added. This list can help understand the projects, whether from an organization's perspective or from a customer's perspective.

Martin and Tate describe projects as Type 1, Type 2, and Type 3 for a degree of definition of a range of projects. These project types are defined primarily on the basis of need for coordination of efforts. Type 1 projects include process improvement, reengineering, and strategic planning. Type 2 projects are typically small projects accomplished by a small team. Type 3 projects require extensive coordination because of their size and complexity.[22]

Although a degree of coordination is required for a project, both internal and external, it is but one parameter for classifying projects. The degree of coordination perhaps relates to project complexity, project, and number of stakeholders.

Many organizations classify projects only by the size—small, medium, and large. These classifications are subjective in that a "large" project for a medium-size company may be a "small" project for a large-size company. Large (major) projects for the U.S. Department of Defense are $1 billion and more. Many organizations do not classify projects by "urgency of need," or have a priority system. It is often assumed that the larger the project, the more important the work.

Priority of a project should be assigned because it relates to delivering benefits to either internal or external customers. Some small projects, such as a software upgrade to fix defects, may have more benefits than a large project for say $50 million. Failure to fix the software defects could potentially have impacts that range across the organization's ability to do business.

---

[22]Paula K. Martin and Karen Tate, "What's Your Type," *PM Network,* Project Management Institute, April 1999, pp. 53–55.

TABLE 3.4 List of Feature Areas for Project Classification

Size
• Dollar value • Number of people on project • Duration (calendar time) • Geographic span (global, multicountry) • Combination of above
Priority
• Urgency of need for business • Urgency of need for customer • Urgency of need to meet market requirements
Degree of risk
• High risk • Medium risk • Low risk
Profitability
• High margin • Medium margin • Low margin • No margin • Loss
Cash flow
• Immediate return on investment • Near-term return on investment • Long-term return on investment
Technology
• Low technology • Medium technology • High technology • Mature technology • Immature technology
Business experience
• Projects that have been done before • Projects that haven't been done before • Projects that have some new work, for which no experience base exists • Projects that no one has done before

TABLE 3.4 List of Feature Areas for Project Classification (*Continued*)

Business base
• Projects that build on core competencies • Projects that build on new core competencies • Project unrelated to core competencies • Projects that leverage core competencies
Project definition
• Undefined projects • Ill-defined projects • Partially defined projects • Fully defined projects
Results (objectives)
• Strategic results • Operation/business results • Part of larger program
Deliverables
• Product • Product and service • Service • Product improvement/upgrade • Product disposal (e.g., nuclear power plant closeout) • Product restoration (repair, renovate) • Product maintenance (facilities maintenance—planned and unplanned) • Emergency operations (fire, life saving, rescue, utilities outages) • Legal services (litigation, felony prosecution, and trials) • Law enforcement (car patrols, foot patrols, accident investigation, and incarceration) • Political campaigns (fund raising, speaking engagements, travel schedule)
Other features
• Engineering • Research • Production with many of the same products • Production with several models of products • One-of-a-kind product • Prototype product (brassboard, breadboard)

Categorization of projects is not an exercise in trivia, but it is an essential understanding of one's business and what comprises the business. Building a business base to support the organization's mission and purposes dictates that project types be defined and the correct type applied to potential business situations. Matching a project type with its requirements should enhance the capability of the organization.

3.13 THE MANAGEMENT OF SMALL PROJECTS

In any organization that is in motion there are usually many small projects that are used to cope with the minor changes that are underway in that organization's products, services, or organizational processes. A small project is one whose performance takes only a few weeks, has an easily defined scope, and has a dollar value between \$5000 and \$50,000. The project team is small, only a few cost centers are involved, and manual methods are used for facilitating the management of the project. Some examples include:

- Reengineering of a production line
- Realignment of a production line
- Development of an information system to support an element of the enterprise
- Revitalization of procurement practices
- Reorganization of an enterprise

The management of small projects is much like managing a large project—except for the degree of complexity involved. There is a routine protocol that should be followed to include (1) identify the need, (2) plan the project, (3) collect relevant information, (4) analyze the data, (5) develop alternative ways to accomplish the project results, and (6) present suitable recommendations. The "work packages" that are involved in each of these protocols are indicated below.

Identify the Need

- Identify the client/sponsor and their perception of the problem.
- Conduct an initial analysis to get an idea of what is involved in the small project.
- Be careful to separate "problems from opportunities."
- Establish tentative objectives and goals for the project.
- Identify the funds that are available for the project.
- Find the initial documentation that describes the problem or opportunity.

Plan the Project

Every small project needs a plan. The essentials of a small project plan are:

- A summary that can be read in a few minutes.
- A list of milestones (goals) identified in such a way that there can be no ambiguity when a goal is achieved.
- A work breakdown structure (WBS) that is sufficiently detailed to provide for the identification of all tasks associated with the project.

- An activity network that shows the sequence of the work packages and how they are related.
- Separate budgets and schedules that are consistent with the work breakdown structure.
- A description of the review process.
- A list of key project team members and associated stakeholders.
- Identification of final objectives, goals, and strategies for the project.
- Identification of what the client or sponsor expects by way of deliverables from the project.
- Identification of and attempt to seek potential answers regarding the key questions surrounding the problem and the project.
- Development of a work plan on how and who will perform tasks.
- Organization of the project team to include identification of individual and collective roles to be carried out by members of the team. The use of an LRC, as described in Chap. 9, is useful here.
- Familiarization with the organization's work authorization process through which funds are transferred for work on the project to an organizational unit within the organization or to an outside vendor.
- Preparation of schedules for the work to be carried out.
- A preliminary outline of the expected final report.

Collect Information

- Use interviews, surveys, or other data collection mechanisms.
- Develop a bibliography of basic information regarding the problem.
- Study the background information.
- Review miscellaneous data and information regarding the problem and the surrounding circumstances or situations.
- Observe activity by the people associated with the problem to discern what is going on.
- Correlate the data and information that have been gathered.
- Use techniques such as work sampling, work flow, and individual and collective behavior by the people associated with the problem.
- As the strategies for the solution of the problem begin to emerge, conduct a preliminary test of these strategies (policies, procedures, processes, methods, techniques, roles, etc.).

Analyze Data

- Classify the data by some common methodology.

- Question what the data appear to be revealing.
- Count, measure, and evaluate the forces and factors that begin to emerge during the analysis of the data.
- Compare data to the objectives and goals that have been established for the project.
- Look for trends, deviations, and other distinct characteristics of the data.
- Correlate different data that have emerged on the project.
- Conduct quantitative and qualitative assessment of the data. Consider using statistical techniques to assess the data.
- Follow your instincts in terms of what the data are revealing—which elements of data are providing meaningful insight into the problem and its solution.

Develop and Evaluate Alternatives

- Identify a few alternatives that might solve the problem.
- Evaluate these alternatives through the use of informal "cost-benefit" analysis to select the one or two that promise a useful solution to the problem.
- Test the one or two alternatives with the client.
- Select a final alternative.
- Develop implementation strategy.

Present Recommendations

- Prepare report.
- Brief client and/or sponsor.
- Rework as needed.
- Submit final report.
- Send a thank-you note (e-mail) to the project team members and other stakeholders who helped bring the project to a successful conclusion.
- Work with the project team members to prepare a "lessons learned" summary of the project and forward it to key stakeholders.

A small project can be managed by using a scaled-down version of most of the concepts, processes, and techniques employed for larger projects. The client that is sponsoring the project and the project stakeholders should be kept informed on the status of the project on a regular basis.[23]

---

[23]Material on small project management is paraphrased from David I. Cleland and Lewis R. Ireland, *Project Manager's Portable Handbook* (New York: McGraw-Hill, 2000), pp. 3.17–3.21.

3.14 TO SUMMARIZE

Some of the major points that have been expressed in this chapter include:

- Project management can be used for a wide variety of purposes. On balance, the reasons for the use of project management can be centered around the following categories: (1) size of the undertaking, (2) unfamiliarity, (3) market change, (4) interdependence, (5) resource sharing, (6) importance of the project, and (7) organizational reputation. Within these categories project management is used:
  - To share resources across organizational units
  - To focus attention on specific customers in specific market segments
  - To integrate systems and subsystems simultaneously or in parallel within independent organizations
  - To manage focused interorganizational efforts from a profit-center perspective
  - To deal with specific ad hoc problems and opportunities
  - To expedite responses to new events in the organization or its environment
  - To accommodate the inherent interdependency within an organizational system
  - To combine several proven methods of organizational design, such as product, functional, and geographic
  - To preserve unity of command, unity of direction, and parity of responsibility and authority for disparate activities
  - To fix accountability within organizations
  - To bring a wide range of experience and viewpoints into focus on tasks, opportunities, and problems
  - To formalize an informal management process such as project engineering
  - To establish a liaison role between organizational units or specialties
  - To test a new organizational strategy without committing to a formal structural reorganization
  - To deal with the magnitude of an undertaking requiring massive input of capital, technology, skills, and resources
  - To manage unique or rare activity
  - To focus effort to maintain an organizational reputation
  - To keep a low-profile, long-term organizational effort alive while awaiting suitable competitive or environmental conditions
  - To facilitate the participation of organizational members in the management process of the enterprise
  - To deal with a new technology which requires pooling of existing resources and capacities
  - To satisfy a customer's need for the unified management of a project-based contract in order to avoid having that customer work with many different functional organizations
  - To meet competition

  - To deal with a task that is bigger than anything the organization is accustomed to handling
  - To promote participative and professional management
- There is a cost associated with dealing with change. That cost typically centers on the use of resources in dealing with the creation of improved and new products, services, and organizational processes.
- Throughout this chapter are examples of how projects are used to deal with change, including the change forced on an organization through a crisis or unexpected event.
- Project management can be used to shut down operations such as in the deactivation of a plant, meet environmental protection standards, or implement other facility modifications where the objective is to change the circumstances under which the enterprise is operating.
- Major projects dealing with the changing of the infrastructure of societies are particularly complex to manage. A couple of examples were provided in this chapter to illustrate this point.
- A few typical challenges to the design and implementation of projects were cited at the end of this chapter.
- Project management meets the need for providing an organizational focus not found in the traditional form of organization. However justified, project management should not be used until the leaders of the organization are committed to its use and are willing to prepare a suitable culture for project management to germinate and grow.
- "Selling" project management to senior managers is a value proposition that focuses on the difference between what is being done today and what can be done through the use of project management.
- Internal selling of project management is trying to change the organization to adopt project management as the management system of choice.
- External selling of project management is providing a service to an organization when that organization does not want to develop project management as its way of doing business for product and service development and delivery as well as organizational process change.
- Using a system to categorize and define the types of projects an organization will use for business is a means of building and understanding the organization's capability.
- Priority of projects equates to urgency of need for the benefits of that project.
- Small projects require disciplined planning and management, but at a lesser scale than large projects.
- Small projects can be used for a variety of work efforts for the organization, such as reengineering, organizational change, and studies.

3.15 ADDITIONAL SOURCES OF INFORMATION

The following additional sources of project management information may be used to complement this chapter's topic material. This material complements and expands on various concepts, practices, and theory of project management as it relates to areas covered here.

- Christopher A. Chung and Abu Md Huda, "Practical Tools for Project Selection," chap. 4 in David I. Cleland (ed.), *Field Guide to Project Management* (New York: Van Nostrand Reinhold, 1997).
- Lynn Crawford, "Winning the Sydney to Hobart: A Case Study in Project Management," and Gerald W. Crabtree, "TAXOL—An Example of 'Fast-Track' Drug Involvement," in David I. Cleland, Karen M. Bursic, Richard J. Puerzer, and Alberto Y. Vlasak, *Project Management Casebook,* Project Management Institute (PMI). (First published in *Proceedings,* PMI Seminar/ Symposium, 1993, pp. 53–59; and *Proceedings,* PMI Seminar/Symposium, 1993, pp. 616–621.)
- Russell D. Archibald, *Managing High-Technology Programs and Projects* (New York: Wiley, 1992). This book provides a wealth of information based on theory and tempered by the author's experiences over an extensive career in project management. Written from the project manager's perspective and designed to provide detailed guidance, this book expounds on proven methods that include checklists for start-up and closeout of projects.
- Kevin Forsberg, Hal Mooz, and Howard Cotterman, *Visualizing Project Management: A Model for Business and Technical Success* (New York: Wiley, 2000). This book details the practices and a proven methodology for performing project management in a variety of industries—most specifically in manufacturing. The concrete examples of "how to" perform various tasks that support project management work. It also addresses the technical performance requirements for projects and provides a model for requirements analysis, which most project management books omit.
- Lynn Crawford and Terry Cooke-Davies, "Enhancing Corporate Performance through Sustainable Project Management Communities," *Proceedings,* Project Management Institute, 1999 Annual Seminars and Symposium, October 10, 1999, 7 pp. This paper looks at the success and failure of projects as they affect organization performance. The authors identify seven prerequisites for an effective project management community.
- George Pitagorsky, "A Scientific Approach to Project Management," *Machine Design,* Cleveland, Ohio, July 26, 2001, pp. 78–83. The author touches on several areas of importance to consider when using project management and the areas in which one may apply the principles of project management. This article also challenges the reader to find new and workable solutions to business problems.

- Janice Thomas, Kam Jugdev, Connie L. Delisle, and Pam Buckle, "Selling Project Management to Senior Executives: What's the Hook?" *Proceedings,* Project Management Institute, Annual Seminars and Symposium, September 7–16, 2000, pp. 827–833. This paper looks at research efforts on "selling" project management to senior executives. The authors make specific points about executive needs, value alignment, and competitive "selling" to executives

3.16 DISCUSSION QUESTIONS

1. What is involved in strategic planning? In strategic implementation?
2. How do projects become the driving force in determining how organizational resources are used?
3. What is meant by the sentence, "Projects are resource-consuming activities"?
4. List and describe some of the major reasons for an organization to use project management.
5. How does an organization know when the size of an undertaking suggests using project management?
6. Discuss situations in your work or school experiences that were permeated by unfamiliarity. Could project management have been used to address the unfamiliarity?
7. What kinds of projects can be crucial to an organization's professional reputation?
8. What types of questions would a manager ask to determine the importance of a project?
9. How does the use of a project management structure affect the culture of a corporation?
10. What kinds of questions are important in determining whether a project supports an organization's strategies and its overall mission, objectives, and goals?
11. How would you "sell" project management within a major corporation, if you work in the organization?
12. What "value statements" would you use to "sell" project management to your boss?

3.17 USER CHECKLIST

1. Is the use of project management in your organization driven by any outside forces? What forces?
2. Is your organization project-driven in any way? In what way?

3. Does your organization recognize when the need for project management arises? Give examples.
4. Does the size of any of your current undertakings warrant using project management?
5. Are any of the ad hoc projects currently being undertaken by your organization fraught with unfamiliarity?
6. Is your organization comfortable in understanding the competitive market in which it works? Is the market dictating the use of project management?
7. Does management recognize and facilitate the large number of interrelationships that can exist between functional departments when each has some role in a project?
8. Are any of your organization's current undertakings crucial to its reputation?
9. How does your organization combine project and functional approaches to management?
10. Does your organization take advantage of emerging opportunities for new products by using a project management team to design and develop innovative ideas?
11. Does your organization promote individual entrepreneurship by supporting the development of creative ideas?
12. Does your organization recognize when not to use project management?

3.18 PRINCIPLES OF PROJECT MANAGEMENT

1. Project management provides a sharp focus on planning, implementation, and control over work of a unique or unfamiliar nature.
2. Project management has three functions: develop and deliver products and services as well as support organization process change.
3. Project management involves greater stakeholder involvement than typical general management approaches.
4. Project management "flattens" an organization's structure, which puts a sharp focus on the work to be accomplished through dedicated efforts of planning, controlling, and implementing.
5. Project management is used to implement the strategic goals through the best use of resources to meet organizational purposes.
6. Project management, when properly implemented in an organization, has an interdependence with strategic objectives.
7. Selling project management requires that the value be sold rather than the features.

8. Selling project management as the system of choice changes the fundamental way business is pursued.
9. Selling project management requires changing a mindset that the current method of work is adequate.
10. Crisis creates opportunities to implement project management.

3.19 PROJECT MANAGEMENT SITUATION—WHEN TO USE PROJECT MANAGEMENT

This book promotes the use of project management as a key competitive advantage for all organizations. It also avoids stating that project management is the only solution for all business needs. Project management, however, is considered a vital part of the strategies for an organization to use when certain conditions exist and when there is a need for rapid planning and execution of critical work to develop and deliver products and services as well as a means of effecting organizational process change.

Knowing when to use project management requires some consideration of the situation and getting the stakeholders to support project management as "the best solution" for a given need. This approach drives the need to understand project management capabilities and when to apply these capabilities. It also drives development of a project management capability within a company.

The situation is that a company with 122 employees provides maintenance services for large residential housing complexes where the residential complexes require common services to retain their operations. Some of the areas that require services are:

- Clean and paint swimming pool; winterize for cold weather.
- Maintain common landscape areas by cutting grass, trimming bushes and hedges, fertilizing plants, trimming trees, and planting replacement ground cover materials.
- Service air conditioning units that serve common areas.
- Service heating units that serve common areas.
- Remove snow during winter months.
- Service plumbing, water, electrical, and gas items when there is a failure.
- Telephone, cable television, and personal items of residents are not serviced.

This company has grown from 7 people approximately 9 years ago to 122 today. The company performs tasks associated with 14 major residential complexes and 6 smaller residential units. Over the years, growth has been steady, but the turnover of technicians has been high. Those who left complained that the amount of work and the erratic schedules hampered their personal lives. Management, consisting of two owners and a general manager, has been frustrated in efforts to smooth out the fluctuations in work load. Further, there are several individuals, some former employees, who are starting companies that will directly compete for the work at residential units.

It is recognized that the high turnover rate with technicians and the emerging competition will probably stop further growth and, if nothing is done, the company will probably lose some of its current customers. A consultant recommends converting to a project-driven enterprise whereby all work is performed as projects by teams. Management is considering this, but doesn't know if it will work because they are uncertain as to how projects could be used.

Concerns facing the company are:

- Is the work to be accomplished compatible with a project approach?
- Can the employees work in teams to perform the different tasks?
- What is the cost to change to a project-driven enterprise?
- What is the cost if the organization does not adopt project management?
- How soon could the organization change and start delivering services under project management?
- How will a change to project management affect the customers? Positively? Negatively?

The company faces a dilemma that it must change. What is the proper direction? Can the company use project management for all of its work or just a part of it? One thing is sure, it must change to something that is successful in delivering services and reducing costs for these services to beat the emerging competition.

3.20 STUDENT/READER ASSIGNMENT

1. Using the project management situation, identify the benefits and shortcomings for an immediate transition to use project management for all services.
2. Considering the nature of the work being done and the knowledge levels of the technicians (probably high school graduates, some tech school graduates), how long would it take to indoctrinate them in the principles of project teams?
3. What do you see as most challenging if the company decides to convert to a project-driven enterprise?
4. Do the customers have any say in converting to a project-driven and delivered service? If so, what are some potential questions that the customers may ask?
5. Are there any potential savings (productivity increase) that you can identify from changing this company to a project-driven company?

CHAPTER 4
THE STRATEGIC CONTEXT OF PROJECTS[1]

"Many of us are like the little boy we met trudging along a country road with a cat-rifle over his shoulder. 'What are you hunting, buddy?' we asked. 'Dunno, sir, I ain't seen it yet.'"

R. LEE SHARPE

4.1 INTRODUCTION

An emerging conviction among those professionals who do research on, publish, and practice project management is the belief that projects are building blocks in the design and execution of organizational strategies. An ongoing and competitive organization has a "stream of projects" flowing through the organization that support changes in operational and strategic initiatives.

In this chapter the strategic relationship of projects to organizational purposes will be considered. A project selection framework will be suggested. An initial look at project planning will be provided, along with a description of the project "owner's" need for participating in the selection and use of projects to support organizational purposes. A key part of this chapter includes the description of a *project management system,* which provides a philosophy and standard for a "systems view" in the management of projects.

4.2 STRATEGIC TRANSITIONS

The most dangerous time for an organization is when old strategies are discarded and new ones are developed to respond to competitive opportunities. The changes that are appearing in the global marketplace have no precedence; survival in today's

[1]Portions of this chapter have been taken from D. I. Cleland, "Measuring Success: The Owner's Viewpoint," *Proceedings,* PMI Seminar/Symposium, Montreal, Quebec, September 20–25, 1986; and D. I. Cleland, "Project Owners: Beware," *Project Management Journal,* December 1986, pp. 83–93.

unforgiving global marketplace requires extraordinary changes in organizational products, services, and the organizational processes needed to identify, conceptualize, develop, produce, and market something of value to the customers. Projects, as building blocks in the design and execution of organizational strategies, provide the means for bringing about realizable changes in products and processes. Senior managers, who have the residual responsibility for the strategic management of the enterprise, can gain valuable insights into both the trajectory of the enterprise and the speed with which the competitive position of the enterprise is being maintained and enhanced. This can be done by conducting a regular review of the status of the "portfolio of projects" in the enterprise.

A belief that projects are building blocks in the design and execution of future strategies for the enterprise means that the organizational planners recognize that preparing for the future on the basis of extrapolations of the past results from a well-understood and predictable platform of past experience is not valuable—and can be dangerous to the health of the enterprise. Although planning based on extrapolation of the past has some value for an ongoing business providing routine products and services, it makes little sense when the enterprise's future is dependent on developing and producing new products and services through revised or new organizational processes. All too often people persist in believing that what has gone on in the past will go on into the future—even while the ground is shifting under their feet. If the enterprise is engaged in a business where competition is characterized by the appearance of unknown, uncertain, or not yet obvious new products and services, especially to the competition, then project-driven strategic planning is needed. Project-based strategic planning assumes that:

- Little may be known of the new product or service but much is assumed about potential customer interest in the forthcoming initiative.
- Decision making on the project during its early conceptual phases is based on what information is available. Assumptions concerning the potential future business success of the innovation are an important source of knowledge on which decisions can be made.
- Assumptions concerning the new venture are systematically converted into meaningful databases as new knowledge concerning the innovation evolves through study by the project team.
- Even after the prototype is developed and field-tested with customers, uncertainty remains as to how well the product or service will do in the competitive marketplace.

One company that has married strategic planning and project management is Blue Cross and Blue Shield of Louisiana (BCBSLA). Linking project management, strategy formulation, and implementation provides for a system of checks and balances for the company. A Corporate Project Administration Group was formed to assist corporate executives in a refinement of the scope of corporate initiatives—and to develop goals and objectives for corporate projects. The group was responsible for working with project managers to develop and execute plans and keep corporate executives informed of the progress being made on the projects. A process was created that tracks corporate initiatives and project performance as they relate to corporate goals and objectives. Monthly status reports are provided for each project initiative. By combining project management and strategic planning the company is better able to select corporate objectives and goals, and initiate and track projects that are related to corporate initiatives.[2]

4.3 IMPLICATIONS OF TECHNOLOGY

Management of an enterprise so that its future is ensured requires that the technology involved in products and/or services and organizational processes is approached from two principal directions: the strategic or long-term perspective and the systems viewpoint. In both directions, projects play a key role. In this chapter these two directions will be woven into a project management philosophy in which projects are building blocks in the design and execution of organizational strategies. A couple of examples of how contemporary organizations deal with projects make the point:

- At Banc One Corporation, one of the fastest-growing and most profitable banks in the United States, 3 percent of the profits has been dedicated to technology R&D. One of the bank's most important technology projects is a new computer system that has dramatically altered the way Banc One branches operate. The system includes the creation of a new credit card processing system. With the assistance of the Dallas-based Electronic Data Systems Corporation, the bank has moved from older mainframe systems to a distributed architecture.[3]
- Sony is probably the most consistently inventive consumer electronics enterprise in the world. It has had hit after hit of high-technology products. Its products have created billion-dollar markets, with devices that have altered people's work and leisure. Sony's portfolio of products ranges from semiconductors, batteries, and recording tapes to video and audio gear for consumers, professionals, computers, communications equipment, and factory robots. Last year the company spent $1.5 billion on research and product development projects—roughly 5.6 percent of revenues. Each year the company sends out 1000 new products—an average of almost 4 a day. Some 200 of these new products are aimed at creating whole new markets, such as the Mini Disc portable digital stereo. Sony founder and honorary chairman Masaru says that the key to success at Sony—and to everything in business, science, and technology—is never to follow the others. In other words, use innovation—the creation of something that does not currently

[2]Philip Diab, "Strategic Planning + Project Management = Competitive Advantage," *PM Network,* Project Management Institute, July 1998, pp. 25–28.

[3]Alice LaPlante, "Shared Destinies: CEOs and CIOs," *Forbes ASAP,* December 7, 1992, pp. 32–42.

exist. Product/project ideas come from many different organizational levels in the company, from the senior managers to the young engineers working in the product design department. Some of Sony's key philosophies are:

- An emphasis on making something out of nothing
- People who are optimistic, open-minded, and wide-ranging in their interests, who move around a lot among product groups
- A belief that having continuous success in the same area makes you believe too much in your own power, which harms your creativity
- A belief that new products come primarily out of a creator's imagination, not from a marketing study
- Occasional use of a "skunk works" project to circumvent the formal project approval process in the company
- Use of competing project teams to work on promising technologies[4]

- In the early 1990s, Boeing invested heavily in new technology so that it could design a commercial aircraft, the 777 twin jet, entirely by computer. It connected 1200 engineers and countless other staffers to 2200 work stations and four mainframes, in Seattle, Philadelphia, the Midwest, and Japan. That technology enabled the aircraft manufacturer to solve virtually every design problem through computer animation—without having to build a prototype—and thereby limit the cost of making design changes down the line. For instance, when a team of engineers discovered a glitch in the jet's wiring, they "fixed" it instantly on their 3-D digital model. The technology cut the design time for the 777 in half.[5]

Projects are essential to the survival and growth of organizations. Failure in project management in an enterprise can keep the organization from accomplishing its mission. The greater the use of projects in accomplishing organizational purposes, the more dependent the organization is on the effective and efficient management of those projects. Projects are a direct means of creating value for the customer in terms of future products and services. The pathway to change will be through projects. Future strategies will entail a portfolio of projects, some of which will survive and lead to new products and/or services and the manufacturing and marketing processes that will beat out the competition. With projects playing such a pivotal role in future strategies, senior managers must approve and maintain surveillance over these projects to determine which ones can make a contribution to the strategic survival of the company. Two authors state:

> The challenge facing senior management seeking to implement revolutionary change within the organization is to manage that change outside the straitjacket of the existing bureaucracy, procedures and norms. Projects and project management help senior management to do precisely that.[6]

---

[4]Brenton R. Schlender, "How Sony Keeps the Magic Going," *Fortune,* February 24, 1992, pp. 75–82.

[5]Paraphrased from Anne Bernasek, "Prosperity," *Fortune,* October 2, 2000, pp. 101–108.

[6]Sergio Pellegrinelli and Cliff Bowman, "Implementing Strategy through Projects," *Long Range Planning,* vol. 27, no. 4, 1994, pp. 125–132.

For the last decade or so, many managers have been preoccupied with the improvement of operations through remedial strategies involving the use of reengineering, benchmarking, TQM, time-based competition, empowerment, team-based organizational designs, continuous improvement, and the so-called learning organization. The use of outsourcing and the "virtual" organization helped eliminate inefficiencies, improve customer satisfaction, and make the enterprise more competitive. In the short run, these remedial strategies helped improve organizational efficiency and effectiveness. But survival in the long-term requires that the enterprise do something that will establish a difference in its products/services and organizational processes that it can preserve in the marketplace. Although current activities are the basic components of today's competitiveness, overall strategic competitive behavior requires that new initiatives are conceptualized, developed, and implemented that will lead to changes in products/services and organizational processes that will ensure future competitiveness. In many cases this means that these new initiatives have to be different from those of the competitors. Few enterprises are able to survive and compete successfully on the basis of current operational capabilities over an extended period. The reason for this is the simple diffusion of new technologies, practices, and best products/services and supporting organizational processes—expressed in a superior way of meeting and exceeding customer expectations. The more a company benchmarks its competitors, the more likely it is that the enterprise and the competitor will become similar. The more a company uses outsourcing as a competitive thrust, the more likely it is that its competitors will copy its strategies and move to an equitable market position.

As rivals imitate each other's operational competitive strategies, the more probable it is that their strategies will converge. Competition becomes a series of behaviors that look similar—and no one competitor can become a big winner. Competition based on operational performance becomes self-defeating, leading to wars of competitive attrition. Unfortunately, many of the "flavors of the year" in the last 10 years have led to diminishing competitive returns. Competition based on continuous improvement reinforced by many of the flavors of reengineering, benchmarking, change management, and so forth, have drawn all too many enterprises into a "me too" mentality that has inhibited true creativity and innovation in creating *strategic* pathways for true competition in strategic performance.

The responsibility for allowing companies to degenerate into competition based on operational improvements clearly rests with the company's leaders. Unfortunately, this means that such leaders have failed to recognize their larger role beyond just operational stewardship, namely, a proactive role in selecting and executing the use of resources to provide a competitive, strategic pathway for the enterprise. Enterprise leaders have to work with the creative and innovative talent in the enterprise's pool of people and define and communicate new directions, allocating resources, making trade-offs through the study of alternatives, and making the hard choices of what to do for the future and—just as important—what not to do by way of committing organizational resources.

A product or process development project is a business venture—the creation of something that does not currently exist but which can provide support to the overall organizational strategy being developed to meet competition. Many projects are found in successful organizations.

4.4 A STREAM OF PROJECTS

An enterprise that is successful has a "stream of projects" flowing through it at all times. When that stream of projects dries up, the organization has reached a stable condition in its competitive environment. In the face of the inevitable change facing the organization, the basis for the firm's decline in its products, services, and processes is laid—and the firm will hobble on but ultimately face liquidation.

In the healthy firm, a variety of different preliminary ideas are fermenting. As these ideas are evaluated, some will fall by the wayside for many reasons: lack of suitable organizational resources, unacceptable development costs, a position too far behind the competition, lack of "strategic fit" with the enterprise's direction, and so on. There is a high mortality rate in these preliminary ideas. Only a small percentage will survive and will be given additional resources for study and evaluation in later stages of their life cycles. Senior managers need to ensure that evaluation techniques are made available and their use known to the people who provide these preliminary innovative ideas. Essentially this means that everyone in the organization needs to know the general basis on which product and process ideas can survive and can be given additional resources for further study. Senior management must create a balance between providing a cultural ambience in the enterprise that encourages people to bring forth innovative product and process ideas and an environment that ensures that rigorous strategic assessment will be done on these emerging ideas to determine their likely strategic fit in the enterprise's future.

For example, Elan Corporation, Plc., whose mission is the development of novel drug absorption systems for therapeutic compounds that provide distinctive benefits for the physician and patient—carrying out all the necessary clinical studies and regulatory work prior to market introduction—follows a fundamental strategy called *mind to market.* To implement this strategy, which brings their products to market through the formulation, clinical testing, registration, and manufacturing phases, project management is used. In the product development area, the company was committed to 56 active projects, utilizing 9 specialized drug delivery technologies in 18 therapeutic categories, which range from cardiovascular and narcotic analgesics to antiemetics and neuropharmacological agents. Research and development is the very essence of the company's business. Its work in R&D ensures a continuing stream of new products and technologies. In the global marketplace, the company currently has new-drug applications or their equivalent filed for 20 products in 30 countries around the world.[7]

[7] *Annual Report,* Elan Corporation, Plc., 1992.

A large retailer's strategy in assessing strategic opportunities is to jump-start a number of small projects at a relatively low cost and then shift the money into the promising ideas as the development work evolves. One example of such a promising project involves the development of electronic shelf tags, which would display pertinent information about a product, including the unit price, price per ounce, sales data, or whatever the company wanted to highlight. No longer would the employees have to change the traditional shelf tags. Another project is under development for a ceiling-mounted scanner to track the number of customers entering and exiting a store, thus alerting personnel that additional sales assistance is needed in specific departments. Another project borrows from just-in-time manufacturing inventory management concepts and processes. Products are shipped to distribution centers only when needed, thus reducing inventory requirements. Suppliers under this new procedure would write their own purchase orders by looking into the retailer's inventory databases and would ship products in time to keep the shelves from becoming bare.

When the use of project management is described in an enterprise, it is easy to think of just one project in the organization. Often we think of a large single dedicated project team led by a project manager who has the proper authority and responsibility needed to do the job. What usually exists after the enterprise has experimented with project management for a while is that several and perhaps many projects are under way, each having its own life-cycle phases. Team members may be working on several different small projects. As the use of project management continues to expand, the matrix organizational design emerges more fully and many projects share common resources provided by the functional entities and appropriate stakeholders. As the growth of project management continues and different projects come and go, there are some unique forces at work. The projects share common resources but will likely have objectives that are not shared with other projects, particularly if a diverse set of customers is involved. As projects start and are closed out or terminated for cause, a new mix in the use of resources comes forth. New projects may have a higher priority than the existing ones. As the competition for resources gets under way in the matrix organization, the opportunities for conflict in the assignment of the resources to the projects will erupt, often requiring senior management to intervene in deciding how the project priorities will impact the priorities for the use of the resources; the opportunity for gamesmanship emerges. Also, having many projects under way provides the opportunities for politics to enter the picture. Sometimes the enterprise will appoint a "manager of projects" who has jurisdiction over the project managers who are acting as a focal point for the projects.

4.5 STRATEGIC RELATIONSHIP OF PROJECTS

Organizational conceptual planning forms the basis for developing a project's scope in supporting the organizational mission. For example, a project plan for facilities design and construction would be a series of engineering documents from which detailed design, estimating, scheduling, cost control, and effective project management will flow. Conceptual planning, while forming the framework of a successful project, is strategic in nature and forms the basis for the following:

- Contributing, through the execution of strategies, to the organizational objectives, goals, and mission
- Standards by which the project can be managed
- Coping with the market and other environmental factors likely to have an impact on the project and the organization

Senior management deficiencies in the organization using project management will probably be echoed in the management of the projects. For example, an audit conducted in the early 1980s of a gas and electric utility that experienced problems with a major capital project found several key deficiencies in that utility, such as:

- Weak basic management processes
- No implementation of the project management concept for major facilities
- Fragmented and overlapping organizational functions
- No focus of authority and accountability[8]

Ford Motor Company is committed to the use of project management in its corporate strategy. To provide consistency in the use of project management, Ford realized during the 1980s that a common project management system was required. To bring about a consistent way to manage projects, a Ford corporate mainframe project management tool selection committee was created. Care was taken to ensure that users would be given a voice in the system selection process. Several key policies were established to both guide and motivate the committee to pursue its work. (1) There was agreement by senior management to accept the recommendations of the committee, assuming that such recommendations were supported by adequate facts. (2) The committee agreed to operate as a cross-functional project team. (3) A schedule was adopted to maintain user interest and enthusiasm; decisions by the committee would be made by consensus. (4) It was recognized that leadership of the committee was an important variable in realizing success of the work under way.[9]

4.6 DETERMINING STRATEGIC FIT

Projects are essential to the survival and growth of organizations. Failure in the management of projects in an organization will impair the ability of the organization to accomplish its mission in an effective and efficient manner. Projects are a direct means of creating value for customers—both customers in the marketplace

[8] Cresap, McCormick, and Paget, Inc., "An Operational and Management Audit of PG&E: Executive Summary," June 1980.

[9] Paraphrased from "Using a Cross-Functional Team at Ford to Select a Corporate PM System," *PM Network*, August 1990, pp. 35–59.

and "in-house" customers, who work together in creating value for the ultimate customer in the marketplace. The pathway to change is through the use of projects that support organizational strategies. Future strategies for organizations entail a portfolio of projects, some of which survive during their emerging life cycle and create value for customers. Because projects play such a pivotal role in the future strategies of organizations, senior managers need to become actively involved in the efficiency and effectiveness with which the stream of projects is managed in the organization. Surveillance over these projects must be maintained by senior managers to provide insight into the probable promise or threat that the projects hold for future competition. In considering these projects, senior managers need to find answers to the following questions:

- Will there be a "customer" for the product or process coming out of the project work?
- Will the project results survive in a contest with the competition?
- Will the project results support a recognized need in the design and execution of organizational strategies?
- Can the organization handle the risk and uncertainty likely to be associated with the project?
- What is the probability of the project's being completed on time, within budget, and at the same time satisfying its technical performance objectives?
- Will the project results provide value to a customer?
- Will the project ultimately provide a satisfactory return on investment to the organization?
- Finally, the bottom line question: Will the project results have a strategic fit in the design and execution of future products and services?

As senior managers conduct a review of the projects under way in organizations, the above questions can serve to guide the review process. As such questions are asked and the appropriate answers are given during the review process, an important message will be sent throughout the organization: Projects are important in the design and execution of our organizational strategies!

The question of the strategic fit of a project is a key judgment challenge for senior executives. Who should make such decisions? Clearly those executives whose organizational products and services will be improved by the successful project outcome should be involved. Senior executives of the enterprise should act as a team in the evaluation of the stream of projects that should flow through the top of the enterprise for assessment and determination of future value. Participative decision making concerning the strategic fit of projects is highly desirable. For some senior executives this can be difficult, particularly if they have been the entrepreneurs who conceptualized the company and put it together. Such founding entrepreneurs tend to dominate the strategic decision making of the organization, reflecting their ability in having created the enterprise through their strategic vision in developing a sense of future needs of products and services.

But senior executives, too, can lose their sense of future vision for the enterprise. Or they can become fixated on favorite development projects that may not make any strategic sense to the organizational mission and goals. For example, in a large computer company the founder's dominance of key project decisions drove out people whose perceptions of a project's strategic worth were contrary to that of the CEO. A new-products development group was abruptly disbanded by the CEO, who had sharp differences of opinion with the group executive over several key projects. This group executive had disagreed with the CEO on a key decision involving continuing development of a computer mainframe project whose financial promise was faint—if potentially attainable at all.

4.7 THE VISION

Projects and organizational strategies start with a vision. A "vision is the art of seeing things invisible to others," according to Jonathan Swift.

The corporate vision statement of Whirlpool Corporation is, "Whirlpool, in its chosen lines of business, will grow with new opportunities and be the leader in an ever-changing global market." Implicit in the statement are commitments to market orientation, leadership, customer satisfaction, and quality.

During the strategic fit review of organizational projects, insight should be gained into which projects are entitled to continue assignment of resources and which are not. Senior managers need to decide; the project manager is an unlikely person to execute the decision. Most project managers are preoccupied with bringing the project to a successful finish, and they cannot be expected to clearly see the project in an objective manner of supporting the enterprise mission. There is a natural tendency for the project manager to see the termination of the project as a failure in the management of the project. Projects are sometimes continued beyond their value to the strategic direction of the organization. The selection of projects to support corporate strategies is important in developing future direction.

4.8 A PROJECT SELECTION FRAMEWORK

In general terms, projects are selected through a filtering process, which considers all alternative projects available to the organization. Figure 4.1 depicts this general filtering process.

A project selection framework is shown in Table 4.1. In the leftmost column is a set of evaluation criteria. The body of the table shows how a proposed new program to begin manufacturing system components in Europe might be evaluated. The following explains the table's components:

- The "criteria weights" in the third column of the table reflect the components' relative importance and serve to permit the evaluation of complex project characteristics within a simple framework. A base weight of 20 is used here for

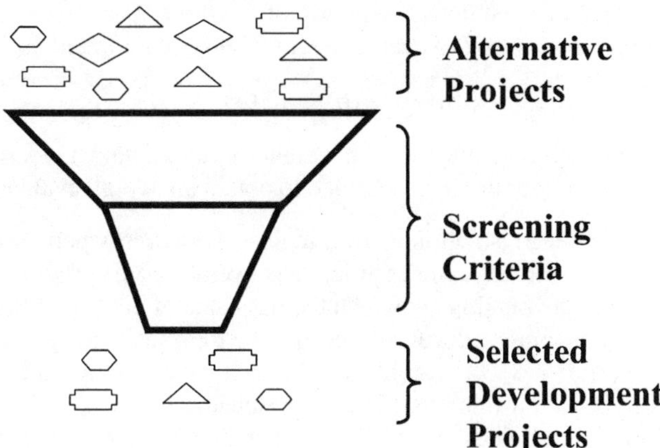

FIGURE 4.1 New project/process development filter.

the major category related to mission, objectives, strategy, and goals. Weights of 10 are applied to the subcategories.

- Within each major category, the 20 "points" are judgmentally distributed to reflect the relative importance of subcategories or some other characteristic of the criterion. For instance, the three stages of *strategy* and the four subgoals are weighted to ensure that earlier stages and goals are treated as more important than later ones. This implicitly reflects the time value of money without requiring a more complex "present value" discounting calculation.
- The first criterion in Table 4.1 is the *"fit" with mission.* The proposal is evaluated to be consistent with both the "product" and "market" elements of the mission and is thereby rated to be "very good," as shown by the 1.0 probability entries at the upper left.
- In terms of *consistency with objectives,* the proposal is rated to have a .2 probability (20 percent chance) of being "very good" in contributing to the ROI element of the objectives, a 60 percent chance of being "good," and a 20 percent chance of being only "fair," as indicated by the probabilities entered into the third row of the table. The proposed project is rated more poorly with respect to the "dividends" and "image" elements.
- The proposal is also evaluated in terms of its expected contribution to each of the three stages of the strategy. In this case, the proposed project is believed to be one that would principally contribute to Stage 2 of the strategy. (Note that only certain assessments may be made in this case, because the stages are mutually exclusive and exhaustive.)
- The proposal is similarly evaluated with respect to the other criteria.
- The overall evaluation is obtained as a weighted score that represents the sum of products of the likelihood (probabilities) and the 8, 6, 4, 2, and 0 arbitrary

level weights that are displayed at the top of the table. For instance, the *consistency with objectives—ROI* expected level weight is calculated as:

$$.2\,(8) + .6\,(6) + .2\,(4) = 6.0$$

This is then multiplied by the criterion weight of 10 to obtain a weighted score of 60. The weighted scores are then summed to obtain an overall evaluation of 610.[10]

Of course, this number in isolation is meaningless. However, when various projects are evaluated in terms of the same criteria, their overall scores provide a reasonable basis for developing the ranking shown on the right side of Table 4.1. Such a ranking can be the basis for resource allocation, because the top-ranked program is presumed to be the most worthy, the second-ranked is the next most worthy, and so forth.

It can readily be seen that such a project selection process will enhance the implementation of the choices made in the strategic planning phase of management. The critical element of this evaluation approach is its use of project selection criteria that relates to the organization's mission, objectives, strategy, and goals and will reflect critical bases of strategy, such as business strengths, weaknesses, comparative advantages, internal consistency, opportunities, and policies.

4.9 PROJECTS AND ORGANIZATIONAL MANAGEMENT

Projects, goals, and objectives must fit together in a synergistic fashion in supporting the enterprise mission. Project success by itself may not contribute to enterprise success. Projects might, early in their life cycle, show promise of contributing to enterprise strategy. A project that continues to support that mission should be permitted to grow in its life cycle. If the project does not provide that support, then a strategic decision faces the senior managers: Can the project be reprogrammed, replanned, and redirected to maintain support of the enterprise mission, or should the project be abandoned?

Project managers cannot make such a strategic decision because they are likely to be preoccupied with bringing the project to a successful finish, and project termination is not their responsibility. Such managers may lack an overall perspective of the project's strategic support of the enterprise mission. Therefore, the decision of what to do about the project must remain with the general manager, who is the project "owner" and has residual responsibility and accountability for the project's role in the enterprise mission and usually puts up the money for the project.

Project success is very dependent upon an appropriate synergy with the enterprise's success. The management of the project and the management of the enterprise depend on a synergistic management approach—planning, organizing, evaluation, and control tied together through an appropriate project-enterprise leadership. This synergy is shown in Fig. 4.2.

---

[10]Adapted from D. I. Cleland and W. R. King, *Systems Analysis and Project Management,* 3d ed. (New York: McGraw-Hill, 1983), pp. 68–70.

TABLE 4.1 Project Selection Model

Program/project evaluation criteria	Sub-category	Criteria weights	Very good (8)	Good (6)	Fair (4)	Poor (2)	Very poor (0)	Expected level weight	Expected weighted score
"Fit" with mission	Product	10	1.0					8.0	80
	Market	10	1.0					8.0	80
	Subtotal	20							
Consistency with objectives	ROI	10	.2	.6	.2			6.0	60
	Dividends	5		.2	.6	.2		4.0	20
	Image	5			.8	.2		3.6	18
	Subtotal	20							
Consistency with strategy	Stage 1	10					1.0	0	0
	Stage 2	7	1.0					8.0	56
	Stage 3	3					1.0	0	0
	Subtotal	20							
Contribution to goals	Goal A	8					1.0	0	0
	Goal B	6	.8	.2				7.6	45.6
	Goal C	4		.8	.2			5.6	22.4
	Goal D	2					1.0	0	0
	Subtotal	20							
Corporate *strength* base		10				.8	.2	1.6	16
Corporate *weakness* avoidance		10				.2	.8	.4	4
Comparative advantage level		10	.7	.3				7.4	74
Internal consistency level		10	1.0					8.0	80
Risk level acceptability		10				.7	.3	1.4	14
Policy guideline consistency		10			1.0			4.0	40
								Total score	610

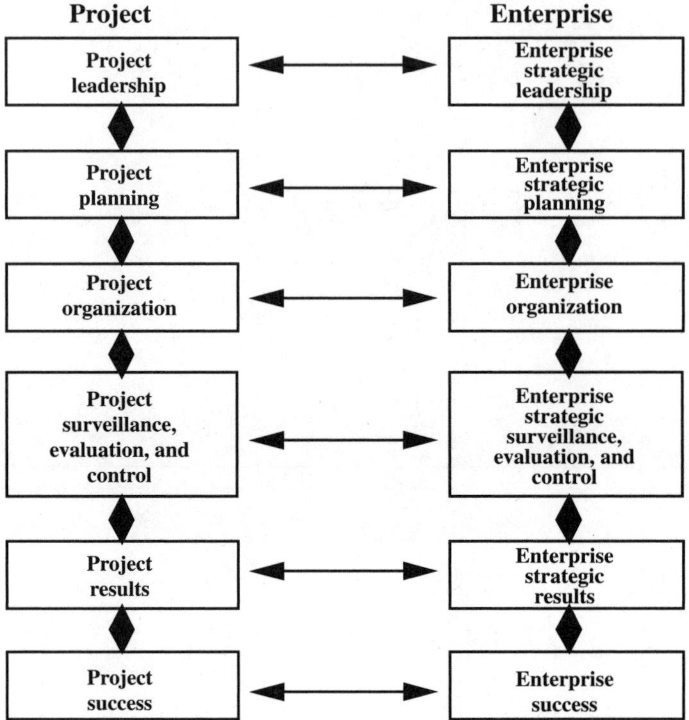

FIGURE 4.2 Project/strategic enterprise synergy. (*Source: David I. Cleland, "Measuring Success: The Owner's Viewpoint,"* Proceedings, *Project Management Institute Seminar/Symposium, Montreal, Quebec, September 20–25, 1996, p. 6.*)

Projects are designed, developed, and produced or constructed for a customer. This customer or project owner may be an internal customer, such as a business unit manager who pays for product development by the enterprise central laboratory. An external customer might be a utility that has contracted with an architectural and engineering firm to design, engineer, and build an electricity generating plant.

Senior managers, who have the responsibility to sense and set the vision for the enterprise, need a means of marshaling the resources of the organization to seek fulfillment of that vision. By having an energetic project management activity in the enterprise, an organizational design and a development strategy are available to assist senior managers in bringing about the changes and synergy to realize the organizational mission, objectives, and goals through a creative and innovative strategy. Leadership of a team of people who can bring the changes needed to the enterprise's posture is essential to the attainment of the enterprise's vision. As additional product and/or service and process projects are added to marshal the enterprise's resources, the strategic direction of the enterprise can be guided to the attainment of the vision. When projects are accepted as the building blocks in the design and execution of organizational purposes, a key strategy has been set in motion to keep the enterprise competitive. Such strategies are dependent on the quality of the leadership in the enterprise.

Compaq Computer Corporation CEO Eckhard Pfeiffer provided the leadership in launching a long-term strategy initiative in that company. This planning effort was launched as soon as the immediate situation at the company was moving adequately toward correction. A comprehensive long-range strategy was developed and put in place. The CEO stated that the focus of the company was clearly on future strategy—a means to totally transform the company.

One of the more important strategic decisions made by the Compaq CEO was to launch development efforts into cut-rate personal computers (PCs). An independent business unit was organized into a project team to develop a low-price machine—a real Compaq.

Revised manufacturing strategies were developed to get costs down at plants in Houston, Singapore, and Scotland. The entire manufacturing process strategy was rethought. An entire system is now built on a single assembly line instead of making the motherboard in one building and the chassis in another. Testing of every subassembly was stopped in favor of testing a sample. All finished systems were still fully tested. Compaq leaned on suppliers to cut prices to bring down overall manufacturing costs.[11] Project planning contributed to the overall planning strategy at Compaq.

4.10 PROJECT PLANNING

Why is project planning so important? The answer is simply because decisions made in the early phases of the project set the direction and force with which the project moves forward as well as the boundaries within which the work of the project team is carried out. As the project moves through its life cycle, the ability to influence the outcome of the project declines. After design of the project done early in the life cycle, the cost of producing the resulting product, as well as the product quality, has been largely determined. Senior managers tend to pay less attention during the early phases of the project than when the product development effort approaches the prototype or market-testing stage. By waiting until later in the life cycle of the project, their influence is limited in the sense that much of the cost of the product has been determined. Design has been completed, and the manufacturing or construction cost has been set early in the project. Senior managers need to become involved as early as possible, and they must be able to intelligently assess the likely market outcome of the product, its development cost, its manufacturing economy, how well it will meet the customers' quality expectations, and the probable strategic fit of the resulting product in the overall strategic management profile of the enterprise. In other words, when senior managers become involved early in the development cycle through regular and intelligent review, they can enjoy the benefits of leverage in the final outcome of the product and its likely

[11]Catherine Arnst and Stephanie Anderson Forest, "Compaq," *Business Week,* November 2, 1992, pp. 146–151.

acceptance in the marketplace. What happens early in the life cycle of the project essentially lays the basis for what is likely to happen in subsequent phases. Because a development project is taking an important step into the unknown—with the hope of creating something that did not previously exist—as much information as possible is needed to predict the possible and probable outcomes. For senior managers to neglect the project early in its life cycle and leave the key decisions solely to the project team is the implicit assumption of a risk that is imprudent from the strategic management perspective of the enterprise. Project planning is discussed more fully in Chap. 11.

Project planning and organizational renewal are linked through the development of organizational strategy. For example, Lawrence A. Bossidy, former CEO of Allied-Signal Company, upon joining the company established ambitious objectives including:

- An 8 percent annual revenue growth
- A total-immersion total-quality program
- A top-to-bottom change in human resources management

A statement of corporate vision and values listed these objectives, developed by the company's top 12 executives to include such things as being "one of the world's premier companies, distinctive and successful," and also the values of satisfying customers, integrity, and teamwork. The vision helped galvanize people. In addition, with these objectives as guidelines, Bossidy chopped $225 million from capital spending, reduced the annual dividend to $1 a share from $1.80, put eight small divisions up for sale, cut 6200 jobs, and combined 10 data processing centers into 2.

The company formed commodity teams—cross-functional project teams of manufacturing, engineering, design, purchasing, and finance in such areas as castings, electronic gears, machine parts, and materials. Each team was responsible for picking the best suppliers in its specialty, with the chosen suppliers getting long-term national contracts. Suppliers were expected to bring down costs for themselves and for Allied-Signal.

Projects are usually paid for by the project owners, key members of project teams. The project owner has the residual responsibility and accountability for managing the project during its life cycle.

4.11 THE PROJECT OWNER'S PARTICIPATION

Project owners cannot leave to others the responsibility for continuously measuring the success of the project, even experienced project management contractors and constructors. Foxhall stated:

> The owner must recognize that he is the key member of the project development team. Only he can select and organize the professional team, define his own needs,

> set his priorities and make final decisions. He cannot delegate these roles, so he must have a sustained presence in project management.[12]

The project owner clearly has responsibility for the efficiency and effectiveness of a contractor involved on the project. This requires a surveillance system to know what the contractor is doing and how well the contractor is performing. For example, one report noted:

> Another essential characteristic of a successful nuclear construction project is a project management approach that shows an understanding and appreciation of the complexities and difficulties of nuclear construction. Such an approach includes adequate financial and staffing support for the project, good planning and scheduling, and close management oversight of the project.[13]

Project owners in the utility industry, driven by the need to better manage projects, have responded by building up personnel and developing improved management systems. Such involvement has enabled the owners to obtain better control over projects and reduce risk.[14]

Every project has (or should have) its owner: the agency or organization that carries the project on its budget and whose strategic plans include the project as an essential building block for future growth or survival. The project owner has the residual responsibility to approve and maintain oversight of the project during its life cycle. The project owner should be more than a corporation or a government agency. Rather, the project owner should be identified by name, an individual recognized as the "personal owner," who assumes managerial oversight of the project as an element of future strategies. Project owners can come from within the organization, such as:

- A senior manager who budgets for a product or process development project
- A division profit center manager who funds an R&D project to support a product improvement program
- A manufacturing manager who is converting a traditional factory to an automated, flexible manufacturing system

Outside project owners usually contract for the project work through architects, engineers, and constructors. The Department of Defense contracts for substantially all the work involved in designing, engineering, and manufacturing weapon systems. In the electric utility industry, many investor-owned utilities do not design and construct their own generating facilities but hire architects, engineers, and constructors to perform most of the work. However, other utilities, such as

---

[12]William B. Foxhall, "Professional Construction Management and Project Administration," *Architectural Record,* March 1972, pp. 57–58.

[13]"Improving Quality and the Assurance of Quality in the Design and Construction of Nuclear Power Plants," NUREG-1055, U.S. Nuclear Regulatory Commission, Washington, D.C., May 1984, pp. 2–17.

[14]Theodore Barry & Associates, "A Survey of Organizational and Contractual Trends in Power Plant Construction," Washington, D.C., March 1979.

Duke Power and Pacific Gas & Electric, perform a substantial portion of the design and construction for major projects in-house.

> To put it simply, the project owner is the one who puts up the money to fund the project. On such a project funder rests the responsibility to see that those funds are used in a prudent and reasonable fashion. This requires adequate assessment of the project risk, project plans, and ongoing monitoring, evaluation, and control of the resources used on the project. Furthermore, an owner's decision to fund a project affects a variety of "stakeholders" who have, or believe they have, a stake in the project and its outcome. In some cases some of these stakeholders will seek legal redress if the project does not meet their particular expectations. Emerging case law establishes that project managers have the legal responsibility for the strategic management of projects.[15]

These stakeholders and their predilections are discussed in Chap. 6.

A landmark study of the design and construction of nuclear power plants found that deep involvement by utilities (owners) in cost, schedule, productivity, and quality considerations contributed to project success as much as close management oversight of the project and the project's contractors.[16]

Project success depends on a commitment by the owner to use contemporaneous project management theory and practice. Support of the enterprise mission comes about through the project owner's effective discharge of her or his strategic planning and management responsibility.

Successful project management depends on senior enterprise management for authority, strategic guidance, and support. Senior managers in turn depend on project managers for timely, cost-effective achievement of project results to support corporate strategy. Project management is a form of "strategic delegation" whereby senior managers delegate to project managers the authority and responsibility to do such things as building capital facilities, introducing new products, conducting research and development, and creating new marketing and production opportunities.

Project management also is a type of strategic management control. Senior managers can use project management as a way to ensure that key strategies are accomplished in an effective manner. A senior manager oversees the strategic direction of the enterprise by providing resources to accomplish the mission, objectives, goals, and strategies. By determining the success or failure of a project, senior management ensures that control systems are instituted to track strategic progress of the enterprise. As project managers make and execute key decisions, these key decisions should be reviewed by senior managers to determine if the decisions are consistent with corporate strategy. Senior enterprise managers commit a serious breach of responsibility and accountability for the manage-

[15]For a more thorough analysis, see Randall L. Speck, "The Buck Stops Here: The Owner's Legal and Practical Responsibility for Strategic Project Management," *Project Management Journal,* September 1988, pp. 45–52.

[16]"Improving Quality and the Assurance of Quality in the Design and Construction of Nuclear Power Plants," NUREG-1055, U.S. Nuclear Regulatory Commission, Washington, D.C., May 1984, pp. 3–15.

ment of the enterprise when they ignore or accept key project decisions without review. When adequate project evaluation is carried out to determine project success, senior managers get information on how effectively enterprise strategies are being implemented.

In order for the owner to do a credible job of measuring project success, several conditions must exist:

- An appropriate organizational design is in place that delineates the formal authority, responsibility, and accountability relationships among the enterprise corporate senior managers, project manager, functional manager, and work package managers.
- Adequate strategic and project planning have been carried out within the enterprise.
- Relevant and timely information is available that gives insight into the project status.
- Adequate management monitoring, evaluation, and control systems exist.
- Contemporary state-of-the-art management techniques are used in the management of the project.
- A supportive cultural ambience exists that facilitates the successful management of projects.

An important part of the strategic management of a project is to carry out such management in the context of a project management system.

4.12 PROJECT MANAGEMENT SYSTEM

Once the mission of the enterprise is established through the operation of a strategic planning system, planning can be extended to select and develop organizational objectives, goals, and strategies. Projects are planned for and implemented through a *project management system* composed of the following subsystems.[17]

The *facilitative organizational subsystem* is the organizational arrangement that is used to superimpose the project teams on the functional structure. The resulting "matrix" organization portrays the formal authority and responsibility patterns and the personal reporting relationships, with the goal of providing an organizational focal point for starting and completing specific projects. Two complementary organizational units tend to emerge in such an organizational context: the project team and the functional units. The *project control subsystem* provides for the selection of performance standards for the project schedule, budget, and technical performance. The subsystem compares actual progress with planned progress, with the initiation of corrective action as required. The rationale for a control subsystem arises out of the need to monitor the various organizational

[17] David I. Cleland, "Defining a Project Management System," *Project Management Quarterly,* vol. 10, no. 4, 1977, pp. 37–40.

units that are performing work on the project in order to deliver results on time and within budget.

The *project management information subsystem* contains the information essential to effective control of the projects. This subsystem may be informal in nature, consisting of periodic meetings with the project participants who report information on the status of their project work, or a formal information retrieval system that provides frequent printouts of what is going on. This subsystem provides the data to enable the project team members to make and implement decisions in the management of the project.

Techniques and methodology is not really a subsystem in the sense that the term is used here. This subsystem is merely a set of techniques and methodologies, such as PERT, CPM, and related scheduling techniques, as well as other management science techniques which can be used to evaluate the risk and uncertainty factors in making project decisions.

The *cultural ambience subsystem* is the subsystem in which project management is practiced in the organization. Much of the nature of the cultural ambience can be described in how the people—the social groups—feel about the way in which project management is being carried out in the organization. The emotional patterns of the social groups, their perceptions, attitudes, prejudices, assumptions, experiences, and values, all go to develop the organization's cultural ambience. This ambience influences how people act and react, how they think and feel, and what they say in the organization, all of which ultimately determines what is taken for socially acceptable behavior in the organization.

The *planning subsystem* recognizes that project control starts with project planning, because the project plan provides the standards against which control procedures and mechanisms are measured. Project planning starts with the development of a *work breakdown structure,* which shows how the total project is broken down into its component parts. Project schedules and budgets are developed, technical performance goals are selected, and organizational authority and responsibility are established for members of the project team. Project planning also involves identifying the material resources needed to support the project during its life cycle.

The *human subsystem* involves just about everything associated with the human element. An understanding of the human subsystem requires some knowledge of sociology, psychology, anthropology, communications, semantics, decision theory, philosophy, leadership, and so on. Motivation is an important consideration in the management of the project team. Project management means working with people to accomplish project objectives and goals. Project managers must find ways of putting themselves into the human subsystem of the project so that the members of the project team trust and are loyal in supporting project purposes. The artful management style that project managers develop and encourage within the peer group in the project may very well determine the success or failure of the project. Leadership is the most important role played by the project manager.

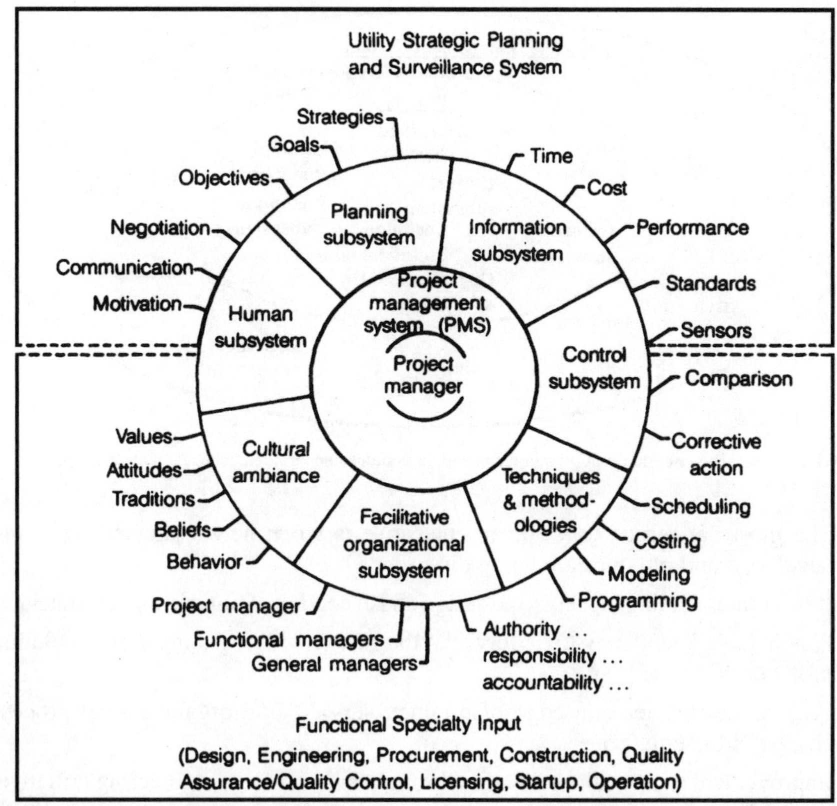

FIGURE 4.3 The project management system. (*Source: Adapted from D. I. Cleland, "Defining a Project Management System,"* Project Management Quarterly, *vol. 110. no. 4, p. 39.*)

Figure 4.3 depicts the project management system in the context of a public utility commission with all its subsystems. The utility owners responsible and accountable for the effective management of the project work through their boards of directors and senior management with the project manager, functional managers, and functional specialists.

In Fig. 4.4, the integrated relationship of a strategic management system and project management system is portrayed. Indeed, the subsystems of a project management system touch all the "choice elements" of a strategic management system.

4.13 TO SUMMARIZE

Some of the major points that have been expressed in this chapter include:

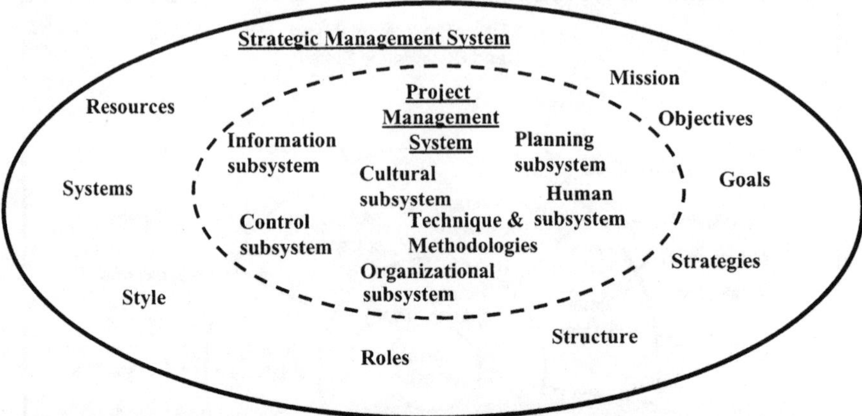

FIGURE 4.4 The integration of project management systems and strategic management systems.

- The most dangerous time for an enterprise is when new strategies are being developed and old ones are being discarded.
- Projects are building blocks in the design and execution of organizational strategies.
- People have a tendency to believe that the future will be a simple extrapolation of the past.
- Projects are the leading edge of product, service, and organizational process change in the enterprise.
- Improvement of operations, such as through the use of a reengineering initiative, has its place. But the enterprise must have strategic projects under way to prepare the organization for its uncertain future.
- Examples were given in the chapter of how contemporary organizations have used projects to change their products, services, and organizational processes.
- A successful enterprise has a "stream of projects" flowing through it all the time. Conversely, an organization that is failing is likely to have few projects under way to allocate resources for future purposes.
- A series of important questions can be asked to determine if an existing or proposed project has a "strategic fit" in the enterprise.
- A simple scoring model was suggested as a useful way to select projects from an inventory of potential projects that exist in the enterprise.
- There is a synergy between projects and the other elements of the enterprise. This synergy is shown in Fig. 4.2.
- A brief introduction to project planning was given along with a promise of more material on project planning contained in Chap. 11.
- The project owner has specific responsibilities in the overall management of the project.

- A project management system is a useful way of depicting the principal subsystems that are involved in the management of a project. Unless all of these subsystems are up and running, there is likely to be a deterioration in the effectiveness and efficiency with which the project is managed.

4.14 ADDITIONAL SOURCES OF INFORMATION

The following additional sources of project management information may be used to complement this chapter's topic material. This material complements and expands on various concepts, practices, and theory of project management as it relates to areas covered here.

- David I. Cleland, "Strategic Planning" and "New Ways to Use Project Teams," chaps. 1 and 29 in David I. Cleland (ed.), *Field Guide to Project Management* (New York: Van Nostrand Reinhold, 1997).
- John Tuman, Jr. and Moses Thompson, "Using Project Management to Create an Entrepreneurial Environment in Czechoslovakia," and Virginia Fairweather, "The Channel Tunnel: Larger than Life, and Late," in David I. Cleland, Karen M. Bursic, Richard J. Puerzer, and Alberto Y. Vlasak, *Project Management Casebook,* Project Management Institute (PMI). (First published in *Proceedings,* PMI Seminar/Symposium, September 1992, pp. 405–409; and *Civil Engineering,* May 1994, pp. 42–46.)
- John Stanley Baumgartner, *Project Management* (Homewood, Ill.: Richard D. Irwin, 1963). This is probably the first book on project management that was published by a commercial publisher. The book's principal focus is on the management processes with which the project manager of DOD and NASA projects had to contend. It is intended primarily for people in companies doing business with the U.S. government. The author also suggests that it may also be of interest to construction and other project-oriented activities. The book's value is in the background that is provided as project management began its emergence as a major building block in management thought and theory.
- Peter W. G. Morris, *The Management of Projects* (London: Thomas Telford, 1994). This book provides an interesting and comprehensive survey of the issues involved in appraising, beginning, and accomplishing any project. It details the experience and lessons learned from the management of projects over the past 50 years. The book concludes with a prediction of how the discipline of project management will likely develop over the next 10 to 20 years. The book is particularly interesting to those people who have a desire to trace the development of project management.
- Janice Thomas, Connie L. Delisle, Kam Jugdev, and Pamela Buckle, "Selling Project Management to Senior Executives: What's the Hook?" *Proceedings,* PMI Research Conference, Paris, France 2000, pp. 431–442. The article

reports on a research project under way under PMI's sponsorship to develop insights—and potential strategies—on how best to convince senior executives to use project management in leading their enterprises. Subsequent articles and research reports will doubtlessly follow from these authors on the subject matter.

- Sergio Pellegrinelli and Cliff Bowman, "Implementing Strategy through Projects," *Long Range Planning,* vol. 27, no. 4, 1994, pp. 125–132. The authors make the point that the challenge facing senior managers who wish to bring revolutionary change within an organization is to manage that change outside the "straitjacket" of the existing bureaucracy. Projects and project management can help senior management to do just that. They also make the important point that the concept of a project has to be understood in a wider sense as a vehicle for achieving change. The authors also state that most strategic initiatives can be conceived and handled as projects—and that the conceptualization and implementation of a strategy usually involves defining and undertaking a range of projects which address a component of the strategy.

4.15 DISCUSSION QUESTIONS

1. Discuss the importance of the strategic management of projects.
2. The chapter described one quantitative method for project selection. What are some other possible methods of project selection? What other factors can be included in the analysis?
3. Why is it important for general managers to take responsibility and accountability for the strategic fit of a project?
4. Discuss criteria for when a project would or would not be a strategic fit.
5. Discuss how a project could be selected for an organization and not be a strategic fit.
6. Discuss the importance of owner participation in measuring and controlling the success of a project.
7. What hinders senior management involvement in large organizational projects?
8. What kinds of questions need to be addressed in order to measure project success?
9. What contemporaneous state-of-the-art management techniques can be used to help control and measure project success?
10. Discuss the responsibilities of project owners with respect to strategic planning and management.
11. Discuss the importance of establishing policies that describe the organizational structure and the authority, responsibility, and accountability of managers within the structure.
12. List and define the various subsystems of the project management system.

4.16 USER CHECKLIST

1. Are the projects within your organization being managed from a strategic perspective? Why or why not?
2. What quantitative and qualitative methods does your organization use for project selection?
3. Does the top management of your organization accept the responsibility for determining the strategic fit of projects?
4. Does the top management of your organization accept the responsibility for monitoring the cost, time, and technical performance objectives of major projects?
5. How do the senior managers of your organization monitor the ongoing progress of major projects?
6. What issues do you see regarding senior managers not monitoring project progress and what is the result?
7. Do the key project managers use state-of-the-art management techniques to control the projects of the organization?
8. Are organizational projects being managed from a project management system perspective?
9. Does the top management of your organization accept the responsibility to develop and implement adequate strategic plans for the enterprise and for projects?
10. Are adequate information systems available to support managers and professionals working on various organizational projects?
11. Does appropriate policy exist that defines the organizational structure and the fixing of authority, responsibility, and accountability of managers at each organizational level?
12. Does the top management of your organization foster an attitude that supports the management of projects?

4.17 PRINCIPLES OF PROJECT MANAGEMENT

1. Projects are building blocks in the design and execution of organizational strategies.
2. Projects provide the means for bringing about realizable changes in products, services, and organizational processes.
3. A regular review of the status of the "portfolio of projects" in an organization provides an excellent assessment of how well the organization is preparing for its future.

4. Failure in project management in the enterprise will impact how well the organization is able to accomplish the "choice elements" in its strategic management strategy.
5. The health of an enterprise can be determined from a review of the "stream of projects" flowing through the organization.
6. The question of the strategic fit of a project is a key judgment challenge facing senior executives.
7. The use of a project selection framework can help in the selection of projects to support the enterprise's strategy.
8. Project success is dependent upon an appropriate synergy with the enterprise's success.
9. The project owner has a key responsibility to maintain surveillance of the status of the projects that are under way to support owner initiatives.
10. Projects are best managed under a project management system philosophy and process.

4.18 PROJECT MANAGEMENT SITUATION—IMPROVEMENT OF PROJECT MANAGEMENT

An electric utility was organized in a very traditional way. Project engineering was carried out in the engineering organization. Any project management that was done was also planned and executed within engineering, a subdivision of the Engineering and Research Department. Although the professionals in the engineering department were excellent project engineering planners and executors, very few of these professionals had any real appreciation of project management—and its broader context in the management of the other functional input areas for a project. Consequently, the quality of project management and leadership was sadly lacking in the company.

Historically, as the company grew, many additional levels were added to the existing bureaucratic structure. Communication between functional "silos" became complex, cumbersome, and slow. Responsibilities were diluted for the overall management of projects for new power plant construction. The cultural ambience of the company could be described as a highly structured, hierarchical enterprise. People tended to focus on the specialization of their functional organizations. Boundaries between the functional organizations became more rigid. As the company grew, its bureaucratic organizational design continued. Fences between functions became higher, and coordination of any particular project became more difficult. Many of the problems in the development and construction of new power plants continued to be difficult.

Senior management, after some major cost and schedule overruns on new power plant construction, concluded that there was no cohesive force to bring together the diverse activities involved in the design, construction, and start-up of new plants. No one individual, other than the CEO, had the authority and responsibility for the building of new plants. Some of the major problems in the management of the acquisition of new power plants included:

- Planning was diffused throughout the organization and no individual had the responsibility for maintaining oversight of the planning for the new plant.
- Although project coordinators had been appointed, the authority of these coordinators was lacking. About all those coordinators could do was to persuade, cajole, or "threaten" the functional people into working in a cooperative fashion in designing, construction, and getting the plant up and running.
- One-way communication prevailed. When a functional element finished its work, the results were "thrown over the organizational wall" to the next function. Concurrent work by the functions on a particular project was limited.
- No person watched the overall project budget. This lack of budget control set the stage for subsequent project cost and schedule overruns.

The stress on the organization and the people was severe. Everyone knew that there must be a better way of dealing with new plant initiatives.

Finally the senior management of the company established a new project management organizational unit. Some of the key actions undertaken by this new organizational unit included:

- An in-depth assessment of the problems and difficulties being encountered in designing, building, and bringing new plants on-line.
- The development of a strategy on how project management concepts and processes could be implemented in the organization.
- The identification of key individual and collective roles in the organization, particularly those that would be concerned with the management of forthcoming projects.
- The appointment of project teams to evaluate and come up with recommended strategies for the improvement of the management of projects in the enterprise.
- A training program for all key people on the concept and process of project management.
- A commitment on the part of the senior managers, to include corporate directors to provide support and resources to develop a project-driven culture in the enterprise.
- A project review strategy for all projects whereby the project's cost, schedule, and technical performance would undergo careful scrutiny.
- A commitment by the senior managers of the enterprise that projects would be considered key building blocks in the design and execution of enterprise strategies.
- A plan to go through a formal assessment process of the efficacy of the emerging project management concept and processes within the enterprise.

4.19 STUDENT/READER ASSIGNMENT

1. Evaluate the strategy initiated by this company for the improvement of project management in the enterprise.
2. What could have been done differently in the improvement of the company's remedial strategy for project management?
3. What would you have done if you had been appointed as the project manager for the remedial strategy for the improvement of project management within the enterprise?
4. What are some of the key issues and considerations to keep in mind when initiating a remedial strategy in a company undergoing a cultural change to emphasize the practice of project management?
5. What project management principles and process could be applied to a situation such as this where an entrenched traditional bureaucracy needs to become a project-driven enterprise, with all the support that a philosophy of project management can provide?

CHAPTER 5
THE BOARD OF DIRECTORS AND MAJOR PROJECTS[1]

"There is plenty of substantive evidence that 'too many corporate boards fail to do their jobs.'"

WALTER J. SALMON
"Crisis Prevention: How to Gear Up Your Board,"
Harvard Business Review, *January–February 1993, p. 68.*

5.1 INTRODUCTION

An important responsibility that managers at all levels of the organization have is to be involved in the decisions to initiate a project, and to maintain surveillance over ongoing projects during their life cycle. The board of directors (BOD), supported by senior managers, needs to be involved in the selection of major projects to support the "choice elements" of the enterprise. Once projects are selected and funded, the responsible managers—to include the BOD—must maintain an ongoing review of how well the project is being managed and the potential results that are promised by the project.

This chapter starts off with looking at how some boards of directors have neglected their responsibilities for the management of major projects, and through their neglect have allowed major problems in the management of the project to develop and endure. Conversely, exemplary BOD behavior, the empowerment of the BOD, and how major projects are to be reviewed are also subjects in this chapter. Finally the chapter looks at the project information needed by the BOD, how project performance audits can be used, how cultural considerations impact the role of the BOD, and what general criteria can be used in the selection of the members of the BOD.

[1]This chapter is an extension of the paper "Capital Projects: The Role of the Board of Directors," presented at the PMI 1988 Annual Seminar/Symposium in San Francisco and published in the 1988 *Proceedings,* pp. 8–12. Appreciation is extended to Attorney Randall L. Speck of Rogovin, Huge & Schiller, Washington, D.C., and Attorney Edward O'Neill of the California Public Utilities Commission, for their helpful guidance in the preparation of this chapter.

5.2 THE NEED FOR BOARDS OF DIRECTORS

Boards of directors have been used in the business community for over 150 years. State general corporation laws require that all business corporations have boards, typically stipulating that the corporation "shall be managed by a board of at least three directors."

Once a project is funded and corporate resources are expanded to design, develop, and construct or manufacture the project, it becomes an important responsibility of the board to maintain surveillance over the efficiency and effectiveness with which corporate strategy is being implemented through the use of major projects. Corporate strategy is clearly a key responsibility of the corporate board of directors. Strategy, according to Chandler, is "the determination of the basic long-term goals and objectives of an enterprise, and the adoption of courses of action and allocation of resources necessary for carrying out these goals."[2] The board of directors cannot work out a company's strategy, but it is the duty of the board to make sure that a company has adequate strategies.[3]

5.3 SURVEILLANCE

By maintaining surveillance over the status of major product and process projects within the enterprise, senior managers—to include members of the board of directors—can gain valuable insight into the effectiveness with which the enterprise is preparing for its future. There are, of course, limits to the number of projects that such managers can monitor. There are, however, certain projects whose outcomes can have major impacts on the organization's future direction. Senior managers should review the adequacy of the planning for these projects and keep abreast of which projects are being executed to further corporate purposes. The projects in which the directors should be particularly interested include:

- New-product, service, and process development projects that have the promise of giving the company a competitive advantage in the marketplace. Projects that contain the possibilities of technological breakthroughs, or significant incremental improvements in products, services, and processes, should be of particular interest to the directors.
- Projects whose execution requires the commitment of substantial resources, such as the building of new facilities, or the development of major supporting organizational resources, such as restructuring or downsizing initiatives.

---

[2]Alfred D. Chandler, Jr., *Strategy and Structure: Chapters in the History of the Industrial Enterprise* (Cambridge, Mass.: MIT Press, 1962), p. 13.

[3]Peter F. Drucker, "The Real Duties of a Director," in *The Changing World of the Executive* (Heinemann, 1985), p. 33.

- Projects that are the outgrowth of a strategic alliance being negotiated for the sharing of resources, results, and rewards with another organizational entity. Research consortia, partnering, and sharing of manufacturing facilities and marketing facilities are some common examples.
- Other projects for supporting the strategic purposes of the firm, such as major cost-reduction initiatives; new major, corporatewide information systems; and investment opportunities.

When the directors accept the concept that projects are building blocks in the design and execution of organizational strategies, the directors gain the use of an important strategic management tool—the inventory of product, service, and process projects under way in the enterprise. As the corporate directors sense competitive changes in the marketplace, or realize that new technological initiatives are coming forth from the research of competitors, or recognize any other major change in the enterprise's future, they need to ask a key question: What projects are under way in this company to meet—and exceed—these competitive threats coming out of the firm's environmental and competition system? If relevant product, service, or process projects are not under way, then the firm's competitive position will be threatened, and projects to position the enterprise to meet these changes need to be undertaken in a forthright manner. If the directors become involved in regular and rigorous review of these important projects, an important message will be sent throughout the company: Projects are important to this company, because it is through projects that we are able to organize our resources to position ourselves for the uncertain future.[4]

The chief executive officer of an aerospace firm assessed its market challenges and keynoted the need for project-related strategies in the following manner:

- Earlier commercialization is increasingly important as a competitive weapon.
- The company tends to revitalize its product and service development process to get the right products to the market quickly and effectively to enhance the company's competitive position.
- Supporting management systems and organizational design alternatives need to be assessed and developed to support the above strategies.

From these policies, assessment strategies were launched to design, develop, and implement project management strategies throughout the enterprise.

Why have some boards overseeing capital projects not carried out their responsibilities? This chapter will attempt to answer this question by examining the role of the board of directors with respect to the strategic management of projects. To gain insight into why many boards seem to have been so ineffectual, the activities of several boards will be discussed. Then some constructive ideas will be offered about what the role, the information required, the actions, and what the background of the board members should be.

---

[4]David I. Cleland, "The Board of Directors and Projects," *PM Network*, January 1991, pp. 6–7.

5.4 SOME BOARD INADEQUACIES

Directors watched seemingly idly as one seemingly invincible corporation after another—from Eastman Kodak and General Motors to IBM, Sears Roebuck, and Westinghouse Electric—faltered and declined. Inadequacy of a board of directors' performance is not limited to for-profit enterprises. The following examples reflect on the board performance in a not-for-profit entity.

After more than a decade of excessive spending, rapid expansion and poor decision making, Allegheny Health Education and Research Foundation (AHERF) filed for bankruptcy on July 21, 1998.[5] At this time, AHERF had accumulated over a billion dollars in debt, representing the largest nonprofit health-care system failure in history. In the ensuing postmortem analysis, the inability of its board of directors to rectify major problems in a timely manner was evident. Following are some examples:

> . . . directors didn't have the time to study all of the documents made ready for every meeting. Sometimes there'd be more than 1,000 pages.

> There weren't a lot of probing questions, and those who did speak up were discouraged from doing so again.

Fortune magazine published an article in their May 14, 2001 issue titled "The Dirty Half-Dozen: America's Worst Boards." The boards that qualify for the magazine's "Hall of Fame" include Coca-Cola, Intel, Pfizer, Target, and Texas Instruments. The boards that are in the "Hall of Shame" include Advanced Micro Devices, Archer Daniels Midland, Maxxam, Occidental Petroleum, and Warnaco.

In a strategy to improve directors' performance, companies are cracking down on the number of directorships board members can hold.[6] The article cites examples of notable executives who serve on seven to nine boards.

But things are changing. There is a quiet revolution going on in American boardrooms—the directors are waking up and taking the job as director more seriously. Unfortunately, there are still many firmly entrenched CEOs and old-line directors who resist modern governance. For most CEOs the reality of global competition has motivated the need for knowledgeable, talented directors to serve as sounding boards and advisers. Some of the enhanced assumption of responsibility on the part of directors has come about from the heightened scrutiny of boards by the press and public, combined with a growing respect—if not fear—of the threat of litigation against directors. With so much stock concentrated in a few large institutions, pension funds, and mutual funds, shareholders are more organized and active. They are making their presence felt and are demanding that the directors exercise genuine involvement and oversight.

---

[5]The story of their demise is described in a six-part series appearing in the *Pittsburgh Post Gazette* (Carpenter, January 24, 1999; Massey, January 17–22, 1999).

[6]See John S. Lubin, "Multiple Seats of Power," *The Wall Street Journal,* January 23, 2001, p. B-1.

In recent years there has been a strong movement for companies adding more outsiders to their boards, but also vastly upgrading their requirements for directors. Figureheads, celebrities, and yes-persons are not wanted. Companies want outside directors who can take an active role in helping guide the company to sustained superior performance. Companies also express a preference for only those active executives that sit on no more than three outside boards.[7]

Some boards have performed in a stellar fashion in their involvement in the shaping and review of strategic plans for the enterprise. The board at Campbell Soup Company is one such board—and it topped *Business Week* magazine's list of the best boards of directors.[8] According to *Business Week,* the best boards, among other things:

- Do an annual assessment of the CEO, conducted in a meeting of independent directors, and link the CEO's remuneration package to specific goals
- Actively participate in the assessment of strategic plans and 1-year operational plans for the enterprise
- Use a governance committee that regularly evaluates board performance and that of the individual directors
- Require each director to own a significant amount of company stock

Further, recommendations by *Business Week* include putting the entire board up for election every year; limiting the number of inside directors; ensuring that the audit, compensation, and nominating committees are composed of independent directors; and, finally, banning interlocking directorships and putting limits on the number of boards on which directors can serve.[9]

The linkage with project management comes about through the board's assessment of the ability of the CEO to provide the leadership of the enterprise through the use of effective strategic management initiatives. Put in the language of management theory, this means that the CEO should provide the environment, resources, and proactive actions to develop and implement the core elements of strategic management: mission, objectives, goals, and strategies. The goals—the milestones for the enterprise—are projects that provide for the design and development of new products, processes, and organizational processes. Thus, to prepare for the enterprise's future means that current projects are the basic building blocks and the means for identifying and integrating resources to develop future initiatives for the enterprise. Major projects, which represent a significant commitment of resources for the enterprise, should be reviewed on a regular basis by the directors. By doing so the directors should have an excellent means for determining if enterprise resources are being committed in meaningful ways to prepare the enterprise for its future.

Corporate difficulties, many of which can be traced to an inactive or inefficient board of directors, are helping shape changes in the governance of corporations. Following some of the past strategic difficulties at IBM, a more proactive board created a "directors and corporate governance committee." Critics of the IBM board pointed out that board members waited too long to hold former IBM CEO John F. Akers accountable for IBM's dismal performance.

---

[7]Anthony Bianco, et al., "The Rush to Quality on Corporate Boards," *Business Week,* March 3, 1997, pp. 34–35.
[8]Richard A. Melcher, "The Best and the Worst Boards," *Business Week,* November 25, 1996, pp. 82–98.
[9]Ibid.

Two studies report that boards are getting smaller—moving to about a dozen members that would enable more ease in discussing issues. Then too, on smaller boards, members would likely take responsibility more personally. Both studies show that boards continue to favor outsiders on the board. More companies are paying directors partly in stock—or at least extending that option to the members—hopefully leading to having the outside directors identify with the shareholders they represent rather than the CEO over whom they maintain oversight. Board reformers continue to push such issues as independent nominating committees, payment in stock, and splitting the job of chairperson and CEO.[10]

It is clear that every time you find a business in trouble, you find a board of directors either unwilling or unable to fulfill its responsibilities.[11] On the Trans-Alaska Pipeline System (TAPS), the individual oil companies that owned the project formed an owner's committee to maintain oversight of the TAPS project. In addition, an owner's construction committee was established to administer the contract with Alyeska, the agent for the owners and their designated project manager. This committee, which was to act much in the manner of a board of directors, did not focus adequately on the strategic decision making on the TAPS project. Its members also improperly intervened in day-to-day operating decisions. A review of the record of this committee indicated little resolution of substantive strategic issues on the project, such as:

- The development of a master strategic plan for the project
- Early integrated life-cycle project planning
- Design and implementation of a project management information system
- Development of an effective control system for the project
- Design of a suitable organization[12]

Too many corporate boards are overpopulated with members of management. Inside directors tend to be committed to the way things have always been done and to their own ideas. Outside directors often have insufficient information about the company, and in too many situations, they receive information concerning the matters scheduled for a board meeting only shortly before the board is convened. In practice, when the CEO encourages board members to meet with senior company managers on a regular basis, outside the formal board meetings, this increases the likelihood that the outside directors will be able to have a fuller grasp of what is really going on in the company.

The nuclear industry was a striking example of the laxity of the directors. In the nuclear industry in the past all too many utilities had boards that neglected to exercise "reasonable and prudent" strategic management in their oversight of nuclear power plant projects. As a result, administrative courts disallowed substantial costs from inclusion in the customer rate base for the utility. In many cases, a failure

[10]Judith H. Dobrzynski, "Corporate Boards May Finally Be Shaping Up," *Business Week,* August 9, 1993, p. 26.

[11]Dan Bayly, "What Is the Board of Directors Good For?" *Long Range Planning,* vol. 19, no. 3, 1986, p. 22.

[12]David I. Cleland, prepared direct closing testimony, Trans-Alaska Pipeline System, Alaska Public Utilities Commission, Federal Energy Regulatory Commission, Washington, D.C., October 19, 1984.

of the board to participate in key decisions set the stage for the major difficulties later in the project's life cycle.

The clear responsibility and accountability of the board of directors can be demonstrated by reviewing a few key litigation conclusions drawn from the nuclear power industry. Although these projects are now history, there are important lessons to be remembered!

- Cincinnati Gas & Electric Company reached a $14 million settlement in a shareholder suit that charged directors and officers with improper disclosure concerning a nuclear power plant.
- The Washington Public Power Supply System (WPPSS) defaulted on interest payments due on $2.5 billion in outstanding bonds in part because of the failure of its directors. Communication at the senior levels of the organization, including that of the board of directors, tended to be "informal, disorganized, and infrequent."
- On the Long Island Lighting Company Shoreham project, the public utility commission determined, "The company should be able to show that its directors . . . were attentive to the project's progress, and aggressively pursued cost containment measures wherever there were reasonable opportunities to do so."[13] Noting the small proportion of board minutes devoted to addressing the Shoreham project, the commission remarked on the "lack of urgency in the board's approach to the project's large cost escalations." The commission also was concerned with the board's "lack of involvement" regarding the critical decision to replace the project's construction management firm. In addition, it found that "prudence dictated that the board carefully examine management's plan and its potential consequences."[14]
- On another nuclear project in the state of Washington, the Washington Utilities and Transportation Commission determined that a number of ominous external occurrences should have caused the officers and directors of the Puget Sound Power and Light Company to call for an in-depth cost-effectiveness study, something they neglected to do.[15] In a separate opinion, one of the commissioners elaborated:

  > It is clear the deficiency extends to the company's board of directors. Board minutes . . . provide no indication that Puget's board either was informed of the magnitude of the problem by management or on its own motion requested management to study the economic consequences of continued investment in the . . . plant.[16]

- In a review of the role of the board of directors of the Diablo Canyon project, an expert witness testified that the board's decisions and actions were either limited or nonexistent in regard to several key decisions and actions. These included the approval of a strategic project plan and the decision that the company acted as its own architect and its own engineering and construction manager.

[13]Long Island Lighting Company, 71 Pub. Util. Rep. 4th 262 (N.Y.P.S.C. 1985).

[14]Ibid., p. 273.

[15]Washington Utilities and Transportation Commission (WUTC) v. Puget Sound Power & Light Co., 62 Pub. Util. Rep. 4th 557 (WUTC 1984).

[16]Ibid., p. 598.

Furthermore, the board did not give proper attention to the choice of a basic organizational design for the project, nor to the implications of the discovery made during the construction of the plant that there was a major earthquake fault in close proximity to the plant. Nor did the directors make a full assessment of the flawed quality assurance and control procedures that led to major design deficiencies in the plant. The board also had too little to say about the selection of a project manager and constructor in the final phases of the plant's construction.[17] At key decision points in that project, the board of directors' role was little more than that of a passive onlooker. The failure of this board to insist upon thorough information and its inaction in the face of serious problems confronting the project were incautious, far from what one would expect a "reasonable and prudent board" to carry out.

In this same case, it was found that from the very outset, the board's role was deficient in overseeing the selection of the plant site. The selection was not even considered by the board but was relegated to the chief executive officer's advisory committee, a top-level executive body whose authority was purely advisory. Although this committee evaluated the site for the nuclear plant, it was done during one of their regular meetings along with 19 other agenda items, allowing only 5 minutes per item on average. Later, during the construction of the plant, an earthquake fault was found offshore, and it caused the Nuclear Regulatory Commission to order a redesign of the plant to bring the plant up to a higher earthquake design configuration. The redesign of the plant and the subsequent reconstruction increased the total cost of the plant by approximately $1.4 billion.

It is clear that the boards of the various nuclear projects mentioned in the preceding discussion could have helped reduce their projects' problems or reduce the threats that faced their projects by careful, informed involvement in key project matters on a regular basis.

5.5 EXEMPLARY BOARD BEHAVIOR

The inadequacies of the boards mentioned up to this point reflect a pattern of inactivity and ignorance concerning the problems and threats that buffeted the projects. It is clear that the boards of the various projects, nuclear and otherwise, could have helped reduce their problems and the associated threats that faced their projects by careful, informed involvement in key matters on a regular basis. This has been done on some nuclear plant projects. For example, the Pennsylvania Power and Light Company's board of directors played an active role on the Susquehanna nuclear plant project, as stated in a letter to one of the authors:

> Our Board of Directors was kept abreast of project activities on a monthly basis. The project issued a monthly report to the Board prior to their meetings. The Project

[17]David I. Cleland, rebuttal testimony, Diablo Canyon project, California Public Utilities Commission, Division of Ratepayer Advocate, Application Nos. 84-06-014 and 85-08-025, San Francisco, June 20, 1988.

> Director was then available at the Board meeting to discuss the report. In addition, for several of the critical construction years, the Board held an expanded meeting at the plant site annually. This permitted Board members to view progress firsthand and permitted additional nuclear topics to be included in the agenda.
>
> The monthly reviews . . . also served as the regular, integrated review of the project by the project manager/project team. These reviews included senior management from our engineer/constructor . . . . Senior representation from the reactor manufacturer was also present when appropriate. These meetings focused on performance and progress and highlighted issues significant to management. The reporting of progress and performance was an integrated team effort.

This plant earned high marks from the Nuclear Regulatory Commission in its latest Systematic Assessment of Licensee Performance (SALP). Susquehanna earned the highest rating possible in 9 of 11 categories and the second-highest rating in the remaining two areas. This gave Susquehanna the second-highest average rating of all nuclear reactors in this country.[18]

There are other examples of good board review. The $2.1 billion Milwaukee Water Pollution Abatement Program initiated a comprehensive review of the status of the projects in that program to be conducted on a monthly basis by the owner's senior managers. The program manager was present to explain the program's status and to answer any questions posed by these senior managers. The senior managers, in turn, kept the board of commissioners of the Milwaukee Metropolitan Sewerage District informed on a regular basis. This complex, high-visibility program, which has held the attention of many stakeholders during its life cycle, finished on schedule and close to the original project budget estimates. The continued review by the senior managers and the commissioners is a major reason this project was successful.

Some corporations have special meetings of the board to deal with major projects in the corporation's strategic plans. Besides providing more concentrated time for discussion on the projects, the social events of such meetings provide the opportunity for the directors to learn about the capability and knowledge of the senior corporate executives as well as something of the credentials of the other directors.

In Fig. 5.1, a conceptual model is offered which notes the key roles to be carried out by an ideal board of directors. As discussed in this chapter, all too often some of these key roles are not effectively carried out by some incumbent board members.

5.6 THE BOARD'S RESPONSIBILITIES

Directors are the representatives of the owners of the corporation. Boards often move glacially in reviewing and approving the strategic management initiatives for the enterprise. Companies can become noncompetitive and dwindle, often without any intervention initiatives encouraged by the directors. Today increasingly

[18]PP&L shareowners' newsletter, July 1, 1988.

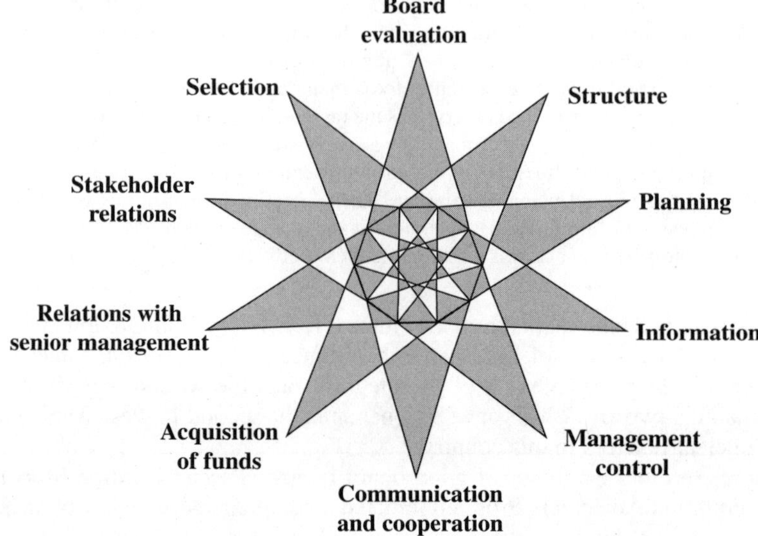

FIGURE 5.1 Conceptual model of roles for the board of directors.

impatient owners, representing such groups as government officials, shareholders, and institutional investors, are taking the lead in evaluating and changing the attitudes of investors.

The corporate governance system may be in place—the senior managers of the enterprise and the directors provide the strategic direction for the enterprise. But these officials, failing to set and follow up on the strategic initiative of the enterprise, often set the game plan unknowingly for corporate decline, leading to failure. Leadership is a critical function for individual and collective roles in discharging the director's functions.[19]

Directors exercise a special kind of management surveillance. Although generally not concerned with short-term operational matters, directors should be alert to any problems and opportunities that are significant to the long-term performance of the company, such as profitability trends, competitive threats, increased costs, loss of future business opportunities, loss of market share, regulatory changes, and quality problems. Observation of any of these problems should alert the directors to the need for an investigation or audit of the company's strategies. Such an audit should determine whether the corporation's strategies have been designed to cope with these difficulties and what the possible and probable long-term impact of the current operations would be.

The directors should expect the corporation's senior managers to manage the organization's resources in a reasonable and prudent manner. Why should any less be expected of the directors themselves? Although their involvement in the corporate affairs is necessarily much broader, the directors need to determine to what extent

[19]Paraphrased from Myron Magnet, "Directors, Wake Up," *Fortune,* June 15, 1992, pp. 85–92.

the senior managers are executing their own responsibilities in the planning, organizing, and control of corporate resources. In other words, the directors are still managers in the truest sense of the word. As managers, they should be expected to perform as any other senior manager, differing only in degree. Because corporate resources are at stake, commissions or omissions at the corporate level are vastly more serious than at lower levels in the organization.

The directors are the "most senior" managers in the corporation; they should set an example for reasonable and prudent management on the part of the corporate senior managers who are concerned with the strategic management and operational effectiveness and efficiency of the corporation. To accomplish this, the directors must demand high performance from the senior managers by ensuring that strategic planning and surveillance are carried out in the corporation, and that efficient and effective operational performance is realized.

The role of senior managers is critical in the success of project management in an enterprise. Because projects are building blocks in the design and execution of strategies for the enterprise, managers at all levels must have an interest and obligation to strategically manage the enterprise. The successful management of the relevant product, service, and organizational process projects means that an appropriate future is likely to come for the enterprise. Some of the common "failures" of those managers who are charged with the responsibility for strategically managing the enterprise include:

- An inappropriate linkage of projects to the strategic direction of the enterprise, resulting in too many of the projects lacking a "strategic fit" after too many organizational resources have been committed to those projects.
- Failure to integrate project development efforts with other development strategies that are under way in the enterprise such as training initiatives, market development, recruitment and training of people, reengineering efforts, and so forth.
- Delaying the decision to establish a project or a family of projects causing "catch-up ball" in getting the project team assigned and the identification of the project resources under way.
- Failure to establish firm and timely project technical performance objectives, which leads to future changes in the project's scope, cost, and schedule.
- Failure to review the project results on a regular basis and to adjust resource input into the projects if needed as a result of the project review.
- Failure to build and maintain alliances with key stakeholders of the project to include visitations to the key stakeholders—like customers and suppliers—to keep them informed and help solidify their continuing support to the project.
- Failure to provide an ongoing training program for the enterprise to include project teams to update knowledge, skills, and attitudes of these people.
- Failure to recognize the motivational considerations of the project teams—and provide those teams with a leadership model for the managers and team members to carry out in the enterprise.

5.7 EMPOWERMENT OF THE BOARD

According to the *Harvard Business Review,* the age of the empowered board of directors is here. Almost all major public corporations recognize that they must make their senior managers more accountable to their stakeholders and strengthen the role of outside directors in development and review of major strategies for their enterprises. Boards need to maintain a certain distance between themselves and the company CEO without turning a meaningful and constructive relationship into an adversarial one.[20]

Empowerment of the board of directors means that outside directors have the competency and independence to monitor the performance of the company and the senior managers—and to influence top management on their role as overseers of the company's fortunes. Outside directors create the opportunity to influence the strategic management of the enterprise and to influence changing corporate leadership if needed to keep the enterprise moving forward in new products, services, and organizational processes. Senior executives should seek the appointment of an empowered board that is truly involved, in an oversight role, in the strategic management of the enterprise. This calls for the board of directors to operate in a new form of teamwork—one in which the individuals work together in their individual and collective roles to improve the operational and strategic performance of the enterprise.

In theory the director's legal authority to govern a company comes from the laws of the state in which the enterprise is incorporated. For example, in Delaware the law states that "The business and affairs of every corporation organized under this charter will be managed by or under the direction of a board of directors." In carrying out their residual responsibility of overseeing corporate management, board members are expected to demonstrate care and loyalty and to exercise judgment in their trustee relationship to the board.

The characteristics of an empowered board include having most of the directors come from outside the enterprise. The board should be small enough to be a cohesive group, with the individual members understanding their reciprocal obligations. A range of business and leadership experience by the board members is required—and the board members should communicate freely when attending board meetings and, more important, in between meetings. If the corporate CEO is also chair of the board, then the outside directors should select a leader from among themselves. Board committees should be made up entirely of outside directors. Finally, board members should receive information about the company's performance in a form that is intelligible to the members.[21]

5.8 THE ROLE OF PROJECTS

In 1968 a landmark study of the practices of senior management in leading industrial corporations noted the responsibilities of directors for project management. The

[20] "Redraw the Line between the Board and the CEO," *Harvard Business Review,* March–April 1995, pp. 153–166.

[21] Paraphrased from Jay W. Lorsh, "Empowering the Board," *Harvard Business Review,* January–February 1995, pp. 107–117.

study was conducted by Paul Holden and several members of the faculty at the Graduate School of Business at Stanford University. Their findings established that project management was an important factor in overall enterprise management. The study further found that the high-level committee (such as the board of directors) was widely used as a valuable organizational design to:

- Establish board policies
- Coordinate line and technical management
- Render collective judgments on the evaluation of corporate undertakings
- Conduct periodic review and monitoring of ongoing programs and projects[22]

To state again—major projects are key building blocks in the design and execution of corporate strategy. This is a fundamental principle all too often missed by key corporate managers and directors. Project management is not recognized for what it is: *a process for the creation of something that does not currently exist but is needed to support future corporate purposes.* When perceptive directors recognize the intertwining of projects and corporate strategies, project management takes on a new significance in the management of the corporation. Unfortunately, some directors have not recognized this fundamental principle. Before the project starts, the board should take action to require that a project plan be developed and presented for its review. Why should a board concern itself with the plan for projects? Several principal reasons are suggested:

- The board needs specific evidence that corporate managers have a planned process for managing projects.
- The project plan provides a performance standard against which project progress can be evaluated as the directors carry out their strategic monitoring, evaluating, and control responsibilities.
- If the project team, project manager, and responsible general managers know that the board will review the project plan, a clear message will reverberate through the organization: This project is important.
- Knowing the project plan can help give the board a reference point for other key corporate decisions which interface with the major project such as recapitalization issues, product introduction plans, and support facilities.
- The evaluation of the project plan and of management's adherence to it allows continual evaluation of key managers.

In some cases a committee of the board, such as the executive committee, is given the authority to act for the full board. Such a delegation without adequate monitoring by the outside directors can have a deleterious effect, particularly if the executive committee's deliberations are not reviewed or are accepted with only minimal questioning. Even with an active and competent executive committee, the board should reserve for itself a regular review of capital projects. Such reviews should

[22]Paul E. Holden, et al., *Top Management* (New York: McGraw-Hill, 1968), pp. 6, 71–74, 108–109.

include discussions of the cost and schedule of the project and future strategies for the resolution of any problems known or anticipated on the project.

The existence of projects in organizations is one clear indication that the organization is changing and is attempting to meet changing future environments. This is a key point not to be missed by senior managers and directors.

5.9 THE ORGANIZATIONAL DESIGN

An important part of corporate strategy is an appropriate organizational design for the implementation of projects. The directors should ensure that an appropriate organizational design is in place for the project. The design should delineate the formal authority, responsibility, and accountability relationships among the senior managers, project manager, functional managers, and work package managers of the enterprise.

Russell D. Archibald, an expert witness evaluating a utility's nuclear plant project organization in a rate case litigation, found serious deficiencies in one utility's project organization and staffing. His findings were:

- The absence of a true project manager
- Inadequate planning and control supporting staff
- Lack of definition of responsibilities and inadequate policies and procedures for fulfilling assigned responsibilities[23]

These deficiencies were found to contribute to schedule delays and cost overruns on this project.

5.10 PROJECT REVIEWS

Directors and senior managers who clearly recognize their responsibilities should feel the need to regularly review projects along with other major organizational activities. Why should the board concern itself with review of the major projects? The board needs specific information that the projects are being designed and developed according to plan and in support of corporate strategies.

Then, too, knowing the status of the projects can give the board a reference point for the review of management actions and recommendations that are interdependent with the other projects and strategies in the organization's strategy. Directors will gain an appreciation of the underpinnings of strategy such as policies, resource commitments, and executive and professional development to support the company's strategies along with its capital projects. By having the directors insist that the company have a strategy and a management philosophy for major project review, another mechanism is in place for facilitating the continuous evaluation of senior managers.

[23]Russell D. Archibald, testimony on project management, Diablo Canyon rate case, California Public Utilities Commission, San Francisco, Exhibit No. 11, 175, March 1987.

Some projects reach the point where their continuation does not make sense for the organization. Because of the vested interest that the project manager and the project team have in the project, they are in the least logical position to recommend termination of the project. But total reliance on the senior managers to do this evaluation is not sufficient because the board is the corporate conscience to make an independent evaluation of where the project stands within corporate strategy. Therefore, both senior managers and directors are the most appropriate decision makers to recommend termination of the project.

How is the project review best done? Here is a prescription to guide directors' surveillance of major projects:

- Accept a philosophy that projects are indeed basic building blocks in the design and execution of corporate strategies, requiring ongoing strategic management and surveillance.
- Conduct a formal review of the strategic plan for the project to determine if appropriate technology is planned and if suitable management systems are in place to keep all the principal managers abreast of the project.
- Require special briefings on the project during key periods of the project's life cycle, such as finalization of design, commitment to construction or prototype manufacturing, design reviews, engineering completion, preliminary customer acceptance, or delivery of the first production unit.
- Go out and "kick the tires." Use plant or construction site visits to observe first-hand what is really happening on the project.
- Insist that the project manager (and the responsible general manager) appear before the board on a regular basis to give a status report on the project.
- Question and question again any funding changes on the project to ascertain what caused the change and what the longer-term impact would be.
- Carefully deliberate on what information the board needs to do its job on capital projects, and relate this information to the major decisions or actions that require board scrutiny.
- If things on the project are not fitting together well, or if major questions and issues are emerging for which answers are not forthcoming, consider a performance audit of the project.

The foregoing list hints at overtones of interference with senior management responsibilities. Perhaps so. But as one reviews some of the major project failures of the recent past, a clear message comes through: Most of these failures can be attributed to the failure of senior management and the board of directors to follow some of the basic "commonsense" prescriptions just outlined. What is the cost of not following these commonsense principles? The answer to this question is imprudent financial performance, delay of effective strategies, waste of corporate resources, and support of a corporate culture that condones poor quality in the management of corporate resources.

During the review of major projects, with the project managers present to answer questions, the review should be structured to focus discussion and debate on the hard questions about the projects. Both the bad news and the good news of the project should get attention. The board should be concerned about the schedule, cost, and technical status of the project, as well as an ongoing assessment of the strategic fit of the project. Does the project continue to occupy a building block in the design and execution of corporate strategies? If not, why not? If there is adverse information about the project, what significance does the information have for the directors in coping with their responsibilities?

This discussion about the need for a regular review of the project implicitly assumes that performance standards exist which provide the basis for reaching a judgment of where the project stands. Experience has shown that such assumptions cannot always be made. If a comprehensive project plan and performance standard for the project do not exist, then monitoring, evaluation, and control of the project are difficult, if not impossible.

Something is added to the discipline of the project team simply because the project is reviewed by the board of directors. When the project team knows that a formal presentation on the project's status will be required by the board, the team will be motivated to do a better job of thinking through the problems and of being prepared with solutions, explanations, or rationales.

What do the directors need to know to adequately review the project? The key to satisfaction of this need is the quality of the information provided to the board.

5.11 INFORMATION FOR THE BOARD

The *Corporate Director's Guidebook* makes the point that "the corporate director should be concerned with the establishment and maintenance of an effective reporting system."[24] A reporting system involving major projects takes the form of a project management information system (PMIS), which contains the intelligence essential to the effective monitoring, evaluation, and control of the project. Corporate directors require such information to determine the efficiency and effectiveness with which corporate resources are being used on the project. Also, the directors need other corporate information relative to the enterprise's forward planning. This includes critical events and issues facing the enterprise that often might have a strong project context such as new products, facilities, and recapitalization strategies. The project's cost, schedule, and technical performance considerations are certainly worthy of a director's ongoing surveillance.

Juran and Louden, in a book published in 1966, addressed the information that the board requires to fulfill its obligation to exercise due diligence and to increase the knowledge that directors have about the company. Juran and Louden spoke of the "philosophy of completeness," regarding information as an essential part of the climate in which the board and management operate. They stated, "Under this

[24]*Corporate Director's Guidebook,* American Bar Association, January 1978, p. 14.

philosophy the rule with respect to information for the board is: *Resolve all doubts in favor of completeness.*"

According to the authors, the practical result of the philosophy of completeness is the advance information package in widespread use in many companies. According to them:

> This package is sent to the directors in advance of each meeting to include the agenda, which is a listing of the topics, which are to be discussed at the meeting. It is not merely a table of contents; it serves also as a kind of notice of what is to come up at the meeting. (By strong implication, anything not on the agenda will be regarded as a surprise.) In some companies the agenda carries notations showing just what actions, if any, the board is being asked to take with respect to each item.[25]

Juran and Louden also recommended that the typical information package for board approval include not only the project proposals on expenditures and actions which are on the list of reserved board powers but also those actions which chart a new course. The reports furnished to the board on the project's status are important tools to help the directors do their job. At the minimum, such reports should contain summary information to help the directors meet their responsibilities: the surveillance of the project's cost, schedule, technical performance objectives, and the probability of continued strategic fit in the enterprise. The project manager has the responsibility to see that the project's status report provides sufficient intelligence for the directors to reach a conclusion about where the project stands.

The typical board meets on a monthly basis. Prior to a meeting, the directors usually are provided with an agenda and appropriate supporting materials for review so that they are able to do their "homework."

Some important things to consider in the use of project-related information for the board include:

- Presenting important issues on the project to the directors before, and not after, corporate senior management has taken a firm position
- Making sure that the directors get any important information before the board meeting in order to make an informed judgment about the project
- Not burying the project information in a stack of corporate information
- Allowing the directors sufficient time to make a decision in which they have confidence
- Making sure there is time at the board meetings for a full discussion of the project with the project manager present to answer questions
- Using the board committees, such as the executive committee and the audit committee, to do detailed analyses and present their recommendations to the full board

[25]Reprinted by permission of the publisher from *The Corporate Director* by J. M. Juran and J. Keith Louden, pp. 257–258. © 1966 AMACOM, a division of the American Management Association, New York. All rights reserved.

5.12 THE PERFORMANCE AUDIT

If the information reported to the board and obtained during the project manager's status report reveals project inadequacies or problems, a performance audit may be in order. Independent performance audits on large projects can provide valuable insight for the board and other corporate managers. An independent performance audit on a project may be defined as an in-depth, process-involving analysis of a project's performance and outlook. The analysis should cover both the technical side of the project and its management. Project performance audits are best made at key points in the project's life cycle or when the project is being buffeted by important problems or changes whose effects may not be fully fathomed. Heyel noted, "Regardless of intent, a failure to investigate independently may be deemed culpable ignorance and a breach of duty to stockholders."[26] Although the full board may order the audit, a subcommittee of the board can make sure the audit is appropriately executed and followed up with the most efficient and productive remedial action.

Although project history is relevant, because past events provide a base from which the project moves forward, the performance audit should not be done to find fault or to debate over past disappointments. Rather, it should use the past to develop a better understanding of how current and future performance on the project can be improved.

On a large water pollution abatement system project, an audit was conducted prior to initiation of detailed planning to turn the project results over to the user. This audit disclosed several contract modification changes that were unduly delayed and that could have an adverse influence on the operational availability of the system. By discovering this delay of changes through the audit, the project manager was able to initiate remedial strategies to get the project back on schedule and meet its operational date.

An independent performance audit appraises results so that the board and its subcommittees can objectively evaluate the need for and extent of remedial strategy and resources required. Failure to conduct an independent performance audit on an ailing or failing project may very well be considered culpable negligence and a breach of duty to the stockholders, leading to legal action.

5.13 CULTURAL CONSIDERATIONS

Over time camaraderie develops among members of the board of directors and the corporate managers. This camaraderie makes it difficult for outside directors to remain objective regarding corporate matters—and makes it unlikely that any director would want to play the role of a spoiler by challenging the board's actions. Yet there would be value in having a forum where consensus could be developed regarding the viewpoints and concerns of the outside directors. One writer on the

[26]Carl Heyel (ed.), *The Encyclopedia of Management,* 3d ed. (New York: Van Nostrand Reinhold, 1982), p. 222.

subject has suggested the appointment of a management advisory committee composed of outside directors only. Through such a collective consensus the role of a spoiler could be played without jeopardy of any individual outside directors. Most public companies have an audit committee and a compensation committee. Having an advisory committee would be an extension of such committees and would bring another fiduciary watchdog to the board of directors' processes.[27]

Directors (and senior managers) influence the culture of the organization, and that culture in turn influences projects. Corporate culture is reflected in the key values held by members of the organization. Managerial and professional behavior are influenced by what the people perceive as the "corporate way of doing things." The value orientation, leadership style, and example set by senior managers greatly influence the behavior of the people.

The attitudes expressed by senior managers can have a significant effect on the organization's culture. Communication by senior management can influence the outcome of the project. Davis noted that senior managers' most important task is to foster a corporate environment that facilitates honest and frank disclosures in dealing with a budget-breaking project. He further notes that discouraging cover-ups and recriminations by senior executives depends on their management style.[28]

A corporation that does not commit itself to comply with government regulations sends an important message throughout the organizational hierarchy. On the other hand, a senior corporate management that takes the lead in developing and promulgating policies that demand full cooperation and disclosure to government bodies will find such policies echoed and enforced throughout the company's organizational structure.

In the nuclear plant construction industry, the Nuclear Regulatory Commission found a direct correlation between the project's success and the utility's view of NRC requirements. More successful utilities tended to view NRC requirements as minimum levels of performance, not maximum, and the utilities strove to achieve increasingly higher, self-imposed goals. This attitude covered all aspects of the project, including quality and quality assurance.[29]

During a performance audit of a large project, it was found that the attitudes, values, beliefs, and behavior demonstrated by senior management of the organization were detrimental to the successful outcome of the project. In an assessment of the corporate culture of this project, it was found that senior management had condoned a culture that contributed to various problems on the project with significant injurious results, such as:

- A lack of candor and openness in dealing with government agencies, particularly the Nuclear Regulatory Commission
- Management leadership which encouraged the destruction of documents that might have negatively affected the company during customer rate litigation

---

[27]John L. Grant, "Shield Outside Directors from Inside Seduction," *The Wall Street Journal,* November 23, 1992.

[28]David Davis, "New Projects: Beware of False Economies," *Harvard Business Review,* March–April 1985, p. 97.

[29]"Improving Quality and the Assurance of Quality in the Design and Construction of Nuclear Power Plants," NUREG-1055, U.S. Nuclear Regulatory Commission, Washington, D.C., May 1984, pp. 2-1 to 2-6.

- A lack of commitment to adequate communications within the company concerning the status of the project
- Not taking a conservative approach to unknown factors in the design and construction of the project
- The general lack of leadership to solve problems on the project in a timely manner
- Reliance on past management philosophies and practices and a failure to recognize the impact of new technology on both the design of the project and the use of contemporaneous project management practices

5.14 SELECTION OF DIRECTORS

Every corporation should have formalized criteria for the election of the directors, including the insider-outsider mix, occupational expertise, and length of tenure. Used as guidelines, such criteria can be varied to accommodate different requirements for the board. Considering the importance of project management to the corporation, the board should include individuals who have had experience in either the management of projects or the senior executive oversight of such projects. If the projects involve new technology, then at least some of the outside directors should have experience in that technology. Directors should be chosen who have experience in the industry or knowledge about the business the corporation pursues. In large integrated corporations, this is difficult, but by careful choice of the directors, a collective understanding of the corporation's business can be known. If the board does not have outside directors with such experience, then the board should request external assistance in the form of project performance audits and consultations to evaluate and question the projects' status.

5.15 TO SUMMARIZE

The major points expressed in this chapter include:

- Corporate strategy is clearly a responsibility of the enterprise's directors.
- Because projects are building blocks in the design and execution of enterprise strategies, the board of directors should be vitally concerned about the status of major product, service, and organizational process projects in the organization.
- Specific policies and philosophies should be established in the enterprise that deal with how the directors carry out their fiduciary responsibilities for the surveillance of major projects.
- In the past, and even today, there are directors whose performance has been inadequate. In the chapter, examples were given of how poorly some directors have performed.
- Some boards of directors have performed their fiduciary duties in a stellar fashion. Examples were given in the chapter of such performance.

- A survey by *Business Week* established key distinctive characteristics of competent boards. The reader should review the results of this survey.
- There is evidence that boards of directors are reducing the number of members, leading to easier assessment and discussion of corporate performance and future strategies.
- Whenever you find a business or a major project in trouble, the cause can likely be traced to a board of directors that was unwilling or unable to fulfill its responsibilities.
- Directors are becoming more liable for lawsuits, which charge them with imprudence in their fiduciary role.
- Examples were given of how inadequate performance by directors on projects in the nuclear power industry adversely impacted the performance of construction projects in that industry.
- Some of the common "failures" of middle and senior managers charged with responsibility for project management were given.
- Boards of directors are becoming more empowered, particularly in increasing the authority of outside directors.
- Regular and rigorous review of the status of major projects is the best way for directors to be kept informed of how corporate strategy is evolving—and how well the enterprise is preparing for its future.
- A prescription for how directors could best review major projects was given. In order to conduct effective reviews, the directors need timely and relevant information on the key projects in the enterprise.
- Project performance audits can be a powerful tool to use in gaining an independent assessment of a project's status.
- Cultural considerations in the enterprise impact the rigor with which projects are managed and reviewed. When the directors participate in regular and rigorous review of key projects, an important message is sent throughout the enterprise.
- In the selection of directors, consideration should be given to the individual's competency in project management.

5.16 ADDITIONAL SOURCES OF INFORMATION

The following additional sources of project management information may be used to complement this chapter's topic material. This material complements and expands on various concepts, practices, and theory of project management as it relates to areas covered here.

- Randall L. Speck, "Legal Considerations for Project Managers," and Kenneth O. Hartley, "The Role of Senior Management," chaps. 13 and 17 in David I.

Cleland (ed.), *Field Guide to Project Management* (New York: Van Nostrand Reinhold, 1997).

- Philip J. Damiani and Robert J. Teachout, "Pittsburgh International Airport Midfield Terminal Energy Facility," in David I. Cleland, Karen M. Bursic, Richard J. Puerzer, and Alberto Y. Vlasak, *Project Management Casebook,* Project Management Institute (PMI). (First published in *Proceedings,* PMI Seminar/Symposium, 1992, pp. 44–50.)
- Paul E. Holden, *Top Management* (New York: McGraw-Hill, 1968). Although this book was published over 30 years ago, its description of the theoretical role of senior managers in an enterprise holds true today. The book was one of the first to recognize the important role that senior managers and directors have in maintaining oversight of the planning for, organization of, and execution of projects in the enterprise.
- Jay A. Conger, *Strategies for Adding Value at the Top* (San Francisco: Jossey-Bass, 2001). In this book the author explores the roles that corporate governance will play in the twenty-first century. Although the book provides little guidance on the role of projects in the design and execution of enterprise strategies, it provides an excellent overview of the major responsibilities of directors. The author proposes that the focus of judging a board's success should move from a shareholder to a *stakeholder* point of view. Conger also examines a board from an organizational effectiveness perspective and proposes a framework that centers on what really influences effective governance strategy.
- Kenneth R. Andrews, "Director's Responsibility for Corporate Strategy," *Harvard Business Review,* November–December 1980. This article points out the key fiduciary responsibilities that directors have for the well-being of the enterprise—to include the obligation to ensure that the senior managers prepare strategic plans for the directors' review. Andrews makes it clear that a board of directors should review corporate strategy periodically to determine its validity, and use it as the reference point for other key board decisions. He further states that key approval decisions on the part of the board should evaluate the risks involved, and share with management the risks associated with its adoption.
- Jay W. Lorsh, "Empowering the Board," *Harvard Business Review,* January–February 1995, pp. 107–117. In this article, the author provides a general assessment of the responsibilities of the board, and suggests strategies that can be used by boards working with senior managers for the empowerment of the board. In addition, the author makes a key point about the key roles to be carried out by board members. He suggests that outside directors should select a chairperson from among themselves and board committees should be made up of outside directors.
- David I. Cleland rebuttal testimony, Diablo Canyon Project, California Public Utilities Commission, Division of Ratepayer Advocate, Applications

Nos. 84-06-014 and 85-08-025, San Francisco, June 20, 1988. This testimony provides expected performance standards for exemplary board behavior regarding the design and construction of a major nuclear power generation plant—and how the project owner neglected these standards. In his testimony, the author found major discrepancies with the manner in which senior managers and board members conducted themselves with regard to the management of this nuclear power plant. Other major problems were found in other areas of this project by expert witnesses. After reading this testimony, and the testimony of other expert witnesses on this project, one can easily have the perception that the project could not have been more badly managed, if the project team, senior managers, and directors had really tried to fail in the management of this project.

5.17 DISCUSSION QUESTIONS

1. What kind of evidence might indicate that a company's board of directors has been inadequate in its monitoring of major project undertakings? Explain.
2. What actions and activities indicate that a company's board of directors has taken an active interest in major projects? Explain.
3. Briefly describe some of the major responsibilities of a board of directors with respect to project management.
4. "Projects are key building blocks in the design and execution of corporate strategy." Explain what is meant by this. What ramifications does this idea have with respect to the responsibilities of a corporate board of directors?
5. Cite and explain some of the reasons for the need for board interest in a project plan.
6. How can a board of directors ensure that the organization design will be effective for accomplishment of corporate strategies and projects?
7. What specific questions should be addressed by the board in project review meetings?
8. What kind of information about a project should be prepared for and presented to the board of directors? Explain.
9. What is the purpose of a performance audit? Under what circumstances might a board of directors consider such an audit? Why?
10. Discuss the effect of corporate culture on organizational performance. What role does the board of directors play in shaping corporate culture?
11. What steps can be taken to protect a board of directors from litigation and subsequent court actions? Explain.
12. Discuss the importance of proper board member selection for organizational effectiveness.

5.18 USER CHECKLIST

1. What evidence indicates the possibility of inadequate attention by the board of directors to the projects within your organization? Does any evidence suggest that your organization's board of directors has been adequately involved in the corporation's major undertakings?
2. What responsibilities do you believe the board of directors should be taking but has not? Explain.
3. How do the major projects within your organization contribute to strategic plans and achievement of objectives and goals? What does this suggest about the need for board involvement?
4. Does your corporation's board of directors receive information about major project plans? What contributions do they make to these plans?
5. What attention has the board of directors of your organization given to the organizational design? What attention is needed?
6. Is the board involved in project review meetings? Why or why not?
7. What questions are addressed by the board of directors with respect to the project's progress? What questions should they be asking?
8. What kind of project status information is presented to the board of directors? Is information presented on a regular basis and in advance of meetings?
9. Under what circumstances might a project audit be needed on a major project in which your organization is involved?
10. What board actions have had an impact on the corporate culture of your organization? Explain.
11. Have any of your organization's projects undergone scrutiny in litigation? How could the company have been better prepared for such litigation?
12. Is your corporate board of directors staffed with knowledgeable, competent members? Why or why not?

5.19 PRINCIPLES OF PROJECT MANAGEMENT

1. The board of directors, along with senior management in an enterprise, has the responsibility to maintain surveillance over the planning for and execution of a project.
2. It is the responsibility of the senior managers of an enterprise to select major projects that should be reviewed by the board of directors.
3. The directors of an enterprise must review, on a periodic basis, the linkage between the strategic management and projects in the enterprise.

4. The board of directors and the senior managers of the enterprise have residual responsibility for the success or failure of a project.
5. In order to maintain effective surveillance over the conduct of a project in an enterprise, the board should approve the project's plan.

5.20 PROJECT MANAGEMENT SITUATION—BOARDS OF DIRECTORS' INADEQUACIES

In the material that follows, some of the key testimony presented by an expert witness on the prudence and reasonableness of the role of the Pacific Gas & Electric Company (PG&E) in the design and construction of the Diablo Canyon nuclear power plant is presented. This testimony, along with other expert witness testimony, was presented during the period when the State of California was involved in evaluating the utility's request for rate charges to offset the cost of the plant's design and construction of approximately $5 billion.

The Diablo Canyon nuclear power plant is located in San Luis Obispo, California. Nearly 20 years elapsed before the plant became operational. It was the first nuclear power plant constructed by PG&E—who did the design engineering for the plant. Other plants that were built by the company were traditional "fossil-fueled" power plants. Excerpts from the expert witness testimony follow:

- "The evidence is clear, however, that neither the Board nor the Executive Committee played any significant role in directing and controlling Diablo until late in the project."
- ". . . at least until 1979, the Board functioned without meaningful formal input of significant Diablo Project data."
- "Because they lacked adequate information, PG&E Directors were unable to take appropriate action in the strategic management of the Diablo Project until late in its history."
- "My further review of all of the meeting minutes of the Board of Directors and Executive Committee cited by PG&E's witnesses indicates that major periods passed during which these senior executive bodies took no action on the Diablo Project."
- ". . . the Board's effectiveness was limited severely due to its failure to insist upon timely, easily understood information on the project."
- "The Major Construction Report (in both its weekly version given to the Executive Committee and the monthly version provided to the Board) was deficient in at least the following significant respects:
    - It did not distinguish between Diablo and other, much less significant jobs.
    - The report provides no basis for comparing planned and actual costs of the job.
    - The report provides no basis for comparing the planned and actual schedule.

  - The report did not identify key problems, events or issues that could affect cost or schedule.
  - The report totally neglected consideration of technical performance, including quality assurance and quality control.
  - The report did not facilitate identification of project trends."
- ". . . PG&E failed to develop an effective project information system for the Diablo Project until 1982."
- "At PG&E, however, such informational material and agenda items were typically not presented to the Board until the outset of the meeting."
- "My review of a number of key strategic decisions and actions on the Diablo Project indicates little, if any, involvement by the Board of Directors and the Executive Committee."
- "These decisions and actions included:
  - Approval of a strategic plan for Diablo;
  - PG&E's decision to act as its own architect, engineer and construction manager (AE/CM);
  - Choice of a basic organizational structure for the project;
  - Assessment of the suitability of the Diablo Canyon site;
  - Assessment of the implications of the Hosgri fault;
  - Full assessment of the implications of the Mirror Image Error; and
  - Selection of Bechtel Power Corporation as Project Completion Manager."
- "*Conclusion:* The most crucial questions in evaluating the reasonableness of the Board of Directors' performance are: (a) what did the Board know, and (b) what action did it take. It is apparent from PG&E's witnesses' testimony and their voluminous exhibits that the Board knew very little about the most significant project the Company has ever undertaken. It is also apparent that the Board was little more than a passive onlooker at key decision-points in the Diablo Project. The PG&E Board's failure to insist upon thorough information and its inaction in the face of various problems were unreasonable."[30]

5.21 STUDENT/READER ASSIGNMENT

1. What overall action should the board of directors of the PG&E Company have taken with regard to this major project when it was initiated?
2. What project management principles were not followed in the management of this project?
3. What do you believe to be the most serious omission in the management of this project by the senior managers and directors of this company?

[30]Rebuttal testimony of David I. Cleland, California Public Utilities Commission, Division of Ratepayer Advocates, San Francisco, California, June 20, 1988.

4. What "philosophy of project management" should the senior managers and board members have followed with respect to the project?
5. The PG&E Company had an excellent "track record" in the design and construction of fossil-fueled power plants. Why did they have major problems on the Diablo Canyon project?

CHAPTER 6
PROJECT STAKEHOLDER MANAGEMENT[1]

"Smile at the claims of long descent."

ALFRED, LORD TENNYSON, 1809–1892

6.1 INTRODUCTION

One of the major concerns coming forth in the management of projects is the recognition and "management" of project stakeholders. These stakeholders are project team members, higher-level managers, and outside organizational entities such as contractors, subcontractors, customers, regulators, financial institutions, and other claimants who have—or believe they have—vested rights in the project. Acceptance of the notion of project stakeholders means that the project has to be managed from an overall perspective of all of the stakeholders—not just the customer(s) and the organization.

This chapter provides a description of generic stakeholders, along with examples of successful and not-so-successful stakeholder management. A model and a process will be suggested that can be used in identifying and understanding project stakeholders, the management of such stakeholders, and how to understand and deal with the likely parochial interests of stakeholders. How to predict stakeholder behavior and how to manage the project from a total stakeholder context will be examined. After reading this chapter the people who are associated with enterprise projects or are engaged in learning how to manage projects should gain important concepts and processes to be added to their overall philosophy of project management.

6.2 WHY MANAGE STAKEHOLDERS?

Stakeholder management is also an important part of the strategic management of organizations. There is abundant literature in the management field that establishes

[1]Portions of this chapter have been paraphrased from David I. Cleland, "Project Stakeholder Management," *Project Management Journal,* September 1986, pp. 36–43. Used by permission.

the need to analyze the enterprise's environment and its stakeholders as part of the strategic management of the enterprise. Full recognition of the role of enterprise stakeholders is a recent phenomenon.

Political, economic, social, legal, technological, and competitive environments affect an enterprise's ability to survive and grow. Project managers need to identify and interact with key institutions and individuals in the project's systems environment. An important part of the management of the project's systems environment is an organized process for identifying and managing the probable stakeholders in that environment. This management process is necessary to determine how the probable stakeholders are likely to react to project decisions, what influence their reaction will carry, and how the stakeholders might interact with each other and with the project's managers and professionals to affect the chances for success of a proposed project strategy.

Cleland and King,[2] Rothschild,[3] Freeman,[4] and Mendelow[5] have presented strategies for dealing with stakeholders in the corporate context. The management of a project's "stakeholders" means that the project is explicitly described in terms of the individuals and institutions that share a stake or an interest in the project. Thus, the project team members, subcontractors, suppliers, and customers, to name a few, invariably are relevant. The impact of project decisions on all of them must be considered in any rational approach to the management of a project. But management must also consider others who have an interest in the project and are, by definition, also stakeholders. These stakeholders are outside the authority of the project manager and often present serious management problems and challenges.

Corporations have always been accountable to their shareholders. Now they are also accountable to their stakeholders. At times, the interests of employees, government, customers, suppliers, creditors, or environmentalists can dominate the interest of the shareholders. In recent years laws have been passed giving stakeholders legal protection, such as found in the use of environmental considerations in making and marketing products. There is a subtle shift from a *shareholder* paradigm to a *stakeholder* paradigm. Project teams are using techniques such as "stakeholder reaction assessment" to identify stakeholder interests.[6]

Some companies have explicit policies that guide their strategies regarding stakeholders. For example, Elan Corporation, Plc., states that "It is our policy to support the communities in the areas in which we are based. In Ireland, we have given significant support to the higher education system, including the new school of pharmacy at Trinity College, Dublin, and to the Michael Smurfit Graduate School of Business at University College, Dublin. Ireland has been good for Elan, and we have enjoyed strong support from its government and development agencies from our early years. In giving something back through the support of higher

---

[2]D. I. Cleland and W. R. King, *Systems Analysis and Project Management,* 3d ed. (New York: McGraw-Hill, 1983).

[3]W. E. Rothschild, *Putting It All Together: A Guide to Strategic Thinking* (New York: AMACOM, 1976).

[4]R. E. Freeman, *Strategic Management—A Stakeholder Approach* (Boston: Pitman, 1984).

[5]Aubrey Mendelow, "Stakeholder Analysis for Strategic Planning and Implementation," in William R. King and David I. Cleland (eds.), *Strategic Planning and Management Handbook* (New York: Van Nostrand Reinhold, 1985).

[6]Leonard J. Brooks, "Higher Stakes," *CA Magazine,* March 1995, pp. 53–56.

education, we are benefiting both Elan and the country from which we will have access to future generations of scientists and managers."[7]

6.3 ORGANIZATIONAL STAKEHOLDERS

Organizational stakeholders have been defined in the context of a business organization. Table 6.1 shows a model of generic organizational claimants (stakeholders) and their claims (stake) for a business organization. The model requires the key managers to develop an appropriate strategy to manage the organization through:

- Identifying appropriate stakeholders
- Specifying the nature of the stakeholder's interest
- Measuring the stakeholder's interest
- Predicting what each stakeholder's future behavior will be to satisfy her or his stake
- Evaluating the impact of the stakeholder's behavior on the project team's latitude in managing the project

The value of using a model like Table 6.1 is to establish a point of departure for developing a model appropriate to a project. It is interesting to know that an environmental group may be concerned about the outcome of a project. But it is vital that the project team have a specific delineation of the various strategies that a stakeholder, such as an environmental group, intends to employ in satisfying that stakeholder's goals and objectives, along with a prediction of the future impact of that stakeholder's actions on the project's outcome. For example, a project manager who must make a recommendation concerning the design of a new plant must be aware of state and local land use, plant design, and tax laws, and the area's likely pattern of growth. The project manager must be aware of the local political climate, availability of a skilled labor force, and public attitudes toward the location of the plant in the community. To put all aspects of the stakeholders together requires an understanding of how to apply the management process in dealing with project stakeholders.

The political side of project management is very real. The project manager who neglects the building and maintenance of alliances with key political stakeholders will soon find indifference or opposition to his or her project. There are several notable examples of projects that were impacted by the political exigencies of the period. The recently completed English Channel tunnel project was first proposed in 1802 and was actually started in 1876. Technologically, the tunnel was possible even in that period with more than a mile of tunnel started at each side of the English Channel. But the project was killed by politics many times—and it was not until 1993 that the tunnel became a reality.

The Interstate Highway System proposed by President Eisenhower was sidelined by politics—and when President Eisenhower adopted a bipartisan approach,

[7]Annual Report 1996, Elan Corporation, Plc., p. 10.

TABLE 6.1 Organizational Claimants and Their Claims

Claimants	Claims
Stockholders	Participate in distribution of profits, additional stock offerings, assets on liquidation; vote of stock, inspection of company books, transfer of stock, election of board of directors, and such additional rights as established in the contract with the corporation.
Creditors	Participate in legal proportion of interest payments due and return on principal from the investment. Security of pledged assets; relative priority in event of liquidation. Participate in certain management and owner prerogatives if certain conditions exist within the company (such as default of interest payments).
Employees	Economic, social, and psychological satisfaction in the place of employment. Freedom from arbitrary and capricious behavior on the part of company officials. Share in fringe benefits, freedom to join union and participate in collective bargaining, individual freedom in offering up services through an employment contract. Adequate working conditions.
Customers	Service provided the product; technical data to use the product; suitable warranties; spare parts to support the product during customer use; R&D leading to product improvement; facilitation of consumer credit.
Suppliers	Continuing source of business; timely consummation of trade credit obligations; professional relationship in contracting for, purchasing, and receiving goods and services.
Governments	Tax (income, property, other), fair competition, and adherence to the letter and intent of public policy dealing with the requirements of "fair and free" competition. Legal obligations for business people (and business organizations) to obey antitrust laws.
Unions	Recognition as the negotiating agent for the employees. Opportunity to perpetuate the union as a participant in the business organization.
Competitors	Norms established by society and the industry for competitive conduct. Business statesmanship on the part of contemporaries.
Local communities	Place of productive and healthful employment in the local community. Participation of the company officials in community affairs, regular employment, fair play, local purchase of reasonable portion of the products of the local community, interest in and support of local government, support of cultural and charity projects.
The general public	Participation in and contribution to the government process of society as a whole; creative communications between government and business units designed for reciprocal understanding; bearing fair portion of the burden of government and society. Fair price for products and advancement of the state of the art in technology, which the product line offers.

Source: D. I. Cleland and W. R. King, *Systems Analysis and Project Management,* 3d ed. (New York: McGraw-Hill, 1983), p. 45.

the project was initiated. The Interstate Highway Act of 1956 provided something for everyone, and that was the secret of political success of the highway system. The Superconducting Supercollider project was killed by stakeholder action. Projects can and do fail because of politics. Some of the lessons learned concerning politics and projects include:

- The story of the project must be told so that all stakeholders understand its rationale and purpose.
- Senior management must be fully behind the project.
- Project managers must sell their project to the stakeholders, particularly those who are "nonbelievers."
- Benefits must be widespread and provide something for all stakeholders.

One author summed up the political challenges in project management very well by stating: "Successful project management means successful political management as well."[8]

Corporate executives are becoming more aware of the need to consider the needs of the stakeholders in their management of the company. For example, NCR, in support of its mission of "creating value for our stakeholders," believes it must first satisfy the legitimate expectations of every person with a stake in the company. NCR attempts to satisfy their stakeholders' expectations by promoting partnerships in which everyone is a winner. The company describes this commitment to its mission in the following way:

> - We believe in building mutually beneficial and enduring relationships with all of our stakeholders, based on conducting business activities with integrity and respect.
> - We take customer satisfaction personally; we are committed to providing superior value in our products and services on a continuing basis. We respect the individuality of each employee and foster an environment in which employees' creativity and productivity are encouraged, recognized, valued and rewarded. We think of our suppliers as partners who share our goal of achieving the highest quality standards and the most consistent level of service. We are committed to being caring and supportive corporate citizens within the worldwide communities in which we operate.
> - We are dedicated [to] creating value for our stakeholders and financial communities by performing in a manner that will enhance the return on their investments.[9]

6.4 PROJECT STAKEHOLDERS

Each project has its own unique set of stakeholders. For example, on the O'Hare Development Program (ODP), a $1.6 billion, 10-year expansion program of Chicago's

[8]Bud Baker, "The Power of Politics: The Fourth Dimension of Managing the Large Public Project," *Proceedings,* Project Management Institute, 25th Annual Seminar/Symposium, Vancouver, Canada, October 17–19, 1994, pp. 830–833.

[9]Courtesy NCR Corporation.

O'Hare International Airport, many different stakeholders were involved. Their involvement is described as follows:

> The City of Chicago is involved on a daily basis at levels from the Mayor's Office to purchasing. Many City departments and other City consultants provide guidance and significant contributions to the ODP. Additional government agencies involved include the Federal Aviation Administration, the Illinois Department of Transportation and the Illinois State Toll Highway Authority. Specialized Architect/Engineer design firms and contractors are selected by the City to execute each project within the Program. Each must be supplied with information, formatted to suit their particular needs and level of participation.[10]

A classic case of stakeholder involvement is found in the Milwaukee Water Pollution Abatement Program (WPAP). In this project, not unlike many others, stakeholders had a major impact on the success of the project. Groundwork was laid for stakeholder involvement in this program through the policy of the Environmental Protection Agency (EPA), which recognizes the need for the citizenry to be involved in the planning of major public works projects and requires a public involvement program on EPA grant–supported projects.

It was necessary to keep the public informed every step of the way on this huge and complex $2.2 billion project to renovate and upgrade the sewage system of Milwaukee and its suburbs. Legislative and judicial actions set the direction of the Water Pollution Abatement Program at Milwaukee. A tight timetable and the involvement of 27 separate municipalities, compounded by the need to undergo massive renovation of an existing sewerage system without disrupting service, added to the complexity of the project. CH2M Hill, an international firm of engineers, planners, economists, and scientists that had been in business for 43 years, and its consortium of principal associate consultants were selected to manage the Milwaukee WPAP.[11]

Successful project management can be carried out only when the responsible managers take into account the potential influence of the project's stakeholders. An important part of the project planning is the identification of all project stakeholders and their relevant stakes in the project. Stakeholder analysis during the planning of the project is particularly useful for the development of strategies to facilitate the "management" of the stakeholders during the life cycle of the project.

Two public works highway projects in Illinois were subjected to effective stakeholder management. The management of stakeholders on these projects involved five essential elements: (1) identification of stakeholders, (2) tiered approach to involvement, (3) active investigation to identify issues of concern or conflict, (4) resolution of concerns and conflicts to an acceptable solution, and (5) formal approval.

---

[10]Paul B. Demkovich, "Goal Achievement through Program Control Systems on the O'Hare Development Program," *Proceedings,* PMI Seminar/Symposium, October 1987, p. 303.

[11]See Henry F. Padgham, "The Milwaukee Water Pollution Abatement Program: Its Stakeholder Management," *PM Network,* April 1991, pp. 6–18.

Some of the stakeholders were obvious: city and county councils, chambers of commerce, and agencies that had regulatory oversight of the highway projects. Others, such as environmental interest groups, neighborhood and historic associations, and business organizations, were less noticeable. Efforts to identify the less visible stakeholders included drives through the area, visits to adjacent businesses, institutions, and residences, and consultation with local representatives. Leading questions that were asked, such as Who cares about this project? and What groups represent the interests of these people? were helpful.

The project team put special effort into forming interest groups that cut across geographic and special interest communities. Meetings were conducted with stakeholders, such as formal briefings for Illinois Department of Transportation officials and Federal Highway Association representatives. Working sessions were held for technical groups composed of representatives of local government, institutions, industries, and agencies. All told, about 65 meetings were held with stakeholder interest groups, for example, informal meetings with interested groups to explain the project, its rationale, its cost, its schedule, and what it would do for the community.

Newsletters and project updates were disseminated on a regular basis. Having the project team take the lead in working with the stakeholder groups and discuss the project increased the comfort levels of the stakeholders. The stakeholders grew to recognize the names and faces of the project team, helping to increase the stakeholders' confidence level. Other strategies that were used to manage the stakeholders and keep them informed included:

- Seeking out and addressing contentious issues promptly to avoid getting blindsided at the formal hearings.
- A series of local drop-in centers to present basic project information and solicit comments.
- Great care in defining the scope, probable cost, and schedule for the stakeholders.
- A constant message: Are there any stakeholder concerns of which the project team is unaware?
- An ongoing willingness to meet with any stakeholders to assure them that their concerns were being considered by the project team.

The result of this proactive stakeholder management strategy: The project team was not taken by surprise by any issue or concerns, nor were the stakeholders surprised about any aspect of the project. The projects were carried out with almost complete acceptance by the affected stakeholders and the general public. Clearly the proactive management of the stakeholders contributed significantly to the project's value.[12]

Project stakeholders include not only the obvious members of the project team but also those principals in the political, economic, social, legal, and technological environments in which the project exists. In some cases the stakeholders will be highly organized and motivated; for example, some environmental groups have been influencing the construction of nuclear power generation plants.

---

[12]Larry Martin and Paula Green, "Gaining Project Acceptance," *Civil Engineering,* August 1995, pp. 51–53.

Because project stakeholder management assumes that success depends on taking into account the potential impact of project decisions on all stakeholders during the entire life of the project, the project team faces a major challenge. In addition to identifying and assessing the impact of project decisions on stakeholders who are subject to the authority of the management, the team must consider how achievement of the project's goals and objectives will affect, or be affected by, stakeholders outside their authority.

The former head of the Bonneville Power Administration in Portland, Oregon, describes the challenges and anxieties involved in making a commitment to public involvement over some company projects and the awesome challenge in making that commitment work. Peter Johnson has become a convert, stating that "public involvement is a tool that today's managers...must understand."[13]

Youker found in his experiences with the World Bank that in reviewing the status of the implementation of its entire portfolio of projects, many of the most important problems of implementation lie in the general environment of the project and are beyond the direct control of the project manager.[14]

Project stakeholders, often called intervenors in the nuclear power plant construction industry, can have a marked influence on a project. At one nuclear power plant, numerous bomb threats over the life of the project lengthened construction schedules, shut down work on select areas, frustrated managers and professionals, and forced more intensive security provisions, including physical searches of people, equipment, and vehicles. Antinuclear blockades and demonstrations impacted productivity. In the fall of 1981, the Abalone Alliance, an antinuclear organization, attempted to blockade the plant. The plant had to pay for housing and feeding the plant operating crew, management staff, National Guard troops, and law enforcement officers. Costs associated with such intervenor action, such as work shutdown and absenteeism because of the physical threats, could not be calculated.

6.5 SOME EXAMPLES OF STAKEHOLDER INFLUENCE

Some recent project management experiences highlight the role of these stakeholders:

- In the investigation of management prudence on the Long Island Lighting Company (LILCO) Shoreham project, Suffolk County, the New York State Consumer Protection Board and the Long Island Citizens in Action (intervenors) argued that the project suffered from pervasive mismanagement throughout its history. The record, in the view of these intervenors, established that approximately

[13]Peter T. Johnson, "How I Turned a Critical Public into Useful Consultants," *Harvard Business Review*, January–February 1993, pp. 56–66.

[14]Robert Youker, "Managing the International Project Management Environment," Management Planning and Control Systems, 5825 Rockmere Drive, Bethesda, MD 20816-2443.

$1.9 billion of Shoreham's cost was expended unnecessarily "as a result of LILCO's mismanagement, imprudence or gross inefficiency."[15]

- One reason that the Supersonic Transport program failed in the United States was that the managers had a narrow view of the essential players and generally dismissed the key and novel role of the environmentalists until it was too late.[16]
- State public utility commissions (PUCs) are key and formidable stakeholders in the design, engineering, construction, and operation of nuclear power generating plants. In past years, state PUCs have prevented the recovery of billions of dollars in generating plant construction costs. Some utilities have been penalized for imprudent spending on nuclear plants; others have been told that their plants were not needed. For example, the Pennsylvania State Public Utility Commission ruled that the Pennsylvania Power and Light Company's newly opened 945-MW $2 billion Susquehanna Unit 2 nuclear plant would provide too much generating capacity for the utility's customers. The utility was allowed to recover only taxes, depreciation, and other operating costs. The Missouri Public Service Commission recently disqualified Union Electric Company from charging ratepayers for $384 million of the $3 billion spent on the new Callaway nuclear plant in central Missouri. The commission cited high labor expenses, improper scheduling of engineering, and "inefficient, imprudent, unreasonable, or unexplained costs" during 4 years of delay.[17]
- In a 1-million-square-foot addition to the New York Hospital, environmental, political, and social challenges existed. The New York City Department of Environmental Protection even required a wildlife preservation plan. Over 45 agencies—"stakeholders"—had to be satisfied even though no public money was involved in the project. Public money was about the only issue missing—air rights, the highway, the river, near-zero work space, historic preservation, sheer size, and other issues had to be considered. The problem was less the outcome of the product—a new hospital addition—and more the process of the conceptualization and management of the project mindful of the key issues involved.[18]
- Diverse stakeholders, or intervenors, are taking active roles in rate-setting case hearings. For example, when the Union Electric Company of St. Louis, Missouri, instituted proceedings for authority to file tariffs increasing rates for electric service, the following parties were granted permission to intervene in the proceedings: 25 cities, the state of Missouri, the Jefferson City school district, the Electric Ratepayers Protection Project, the Missouri Coalition for the Environment, the Missouri Public Interest Research Group, Laclede Gas Company, Missouri Limestone Producers, Dundee Cement Company, LP Gas Association,

[15]Recommended decision by administrative law judges William C. Levey and Thomas R. Matias, Long Island Lighting Company–Shoreham Prudence Investigation, case no. 27563, State of New York Public Service Commission, March 13, 1985, p. 57.

[16]Mel Horwitch, "The Convergence Factor for Successful Large-Scale Programs: The American Synfuels Experience as a Case in Point," in D. I. Cleland (ed.), *Matrix Management Systems Handbook* (New York: Van Nostrand Reinhold, 1984).

[17]William Glasgall, "The Utilities' Pleas Falling on Deaf Ears," *Business Week,* June 17, 1985, p. 113.

[18]Nadine M. Post, "And a Highway Runs through It," *ENR,* August 7, 1995, pp. 24–28.

Missouri Retailers Association, the Metropolitan St. Louis Sewer District, and the industrial intervenors—American Can Company, Anheuser Busch, Inc., Chrysler Corporation, Ford Motor Company, General Motors Corporation, Mallinckrout, Inc., McDonnell Douglas Corporation, Monsanto Company, National Can Corporation, Nooter Corporation, PPG Industries, Inc., Pea Ridge Iron Ore Company, River Cement Company, and St. Joe Minerals Corporation (Monsanto et al.).[19]

- The Nuclear Regulatory Commission is a proactive stakeholder in the management of nuclear power plant projects. Its principal interest is the licensing of nuclear plants to ensure quality assurance, safeguards, inspection, and proper operation. Its influence in the industry is substantial. In addition to licensing individual plants, the NRC conducts studies in the design, engineering, and licensing of plants. In 1984 it published a landmark study of existing and alternative programs for improving quality and the assurance of quality in the design and construction of commercial nuclear power plants.[20]
- Competitors are key stakeholders, particularly during the competitive phase before the architect and engineer, project manager, or constructor firm is selected during a source selection process. During this competitive phase, an in-depth analysis of competitors is essential to winning a contract. The business literature contains descriptions on how to access the competition.[21] A potential winning contract can become a loser if the competition is ignored.
- Stakeholder management includes very favorable situations when companies are creating wealth for their stockholders at a phenomenal rate. *The Wall Street Journal* reports that Royal Dutch/Shell Group has a huge problem while generating profits of approximately $1.5 million per hour and sitting on more than $11 billion in the bank. This creates a predicament with its stockholders—primary stakeholders—because they are looking for growth as well as current profits. Stockholders want the money put to use for increased benefits over the long haul. Shell has used some of the money to reduce debt and some to buy back stock. Efforts have even been made to acquire small, less successful energy companies. The problem still remains that there is too much cash that is not working for the stockholders.

  Shell, like other energy companies, would like to develop more oil fields, but is constrained by U.S. economic sanctions against countries with huge oil reserves. Such countries as Libya, Iraq, and Iran are included in these biggest oil producers. Areas such as the North Sea and the United States are experiencing a decline in oil production. Other countries experiencing political turmoil, such as Indonesia and Nigeria, present challenges to operating and managing oil production. There are other countries that do not meet the expectations of energy companies and these opportunities are currently being deferred.

---

[19]Case Nos. ER-85-160 and EO-85-17, State of Missouri Public Service Commission, Jefferson City, March 29, 1985.

[20]W. Altman, T. Ankrum, and W. Brach, "Improving Quality and the Assurance of Quality in the Design and Construction of Nuclear Power Plants," NUREG-1055, U.S. Nuclear Regulatory Commission, Washington, D.C., May 1984.

[21]Richard Eells and Peter Nehemkis, *Corporate Intelligence and Espionage* (New York: Macmillan, 1984).

Energy companies are holding nearly $40 billion in cash reserves that cannot be spent to extend the oil production—the area where this reserve of money was generated. Efforts to expand into new fields of work unrelated to energy production have caused stockholder reactions. Companies are being "forced" to stay within their core business areas by stakeholders. To appease stakeholders, including shareholders, Shell points out that the oil business is extremely cyclical, the top of the cycle has lots of expenditures, and opportunities would not be passed up. The stock buyback is a means of preventing too much cash from accumulating, but is not a means of generating more in the oil-producing business.

Throughout this report in *The Wall Street Journal,* there are examples of stakeholders for oil exploration projects. Stockholders want to ensure that their money grows through investments rather than being held in a bank. Different politics for each of the oil-producing countries brings about challenges to management in view of government instability or uncertainty. Executives of the oil companies must also be considered stakeholders, although not specifically addressed as such. Missing and perhaps the largest group of stakeholders is the consumer, both in the United States and overseas.[22]

6.6 SOME EXAMPLES OF SUCCESSFUL STAKEHOLDER MANAGEMENT

There are some excellent examples of successful stakeholder management:

- Care was taken during the design and construction of the Hackensack Meadowlands sports complex to develop cooperation among the groups concerned with environmental impact, transportation, development, and construction.
- On the James Bay project special effort was made to stay sensitive to social, economic, and ecological pressures.[23]
- James Webb and his colleagues at NASA were adept at stakeholder management during the Apollo program. NASA gained the support not only of the aerospace industry and related constituencies, but also of the educational community, the basic sciences, and the weather forecaster profession.[24]
- The 12.5-mile, $490 million highway project through Glenwood Canyon in Colorado is one of the most expensive nonurban sections of the interstate system. This project has no operations component. As such, there is no added level of liability/risk to markets or investors upon its completion. It required more than a decade of planning and 12 years to construct. The project involved an unprecedented degree of cooperation among the project team, environmentalists, and tourists to create a major highway that preserved and even enhanced one of the

[22]Paraphrased from Christopher Cooper and Thaddeus Herrick, "Pumping Money," *The Wall Street Journal,* July 30, 2001, pp. A1, A8.

[23]Peter G. Behr, "James Bay Design and Construction Management," ASCE Engineering Issues, *Journal of Professional Activities,* April 1978.

[24]E. Ginsburg, J. W. Kuhn, and J. Schnee, *Economic Impact of Large Public Programs: The Nash Experience* (Salt Lake City, Utah: Olympus Publishing, 1976).

nation's premier natural settings. The construction of the highway through a scenic gorge overcame fierce initial opposition, a wide variety of design changes and physical constraints, plus remarkable cooperation in creating a four-lane highway that even the environmentalists love.[25]

- Bechtel planned, designed, engineered, and managed the procurement, right-of-way acquisition, and construction of a second gas pipeline extending 875 miles from Canada into central California. This included the construction of a new compressor station and the retrofit of 17 compressor stations and three major meter stations at a cost of approximately $1.6 billion. This new pipeline parallels the first completed in the early 1960s. Throughout the pipeline expansion, concern about the wide range of environmental factors was paramount. Careful planning by Bechtel resulted in the development of extensive safeguarding of environmental factors on the pipeline. Certain measures dealt with the control of erosion, noxious weeds, hazardous material, and construction noise, as well as extensive training for all personnel on environmental awareness of work practices.[26]

6.7 PROJECT STAKEHOLDER MANAGEMENT (PSM) PROCESS

The principal justification for adopting a PSM perspective springs from the enormous influence that key external stakeholders can exert. Arguably, the extent to which the project achieves its goals and objectives is influenced by the strategies pursued by key stakeholders. Stakeholder management leading to stakeholder cooperation enhances project objective achievement, while stakeholder neglect hinders it.

In working with project managers to develop a project strategy, which encompasses a PSM philosophy, the following basic premises can serve as guides for the development of a PSM process:

- PSM is essential for ensuring success in managing projects.
- A formal approach is required for performing a PSM process. Multiyear projects are subject to so much change that informal means of PSM are inadequate. Reliance on informal or hit-or-miss methods for obtaining PSM information is ineffective for managing the issues that can come out of projects.
- PSM should provide the project team with adequate intelligence for the selection of realistic options in the management of project stakeholders.
- Information on project stakeholders can be gained from a variety of sources, some of which might superficially seem to be unprofitable.

PSM is designed to encourage the use of proactive project management for curtailing stakeholder activities that might adversely affect the project and for facilitating the project team's ability to take advantage of opportunities to encourage stakeholder

[25]John Pendergast, "Pioneer Highway," *Civil Engineering,* July 1993, pp. 36–39.
[26]Gary Walker and John Myrick, "Doubling a Pipeline," *Civil Engineering,* January 1994, pp. 50–52.

support of project purposes. These objectives can be achieved only by integrating stakeholder perspectives into the project's formulation processes and developing a PSM strategy. The project manager is then in a better position to influence the actions of the stakeholders on project outcome. Some objectives for PSM might be as follows:

- Ensure the availability of timely, credible, and comprehensive information of the capabilities and the options open to each stakeholder.
- Continue to identify the probable strategies of the stakeholders.
- Determine how key stakeholders' strategies might affect current project interests.
- Continuously monitor and provide comprehensive information about probable actions in the project stakeholder environment that might have an impact on the interests of the project.
- Organize the collection, analysis, and dissemination of stakeholder information for the project team.

Failure to recognize or cooperate with adverse stakeholders may well hinder a successful project outcome. Indeed, strong and vociferous adverse stakeholders can force their particular interest on the project manager at any time, perhaps at the time least convenient to the project. PSM is thus a necessity that allows the project manager to set the timetable to maintain better control. A proactive PSM process is designed to help the project team develop the best possible strategies.

6.8 PLANNING STAKEHOLDER MANAGEMENT

Developing a strategy to manage the stakeholders starts with putting forth a few key questions:

- Who are the project stakeholders—both primary and secondary?
- What stake, right, or claim do they have in the project?
- What opportunities and challenges do the stakeholders pose for the project team?
- What obligations or responsibilities does the project team have toward its stakeholders?
- What are the strengths, weaknesses, and probable strategies that the stakeholders might employ to realize their objectives?
- What resources are at the stakeholders' disposal to implement their strategies?
- Do any of these factors give the stakeholders a distinctly favorable position in influencing the project outcome?
- What strategies should the project team develop and implement to deal with the opportunities and challenges presented by the stakeholders?
- How will the project team know if it is successfully "managing" the project stakeholders?

6.9 A MODEL OF THE PSM PROCESS

The PSM process consists of executing the management functions of planning, organizing, motivating, directing, and controlling the resources used to cope with external stakeholders' strategies. These functions are interlocked and repetitive; the emergence of new stakeholders might require the reinitiation of these functions at any time during the life cycle of the project. This management process is continuous, adaptable to new stakeholder threats and promises and to changing strategies of existing stakeholders. Putting the notion of stakeholder management on a project life-cycle basis emphasizes the need to be aware of stakeholder influence at all times.

The management process for the stakeholders consists of the phases depicted in Fig. 6.1 and discussed below.

6.10 IDENTIFICATION OF STAKEHOLDERS

The identification of stakeholders must go beyond the internal stakeholders. Internal stakeholders must, of course, be taken into account in the development of project strategies. Their influence is usually supportive of project strategies

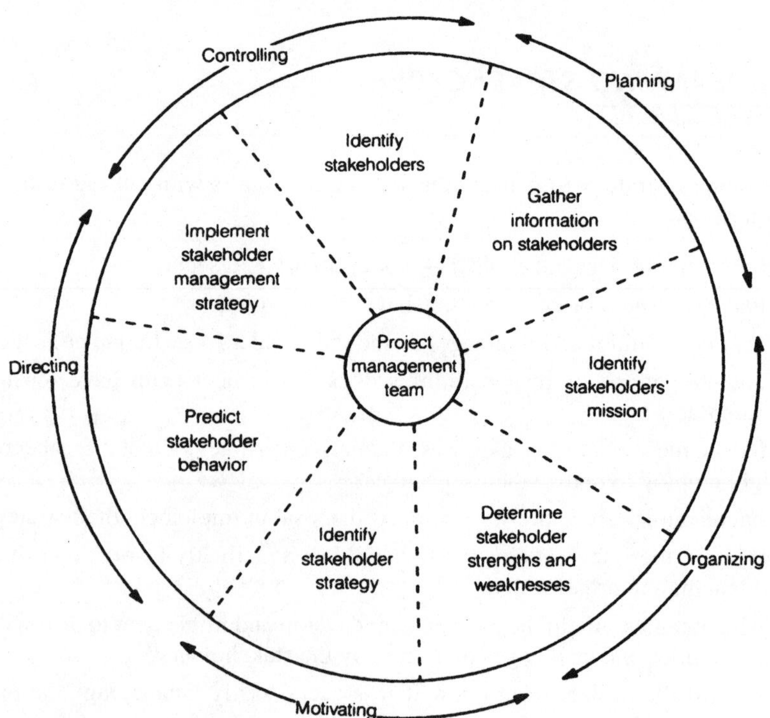

FIGURE 6.1 Project stakeholder management process.

because internal stakeholders are an integral part of the project team. A prudent project manager would ensure that these internal stakeholders play an important and supportive role in the design and development of project strategies. Such a supportive role is usually forthcoming because the project manager has some degree of authority and influence over these individuals.

Stakeholders are persons or groups that have, or claim, ownership, rights, or interests in a project and its activities: past, present, or future. Primary stakeholders are those persons and groups that have a legal contractual relationship to the project. Such stakeholders include the project owner, suppliers, functional groups, investors, and those from the public domain such as communities and institutions that provide infrastructures and markets, whose laws and regulations must be obeyed and to whom taxes and other obligations are owed. Secondary stakeholders are defined as those who influence or affect, or are influenced or affected by, the project but are not regularly engaged in transactions with the project and may not be essential for the project's survival.[27] The media and special interest groups are secondary stakeholders under this definition. These stakeholders have the capacity to mobilize public opinion in favor of or in opposition to the project's purposes and performance.

The management of a project inevitably entails bringing into the picture those persons and groups that have both contractual interests and vested interests in the management of the project as well as its outcome. These persons and groups come from a wide variety of organizational settings to include:

- Senior organizational managers including corporate directors, general managers, functional managers, project managers, work package managers, and project team members
- Customers (users), suppliers, contractors, subcontractors, and vendors
- Local, state, and federal agencies and commissions and judicial, legislative, and executive organizations
- Employees, private citizens, tourists, and families of employees
- Creditors and shareholders
- Social organizations, political organizations, environmentalists, "intervenor" groups such as the Sierra Club, and consumer groups
- Competitors
- Local communities and the general public
- Professional organizations, trade associations, and unions
- Institutions such as schools, universities, hospitals, churches, chambers of commerce, civic groups, minority groups, activists, and the American Civil Liberties Union
- News media

[27]The idea of primary and secondary stakeholders was first expressed in David I. Cleland, "Project Stakeholder Management," *Project Management Journal,* September 1986, pp. 36–43. The concept of project stakeholders was further extended in David I. Cleland, "Project Stakeholder Community—A Revisit," in Jeffrey Pinto (ed.), *Project Management Institute's Project Management Handbook* (San Francisco: Jossey-Bass, 1998).

A model of the project stakeholders is shown in Fig. 6.2. In a sense the secondary stakeholder "organization" is a virtual organization—one that exists in essence or effect though not in actual fact, form, or name. The stakeholder virtual organization is an underlying entity that lurks under the surface—a sort of potential organization that exists between the lines and structure of the formal organizational entity. Although not existing in actual fact, form, or name, the secondary stakeholders that comprise the virtual organization can exert a powerful influence over the project's planning and outcome.

6.11 PRIMARY STAKEHOLDERS

Primary stakeholders are those persons or groups on the project team who have a contractual or legal obligation to the project team and have the responsibility and authority to manage and commit resources according to schedule, cost, and

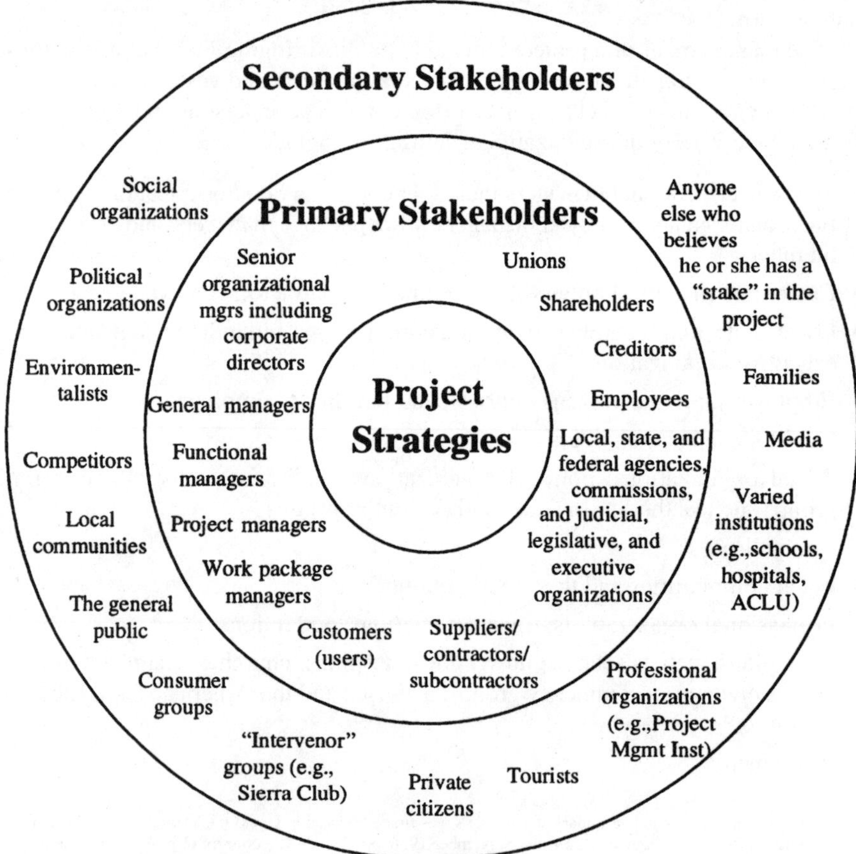

FIGURE 6.2 The project stakeholders.

technical performance objectives. Such stakeholders have direct strategic and operational roles through participating in the design, engineering, development, and construction (production), and after-sales logistic support of the project outcomes. In other words, the primary stakeholders belong to the project team and its supporting organizational infrastructure: functional managers, general managers, senior managers, customer and supplier officials, and so forth. These primary stakeholders have the residual authority and responsibility to use resources to support the project objectives. The key authority and responsibilities of these stakeholders include:

- Providing leadership to the project team.
- Allocating resources to be used in the design, development, and construction (production) of the project results.
- Building and maintaining relationships with all stakeholders.
- Managing the decision context in the design and execution of strategies to commit project resources.
- Leading by example to set the cultural ambience of the project, which brings out the best of people in providing high-quality professional resources to the benefit of the project.
- Maintaining ongoing and effective oversight of the project's progress in meeting its schedule, cost, and technical performance objectives, and where necessary, instituting reprogramming and reallocation of resources as required to keep the project on track.
- Periodically assessing the efficiency and effectiveness of the project team in doing the job for which it has authority and responsibility.

6.12 SECONDARY STAKEHOLDERS

Secondary stakeholders are those who have no formal contractual relationship to the project but can have a strong interest in what is going on regarding the project. These stakeholders belong to an informal project stakeholder organization. They include social organizations, competitors, local communities, the general public, consumer groups, private citizens, intervenor groups, professional organizations, the media, families, and varied institutions such as schools, universities, hospitals, churches, civic groups, and so forth (see Fig. 6.2).

The management of secondary stakeholders poses significant challenges for the project manager and other enterprise managers as well, because these managers have no legal authority or contractual relationship with those stakeholders. These secondary stakeholders can exert extraordinary influence over the project, supporting or working against the project and its outcome. "Management" of these secondary stakeholders can be particularly arduous, because no formal relationship exists with them. Consequently, the authority that the managers use is limited to their de facto authority—interpersonal capabilities, knowledge, persuasive powers, political savvy, expertise—in general, their ability to work with and influence these secondary stakeholders. Some of the more important characteristics of these secondary stakeholders include:

- There are no limits to where they can go and with whom they can talk to influence the project.
- Their interests may be real—or are perceived to be real—because the project and its results may impinge in their "territory."
- Their "membership" on the project team is ad hoc—they stay as long as it makes sense to them in gaining some advantage or objectives involving the project.
- They may team with other stakeholders on a permanent or ad hoc basis in pursuing common interests for or against the project's purposes.
- The power they exercise over the project can take many forms, such as political influence, legal actions (such as court injunctions), emotional appeal, media support, social pressure, local community resistance, use of expert witnesses, or even scare tactics.
- They have a choice of whether or not to accept responsibility for their strategies and actions.

In managing the project stakeholders, the legal authority that has been delegated to the project manager is helpful but not enough to motivate people to give that extra commitment to carry out successfully the work required in dealing with the project stakeholders. Through de facto authority—that which comes to the project manager and the team members through competency—reciprocal confidence and commitment can be built with stakeholders to enhance the chances of gaining the loyalty and commitment of the stakeholders. For the project manager this is an exercise in the development and use of personal power in working successfully with project stakeholders. The development of this personal power requires competence, and also a great deal of energy and capacity to keep all stakeholders of the project moving in the right direction. When the team members feel right about the relationships that exist with their stakeholders, there is a greater likelihood of success.

6.13 GATHERING STAKEHOLDER INFORMATION

Gathering information about the project stakeholders is similar to collecting information on competitors.[28] To systematize the development of the stakeholder information means that questions such as the following need to be considered:

- What needs to be known about the stakeholder?
- Where and how can the information be obtained?
- Who will have responsibility for the gathering, analysis, and interpretation of the information?

[28] The techniques described here are paraphrased in part from W. R. King and D. I. Cleland, *Strategic Planning and Policy* (New York: Van Nostrand Reinhold, 1986), chap. 11, pp. 246–270.

- How and to whom will the information be distributed?
- Who will use the information to make decisions?
- How can the information be protected from "leaking" or misuse?

Some of the information collected on the project's external stakeholders may include sensitive material. One cannot conclude that all such stakeholders will operate in an ethical fashion. Consequently, all information collected should be assumed to be sensitive until proved otherwise and protected accordingly. This suggests the need for a security system patterned after a company's business intelligence system. Some information should be available only on a need-to-know basis, whereas some should be available to all interested parties.

The following precautions should be considered in planning for a PSM information system:

- One individual responsible for security
- Internal checks and balances
- Document classification and control such as periodic inventory, constant record of whereabouts, and prompt return
- Locked files and desks
- Supervised shredding or burning of documents no longer useful
- Strict security of offices containing sensitive information

Information on the stakeholders is available from a wide variety of sources.[29] In obtaining such information, the highest standards of ethical conduct should be followed. The potential sources of stakeholder information and the uses to which such information can be put are so numerous that it would not be practical to list all sources and uses here. The following sources are representative and can be augmented according to a particular project's needs:

- Project team members
- Key managers
- Business periodicals, such as *The Wall Street Journal, Fortune, Business Week,* and *Forbes*
- Business reference services—*Moody's Industrial Manual, Value Line Investment Security,* and others
- Professional associations and trade associations
- Customers and users
- Suppliers
- Local press
- Trade press
- Annual corporate reports

[29]Ibid.

- Articles and papers presented at professional meetings
- Public meetings
- Government sources[30]
- Internet

Once the information has been collected, it must be analyzed and interpreted by the substantive experts. The project manager should draw on the company's professional personnel for help in doing this analysis. Once the analysis has been completed, the specific target of the stakeholder's mission can be determined.

6.14 IDENTIFICATION OF STAKEHOLDER MISSION

Once the stakeholders have been identified and information gathered about them, analyze the information to determine the nature of their mission or stake. This stake may be a key building block in the stakeholder's strategy. For example, the Nuclear Regulatory Commission manages the licensing of nuclear power plants to promote the safe and peaceful commercial use of the atom. A useful technique to better understand the nature of the external stakeholders' claim in the project is to categorize their stake as supportive or as adverse to the project. It is in the best interest of the project manager to keep the supportive stakeholders well informed of the project's status. Deal carefully with the potentially adverse stakeholders. Information on these stakeholders should be handled on a need-to-know basis because if such information is available to adversarial stakeholders on the project, it can be used against the project. However, communication channels with these stakeholders should be kept open, because this is critical to getting the project point of view across. Adversarial stakeholders will find ways to get information on the project from other sources, which can be erroneous or incomplete, giving the opportunity for misunderstanding and further adversarial behavior.

6.15 DETERMINING STAKEHOLDER STRENGTHS AND WEAKNESSES

Once the stakeholders' mission is understood, then the stakeholders' strengths and weaknesses should be evaluated. An assessment of stakeholders' strengths and weaknesses is a prerequisite to understanding the success of their strategies. Such analysis is found in nearly all prescriptions for a strategic planning process.[31] This process consists of the development of a summary of the most important strengths on which the stakeholders base their strategy and the most significant weaknesses they will avoid in pursuing their interests on the project. Identifying five or six

[30]Ibid.
[31]W. E. Rothschild, *Putting It All Together.*

strengths and weaknesses of a stakeholder should provide a sufficient database on which to make a judgment about the efficacy of a stakeholder's strategy.

An adversary stakeholder's strength may be based on such factors as:

- The availability and effective use of resources
- Political alliances
- Public support
- Quality of strategies
- Dedication of members

Accordingly, an adversary stakeholder's weaknesses may emanate from:

- Lack of political support
- Disorganization
- Lack of coherent strategy
- Uncommitted, scattered membership
- Unproductive use of resources

Once these factors have been developed, each proposed strategy for coping with the stakeholders can be tested by answering the following questions:

- Does this strategy adequately cope with a strength of the stakeholder?
- Does this strategy take advantage of an adversary stakeholder's weakness?
- What is the relative contribution of a particular stakeholder's strength in countering the project strategy?
- Does the adversary stakeholder's weakness detract from the successful implementation of his or her strategy? If so, can the project manager develop a counterstrategy that will benefit the project?

6.16 IDENTIFICATION OF STAKEHOLDER STRATEGY

For a proposed strategy to be successful, it should be built on a philosophy that recognizes the value of going through a specific strength-weakness analysis to develop project strategy. This can be done, however, only if there is a full understanding of the stakeholder's strategy.

A stakeholder strategy is a series of prescriptions that provide the means and set the general direction for accomplishing stakeholder goals, objectives, and mission. These prescriptions stipulate what resource allocations are required; why, when, and where they will be required; and how they will be used. These resource allocations include plans for using resources, policies and procedures to be employed, and tactics used to accomplish the stakeholder's purposes.

6.17 PREDICTION OF STAKEHOLDER BEHAVIOR

On the basis of an understanding of external stakeholder strategy, the project team can proceed to predict stakeholder behavior in implementing strategy. How will the stakeholder use resources to affect the project? Will an intervenor stakeholder picket the construction site or attempt to use the courts to delay or stop the project? Will a petition be circulated to stop further construction? Will an attempt be made to influence future legislation? These are the kinds of questions, when properly asked and answered, that provide a basis for the project team to develop specific countervailing strategies to deal with adversary stakeholder influence.

In some cases, a stakeholder will provide help to another stakeholder. For example, a group of dedicated nuclear advocates formed an industry association to ensure the nuclear operating safety that the Nuclear Regulatory Commission could not provide. This association, the Institute of Nuclear Power Operations (INPO), is dedicated to improving the safety of nuclear plants. INPO sets safety standards and goals, evaluates plant safety, and provides troubleshooting assistance to its sponsors. INPO oversees the training of plant operators and supervisors. In its role as a stakeholder of nuclear power, INPO works closely with the Nuclear Regulatory Commission. If INPO finds areas for improvement in a utility's operation, it is the utility that alerts the Nuclear Regulatory Commission.[32]

To better predict stakeholder behavior, the project team should take the lead in analyzing the probable impact of the stakeholder on a project. A step-by-step approach for analyzing such impact on a project would consist of the following, depicted in Fig. 6.3 and described below.

First, identify and define each potential strategic issue in sufficient detail to ascertain its relevance for the project. Next, determine the several key factors that underlie each issue and the forces that have caused that issue to emerge. These forces usually can be categorized into political, social, economic, technological, competitive, or legal forces.

Then, identify the key stakeholders that have, or might feel that they have, a vested interest in the project. Remember that several different stakeholders may share a vested interest in one strategic issue. Stakeholders usually perceive a vested interest in a strategic issue because of:

- *Mission relevancy.* The issue is directly related to the mission of the group. For example, members of the Sierra Club see the potential adverse effect of a nuclear power plant project on the environment.
- *Economic interest.* The stakeholders have an economic interest in the strategic issue. A union would be vitally interested in the wage rates paid at a project construction site.
- *Legal right.* A stakeholder has a legal right in the issue, such as the Nuclear Regulatory Commission, which has the power to grant operating licenses for nuclear generating plants.

[32]For more on the role of INPO see James Cook, "INPO's Race against Time," *Forbes*, February 24, 1986, pp. 54–55.

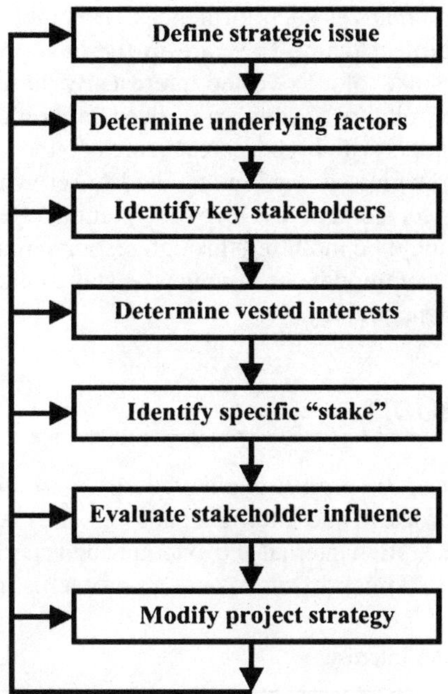

FIGURE 6.3 Stakeholder impact evaluation process.

- *Political support.* Stakeholders see the issue as one in which they believe there is a need to maintain a political constituency. A state legislator would be concerned about the transportation of toxic wastes from a power plant to a repository site within the state or the transportation of wastes across the state.
- *Health and safety.* The issue is related to the personal health and safety of the group. Project construction site workers are vitally interested (or should be) in the working conditions at the site.
- *Lifestyle.* The issue is related to the lifestyle or values enjoyed by the group. Sports groups are interested in the potential pollution of industrial waste in the forests and waterways.
- *Opportunism.* The issue is one that the group can rally others around, with the goal of increasing the group's political power at the expense of the project.
- *Survival.* The issue is linked to the reason for existence of a group of stakeholders. For example, members of the investment community see clearly the financial risks of nuclear plant construction today, considering the uncertainty in the licensing of a nuclear power plant.[33]

Once the stakeholders have been identified, clarify the specific stake held by each. Then judge how much influence each stakeholder might have on the project and its

[33]Paraphrased from Edith Weiner and Arnold Brown, "Stakeholder Analysis for Effective Issues Management," *Planning Review,* May 1986, pp. 27–31.

outcome. Table 6.2 summarizes such influences. This table should be completed by members of the project team. They are in the best position to identify the probable impact of a stakeholder's vested interest. By perusing the table, a manager can get a summary picture of which stakeholders should be "managed" by the project team. Stakeholders with high interest scores on the table should be studied carefully and their strategies and actions tracked to see what effect such actions might have on the project's outcome. Once the potential effect is determined, then the project strategy should be modified through resource reallocation, replanning, or programming to accommodate or counter the stakeholder's actions through a stakeholder management strategy.

6.18 PROJECT AUDIT

An independent audit of the project conducted on a periodic basis will also help the project team get informed and intelligent answers they need on strategic issues and stakeholder interests. Both internal and external audits performed by third parties to analyze the project's strengths, weaknesses, problems, and opportunities can

TABLE 6.2 Stakeholder Interests

	Stakeholders											
Stakeholder interest	1	2	3	4	5	6	7	8	9	10	11	12
Mission relevance												
Economic interest												
Legal right												
Political support												
Health and safety												
Lifestyle												
Opportunistic competitive survival												
Vested interest: H—high M—medium L—low												

shed light on how well the stakeholders are being managed. There is a symbiotic relationship between the project and its stakeholders. The project cannot exist without its stakeholders; conversely, the stakeholders rely to some extent on the project for their existence.

6.19 IMPLEMENTING STAKEHOLDER MANAGEMENT STRATEGY

The final step depicted in Fig. 6.1 in managing the project stakeholders is to develop implementation strategies for dealing with them. An organizational policy, which stipulates that stakeholders will be actively managed, is an important first step of such implementation strategies. Once this important step has been taken, additional policies, action plans, procedures, and allocation of supporting resources can be made to make stakeholder management an ongoing activity. Once implementation strategies are operational, the project team has to:

- Ensure that the key managers and professionals fully appreciate the potential impact that both supportive and adversarial stakeholders can have on the project outcome.
- Manage the project review meetings so that stakeholder assessment is an integral part of determining the project status.
- Maintain contact with key external stakeholders to improve the chances of determining stakeholders' perception of the project and their probable strategies.
- Ensure an explicit evaluation of probable stakeholder response to major project decisions.
- Provide an ongoing, up-to-date report on stakeholder status to key managers and professionals for use in developing and implementing project strategy.
- Provide a suitable security system to protect sensitive project information that might be used by adversarial stakeholders to the detriment of the project.

Henry F. Padgham, former president and chairman of the Project Management Institute, who has managed many successful large projects, believes, "Project management today demands that we pay attention to all who have a stake in our projects."[34]

6.20 TO SUMMARIZE

The major points expressed in this chapter include:

- Stakeholder management is an important part of the management of an enterprise and of the management of a project.

[34]Henry F. Padgham, *PM Network,* April 1991, p. 18.

- In recent years there has been an increase in laws that give stakeholders legal rights.
- Political stakeholders can have a major impact on a project. Examples were given of such impact.
- Project stakeholders can be managed. An important part of that management is to keep them informed of the status of the project, from its inception through to operational use.
- Stakeholders in the nuclear power industry have been particularly aggressive and are often called "intervenors."
- A project stakeholder management process was suggested, which provides a paradigm for applying management theory and practice to project stakeholders.
- Project stakeholders have been classified into two types: primary and secondary.
- Primary stakeholders are those persons or groups that have a legal contractual relationship to the project.
- Secondary stakeholders are those persons who influence or affect or are influenced or affected by the project, but are not regularly engaged in transactions with the project and may not be essential for the project's survival.
- Secondary stakeholders may be thought of as comprising a "virtual" organization that can exert powerful influence over the project's planning and outcome.
- Secondary stakeholders have a great deal of freedom in influencing the project—or ignoring the project and its outcome.
- Information on project stakeholders is available from a wide variety of sources.
- A project audit should include an assessment of the stakeholders to include their degree of satisfaction or dissatisfaction with the project and its planned and actual results.

6.21 ADDITIONAL SOURCES OF INFORMATION

The following additional sources of project management information may be used to complement this chapter's topic material. This material complements and expands on various concepts, practices, and theory of project management as it relates to areas covered here.

- R. Max Wideman, "How to Promote Projects to Stakeholders," chap. 15 in David I. Cleland (ed.), *Field Guide to Project Management* (New York: Van Nostrand Reinhold, 1997).
- E. Payson Willard, "The Demise of the Superconducting Supercollider: Strong Politics or Weak Management?" in David I. Cleland, Karen M. Bursic, Richard J. Puerzer, and Alberto Y. Vlasak (eds.), *The Project Management Casebook,* Project Management Institute. (First published in *Proceedings,* Project Management Seminar/Symposium, 1994, pp. 1–7.)

- Robert J. Graham and Randall L. Englund (contributor), *Creating an Environment for Successful Projects: The Quest to Manage Project Management* (San Francisco: Jossey-Bass, 1997). This book relates experiences of two consultants, who identify the need for upper management to create and sustain an environment that is supportive of project management. The authors assert that such an environment is essential to an organization's growth.
- Robert K. Wysocki, James P. Lewis, and Doug Decarlo (contributor), *The World Class Project Manager: A Professional Development Guide* (Cambridge, Mass.: Perseus Books, 2001). This book focuses on the project manager, who requires the knowledge, skills, and ability to deal with clients (stakeholders). The authors state the project manager must possess superior skills and competencies to survive in the modern project environment. Included in the book are several exercises and self-assessment tools to assist the reader in developing the proper skills.
- Cait Murphy, "How to Fix the Air Traffic Mess," *Fortune,* June 25, 2001, pp. 116–122. This article discusses the start-up airlines and their chances of success in a world dominated by large, established companies. Moreover, the article looks at the intricate weave of multiple interests in airport operation and change. San Francisco, for example, requires the approval of 31 different agencies just to reconfigure a runway. Building a new runway takes considerably more time and interaction with stakeholders.
- Jim Carlton, "Saga of the Santa Lucia Preserve Nears a Close," *The Wall Street Journal,* February 28, 2001, p. B16. This article shows the extent of planning and coordination required to successfully construct a "green housing project" that meets stakeholders' (environmentalists') requirements. The time and energy consumed gave this project the best chance of success and, when complete, proved that all the work was worth it.
- Anonymous, "Satisfied Customers Equal Business Success," *Contractor,* vol. 48, issue 6, Newton, Mass., June 2000, pp. 7, 38. This article addresses the importance of the customer as a major stakeholder in projects. The results of a study show the emphasis that is being placed on customer and client relationships to ensure satisfaction with the end product.

6.22 DISCUSSION QUESTIONS

1. What is meant by a *project stakeholder?*
2. Describe a project management situation from your work or school experience, and list the project stakeholders.
3. Discuss the importance of keeping all project stakeholders informed on the issues relevant to them with respect to projects.
4. In the nuclear power plant described in the chapter, what could the project managers have done to prevent intervenors from disrupting the construction of the plant?

5. Why is it important for project leaders to develop a project stakeholder management (PSM) process? Discuss stakeholders' potential impact on the attainment of project objectives and goals.
6. List and discuss the objectives of PSM as described in the chapter.
7. List and describe the steps in the PSM process.
8. List some sources of information on project stakeholders.
9. What questions must management address to assess the potential impact of an adversarial stakeholder?
10. What factors indicate a vested interest by a stakeholder in a strategic issue of a project?
11. What additional steps must management take once stakeholders and their potential impacts have been identified?
12. What factors of organizational culture contribute to effective management of stakeholders?

6.23 USER CHECKLIST

1. Does your organization continually seek to identify project stakeholders? In what ways?
2. How does your organization manage the interrelationships among project stakeholders? Do any written policies exist that assist in the management of stakeholders?
3. In what ways does your organization seek to manage intervenors?
4. What stakeholder impacts are typical in your organization?
5. Describe a recent project in your organization that was successful in the management of stakeholders. What led to this success?
6. Describe your organizational philosophy and attitudes toward the PSM process.
7. Are there any formal ways that the project managers in your organization accept responsibility for the PSM process?
8. In what ways do project managers go beyond identification in assessing stakeholder impact?
9. What sources are used or can be used to gather information on the project stakeholders?
10. Do project managers attempt to predict stakeholder behavior? In what ways?
11. Are the project stakeholder issues addressed in project audits? What questions are asked or can be asked to help the project team identify and control strategic issues?
12. What proactive measures are taken to ensure continual management of stakeholders? How can the top managers of your organization support the PSM process?

6.24 PRINCIPLES OF PROJECT MANAGEMENT

1. Stakeholder management is a critical part of achieving successful projects.
2. Stakeholder management must be a concentrated effort that is built around a formal process.
3. Stakeholders can positively or negatively affect the progress of a project, depending upon the management of their interests and concerns.
4. Anticipating stakeholder reaction and planning to preempt or respond to actions can materially add to the project's value.

6.25 PROJECT MANAGEMENT SITUATION—STAKEHOLDER INITIATIVES

Projects of all sizes have stakeholders; some are directly involved in the project's work, whereas others have concerns that may be directly or indirectly related to the product of the project. Those directly involved are typically called "primary stakeholders," and they are concerned with the progress of the project, which includes all aspects of ensuring that the project is successful. Those indirectly involved are typically called "secondary stakeholders," and they are concerned with making some change to the direction that the project is taking.

The project's process may have adverse impacts on such items as quality of life. Construction crews operating heavy equipment to build a road can disrupt traffic flow, create loud noises, create dust clouds that carry to housing areas, disrupt the normal flow of drain water during storms, and mar the topsoil sufficiently to cause erosion. The trucks may be speeding through residential areas and posing a hazard to children playing near streets.

On the other hand, there is the product of the project. Construction of new homes in an area can negatively impact the local infrastructure through overloads to the existing system. New homes require water, electricity, telephones, sewage disposal, and gas. In addition, the new families will place new requirements on schools, stores, roads, libraries, fire departments, and other public facilities.

Stakeholders for the road project would want the road, but perhaps object to the inconveniences and hazards associated with building the road. Stakeholders would in this instance have two purposes: (1) improve their road system for a smoother flow of traffic and (2) cause the construction crew to exercise caution in how the work is performed. In this situation, one stakeholder could have two different views and support both in public hearings.

Stakeholders in the construction of new homes could be individual residents in the area, for example, trying to restrict expansion so their existing homes continue to rise in value. They could also be viewing the additional families as placing a burden on the public facilities that are already overcrowded. Stakeholder opposition to the new homes may be based on facts or the stakeholder's perception—in either case, the opposition can be disruptive to the home construction.

Stakeholder opposition to a project is not necessarily bad. Some opposition may cause change of plans that were not well developed with a full range of facts. Project managers should assess the facts of any opposition first before rejecting a stakeholder's position. The stakeholder may have a different approach that will assist the project in being better.

6.26 STUDENT/READER ASSIGNMENT

1. On the basis of the project management situation, what opposition would you take as a stakeholder in the building of residential homes? What actions would you take to prevent the homes from being constructed when there is insufficient water at this time to serve the existing homes?
2. As a project manager, what would you do to avoid conflicting information from being "leaked" to potentially hostile stakeholders in the community?
3. A nuclear power plant is to be constructed in your location. It has been approved as a safe, environmentally friendly design, but there are rumors that it would possibly vent radiation into the atmosphere. What action do you propose to resolve this apparent conflict in information?
4. Your company is proposing to build a new car that has less pollution emitted from exhaust fumes. The car is extremely light and would not fare well in a collision with an SUV. What type of information would you release to counter claims of this being an unsafe automobile?
5. You are a candidate project manager for a sensitive project that is expected to have many primary and secondary stakeholders. What is the process you would use to keep these stakeholders informed?

CHAPTER 7
STRATEGIC ISSUES IN PROJECT MANAGEMENT[1]

"Every advantage...is judged in the light of the final issue."
DEMOSTHENES, 384–322 B.C.

7.1 INTRODUCTION

In the management of a project, there are likely to be issues or contentions that can have a significant impact on what purposes the project fulfills—and how the project should be managed. Often project success or failure rides on these issues—and how they have been adequately considered during the planning for and execution of projects. The notion of strategic issues in the management of projects is another area of consideration that broadens the role of the project manager and his or her team members.

In this chapter, the nuclear power industry will be used to provide some representative examples of what is meant by strategic project "issues." How to identify issues, how to analyze the significance of issues, and how to manage project strategic issues will be suggested. Once the project team has identified the issue and analyzed its real or potential impact on the project, strategies can be developed and executed to deal with those issues that might have an important impact on the management of the project as well as its outcome. Also some insight into why projects succeed and why they fail because of strategic issues will be provided.

7.2 WHAT ARE STRATEGIC ISSUES?

The concept of "strategic issues" has emerged as a way to identify and manage factors and forces that can significantly affect an organization's future strategies and tactics. The importance of strategic issues has therefore appeared in the literature primarily in the context of the strategic management of an organization.

[1]David I. Cleland, "Strategic Issues in Project Management," *Project Management Journal,* March 1989.

King has put forth the notion of strategic issue management as an integral element of the strategic management of organizations,[2] and Brown also has dealt with strategic issues in the management of organizations.[3]

This chapter describes an approach to the assessment and management of strategic issues facing project teams as well as some strategic issues that have had an impact on contemporary projects. Project owners need to be aware of the possible and probable impacts of strategic issues. The project team leader has the primary responsibility to focus the owner's resources in order to deal with project strategic issues. The authors suggest three key aspects of strategic issue management: a need to be aware of strategic issues facing a project, an approach for the assessment of the strategic issues, and a technique for the management of strategic issues.

7.3 SOME EXAMPLES

Sometimes the existence of strategic issues in an industry fosters the use of project management techniques in a fashion not previously used. For example, intense foreign competition in the U.S. automobile industry has prompted U.S. automobile manufacturers to develop innovations in the design of their cars. Cutting costs and cutting car design–development time are other key strategic issues facing U.S. producers. Their response to the need to reduce the time it takes to manufacture a car has, in part, been to use project management techniques in the form of an organizational alignment and a process of engineering manufacturing called simultaneous engineering or use of product design teams. The result: shorter car model product-development cycles with consequent cost savings, improved quality, and a more competitive product in the world car market.

When the Japanese automaker Nissan considered building a plant in the United States, it recognized that a strategic issue facing that project was the adaptability of the local community and the workers to the Nissan culture. By carefully selecting their employees and using exchange trips to Japan, and by orientation sessions at the plant in Tennessee, the Japanese managers were able to resolve this strategic issue, resulting in a successful production facility characterized by model employee-management relations.

Jaafari discusses the strategic issues in the management of macroprojects in Australia by first looking at the typical pattern of managerial relationships that occur and must be administered in such macroprojects. These occur between:

- Each participating owner and the joint venture or company acting as the collective body for owners (herein referred to as the owner)
- The owner and the government(s)
- The owner and the lenders
- The owner and purchasers of the end product(s)

[2]William R. King, "Strategic Issue Management," chap. 15 in W. R. King and D. I. Cleland (eds.), *Strategic Planning and Management Handbook* (New York: Van Nostrand Reinhold, 1986).

[3]J. K. Brown, "This Business of Issues: Coping with the Company's Environment," *The Conference Board Report,* no. 758, 1979.

- The owner and insurer/underwriters
- The owner and project manager or engineer-constructor
- The owner and constructors/suppliers and fabricators
- The owner and the designer[4]

These relationships emerge as the project stakeholders are identified and the nature of their stake is determined. Stakeholders are those persons or organizations that have, or claim to have, an interest or share in the project undertaking. Strategic issues can arise from many different stakeholder groups: customers, suppliers, the public, government, intervenors, and so forth.

In a project, a strategic issue is a condition of pressure, either internal or external, that will have a significant effect on one or more factors of the project, such as its financing, design, engineering, construction, and operation.[5] Some examples of the way that contemporary projects have faced strategic issues follow.

On the U.S. Supersonic Transport Program, the managers had too narrow a view of the essential players or stakeholders and generally dismissed the impact of the environment-related strategic issues surrounding the program until it was too late. Environmentalists, working through their political networks, succeeded in stopping the U.S. supersonic program.[6]

The life cycle of the Tennessee–Tombigbee Waterway provides insight into the negative role that strategic issues can play.[7] On this waterway project, strategic issues played a role in the consideration of funding for this project over many decades. Political considerations, lawsuits, environmental factors, and social factors delayed approval and construction of the project for extended periods. Although the actual construction of this waterway took almost 14 years, the waterway was 175 years in the making. As far back as 1810, the citizens of Knox County in Tennessee petitioned Congress to provide a waterway to Mobile Bay. Congress finally authorized the first federal study in 1974, but the project was delayed through 22 presidential administrations, 55 terms of Congress, 8 major studies and restudies, and 2 major lawsuits. This waterway is one of the largest civil works projects ever designed and built by the Army Corps of Engineers. About 234 miles long, the project cost $2 billion and required more than 114 major contracts during its construction period.

In contrast to the handling of the Tennessee–Tombigbee Waterway, in the Midwest a Water Pollution Abatement Program costing approximately $2.5 million successfully faced challenging strategic issues at the outset and during the early years of the program. The development of a master plan for the project included the development of appropriate environmental impact statements. This master plan could not be changed without court and Environmental Protection Agency (EPA) approval. Because the funding for the project included EPA federal grants, state

[4]A. Jaafari, "Strategic Issues in Formulation and Management of Macroprojects in Australia," *International Journal of Project Management,* vol. 4, no. 2, May 1986.

[5]Definition derived from W. R. King and D. I. Cleland (eds.), *Strategic Planning and Management Handbook* (New York: Van Nostrand Reinhold, 1986), chaps. 1, 4, and 15.

[6]Mel Horwitch, *Clipped Wings: The American SST Conflict* (Cambridge, Mass.: The MIT Press, 1982).

[7]Paraphrased from General Kenneth McIntyre, *The Tennessee–Tombigbee Waterway,* Stone & Webster Engineering Corporation, Boston. Paper presented at the Larger Scale Programs Institute, Colloquium on Research Priorities for Large Scale Programs, Austin, Tex., March 1985. This project is also described in Chap. 1.

grants, general obligation bonds, and tax district levies, the courts became involved in the planning and execution of the project. The Army Corps of Engineers reviewed all construction contract documents before bidding, reviewed all change orders to the construction contracts, reviewed completed construction, and audited contract administration procedures. All the work that received federal grant participation ultimately was audited by the EPA and Army Corps of Engineers, as well as state and local auditors. In addition, the General Accounting Office conducted periodic reviews of the project. All these stakeholder groups became involved in the legal and regulatory strategic issues that arose on this project. Successful management of this project included the management of not only the project team, but also the project stakeholders and the strategic issues that faced this project throughout its life cycle.

Sometimes a strategic issue arises from the attitudes of employees. For example, CEO George Fisher's key strategies for turning around Eastman Kodak included a three-phase plan: first, reconfigure Kodak by selling all businesses unrelated to photography, repay most of the debts, and separate the embryonic digital–electronic imaging operations from the traditional chemistry-based silver halide photography division; second, set strict financial goals that included achieving virtual perfection in manufacturing quality; and third, require accelerated growth initiatives. In all of this CEO Fisher was convinced that his most urgent task was to eliminate resistance to change from employees.[8]

Strategic issues can emerge at any time during a project's life cycle. The following is an illustration of how costly it can be to ignore them. On a large nuclear power plant project, an offshore earthquake fault was discovered only a few miles from the plant site. This occurred midway through the project's life cycle. Although the discovery of this fault was obviously a significant strategic issue, there was little evidence that the senior managers of the owner organization demanded and received a "satisfactory accounting" or made any in-depth inquiry to determine its full ramifications. The potential strategic implications of the fault should have prompted the corporate board of directors to do the following:

- Ask for an immediate, in-depth study of its possible and probable effects on the design of the plant.
- Acknowledge the need to forthrightly resolve the effects of the earthquake fault on the seismic design of the plant.
- Order a full-scale audit of the current status of the plant.

The project owner was not able to provide any evidence that the board of directors or the executive committee of the board considered the available options of

- Withdrawing its license application or stopping work
- Significantly reducing work at the site pending a full-scale investigation of the implications of the fault
- Accelerating offshore investigations to speed resolution of any questions that might have been raised

[8] "Focus Kodak?" *Fortune,* January 13, 1997, pp. 77–79.

There was no evidence that the board of directors considered any options other than that of continuing work, so that after the plant was nearly completed, the board members were faced with the enormous costly problem of redesigning the plant so that it could function safely in spite of its poor location.[9] Public concern over the seismic-geologic potential safety of this plant was expressed through the organized efforts of several intervenor or stakeholder groups acting through the courts to require reassessment, or even cancellation, of the plant.

The successful completion of any substantial public works project is dependent upon the recognition and management of strategic issues surrounding the social, political, legal, and economic aspects of the project as well as the cost, schedule, and technical performance aspects. On these public works the project can expect to encounter strategic issues such as:

- Land acquisition challenges
- Environmental impacts
- Political support or uncertainty
- Advocacy usually related to who conceives, champions, and nurtures the project and provides ongoing maneuvering to keep the project alive and well—a task partially fulfilled by the project managers
- Intervenors ranging from such organizations as local newspapers to vested interest groups such as the Sierra Club
- Competitors who would like to see the project fail so they could pick up some of or all the action

One of the major strategic issues facing the United States and other nations as well is the development of alternative means for generating electrical power. The energy crisis of 1974 pointed out the imprudence of depending on oil and gas as the principal fuels for generating electric power. Today, that crisis seems to be part of our forgotten past—but it has not gone away. Limited research is being carried out in projects leading to the development of alternative means of producing electrical power. In the judgment of the author, another energy crisis is forthcoming. It is not a question of whether such a crisis will emerge—it is a question of when. Although many people will disagree with the author's opinion in this regard, what if such a crisis does come forth? What alternative means for generating electrical power will be available? One such alternative source is nuclear power. But this source is not acceptable to most people because of the history at the Three Mile Island facility and the experiences at Chernobyl. Then, too, the poor management of the construction of nuclear power plants in the United States causes a lot of concern about whether or not the design and construction of plants in the future would do any better.

Jack Welch, General Electric Company's CEO for nearly 20 years, forged an entrepreneurial culture that kept the company at the forefront of U.S. industries. He once observed, "Managing success is a tough job. There's a very fine line

---

[9]Paraphrased from testimony submitted by David I. Cleland during litigation on the Diablo Canyon project, California Public Utilities Commission, Division of Ratepayer Advocate, application nos. 84-06-014 and 85-08-025, June 1988.

between self-confidence and arrogance. Success often breeds both, along with a reluctance to change." When Welch attempted to merge Honeywell with GE, this attitude seemed to be more than self-confidence and the merger met with failure.

Welch did not adequately assess the influence of the European Commission, and specifically the Commission's top antitrust official. Welch was confident of the outcome of the GE-Honeywell merger because of his successes in 1700 other mergers during his 20-year tenure. What was not considered was

- A growing sense of rivalry between the European and the U.S. aerospace companies
- Cultural sensitivities
- Tough top antitrust officials in Europe with a reputation for challenging large mergers
- A perceived arrogance on the part of GE by the Europeans
- European fears that GE would dominate the aircraft maintenance, repair, and overhaul operations in Europe

Observers of the situation attribute Welch's attitude and lack of understanding of the European culture, including the tough stand taken on large mergers in Europe. This attempted merger, initiated just prior to Welch's planned retirement, places a stain on his otherwise brilliant career and demonstrates that 1700 successes in the past do not assure success when fundamental areas are ignored.[10]

According to *The Wall Street Journal,* after years in the wilderness, the nuclear power industry is back on the march. The U.S. government's statements in early 2001 that nuclear power is an essential part of the energy mix for the United States have caused rethinking about new power plants. The Nuclear Regulatory Commission is expecting to receive applications soon for permission to build new power plants—the last application was in 1973.

Although Exelon executives and other nuclear energy backers maintain that they can build reactors that cost less and are far safer than the 1960s- and 1970s-era plants around the United States, it is uncertain whether the public accepts their argument. "We're not looking at this as a nuclear revival, but as a relapse," says Paul Gunter, head of the nonprofit Nuclear Information and Resource Service in Washington, D.C. He believes that the nuclear industry, which suffered from the gargantuan cost overruns as well as a handful of highly publicized safety problems like the Three Mile Island near-meltdown, represents the "biggest managerial disaster" in U.S. history. Gunter has noted

> Reactor safety isn't the only outstanding issue standing in the way of the nuclear comeback…, there's still no approved plan governing the disposal of radioactive waste.
>
> The Price Anderson Act, which protects the nuclear industry against unlimited liability in the event of a nuclear accident, expires in August 2002. Unless it is renewed, say industry executives, it is unlikely any company would build a new plant.

---

[10]Paraphrased from Anthony L. Velocci, Jr., "GE's Own Arrogance Thwarted Bid to Acquire Honeywell," *Aviation Week & Space Technology,* July 9, 2001.

Exelon believes that all these obstacles are minor compared with the benefits of widespread deployment of new-technology nukes that would make the nation more energy self-sufficient and reduce air-pollution emissions. Another plus is that the Nuclear Energy Commission has been streamlining its application approval process since the mid-1980s.[11]

Considerable emotion is involved whenever the subject of nuclear power comes up. Indeed, the potential and the problems for building a case for nuclear power are surrounded by formidable "strategic issues" that are shown in the material that follows in this chapter. Until these strategic issues are resolved, it is highly unlikely that any future program for starting nuclear power projects in the United States could be launched. In the material that follows, the strategic issues involved in nuclear power are discussed as a paradigm for how strategic issues can impact projects in this field.

7.4 AN APPLICATION OF THE CONCEPT OF STRATEGIC ISSUES: NUCLEAR CONSTRUCTION INDUSTRY

Strategic issues vary depending on the industry and the circumstances of a particular project. In the material that follows, the nuclear plant construction industry is used to illustrate the concept of strategic issues as applied to a select industry. This industry has been chosen because of the many strategic issues that have faced the industry—issues which relate to a particular project as well as to the many generic issues that confront project owners, managers, constructors, designers, regulators, investors, local communities, consumers, and other vested stakeholder groups.

A project that has as long a life cycle as a nuclear power generating plant will be affected by many issues (some of them linked) that are truly strategic in nature. For example, the typical strategic issues that a nuclear power plant project faces today include:

- Licensability
- Passive safety
- Power costs
- Reliability of generating system
- Nuclear fuel reprocessing
- Waste management
- Capital investment
- Public perception
- Advocacy

[11]Paraphrased from Rebecca Smith, "Nuclear Power: Revival or Relapse?" *The Wall Street Journal,* May 2, 2001, pp. B1, B4.

- Environment
- Safeguards[12]

The U.S. nuclear power industry has had extraordinary challenges in the past such as uncertain licensing procedures, project cost and schedule control problems, quality assurance disputes, intervenor actions, and other conditions that are strategic issues to be dealt with by a project team in managing a nuclear power plant project. A discussion of these issues follows.

Licensability

All U.S. nuclear plants, to be licensed, must meet federal codes and standards as well as the nuclear regulatory guides for the particular design. But many of these codes, standards, and guides are not applicable to a new concept and design that have not been licensed previously. The first of a kind becomes precedent-setting and will receive a commensurate amount of attention from the Nuclear Regulatory Commission (NRC) staff—so much so that joint groups will be set up with representation from the Department of Energy (DOE), NRC, and a bevy of consultant experts to answer the hundreds of questions posed by the NRC staff and to draft appropriate revisions to the existing federal codes and regulations as well as to set up future guides for the new concept.

This strategic issue can take years to resolve when one includes the judicial, state, and local hearing processes that a nuclear plant must face. The lack of firm and predictable policy emanating from the NRC now adds to the risk and uncertainty involved in the management of this strategic issue. Such issues and uncertainties are reflected in the increased costs and schedules for the project. The challenge facing the NRC is forthright—remove the uncertainty of the current licensing process that exists today. The NRC that licenses the plant and the state and local governments that conduct hearings to ascertain the proper allocation of costs for the utility's rate base are key stakeholders in the project.

Passive Safety

All the commercial reactors built and operated in the United States today require the activation, within a prescribed period, of an auxiliary shutdown system, either automatic or manual. At present, if one allows the reactor to operate without adding reactivity (a process similar to adding coal to a fire) and assuming that the cooling systems remain effective (the pumps operate, the valves open and close on cue, the heat exchangers transfer heat, etc.), the reactor should eventually bring the auxiliary system into operation. The difficulty comes when the auxiliary system cannot halt or lower the reactivity (like removing coal from the fire) and/or maintain the effectiveness of the cooling systems.

---

[12]These strategic issues were developed during the conduct of a research project by D. I. Cleland and D. F. Kocaoglu, *The Design of a Strategic Management System for Reactor Systems, Development and Technology* (Argonne, Ill.: Department of Energy), with the assistance of A. N. Tardiff and C. E. Klotz of the Argonne National Laboratory.

Passive safety, as it relates to a nuclear power plant, refers to the plant's ability to take advantage of inherent, natural characteristics to move itself into a safe condition without the need to activate an automatic auxiliary safety system or a set of predetermined operator procedures to do the same.

Passive safety is the dominant strategic issue facing the nuclear power generating industry today. This issue is both social (it would help overcome fear of nuclear power) and technical (design and operating considerations). The nuclear accidents at Three Mile Island and Chernobyl have intensified the search for a nuclear power plant that promises passive safety. Nuclear vendors and utility companies are the key stakeholders interested in passive safety.

Power Costs

The components of power costs are capital costs, operations and maintenance (O&M), and fuel costs. For a typical nuclear power plant, the capital cost component is four times the O&M cost, which is approximately equal to the fuel cost. Hence it is evident that capital cost is the most significant component. One of the significant factors leading to the current hiatus in orders for new nuclear power plants is that these plants are extremely capital-intensive and have relatively low fuel costs. Coal- and oil-fired plants have a relatively low capital cost component, but their fuel costs are extremely high.

Construction times for many recent U.S. nuclear plants have exceeded 10 years. The U.S. licensing and judicial procedures have accounted for much of the delay, but other factors, such as imprudent project management, also have taken their toll. Whatever the reasons, the delays have an extraordinary impact on the resultant capital investment in these plants even before they have produced 1 kWh of electricity. The interest paid on the capital to build the plant commonly is greater than 50 percent of the capital investment in the plant. As a result, there has been an inordinate increase in the capital cost component so that nuclear power has now lost its competitive power cost edge over its closest competitor, coal.

Utilities, nuclear reactor manufacturers, architectural and engineering firms, plant constructors, and state regulatory commissions are the principal stakeholders concerned with the strategic issue of power costs.

Reliability of Generating System

The reliability of a nuclear power plant must be extremely high, particularly in the safety systems and components. There are reliability differences from one model to another; that is, one might have fewer moving parts, fewer systems, fewer components, and fewer things to go wrong.

Plants designed and constructed under stringent quality assurance controls will be more reliable than plants where the quality standards have been relaxed. Concepts that utilize more factory-built than on-site fabricated and assembled systems tend to be more reliable because quality assurance can be applied more easily at the factory. Gravity- and natural circulation–dependent systems tend to be more reliable than forced-circulation systems. The importance of these and more reliable approaches to a nuclear power plant cannot be overemphasized, particularly in view of Three Mile Island, Chernobyl, and the resultant skeptical public attitude toward nuclear power. Utilities are the principal stakeholders here.

Nuclear Fuel Reprocessing

Commercial nuclear fuel reprocessing in the United States is limited. Instead, the U.S. government has agreed, for a price, to accept the spent fuel from U.S. reactors for long-term storage. Europe and Japan, however, have viable programs to recover for future use the nuclear fissionable fuel from spent fuel assemblies. Any concept, such as the breeder, which requires reprocessing technology, must carry the burden of developing this technology as well as the nuclear proliferation stigma attached to it. Thus any future nuclear plant in the United States may require the arrival of a liquid-metal reactor technology that provides for the use of reprocessed fuel. The time frame for such fuel reprocessing capability is circa 2040 by best current estimates. Utilities and reactor manufacturers are the principal stakeholders.

Waste Management

Public reaction to shipments of nuclear waste is becoming increasingly severe. Therefore, minimum waste streams and minimum movement of such wastes outside the plant boundaries are advisable. The waste disposal program conceived and managed by the U.S. government and the nuclear power industry to store radioactive fuel safely is being challenged under public pressure. Unreasonable management and cost overruns aside, one of the biggest issues for the nuclear power industry is what to do with over 1500 metric tons of lethal atomic waste that it produces each year. Utilities, states where storage sites are located, and the general public are vested stakeholders in this strategic issue.

Capital Investment

Closely akin to the strategic issue of power costs are the financial exposure and risks that investors of nuclear power plants have experienced over the last several years.

There have been awesome financial implications for all too many nuclear plants. One of the most sobering has been the experience of the Washington Public Power Supply System (WPPSS). WPPSS's default on interest payments due on $2.5 billion in outstanding bonds can be laid on the failure of WPPSS management. Management style in WPPSS did not keep pace with the growing size and complexity of the organization. Communication at senior levels of the organization, including the board of directors, tended to be "informal, disorganized, and infrequent."[13] To renew support of nuclear power in the financial communities, it is

[13]James Leigland, "WPPSS: Some Basic Lessons for Public Enterprise Managers," *California Management Review,* Winter 1987, pp. 78–88.

important that the current conditions change along the lines noted in the discussion of the power cost issue in this article. Investment institutions are the principal stakeholders as well as the state public utility groups that must rule on the acceptability of a capital investment cost into the utility's rate base.

Public Perception

Table 7.1 demonstrates the strategic issue of public perception. The experts rank nuclear power 20th in the list of high-risk items, whereas the other groups rank it first or close to first. Note that x rays and nonnuclear electric power fall into the same pattern. When the United States converted from direct current to alternating current in the early 1920s, a similar negative public reaction resulted. Some extensive innovative technical, social, and managerial approaches must be developed and implemented to change perceptions.

Aggravated by the nuclear accidents at Three Mile Island and Chernobyl, the increasingly negative public perception of nuclear power and its associated risks has made this strategic issue more acute, and the need for government research programs more pressing.

Advocacy

Not many government interest research programs can proceed through the government bureaucracy without a strong advocate who can gain substantial support for the program. The base of support must be broad and must include, as is the case with the research in the Advanced Reactor Development Program, key individuals within the DOE, the White House, the Office of Management and Budget, Congress and its staff offices, the nuclear community (the stakeholders), the scientific community (National Science Foundation, National Academy of Sciences, certain universities), the financial community, and others. With such backing, the public generally supports the program. A single-effect advocate also can be an essential ingredient. Military aircraft and the aircraft carrier had Billy Mitchell; the nuclear submarine fleet had Hyman Rickover; the space program had Werner von Braun—the list of successful efforts led by able champions is long. Thus a reactor manufacturer who contemplates obtaining government funds to research advanced nuclear reactors should determine what advocacy exists for such research, both in the government and in the corporation.

Environment

From an environmental viewpoint, the nuclear advocates had essentially convinced the general public that nuclear power plants were environmentally benign—until the media convinced the public otherwise after the Three Mile Island incident. The Chernobyl incident reinforced the sense that nuclear power was a serious threat to the environment and to life itself. Certainly, the environmental impact of the Chernobyl accident on its surrounding environment is as yet uncertain.

TABLE 7.1 High Risk: A Matter of Perception

Four groups rank "what's dangerous and what's not."
People were asked to "consider the risk of dying as a consequence of this activity or technology." (1 has the most consequence and 30 the least.)

Activity or technology	Experts	League of Women Voters	College students	Civic club members
Motor vehicles	1	2	5	3
Smoking	2	4	3	4
Alcoholic beverages	3	6	7	5
Handguns	4	3	2	1
Surgery	5	10	11	9
Motorcycles	6	5	6	2
x rays	7	22	17	24
Pesticides	8	9	4	15
Electrical power (nonnuclear)	9	18	19	19
Swimming	10	19	30	17
Contraceptives	11	20	9	22
General (private) aviation	12	7	15	11
Large construction	13	12	14	13
Food preservatives	14	25	12	28
Bicycles	15	16	24	14
Commercial aviation	16	17	16	18
Police work	17	8	8	7
Fire fighting	18	11	10	6
Railroads	19	24	23	20
Nuclear power	20	1	1	8
Food coloring	21	26	20	30
Home appliances	22	29	27	27
Hunting	23	13	18	10
Prescription antibiotics	24	28	21	26
Vaccinations	25	30	29	29
Spray cans	26	14	13	23
High school and college football	27	23	26	21
Power mowers	28	27	28	25
Mountain climbing	29	15	22	12
Skiing	30	21	25	16

Source: Decision Research, Eugene, Ore. (© *The Washington Post,* May 21, 1986).

Reassuring the public that there will be no future Chernobyl-type accidents will be no easy task. Much work must be done to convince people that such an accident cannot occur in the United States. This certainly must be convincingly transmitted to the stakeholders who are the potential owners of nuclear power plants, the administration, Congress, and above all the general public itself. A most environmentally benign and inherently safe nuclear plant would go a long way to settling this issue. Unfortunately, such a plant may be decades away.

Safeguards

The objective of nuclear safeguards is to keep fissionable material out of unauthorized hands. A nuclear plant security system that does this better than another should have a competitive edge. For example, if throughout the fuel cycle of a particular plant, the plant configuration prevents the fissionable fuel from being deployed and used as source material for a weapon, then one could say that the plant is proliferation-proof.

7.5 MANAGING PROJECT STRATEGIC ISSUES

Project strategic issues often are nebulous, defying management in the literal sense of the word. It is important that the project team identifies the strategic issues the project faces and deals with them in terms of how they may affect the outcome of the project. In the assessment of the issues, some may be set aside as not having a significant impact on the project. These would not be reacted to but would be monitored to see if any changes occur that could affect the project. Of course, some significant issues may not be subject to the influence of the project team.

The early identification of issues is important so that there can be an early decision on how issues are to be handled. An issue tends to go through a life cycle such as described in General Electric's approach to public issues, where phases of *conversion, contention, legislation,* and *regulation* are discussed.[14]

A useful technique to identify strategic issues facing a project is to keep a running tally of all issues that face the project and then take time to have the project team discuss these issues to see which ones are *operational* (short term) and which are *strategic* (in the manner described in this book). Once the project team has been acquainted with the notion of strategic issues, each member should be encouraged to note any emerging issues for discussion and review at one of the regular project team meetings. During this meeting all issues should be reviewed, selecting those that appear to be strategic and assigning a member of the team to follow the issue and keep the project team aware of it and its implications on the project's future. More serious issues may require the appointment of an investigative subproject team that will report back to the full team. An example of how one project team's awareness of strategic issues early in the project's life cycle proved useful appears below.

A kickoff meeting of the project team and the senior managers from the owner's organization was held to get the project team organized and to start preliminary project planning. During this 3-day meeting a tally was made of issues known to

[14]J. K. Brown, op. cit.

impact or have potential future impact on the project. Some issues were determined to be truly strategic, and the group decided to track them to determine their significance. If such a tally had not been done and if the preliminary discussions had not been carried out, it is highly probable that some of the more important issues might not have surfaced until the project was into its life cycle. By then an orderly and timely resolution of some of these issues would have been difficult, if not impossible. This suggests that an important part of any project review meeting is to discuss and update the current project issues to see which ones might be added. By the same process, those issues judged as no longer important could be put aside.

The project team requires a philosophy on how to manage strategic issues. A phased approach is suggested as portrayed in Fig. 7.1. These phases are discussed below.

7.6 ISSUE IDENTIFICATION

Identifying some of the issues often can come about during the selection of the project to support the organizational strategy. During the selection process the following criteria can be addressed to determine if the project truly supports organizational strategy:

- Does the project support a strength that the enterprise holds?
- Does it avoid a dependence on something that is a weakness of the enterprise?
- Does the project support an organizational need?

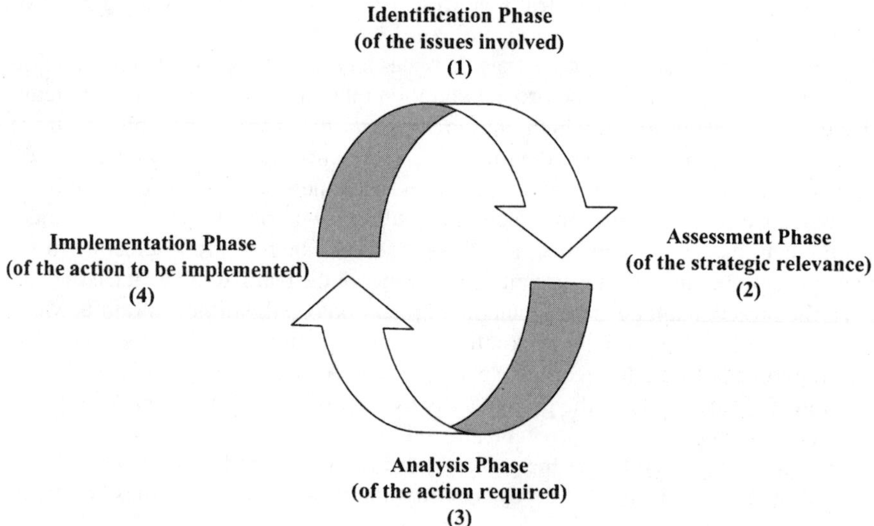

FIGURE 7.1 An approach for the management of strategic issues.

- Is there a customer who is willing to pay for the project?
- Can the project owner assume the risk that is involved in the project?
- Are the resources and management skills available to bring the project to completion on time and within budget?[15]

As the decision makers seek the answers to these questions, there will be some strategic issues that emerge naturally. Other issues can be identified by the project team during its planning, evaluation, and control meetings.

For example, during a customer review of a bid package for a new weapon system, an aerospace contractor's project proposal team discovered that the customer had serious doubts about the contractor's cost-estimating ability. This concern prompted the contractor to engage a consultant to conduct a survey of its customers to assess its image in two general areas: product image (price, quality, reliability, etc.) and organizational image (quality of personnel, responsiveness, integrity, etc.). Both structured and unstructured personnel interviews were conducted with key customer personnel. One significant outcome of this image survey was the perception by key customer personnel that the contractor's cost estimates were far too conservative, invariably resulting in excessive cost overruns. The contractor's key executives were shocked by the customer's perceptions of its cost performance credibility. This matter of credibility immediately became an urgent strategic issue within the contractor's organization. A task force was formed to investigate the issue and recommend a strategy on how to deal with it. In their deliberations the task force found that the contractor's cost performance was in fact quite credible, and that the perception held by the customer's key people was not valid. Consequently, the contractor mounted an advertising and indoctrination program to change the customer's viewpoint by working through the field marketing people and by visiting the customer's offices to present the actual facts on contractor cost performance. The result was a resolution of the strategic issue in the contractor's favor. Had the project proposal team not been alerted to this potential strategic issue, the contractor may well have lost future government contracts.

By maintaining close contact with the customer, an opportunity is provided to identify issues that can have an impact on the project. Another technique is to examine the stakeholders on the project to see if the nature of their claims suggests any strategic issues.[16] As each stakeholder group is reviewed, the following questions should be addressed:

- What claims do the stakeholders have in the project?
- How might the claims affect the outcome of the project?
- What resources and influence do the stakeholders have to push the satisfaction of their claims?
- Can the project live with the stakeholder's purposes and motivation?

---

[15]Paraphrased from D. I. Cleland and W. R. King, *Systems Analysis and Project Management,* 3d ed. (New York: McGraw-Hill, 1985), pp. 67–68.

[16]D. I. Cleland, "Project Stakeholder Management," *Project Management Journal,* vol. 1, no. 4, September 1986, pp. 36–44.

- Can the outcome of the stakeholder's claim on the project be predicted?
- What can the project team do about these claims?

Other techniques can be used such as the *nominal group technique*[17] or brainstorming to aid in the identification of issues. Perhaps the best way to identify issues is to ensure that the project team is well organized, well managed, and well aware of the larger systems context (economic, political, social, technological, and competitive) of the project. If the team meets these conditions, there is a better likelihood that most of the important and relevant strategic issues will surface.

7.7 ASSESSMENT OF AN ISSUE

The act of assessing an issue entails judging its importance in terms of its impact on the project. King has suggested four criteria for first assessing an issue as strategic and then moving to subsequent states of management of the issue:[18]

- Strategic relevance
- Actionability
- Criticality
- Urgency

The *strategic relevance* of an issue relates to whether it will have a long-term impact (more than 1 year) on the project. Most of the strategic issues mentioned earlier in this chapter could be considered to be strategically relevant, such as licensability, passive safety, and power costs. Strategic relevance addresses the question: Will this strategic issue influence the project strategy or the likely consequences of the strategies that are being followed on the project? If an issue is strategy-relevant, then the project manager has two basic courses of action: Try to live with the issue's impact, or do something about the issue.

But some strategic issues will be beyond the authority and resources of the project manager to resolve. In such situations a third course is open to the project manager: Elevate the issue to senior managers for their analysis and possible evaluation. Even though senior managers are aware of the issue, the project manager retains residual responsibility to see that the issue is "tracked" and given due attention.

The *actionability* of a project issue deals with the capability of the project team and the enterprise to do something about the issue. For example, the issue of licensability of a new nuclear power plant is critical to the decision of whether to

[17] The process is explained in A. H. Van de Ven and H. L. Delbecq, "Nominal versus Interacting Group Processes for Committee Decision Making," *Academy of Management Journal,* vol. 14, no. 2, 1971.

[18] W. R. King, "Strategic Issue Management," in W. R. King and D. I. Cleland (eds.), *Strategic Planning and Management Handbook* (New York: Van Nostrand Reinhold, 1986), pp. 252–264.

fund such a plant. A company can help resolve the licensability of nuclear power plants by participating with the industry's groups that are trying to influence the Nuclear Regulatory Commission either directly or through congressional persuasion to do something about the uncertainties related to licensing. Such participation would be useful in influencing the strategic issue as well as for keeping informed about the status of the issue. The related strategic issue of funding support for a power-generating plant would be an issue that the enterprise would actively try to resolve by working with investment bankers in the financial community.

A project may face strategic issues about which little can be done. Keeping track of the issue and considering its potential impact on project decisions may be the only realistic action the team can take. Key project managers should always be aware that there are issues that may be beyond their influence.

The *criticality* of an issue is the determined impact that the issue can have on the project's outcome. The issue of growing congressional disenchantment with the U.S. Supersonic Transport Program arose from the concern of the environmentalists over the sonic boom problem. Proactive environmental groups along with the general public exerted political influence, which contributed to the termination of that program. Project advocates recognized too late that the sonic boom controversy was the critical fulcrum for the environmentalists to use for their public and congressional support. If a preliminary analysis of an issue indicates it is noncritical, then the issue should be monitored and periodically evaluated to see if its status has changed.

The *urgency* of an issue has to do with the time period in which something needs to be done. All else being equal, if an issue should be dealt with immediately, it must take precedence over other issues. Urgent issues emerging during the project planning should be considered a "work package" in the management of the project. Someone should be designated as the issue work package manager to look after the issue, particularly during its urgency status.

The accident at the Three Mile Island nuclear plant and the subsequent uncertainties over plant design and licensing posed serious and urgent strategic issues for all nuclear plants in the design and construction phases of their life cycle. Although most project managers would have considered this an urgent issue, there were limits as to what could be done, except to track the issue and try to influence the NRC and other government agencies through the industry's societies and political contacts.

H. Ross Perot's controversial contract for a project with the U.S. Postal Service faced a strategic issue soon after the award was announced. The contract immediately drew fire because it was awarded without competitive bidding. The General Accounting Office began an investigation, followed by a U.S. Senate resolution requesting that the contract be put on hold pending further study. The General Services Administration's Board of Contract Appeals nullified the contract. It is not known if the lack of competitive bidding with the U.S. Postal Service was ever considered a possible "strategic issue" by the Perot Systems Corporation team. But that is the way things have turned out. It has become an issue with considerable urgency affecting a major project for that corporation.

7.8 ANALYSIS OF ACTION

Identification and assessment of an issue are not enough; the issue has to be managed so that its adverse effect on the project is minimized and its potential benefit is maximized. The issue work package manager is in charge of collecting information, tracking the project, and ensuring that the issue remains visible to the project team. That manager should also coordinate decisions made and implemented regarding the issue.

In the analysis of action required to deal with an issue, seeking answers to a series of questions like the following can be helpful:

- What will be the probable effect of the issue in terms of impact on the project's schedule, cost, and technical performance and the owner's strategy?
- Who are the principal stakeholders who have an interest in the project? What will be the impact on their probable strategy?
- How influential are these stakeholders?
- What strategy should the project team develop to deal with these issues?
- What might be the real cost in relation to the apparent cost to the project owner, and will other projects being funded by the project owner be affected?
- What specific action will be required, and what will it cost the project owner?

The action developed to deal with the issue may, at the minimum, consist of simply monitoring the issue and giving status reports to the project team. Some issues, however, may require a more aggressive approach. The issue work package manager may find it useful to think of the issue as having a life cycle, with such phases as conception, definition, production, operations, and termination, and to identify the key actions to be considered and accomplished during each phase. The manager should be specific and should stipulate what will be done, when it will be done, how to do it, where, and who will be in charge of implementing the action leading to resolution of the issue.

7.9 IMPLEMENTATION

However it is dealt with, the resolution of an issue or the mitigation of its effects requires that a *project plan of action* be developed and implemented. Indeed, the resolution of a strategic issue can be dealt with as a miniproject requiring the execution of the management functions—planning, organizing, motivating, direction, and control—and all these functions entail some degree of work breakdown analysis, scheduling, cost estimating, matrix responsibility, information systems, design of monitoring and control, and so on. What resources are to be used to resolve the issue and who should take the leadership role in resolving that issue and the crucial questions to be answered should be considered.

The potential for the success or failure of a project can have strategic issue implications. In the material that follows, a brief review of some of the reasons for project success and failure is given to remind the leader that such issues are very real and should be considered during the project's life cycle.

7.10 STRATEGIC ISSUES OF PROJECT SUCCESS AND FAILURE

In 1945, Mayo observed that the United States is technically competent, but we have considerable social incompetence.[19] Another perspective is offered by two researchers who found that for the overwhelming majority of failed projects there was not a single technological issue to explain the failure—rather, it was sociological in nature.[20] Thamhain and Wilemon found in their research that newer project management approaches require more extensive human skills and competence. Some of the skills they found that were associated with building multidisciplinary teams involved motivating staff people, developing a healthy work climate, managing conflict, and communicating effectively at all levels.[21]

What are the key critical factors in successful projects? One study identified a general set of critical success factors that could be applied to any project regardless of its characteristics or development methodology. It was recognized that timing was evident in the factors. The top factors identified were those concerned with establishing adequate planning to include goals and the general philosophy and mission of the project, as well as the commitment to these goals. Factors associated with user involvement tended to take up the middle of the rankings. Finally, at the low end of the factors were the project management and technical factors to include the selling of the system, the monitoring of progress, and troubleshooting activities usually found during project development activities.[22]

Another study about project success formulates a conceptual framework to assess the impact of success factors on the success criteria of a technology transfer project. The study drew from a sample of 40 automation industry firms and 48 successful and not so successful cases. The dominant importance of involvement for the success of a technology transfer project became evident. Technical characteristics are the second most important factor in determining project success. One key outcome of this study was a list of policy implications and recommendations to enhance the technology transfer process.[23]

---

[19]E. Mayo, *The Social Problems of an Industrial Civilization* (Boston: The Graduate School of Business, 1945).

[20]Tom DeMarco and Timothy Lister, *Peopleware: Production Projects and Teams* (New York: Dorset House, 1988).

[21]H. J. Thamhain and D. L. Wilemon, "Developing Project/Program Managers," *Proceedings,* PMI Seminar/Symposium, Toronto, Ontario, October 1982, p. II-B.2.

[22]James J. Jiang, Gary Klein, and Joseph Balloun, "Ranking of System Implementation Success Factors," *Project Management Journal,* December 1996, pp. 49–53.

[23]Raykun R. Tan, "Success Criteria and Success Factors for External Technology Transfer Projects," *Project Management Journal,* June 1966, pp. 45–55.

Florida Power and Light management identified what it thought to be the 10 most important factors in completing the St. Lucie 2 Nuclear Power Plant essentially on schedule, within cost, and without major quality-related problems:

- Management commitment
- A realistic and firm schedule
- Clear decision-making authority
- Flexible project control tools
- Teamwork
- Maintaining engineering ahead of construction
- Early start-up involvement
- Organizational flexibility
- Ongoing critique of the project
- Close coordination with the Nuclear Regulatory Commission[24]

The potential for a project failure or success usually has strong overtones of a strategic issue. If a product development project fails, the strategic viability of the enterprise can be threatened. Conversely, if a project succeeds, a significant contribution to the future viability of the enterprise has been made. Project managers, team members, and senior managers should be aware of how strategic issues can impact the success or failure of a project.

7.11 PORTFOLIO MANAGEMENT FOR PROJECTS

The design and implementation of a portfolio for projects is a strategic issue within an enterprise. Portfolio management of projects is a strategy that moves the selection and implementation of projects from a random process to one with structure and discipline. Project portfolio management is used to align projects with strategic goals and objectives for a more effective and efficient organization.

Portfolio management is often viewed in the financial community as a balance between the investments through stocks, bonds, and cash. Within the stocks, there are a wealth of categories that include high, medium, and low risk; speculative and stable stocks; emerging companies and mature companies; and dividends and capital realization. Similarly, projects may be categorized into areas that have specific parametric variables, which permit characterization and classification of projects.

Many project may be selected on the basis of criteria that are not in alignment with the organization's purpose or mission or goals and objectives, or the strategy for the organization's growth. Projects are sometimes selected on the basis of

[24]*Improving Quality and the Assurance of Quality in the Design and Construction of Nuclear Power Plants,* NUREG-1044, U.S. Nuclear Regulatory Commission, Washington, D.C., May 1984.

profitability estimates alone—without regard for any of the multiple factors that can affect the outcome or for an alignment with the organization's strategic direction. Random selection of projects without consideration for the enterprise's strategic direction can have adverse consequences.

To ensure the best outcome for the organization, projects need to be reviewed and assessed for their contributions to the organization and for future business. Ideally, there would be a balance of projects that includes such considerations as risk, profitability, size, and strategic fit.

Project Portfolio

What is the proper balance of projects defined by their multiple characteristics for an organization? If projects are the building blocks to the future of an organization, then an organization needs to understand the types and characteristics of projects in order to select the best ones for success. Selecting the wrong projects can have the effect of diverting funds and resources in a direction that limits organizational growth.

Following are some examples of types of projects that have been observed:

- An energy transport company selected most of its projects on the basis of repairing or upgrading the existing natural gas lines. This focus on repairing and upgrading the existing system proved to be most effective because the natural gas was transported to the consumers more efficiently and effectively. A secondary purpose for projects was to serve new customers. Note that if current customers were poorly served, new customers would only degrade the existing service.
- An insurance company selected software projects on the basis of perceived need for new efficiencies to better serve its customers. Many of the projects were changes to existing practices/procedures that were not fully coordinated; that is, work was being accomplished on top of work to change an old system. The larger picture was not considered when projects were approved, which suboptimized the organization's resultant capabilities derived from the individual projects. The organization did struggle through many duplicate efforts and sometimes conflicting outcomes.
- A telecommunication company's future was dependent upon what the sales force could sell to consumers. Each salesperson targeted different aspects of telecommunication services and "sold" projects on the basis of expected revenue. The cost of the projects was optimistically low and the actual revenue estimate way optimistically high. Less than 1 of 18, or 5 percent, of the projects exceeded revenue, and failures (expenditures exceeded revenue) were many. Project selection was dependent upon a salesperson's business cases—which were typically optimistic—with the salesperson being rewarded for a sale rather than a successful project.
- A major manufacturer of paper products selected and approved projects on the basis of availability of budget and the implied need for the projects. Although all the projects appeared to be valid requirements, several projects were implemented

for products that were not needed until several months after scheduled delivery. This resulted in many projects consuming resources early and the end result being products that could not be used for several months following delivery.

Project Selection Criteria

Ideally, an organization will select projects that align with the strategic goals and ones that build on current capability. Each project selected and implemented should be a building block that promotes the organization's purpose and that positions the organization for an improved future capability. Any project not aligned with the strategic goals may well detract from the organization's purpose and delay its growth.

Each organization should develop a model for assessing project fit within the strategic goals and objectives. This model could, perhaps, be a part of the strategic plan for the organization. A model would construct a generic combination of "best-fit" projects to guide senior management in approving new project work that aligns with the strategic direction. See Chap. 4 for additional information on project selection.

Some considerations for constructing an organization's model for a portfolio of projects follow:

- Projects are typically approved for the purpose of developing and delivering new products and services or effecting organizational change. These projects should be used in a balanced manner to advance the organization. In these turbulent times, an organization needs to focus on rapid delivery of products and services as well as changing the organization's processes to obtain more productivity for future work.
- Projects need to have a balance between providing products and services today and as future building blocks to position for tomorrow's work. For example, an organization might have an R&D program using projects to redesign or develop new products in anticipation of retiring some existing product lines.
- Projects are the means for action and implementing strategic plans. Well-designed and well-implemented projects can contribute to an organization's success today and in the future.
- Core competencies drive which projects should be selected and successfully implemented. Making a decision to conduct a project outside of one's core competencies adds risk and introduces a level of risk that may be unacceptable for the organization.

What are the variables and characteristics that an organization needs to consider for developing a balanced portfolio of projects? A selection model allows consistent application of criteria for an informed project selection process that can optimize and support the strategic goals and objectives. Random selection of apparently attractive projects typically results in less than optimum results.

All projects are "discretionary" in their implementation. Some will argue that a few projects are "must have" types and cannot be avoided. In viewing all projects, the senior managers making the selection have discretion of either delaying, not selecting, or implementing a project. The consequence of the decision may dictate a certain action based on the enterprise's goals. However, objective selection criteria should provide a better understanding of the project under consideration and its promised benefits.

Table 7.2 gives several critical aspects of a project that need to be considered in context with the strategic goals and objectives of the organization. Some of the items may be more important to an organization than others. Only the organization can determine the importance and weight of each item as it fits with their business and strategy.

TABLE 7.2 Elements of a Project Selection Process

Variable*	Comments
Profit margin	Organizations need to earn a profit on their project work when the resultant outcome is a product or service being sold. Organizations typically set a profit goal for projects on the basis of factors such as risk, degree of difficulty to complete the project, the type of work, and whether there is a usable by-product of the project.
Project risk	Organizations must assess the risk of the project. The risk may be whether the product will meet market expectations or whether the project can be completed within established goals for cost, schedule, and technical performance.
Process change	Organizations may use projects to optimize their organizational processes. Using projects to upgrade or establish new processes may be the most cost- and time-effective method.
Resources	Human and nonhuman resources need to be considered. Human resources may or may not have the requisite qualifications to perform on a project. Special material resources or tools may or may not be available to complete the project. In some instances, current human resources may be insufficient in quantity to complete the project within the required time frame.
Financial considerations	Cash flow may be negatively impacted by large initial expenditures on a project. The cost of labor or outsourcing of work can also have major financial impacts for an organization.
Building block	Question whether the project is a building block for the organization by further development of core competencies and contributing to the organization's success. Or is the project something that neither contributes to building the organization nor is within the organization's overall purpose?
By-product	Question whether there are by-products that can be used in subsequent projects or whether by-products may be used to enhance the organization's future capabilities.
Technology	Question whether the technology is one that the organization understands and is building a business based on it. Also the degree of maturity of a technology or whether the technology is to be developed is key to the project selection.

TABLE 7.2 Elements of a Project Selection Process (*Continued*)

Variable*	Comments
Project duration	Question whether the project duration fits into the normal work arrangements and whether there are only long- or short-duration projects in the organization.
Size (relative to organization)	Question whether the size (dollar, resource, duration) is right for the organization. Organizational structure may find the project to be too large for the capability or too small for the type of management used.
Corporate image	Question the image that the corporation will get when taking on a project.
High competition	Question the degree of competition for the project or the product and whether this project is a declining market area.
Client	Question whether the new project is for an existing client or a new one. Determine whether the organization's business is centered around one or two clients where any loss of a client would have a major impact on the organization.
Life-cycle phases	Question whether the project's life-cycle phase provides for continuity of work or whether there is interrupted flow of work. Interrupted flow of work typically requires more resources and costs more.
Core competency	Question whether the project is within the organization's established core competencies or whether it is initiating a new core competency. Existing competencies are easier to perform than starting new ones.
Urgency of need	Question the urgency of need to determine whether delivery is possible within the time frame desired. Also, determine whether there will be resources available to complete the project.
R&D	Question R&D projects to see if there are too many/too few and whether their focus is on the right/wrong areas. Determine whether new projects duplicate effort or lead to enhanced or new projects.

*These variables are purposely in a random order. This is designed to show that the table has not placed any weight on a variable or that a variable may apply to an organization's strategy. Further, organizations must develop their respective model and may use any or all of the above variables.

Once project selection criteria are established for the type of projects determined by an organization's portfolio of needs, a review of all projects can be made. Using a model of what is ideal for the organization, one can then compare the differences to establish a direction for change. Seldom will the model replicate the actual situation.

An example of project selection criteria in a portfolio has been developed on the basis of a generic situation. It does not directly represent an organization, but serves to show what might be used in an organization to define the characteristics of desirable projects. Criteria for a portfolio of projects may have some of the characteristics shown and explained in Table 7.3.

An example of what an ideal organization's project portfolio might be is depicted in Table 7.4. Although this is an example, it provides the reader an idea of what to look for in a portfolio of projects.

TABLE 7.3 Organizational Project Portfolio Selection (Example)

Characteristic	Criteria
Project size	Project mix will consist of small (less than $10K), medium (between $10K and $100K), and large (more than $100K) projects • Small projects will represent 50 percent of the business. • Medium projects will represent 40 percent of the business. • Large projects will represent 10 percent of the business.
Core competencies	Projects will be selected on the basis of fit into one of the organization's competencies. Exception to this will require approval of the board of directors.
Business risk	Projects (products) will have a high degree of success (90 percent or more) before being selected.
Project risk	Project goals will have a 70 percent or greater chance of success.
Technology	Projects requiring new technology will be compatible with existing core competency growth plans.
Profitability	Projects will have an expected profitability of greater than 15 percent.

TABLE 7.4 Organizational Project Portfolio (Example)

Item	Target	Actual
Small-sized projects	32	27
Medium-sized projects	12	17
Large-sized projects	4	7
High-risk projects	2	3
Medium-risk projects	6	0
Low-risk projects	40	45
Projects related to competencies	48	29
Number of project customers	>12	7
High-technology projects	2	0
R&D projects	5	3

Table 7.4 depicts information on the basis of what the organization sets as its target goal. Some discussion is required to show how the portfolio is out of balance and needs some future fine tuning.

- Project size shows a tendency to creep toward the larger projects. This would need a review to determine whether the organization can manage 48 projects at one time with 6 projects moving into the next category (excess of 5 in the medium category and 1 in the large category).
- Project risk exceeds the high category by one, whereas five additional projects are in the low category. If the high-risk projects are all large projects, this may have some impact on the organization—assuming that there will be some risk events that could have significant negative impacts.
- Project-competency connection seems to be out of balance with the goals. Slightly more than one-half of the projects are within the organization's core competencies. This would raise the question: What is the correct mix of core competencies for the organization? When the projects outside the core competencies are examined, one may find a significant disconnect with the business purpose.
- Number of project customers is not meeting the target goal. This needs a review and determination if this could lead to problems in the future.
- There are no high-technology projects. This may indicate a stagnation of technology growth to meet the marketplace or it may indicate a change in the business.
- R&D projects are not maintaining pace with the target. Possible causes are that resources are being used on existing projects or there is no need for R&D this year.

Reviewing Project Portfolio Management

Project portfolio management adds a dimension to an organization's capability and plan for growth. It first establishes criteria for selecting a project—whether internal or bidding on project work for clients. These criteria guide the organization to a model that supports the strategic goals and objectives while optimizing the balance of various characteristics of projects. Senior managers charged with selecting projects have an objective means of making informed decisions when using the selection criteria.

Building a model of what the organization should have in terms of projects provides a means of assessing the balance of the types of projects. One would not want all high-risk projects because of the potential for driving the organization into bankruptcy. The same goes for having all low-risk projects that typically have small profit margins, which will not support the organization's growth potential.

Managing by project portfolios provides high visibility to the organization's total projects without focusing on just one project at a time. It seeks to balance the many characteristics of projects to optimize the organization through alignment with strategic goals and objectives. A portfolio of projects improves any project-driven organization through improved visibility.

7.12 TO SUMMARIZE

The major points expressed in this chapter include:

- In a project, a strategic issue is a condition of pressure, either internal or external, that will have a significant effect on one or more factors of the project, such as its financing, design, engineering, construction, and operation.
- Examples were given of how strategic issues impacted some of the projects in the past and in contemporary times.
- Sometimes a strategic issue is subtle, as in the case of Eastman Kodak, where the CEO felt that his most urgent task was to eliminate the resistance to change from employees.
- A major issue facing the United States and other nations is continued dependence on oil and coal for the production of electrical power.
- The principal strategic issues facing potential projects for the construction of nuclear power plants were presented to keynote the importance of effectively managing the strategic issues in a project.
- Strategic issues on a project can arise from within the enterprise, and from outside, such as major concerns that the project stakeholders have about the project.
- A paradigm was suggested on how best to manage the strategic issues facing a project.
- Project success and project failure were suggested as major strategic issues that face a project.
- Prescriptions were given on how project strategy could be managed to increase the likelihood of project success or project failure in the conduct of a project.
- Potential strategic issues likely to impact a project during its life cycle should be identified during the early planning stages for the project.
- The matter of strategic issues should be given due consideration during the life cycle of a project.
- Project portfolio managers adds value to any organization by understanding the types of projects selected and maintaining a balance of projects consistent with the strategic goals and objectives.
- Project portfolio management highlights adverse trends for an organization when too much risk is accepted through individual projects.

7.13 ADDITIONAL SOURCES OF INFORMATION

The following additional sources of project management information may be used to complement this chapter's topic material. This material complements and expands on various concepts, practices, and theory of project management as it relates to areas covered here.

- Kim LaScola Needy and Kimberly L. Petri, "Keeping the Lid on Project Costs," and Lewis R. Ireland, "Total Customer Satisfaction," chaps. 9 and 26 in David I. Cleland (ed.), *Field Guide to Project Management* (New York: Van Nostrand Reinhold, 1997).
- Regula A. Brunies and Ross Brophy, "Minimizing Construction Claims under the Project Management Concept," and Randall L. Speck, "The Legal Standards for 'Prudent' Project Management," in David I. Cleland, Karen M. Bursic, Richard J. Puerzer, and Alberto Y. Vlasak, *Project Management Casebook,* Project Management Institute (PMI). (Originally published in *Proceedings,* PMI Seminar/Symposium, Montreal, Canada, 1986, pp. 198–212; *Proceedings,* PMI Seminar/Symposium, Milwaukee, Wis., 1987, pp. 566–576.)
- John D. Sterman, *Business Dynamics: Systems Thinking and Modeling for a Complex World* (Burr Ridge, Ill.: McGraw-Hill, 2000). It is not enough to identify the strategic issues in projects, but one must be able to solve those issues through analysis and application of working models. This book provides the basis for strategic thinking in the context of business, engineering, and social and physical sciences.
- Stephen G. Haines, *The Systems Thinking Approach to Strategic Planning* (Boca Raton, Fla., Saint Lucie Press, 2000). This book is a practical application to systems thinking and improves on the systems thinking concept first introduced by Peter Senge in the *Fifth Discipline* (Doubleday, 1990). This book focuses on planning strategies and the change management process in support of customer satisfaction.
- Anonymous, "Seattle Light Rail in Question," *Railway Age,* June 2001. This article describes the cost overrun and schedule delay of the Seattle Light Rail Transport System. It introduces the issue of cost and schedule growth by approximately one-third and the challenges associated with this project.
- Anonymous, "Thinking Outside the Box," *Chain Store Age,* May 2001. This article addresses numerous issues that face organizations when considering doing business in a particular state, county, or municipality. If a project was being planned for a particular location, the list of challenges to good business would be invaluable. With projects bridging several communities, such as a telecommunication tower project, the issues described would apply and require resolution for the best outcome.
- William R. Bigler, "The New Science of Strategy Execution: How Incumbents Become Fast, Sleek Wealth Creators," *Strategy & Leadership,* May–June 2001, pp. 29–34. This article focuses on the delays in implementing strategies for firms and the resultant chaos. Discussion centers on the identification of opportunities and rapid implementation to achieve the desired outcome.

- Robert Buttrick, *The Interactive Project Workout: Reap Reward from All the Business Projects,* 2d ed. (Englewood Cliffs, N.J.: Prentice Hall, 2000). This book covers a variety of subjects for managing projects; most notable is the framework of projects. One chapter is devoted to managing a portfolio of projects. Other chapters deal with reviewing the project and ensuring that it follows generally accepted practices for project management.

7.14 DISCUSSION QUESTIONS

1. Define a strategic issue.
2. Select a project management situation from your work or school experience, and list the strategic issues.
3. What methods might project managers employ to identify the strategic issues of a project?
4. What approaches can be used by project leaders to assess the impact of a strategic issue?
5. What management techniques can be used to address strategic issues?
6. Why is project success or failure considered to be a strategic issue?
7. What roles do environmental issues play in projects such as power plants and other major construction projects?
8. List and define the elements of the phase approach to dealing with strategic issues.
9. In identifying the strategic issues of a project, management can ask questions pertaining to the project stakeholders. What kinds of questions should be asked?
10. What is meant by the strategic relevance of an issue?
11. How can management assess the criticality and urgency of a strategic issue?
12. How can managers ensure that project team members are aware of and understand the project strategic issues?
13. What trends can be recognized when project portfolio management categorizes projects?
14. Discuss the use of project portfolio management when an organization has not fully developed and announced its strategic goals and objectives.
15. What advantages do you see for using project portfolio management in your organization?
16. How does project portfolio management affect the allocation of resources in your organization?

7.15 USER CHECKLIST

1. Do the project managers of your organization understand the concept of strategic issues? How do they manifest this understanding in managing projects?
2. Do any formal methods exist in your organization for strategic issue management? What are they? How are they used?
3. Do the project managers of your organization attempt to identify project interfaces that can seriously impact the outcome of a project? Explain.
4. Does top management use any postproject appraisals to help uncover strategic issue-related problems? Does management see the value in post-project appraisals?
5. Does the management of your organization recognize the importance of understanding public perception? In what ways do project managers control public perception?
6. Are there any outside advocates that can be or are effective in altering public opinion in favor of your organization's projects?
7. Do project managers assess the environmental impacts of projects? In what ways?
8. Could the phase approach to managing strategic issues be used effectively in your organization? How?
9. Are current project managers kept informed of the factors likely to impact project success or failure?
10. Does management seek to identify the relevant issues for each project stakeholder?
11. Does management identify the strategic relevance of each issue and determine the actionability, criticality, and urgency? In what ways is this done? What other methods could be used?
12. Are project team members made aware of strategic issues? How? Do they then attempt to monitor these issues as they relate to their own work packages?

7.16 PRINCIPLES OF PROJECT MANAGEMENT

1. Strategic issues may adversely impact project success through inadequate attention being paid to the issues.
2. Large, long-term projects have a great potential for cost overruns and schedule delays.
3. Relationships between the project owner and others are critical to the success of the project.

4. Large projects have diverse stakeholders who do not agree on the risk of the project or the benefits of its product.
5. Anticipating and addressing strategic issues materially improves the chance for successful projects.

7.17 PROJECT MANAGEMENT SITUATION—SOME STRATEGIC ISSUES

When an organization changes from its present type of work to a new competency, there are typically uncharted courses taken. The obvious impediments to change are easy to identify and resolve through some routine planning or contingency action. It is the unanticipated issues that will create problems—emerging when other activities are taking the time and resources or slowly materializing in a fashion so that it is difficult to characterize the issue.

Anticipating issues for a major change of direction for an organization may be unsuccessful for several reasons. There may not be someone knowledgeable available to review the plans and identify potential problems. Also, the organization's plan for change may have weak goals that are unclear or not understood. This anticipated change is probably a new venture for the organization that is based more on a desire to reposition than the facts regarding difficulty for the change.

One of the more challenging situations is to assume that because the competition has made a similar change this organization can also transition to the new position with relative ease. It must be remembered that the competition will neither share the difficulty of the transition nor share information on the success of their transition. It is not in a competitor's interest to help another position to compete for a part of the market.

Strategic issues may be a part of the operating environment, such as performing work in a foreign country using indigenous unskilled labor or conforming with laws or customs of the foreign country, that must be considered during the decision-making process. The best method of meeting strategic issues and resolving them through various means is to be well armed with information.

Issues that affect relationships in a foreign country may be derived from religious followings, language barriers and language translations, work ethics, local and federal laws, customs of the population, labor unions, trade barriers, and human skill levels. It takes an expert on the country to identify these differences from what is current practice in the United States. Some issues may arise from what one would assume is a favorable situation for the local population.

For example, apparently favorable situations such as employment and good wages can cause competition among those seeking jobs and possible sabotage of the work by those not employed. Some countries require that groups of people be hired for jobs—rather than the typical model of hiring one person at a time. These same groups may require that the leader be paid, but that he will not be required to work or supervise. If work instructions are given in English, the worker may often use lack of communication as the excuse for poor performance.

Initiating a venture that can develop strategic issues can be difficult to complete without thorough planning with a lot of information and a process to handle emerging issues. Failed projects may be the outcome when the organization is overcome by issues that seem to have no immediate answer.

7.18 STUDENT/READER ASSIGNMENT

1. You are the project manager for a light manufacturing firm and it has been decided that the work can be done at less overall cost if the items are manufactured in Dalian, China. What issues do you anticipate with manufacturing the products in China?
2. A competitor has a major coffee-growing effort in Ethiopia and seems to be doing well. Management has decided to invest in a project in Ethiopia start a major wool-growing effort that will provide cheap wool for the world market. What issues do you see in this venture?
3. Your organization, an experienced mine operator in the United States, has been invited to participate in a major mining operation in northern Canada. You have been tasked with identifying any issues associated with partnering on the new mine. What are the issues?
4. Your organization has been awarded a contract to build an airfield in the Sudan. All construction equipment will be transported to the location. One of the issues is that heavy equipment operators are not available from the indigenous labor force. What are some of the means to resolve this issue?
5. While working in a foreign country, you identify the issue that the computers are not working properly because of the difference in frequency of the local electricity (50-Hz current versus 60-Hz current in the United States). Your project management scheduling tools as well as clocks are not working properly. What action should you take to resolve this issue?

P • A • R • T • 3

ORGANIZATIONAL DESIGN FOR PROJECT MANAGEMENT

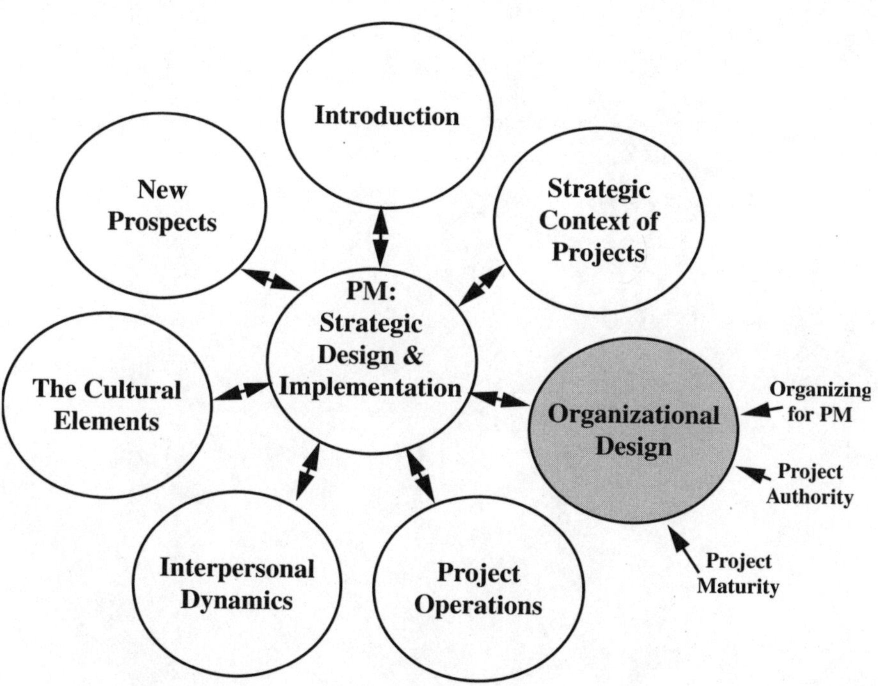

CHAPTER 8
ORGANIZING FOR PROJECT MANAGEMENT

"...to our worship of quantity and indifference to quality, to our unthinking devotion to organization, standardization...."

DANIEL GREGORY MASON, 1873–1960

8.1 INTRODUCTION

How many times has the reader been in situations in which the lack of a clear organizational design has created problems? The authors suggest that the lack of a clear organizational design has created manifold problems in the attainment of the organizational "choice elements," and has been an ongoing source of frustration in all too many organizations. In the project-driven organization, special care has to be given in designing and implementing suitable organizational models to provide the basis for the delegation of authority and responsibility.

In this chapter, a citation of probable organizational deficiencies that do not adequately describe individual and collective roles in the organization will be noted. Then the various forms of project organizations will be presented to include the controversial "matrix" design that has long occupied the attention of theorists and professionals. The relative authority of the project managers vis-à-vis the functional managers will be portrayed, along with a description of the basic project-functional interface typical of the project-driven organizational unit. Organizational networking, the role of the project management office, and administration will be provided.

8.2 PROJECT-DRIVEN ORGANIZATION

This chapter examines the project-driven organization, including its alternatives in coping with the use of cross-functional teams that are characteristic of the matrix organization. Some suggestions are offered on how to deal with the matrix organization. A project-driven organization has the following characteristics:

- It has an organizational design that defines the use of a "matrix organization" structure which provides a focus for the management of projects.
- It supports the linkage of projects as building blocks in the design and execution of organizational strategies.
- The organization has senior executives that are committed to the use of projects in the design and execution of enterprise strategies.
- It constantly reinforces the role of project management as the means for dealing with product, service, and process changes in the enterprise—carried out through the management of a project portfolio that prepares the organization for the future—as if that future mattered.
- An explicit culture of project management is evident in the organization, and means are constantly used to reinforce this culture.
- Project management experience has become an explicit consideration for promotion to higher levels of responsibility in the enterprise.
- Teamwork has become a key characteristic of the culture of the enterprise.
- The management of stakeholders is a key task of the project team's endeavors.
- Project management is a core business process in the enterprise.

8.3 SELF-MANAGEMENT IN ORGANIZATIONS

One of the biggest trends of all in the field of management is the rethinking and reevaluation that is under way in the application of management theory reflected in the major changes in the organizational design and empowerment strategies used today. Many organizations of all kinds are starting to abandon the revered "chain of command" where authority and responsibility were placed, in favor of empowering employees to "manage themselves." In order for the employees to manage themselves, an organizational design is needed. The use of the alternative team organizational structure has provided the basis for such an organizational design.

As employees serve on alternative teams and are properly empowered to do so, they are freed of being closely directed and controlled, as was the case in most traditional organizations where first-level supervisors and other managers often exercised "command and control" over their subordinates. In modern organizations that use alternative teams, these traditional supervisors are gone and even if still around, they carry out different strategies in their new role of maintaining oversight of the employees. Such new supervisors become teachers, mentors, facilitators, coaches, and the like, where they work with the teams rather than supervising them in the traditional sense. Employees are free on the teams to figure out how to get the job done without central planning and control. Sometimes this new paradigm is called "self-management and organization."

The new organizational design embracing teams is akin to the biological world, where uncontrolled environments and actions produce remarkable results of efficiency and effectiveness through a process of self-management and adaptation. In such environments the teams are encouraged to design and execute their own strategies, to experiment, and to seek information and assistance wherever necessary to include organizational members, suppliers, customers, and other stakeholders. New streams of performance information are created and disseminated so that team members as well as other organizational stakeholders can see what is going on and what is working positively for the enterprise's goals and objectives.

Employees serving on these teams express new dimensions of eagerness and enthusiasm, anxious to take on new dual responsibilities: first, discharge of their obligations for the technical work needed by the enterprise, and second, the management of that technical work. For example, in a furniture factory in rural Virginia, productivity soared after workers took over production scheduling and problem solving. At a refinery in the Middle West, those who traditionally had carried out specific work procedures came up with their own policies and procedures, resulting in huge gains in output. Even the process of strategic planning has been changed in some companies through the reduction of staff people to do the planning and dependence on teams of employees to keep in touch with major stakeholders such as customers and suppliers to discern trends that need to be factored into the overall strategic planning being carried out for the enterprise. Yet, there have been some real contemporary organizational deficiencies.

8.4 ORGANIZATIONAL DEFICIENCIES

Here are a few examples of how deficiencies in the organizational design affect project success (and failure):

- On the Shoreham Generating Plant project of the Long Island Lighting Company, the organizational arrangement left lines of authority and responsibility blurred and unclear from the start. The lack of adequate organization was a major deficiency that significantly prejudiced the utility's ability to manage the project. Over the life of the project, despite repeated complaints about role confusion and tangled lines of authority and unclear accountability, the senior managers of the utility failed to create an organizational framework that allowed its managers to direct and manage the construction of the plant efficiently.[1]
- An investigation of the Trans-Alaska Pipeline System (TAPS) project indicated that organizational structure significantly influenced project performance.[2]

[1]Paraphrased from Recommended Decision by Administrative Law Judges William C. Levey and Thomas R. Matias, *Long Island Lighting Company–Shoreham Prudence Investigation,* case no. 27563, State of New York Public Service Commission, March 13, 1985.

[2]T. F. Lenzner, *The Management, Planning and Construction of the Trans-Alaska Pipeline System* (Anchorage, Alaska: Pipeline Commission, 1977).

- A Rand Corporation study of "new technology" process plant construction found that the most prominently mentioned management-related reason for increased costs was "diffused decision-making responsibility for a project."[3]
- The fatal launch of *Challenger* is an example of some difficulties that had their genesis in a faulty organizational design. NASA's leaders were preoccupied with raising money for NASA from Congress. The organizational components of NASA were supposed to work together, but the Marshall, Kennedy, and Johnson Space Centers behaved more like baronies, not communicating with each other or with the top of NASA. The flow of information up and down the NASA hierarchy was, according to *Fortune* magazine, as flawed as the now notorious "O" rings.[4] The Marshall Space Center had an ambiguous chain of command with a reporting relationship to the Johnson Space Center in Houston, but not under Johnson's management control. The Marshall Center also reports to the Office of Space Flight at NASA headquarters and in theory cooperates closely with the Kennedy Center in Florida. However, the anomalies in the organizational reporting relationships were further blurred by cultural factors, which allowed jealousy and rivalry to exist among Marshall, Johnson, and Kennedy Centers. Also, there was resistance to NASA headquarters' oversight of their operations.[5]

Project management has led the way in the formalization of the erosion and crossing of organizational boundaries. In today's competitive world, the crossing of many boundaries—functional, geographic, organizational—is showing promise of becoming a way of life. Jack Welch, CEO of General Electric, says that "to create what we call 'boundaryless' companies, we no longer have the time to climb over barriers between functions like engineering and marketing, or between people—hourly, salaried, management, and the like. Geographic barriers must evaporate. We've got to simply delegate more and simply trust more. We need to drive self-confidence deep into the organization. We have to convince our managers that their role is not to control people and stay on top of things, but rather to guide, energize, and excite."[6] Surely project managers, who have had to survive in boundaryless organizational designs, are well equipped to provide leadership in reaching the boundaryless companies envisioned by Jack Welch.

Some of the shortcomings of traditional organizational hierarchies and organizational design follow:

- Formal enterprise hierarchies tend to be slow, inflexible, and bureaucratic.
- Formal structures can create barriers between the enterprise and the customer.
- It is difficult for enterpreneurship to flourish in a bureaucracy, particularly in the front-line operating units.
- Structure is only one part, a small part, of systems-oriented organizational change.

[3]Rand Corporation, *A Review of Cost Estimation in New Technologies: Implications for Energy Process Plants,* Santa Monica, Calif., July 1978.

[4]Michael Brody, "NASA's Challenge: Ending Isolation at the Top," *Fortune,* May 12, 1986.

[5]Ibid.

[6]1992 General Electric Annual Report.

- A traditional organization looks more to the higher headquarters and less to the customer, suppliers, and other key stakeholders.
- The inadequacies of innovation in the traditional structure gave rise to the creation of "skunk works" where innovation could thrive unimpeded by traditional bureaucracy.
- Scant attention was given to horizontal processes within organizations, and to dealing with outside stakeholders.
- Organizations were viewed more as structures than as processes.
- The successful traditionally organized enterprises were able to foster a culture that encouraged bottom-up ideas and initiatives.

Project management has changed some of these shortcomings.

8.5 THE PROJECT ORGANIZATION

The term project organization is used to denote an interorganizational team pulled together for a specific purpose. Personnel are drawn from the organization's functional units to perform a specific task; the organization is temporary, built around the purpose to be accomplished rather than on the basis of functional similarity, process, product, or other traditional bases. When such a team is assembled and superimposed on the existing structure, a matrix organization is formed. The matrix organization encompasses the complementary functional and project units. Figure 8.1 is a model of the matrix organization.

Before we examine the matrix organization, a brief review of other means of organizing is needed. Organizational theorists have developed various ways of dividing the organization into subunits to improve efficiency and to decentralize authority, responsibility, and accountability through a process of departmentalization, with the objective of arriving at an orderly arrangement of the interdependent parts of the organization. Departmentalization is integral to the delegation process. The most widely used system of departmentalization includes:

- *Functional* departmentalization, where the organizational units are based on distinct common specialties such as finance, engineering, and manufacturing
- *Product* departmentalization, by organizing into distinct units responsible for a major product or product line
- *Customer* departmentalization, where organizational units are designated around customer groups such as the Department of Defense
- *Territorial* departmentalization, with people located in units based on geographic lines, for example, western U.S. marketing zone
- *Process* departmentalization, where the human and other resources are based on a flow of work such as an oil refinery

In the late 1950s and early 1960s, these traditional forms of organizing resources were proving inadequate to cope with the need to integrate the disparate organizational

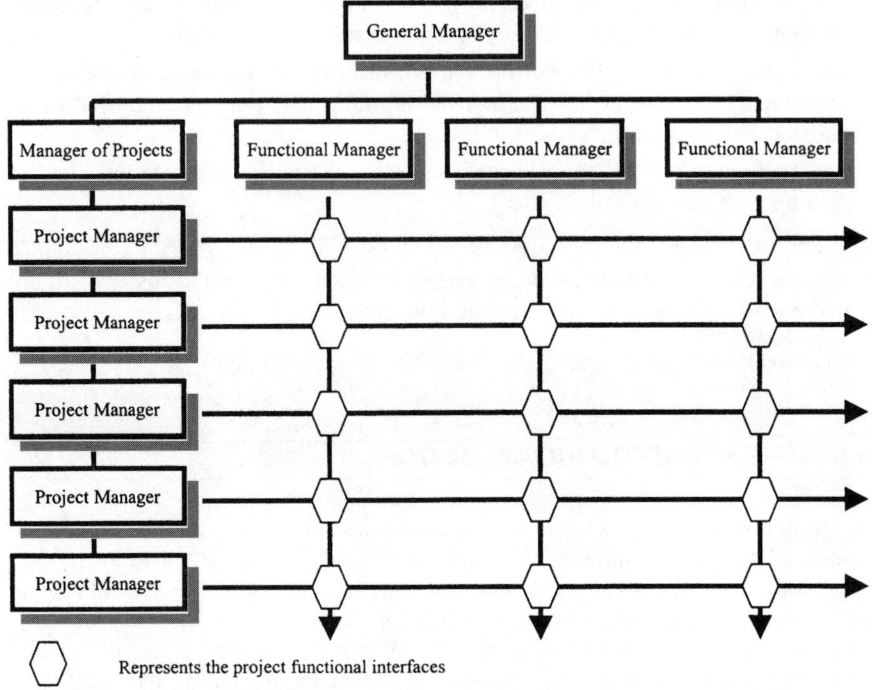

FIGURE 8.1 A basic project management matrix organization.

activities required of such ad hoc endeavors as a project. Experimentation with alternative, more flexible forms of organization was used to meet the demands of the evolving, dynamic "projects" business. The result was a blend of the functional structure and a designated focal point for managing a project within an enterprise. The *project-driven matrix organization* evolved as essentially a network of interactions between a project team and the traditional functional elements of an organization.

8.6 VARIOUS FORMS OF THE PROJECT ORGANIZATION

A variety of project-driven organizational forms exists. At one extreme is the *pure project organization,* where the project manager is given full authority to run a project as if it were a one-product company; at the other extreme is the pure functional organizational department on a traditional basis, reflecting the traditional hierarchy. A functional department is a hierarchical organization where each employee normally has only one superior. Employees are grouped on the basis of the functions to be carried out, such as marketing, engineering, production, finance, and so forth. In a functional organization the managers provide for the use and integration of resources, maintain oversight over the use of such resources, and assign people to jobs on the basis of the need to provide resources to support department needs. Sometimes project managers are assigned to manage projects which are carried out within the functional department—but with common resource support from some other departments such as human resources, finance, or procurement support. Functional groups are becoming more and more specialized resource pools with special capability to nurture the state-of-the-art expertise needed to support project purposes. In the middle lies a variety of project-functional combinations of the matrix organization. Each of these forms has certain advantages and disadvantages; no one form is best for all projects, or even best for one project throughout its entire life cycle. The essence of project organization is flexibility. The project can be built around the organizational strategy; as the strategy changes, so must the focus of the organization.

In one study of the significance of project management structure on the success of 546 development projects, it was found that projects relying on the functional organization or a functional matrix were less successful than those that used a balanced matrix, project matrix, or project team. The project matrix outperformed the balanced matrix in meeting schedule and outperformed the project team in controlling cost.

Basic definitions of the types of structures in this study were similar to the types described in research by Larson and Gobeli:

- *Functional organization.* The project is divided up and assigned to relevant functional areas with coordination being carried out by functional and upper levels of management.
- *Functional matrix.* A person is designated to oversee the project across different functional areas.
- *Balanced matrix.* A person is assigned to oversee the project and interacts on an equal basis with functional managers.
- *Project matrix.* A manager is assigned to oversee the project and is responsible for completion of the project.
- *Project team.* A manager is put in charge of a core group of personnel from several functional areas who are assigned to the project on a full-time basis.[7]

Earlier in this chapter the term matrix was introduced. In the material that follows, a detailed examination of the matrix organizational design will be made. Before such examination is done, a brief review of the *pure* project organizational approach is needed to set the stage for the explanation of the matrix design.

In the pure approach, the project is truly like a minicompany. The project team is independent of major support from any major functional units or departments. Minor functional support in such matters as industrial relations, payroll, and public relations is provided by a functional element that takes care of the entire organization.

---

[7]Erik W. Larson and David H. Gobeli, "Significance of Project Management Structure on Development Success," *IEEE Transactions on Engineering Management,* vol. 36, no. 2, May 1989, pp. 119–125.

The major advantage of the pure project organization is that it provides complete line authority over the project personnel; the project participants work directly for the project manager, with the chief executive (or some other general manager) in the main line of authority. One of the strongest disadvantages of this type of organization is that the cost is increased because of duplication of effort and facilities. In addition, because there would be no reservoir of specialists in a functional element, there might be a tendency to retain personnel on the project long after they were needed. A functional group is needed to look toward the future and work to improve the company's technical functional capability for new projects.

The matrix organizational design is a compromise between the hierarchical structure of the traditional functional organization and the project team design where an explicit sharing of authority, responsibility, and accountability is carried out. As the matrix organizational design has continued to evolve, the use of self-managed work teams has also evolved, resulting in enhanced empowerment of teams as elements of enterprise strategy.

8.7 THE MATRIX ORGANIZATION

A mixed project and functional structure, or matrix organization, is desirable for managing certain projects within desired cost, schedule, and performance standards. The mixture can lie anywhere between the pure project and the pure functional extremes, the exact structure being determined by the particular project requirements.

The matrix organizational design emerged in the early 1960s as an alternative to the traditional means of organizing people serving on project teams. The matrix enjoyed popularity in the 1970s and early 1980s. Original concepts of the matrix organizational design emphasized the individual and collective roles of members of the project team. In some cases, companies went too far in trying to escalate the matrix organizational design throughout the breadth and length of the organization. Texas Instruments pulled back from the extensive matrix organizational design, citing it as one of the key reasons for the firm's economic decline.[8] Xerox Corporation reportedly abandoned the matrix form, claiming that it created a deterrent to product development.[9] Other signs of disenchantment with the matrix organizational design appeared. One of the more assertive statements was offered by Peters and Waterman in their book *In Search of Excellence.* They claim that the matrix was complicated and ultimately an unworkable structure, which "degenerates into anarchy and rapidly becomes bureaucratic and noncreative."[10] Yet, in spite of its challenges the matrix design continues to gain advocates.

The growing use of the matrix organizational design has provided legitimacy to the use of alternative horizontal organizational designs that complement the earlier traditional organizational structure aligned along functional lines. The use of alternative horizontal organizational designs is supported by several basic ideas. First

[8] "An About Face in TI's Culture," *Business Week,* July 5, 1982, pp. 21–24.

[9] "How Xerox Speeds Up the Birth of New Products," *Business Week,* March 19, 1984, pp. 58–59.

[10] Tom Peters and Robert Waterman, *In Search of Excellence* (New York: Harper and Row, 1982), p. 49.

is organizing the enterprise around emerging projects and organizational processes such as order entry, inventory management, and information management. Instead of creating the enterprise around functions or departments, it is built around key processes required for the delivery of value to customers. An individual is assigned as an owner of each process, and a project manager or process manager is appointed to maintain oversight over the development and management of the process as appropriate.

Advocates of the matrix organizational design offer many reasons for its efficiency and flexibility in marshaling and using the resources to support a project. Critics are quick to point out that the matrix arrangement is cumbersome, costly, and difficult to understand. As mentioned earlier, Larson and Gobeli offer a description of the different forms of matrix design in terms of the relative influence of the project and functional managers, that is, the functional matrix, the balanced matrix, the project matrix, and the project team. They also offer an insightful description of the advantages and disadvantages of the different matrix structures. They conclude that although the matrix has its disadvantages in terms of being cumbersome, chaotic, and anarchical, its popularity is not diminishing, but rather is the dominant mode for completing development projects.[11] Follow-on research by Larson and Gobeli leads them to conclude that different management structures can be applied at different phases of the project life cycle, and that there is no one best way to organize the project team except that the functional matrix and the functional organizational design for managing projects are less effective than a form that provides strong project leadership.[12]

Prescription of the expected formal individual and collective roles to be expected in the matrix organization is needed. Table 8.1 suggests a boilerplate model that can be used as a guide to such formal prescription. The use of the linear responsibility charting technique outlined in Chap. 9 is a productive way to develop these roles and in that development educate the people as to how they should operate in the matrix organization.

TABLE 8.1 Organizational Design for Project Management

Project manager	Functional manager
• What is to be done?	• How will the task be done?
• When will the task be done?	• Where will the task be done?
• Why will the task be done?	• Who will do the task?
• How much money is available to do the task?	• How well has the functional input been integrated into the project?
• How well has the total project been done?	

[11]Erik W. Larson and David H. Gobeli, "Matrix Management: Contradictions and Insights," *California Management Review,* Summer 1987, pp. 126–138.

[12]Erik W. Larson and David H. Gobeli, "Organizing for Product Development Projects," *Journal of Product Innovation Management,* vol. 5, 1988, pp. 180–190.

Many roles are carried out by the project manager. These roles arise at different times during the life of the project. At the beginning of the project the following roles are likely to be played out:

- A strategist provides leadership for the design and development of a project plan.
- A recruiter obtains the best possible talent to serve on the project team.
- A negotiator garners high-quality resources for the team.
- A visionary finds and communicates a vision to the project team and to other stakeholders.
- A designer maintains oversight over the design of the organizational structure for the project and the configuration of the anticipated project results to include all supporting systems.

During the execution of the project the project manager, in addition to continual reinforcement of these roles, executes additional roles as follows:

- A mentor who provides counseling and consultation to members of the project team when required.
- A coach who instructs and trains the team performers in the fundamentals of project management.
- An integrator who forms the project resources into a product, service, or process.
- An expediter who keeps people and other resources moving on the project.
- A conflict manager who helps resolve the conflicts over the use of resources that naturally arise during the life of the project.
- An influencer who sways stakeholders to support the project purposes.
- A decision maker who works with the project stakeholders in the removal of uncertainty concerning how resources will be used on the project.
- Finally, a diplomat who builds and maintains alliances with project stakeholders for the continuing support of the project and its role in the operational and strategic management of the enterprise.

Matrix organizational designs emerged to deal with the enigma and perceived inconsistency of having two or more "bosses"—a reflection of the fascination that conventional wisdom held concerning the impropriety of violating Fayol's principle of "unity of command." In today's team-driven organizations, authority-responsibility-accountability relationships are complex, everchanging, and based as much on individual (or group) ability to influence other people as on the formal authority of a defined organizational position. Given these considerations, what is the general nature of the matrix organizational design? Several observations can be offered:

- A formal matrix organizational design should be described along the demarcation suggested in Table 8.1. This formal design should not be inflexible, but should be offered as a way in which the authority-responsibility-accountability patterns should normally operate.

- The ability to influence other people through the continued demonstration of one's knowledge, skills, and attitudes is the final determining factor in achieving successful integration of individual and collective roles in the matrix design. However, one could make much the same statement about a management position in any type of organization.
- The growing use of alternative forms of teams in contemporary organizations will continue to make the matrix organizational form more acceptable and more flexible and will provide for bringing a philosophy of bringing people together regardless of their "home" organization into a focus to accomplish organizational purposes.
- *Matrix,* then, is more a state of mind to encourage people to work together to create value for themselves and for the organization.
- As an organization works in the matrix context, the structural form of the matrix will tend to erode and become institutionalized into the overall manner in which people relate to each other in their individual and collective roles. In such organizations, matrix is described as "simply the way we do things around here," truly a key element in the organization's culture.
- In the matrix organization people relationships work in many directions and are usually dominated by those relationships with (1) team members, (2) functional personnel, (3) upper-level management, and (4) internal and external stakeholders.
- When the company is organized into traditional functional departments, everything involving the project runs the risk of falling between the cracks in the organization structure. By having a project team appointed, a major step has been taken to put together the functional horizontal pieces of the ad hoc effort into a unified, harmonious whole. Projects that cross functional boundaries tend to be orphans, because they lack someone to act as a "champion" for pulling together their functional parts. Because projects require difficult, time-consuming work directed to communicating the need to synergize the project, the work to support the project can easily be deferred or be entirely forgotten. But when a project manager has been appointed, a champion should come into play to integrate the parts of the project into a synergistic whole.[13]

From the project manager's perspective an understanding of the role of the knowledge of the functional manager is important.

8.8 FUNCTIONAL AREA KNOWLEDGE

Because members of a project team come from different functions that are required on the project, questions can arise concerning the degree of knowledge and skill required of the project manager in leading the project team. Indeed, the same

[13]Benson P. Shapiro et al., "Staple Yourself to an Order," *Harvard Business Review,* July–August 1992, pp. 113–122.

questions can arise when a "general manager" assumes the leadership of an organization that consists of different functions and activities. A few guidelines are suggested that can help clarify the leadership required of project managers—and general managers—in this respect.

Each functional manager is required to have the knowledge and skills necessary to provide the leadership of the functional technology involved. For example, the manager of an engineering design functional element would be expected to have sufficient knowledge and skills to command the technology involved in support of the enterprise's mission. A project manager, or a general manager, would depend on that individual to develop the strategies and oversee the application of that function in the organization. A project manager, or a general manager, would be expected to have only the knowledge and skills in those disciplines that support the project or enterprise to the following extent:

- Be able to ask the right questions and know if the right answer is being given regarding the discipline. A command of the functional area would not be necessary.
- Have a general understanding of the function and the role that such functions have in regard to the overall organizational effort under way.
- Be able to know if the individual that has the responsibility for the function is able to make the contributions necessary to support the overall activity.
- Be able to define and understand the general part that the functions play in the overall activity.
- Have an oversight perspective of a conceptual model of what the function is expected to carry out.
- Be comfortable with the knowledge and skills of the team member representing the function, and be able to trust the individual's competencies in supporting the overall activity.
- If required, be able to select additional functional people to support the project, perhaps in a consulting capacity.

An individual who is an expert in some specialty or function, and assumes responsibility as a general or project manager, makes an important step toward obsolescence in that field. If that individual tries to keep abreast of that function, and wants to continue to "command" that function, erosion of the required general manager or project manager knowledge and skills can occur.

8.9 FOCUS OF THE MATRIX DESIGN

Managers should heed the advice given more than 30 years ago by one of the early writers in the then-emerging field of project management. Middleton offered the advice that neither the role of the project manager nor that of the functional manager should dominate in using project management. He further briefly described the general relative roles of these managers and charged top management with the responsibility for resolving the conflicts between them.[14] Middleton's advice is particularly appropriate to those managers who are considering the use of the matrix organization.

The matrix form of organization demands attention, for many managers do not have a clear, consistent concept of what it means. Although the matrix is used in a wide variety of different organizations, there is not a full understanding of its structure, processes, and impact on the parent organizational system. In its basic form a matrix organization is a network of interfaces between a project team and the functional elements of an organization. As additional project teams are laid across an organization's functional structure, more interfaces come into existence. The authority, responsibility, and accountability patterns found in these interfaces are delineated in subsequent portions of this chapter.

In its most elementary form, a matrix organization looks like the model in Fig. 8.2, where the interface of the project and functional elements comes about. The interface of these elements centers on the project work packages. The underlying concept of the work package is simply that of management by objectives and the decentralization of authority, responsibility, and accountability. Implementation of the project requires that the total job be broken down into components (hardware, software, and services) and that these components be further broken down into assignable work packages. Each work package is basically a "bundle of skills" that

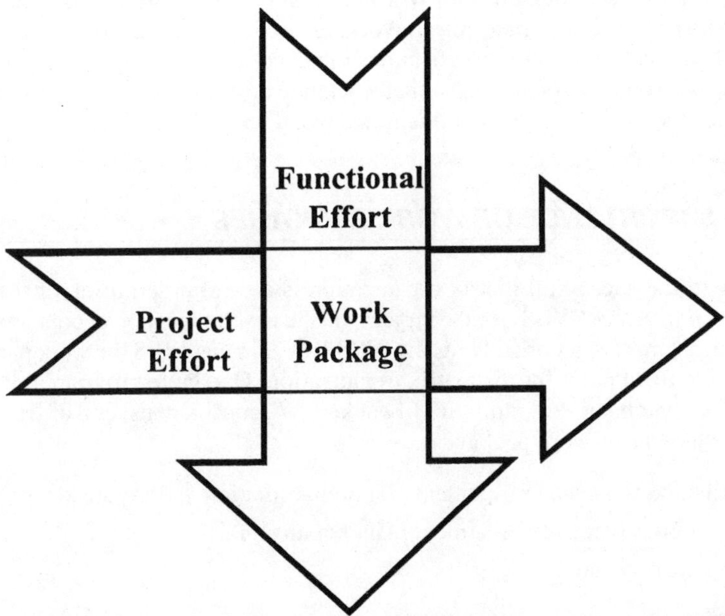

FIGURE 8.2 Interfaces of the project and functional effort around the project work packages.

[14] C. J. Middleton, "How to Set Up a Project Organization," *Harvard Business Review,* April 1967, p. 82.

an individual or individuals have to perform in the organization. A work package is negotiated with, and assigned to, a specific manager or professional. The individual who accepts the work package agrees to specific objectives and goals that are measurable, and to detailed task descriptions, specifications, milestones, budget for the work package, and so forth. This work package manager or professional is then held fully responsible for the work package meeting its objective on time and within budget.

The underlying premise of the matrix organizational form is that project objectives can best be reached if the organization's resources can be directly oriented toward those objectives without regard to traditional organizational structures and constraints. The organizational form of the matrix is used as a means to an end; it can be readily adapted to a changing environment. As the organizational need for new projects changes, the matrix structure tends to be fluid. Because organizations are organized around specific projects, the matrix is in a constant state of flux as projects are completed and resources are deployed to new or other current projects.

When the matrix organizational design is used in the management of projects, there will be modifications in the existing patterns of authority-responsibility-accountability. Reporting relationships will be modified, members of the project team will take on added authority and responsibility for the project work packages, and the role of the functional manager will be modified. The cultural changes coming out of the matrix organization will have reverberations throughout the enterprise as people take on new roles. Working across organizations to deal with stakeholders will also cause modifications in the culture and project teams need to recognize this and accept the reality that what they do as members of the focal point to manage the project can have influences beyond the project itself.[15]

8.10 IMPORTANCE OF WORK PACKAGES

The key to the successful matrix organization is a careful definition of the *work breakdown structure* (WBS) for the project and the development of an organizational structure that most appropriately fits the WBS.[16] Within that WBS the work packages provide the focal point for the matrix organization. One large program, the Water Pollution Abatement Program[17] in Milwaukee, Wisconsin, consisted of five major project elements or work packages:

- Jones Island wastewater treatment plant rehabilitation and expansion
- South Shore wastewater treatment plant expansion
- Conveyance systems

[15]For further insight into the cultural issues of project management see Chap. 19 and Patrick Brown, Sheila Grove, Richard Kelly, and Satyendra Rana, "Is Cultural Change Important in Your Project?" *PM Network*, January 1997, pp. 48–51.

[16]The concept of the WBS and the work package is discussed in Chap. 11.

[17]The program was defined as the entire undertaking in this effort, a $2.2 billion effort that consisted of many projects.

- Solids disposal
- Hydraulics and controls

8.11 THE PROJECT-FUNCTIONAL INTERFACE

Projects are essentially horizontal; the functional organization, as exemplified by the traditional organization chart, is vertical. The basic dichotomy found in matrix organizations centers around a project-functional interface reflected in Table 8.1. The syntax of the statements in that table is to provide a simple set of key words, as indicated by italics in the following list:

> The "demarcation" described in Table 8.1 is a very broad way of portraying the authority and responsibility relationships in the matrix organization which can be used as a point of departure to develop an understanding of the *web of relationships* found in the matrix organization.
>
> The interface clearly describes how project managers accomplish project ends by managing relationships within the total organization. There are few things project managers can do alone. They must rely on the support and cooperation of other people within the organization. They must look to functional managers for specific support. Indeed, project managers *get things done by working through others* in the classic sense of the phrase, which is often used as a definition of successful management.
>
> This managing of organizational relationships is three-dimensional. Upward, project managers must relate to their boss, who is either a general manager or a *manager of projects.* Horizontally, they relate to members of their project team. Diagonally, they relate to functional managers and to representatives of other organizations, for example, the customers.
>
> Managing these sets of relationships is a most demanding task. It is nearly impossible if care has not been taken to describe the formal authority and responsibility relationships that are expected within the organization. This means making explicit the network of relationships that project managers have in each of the three dimensions. To whom do they have to relate? What are the key relationships? What is the work breakdown structure around which action is expected? Who works for whom?[18]

The matrix provides a sound basis for balancing the use of human resources and skills within the total organization as people are shifted from one project to another. A project can be viewed as a small business within a larger enterprise whose ultimate goal is to go out of business when the project is terminated. Hence, as the enterprise has a stream of projects that is flowing through the organization,

[18]David I. Cleland and William R. King, *Systems Analysis and Project Management,* 3d ed. (New York: McGraw-Hill, 1983), p. 351.

each project is in a different phase of its life cycle. The opportunity exists for the general manager to balance human resources in the organization and apply these resources where necessary to keep the stream of projects flowing freely and effectively in the organization.

The key to making the matrix work effectively is to recognize the complementary roles that exist and to carefully delineate the relative authority, responsibility, and accountability for the people filling these roles.

The characteristics of an effective matrix organization include the following:

- First, appropriate empowerment to include appropriate documentation is instituted so that authority, responsibility, and accountability are shared as team members perform their individual and collective roles on the project team.
- Second, lines and methods of communication are well established, and people understand and accept their obligations to communicate freely information team members and stakeholders need.
- Third, functional managers accept and are committed to the matrix organizational design and are committed to the provisioning of functional resources to support the project needs.
- Fourth, there is an explicit understanding of the interdependent roles of the functional managers and the project managers working together to support the project purposes.
- Finally, there is a prevailing culture in the enterprise that basically supports the organizational design of the matrix as a way of sharing scarce resources in the enterprise. People agree that the matrix reflects "simply the way that we do things around here."

8.12 A CONTROVERSIAL DESIGN

The matrix organizational design and the matrix organizational concept have had problems and abuses. Part of the problem in the use of the matrix design has been characterized as caused by, or corrected through, a weak or strong matrix design. A weak matrix is one characterized by the following:

- A failure on the part of key participants to understand the basic principles and roles involved in the matrix
- An inherent suspicion and distrust of any organizational design that departs from the management principle of *unity of command* in which one individual is expected to receive orders and direction from only one individual
- Functional managers who feel threatened by an apparent superiority of the project objectives and goals over those of the functional entity
- A failure on the part of senior management to see to it that some basic documentation is prepared to describe the formal and reciprocal roles of the key

managers involved on the project: the project manager, functional managers, and work package managers

- A lack of appreciation on the part of the project manager and key staff to understand and respect the role of the functional professionals and their authorities and responsibilities in the management of the project
- Poor selection of project and functional managers
- The project manager who sees his or her role as simply a coordinator rather than as a manager in the truest sense of the word
- A project manager who fails to understand the many stakeholders on the project, even those outside the parent organization, who have to be "managed" to fulfill the project ends
- Lack of trust, integrity, loyalty, and commitment on the part of the project team members
- Failure to develop and maintain the project team
- Putting the functional managers on report to senior managers rather than working out the conflict and challenges that are bound to occur in the management of the project
- Indecisiveness on the part of the project manager who would rather defer decisions to the senior managers than make as many decisions as possible on the project, referring only those that must be made by the senior executives

Conversely, a *strong* matrix exhibits these characteristics:

- Care has been taken by senior managers to define the individual and collective authority-responsibility roles of the project manager, functional managers, and work package managers.
- The project manager and the other key managers feel a strong sense of personal ownership and responsibility for their work and are willing to share ownership and responsibility, their resources, and the rewards to be gained from the successful projects.
- The project manager is given full authority and responsibility and is expected to exercise managerial prerogatives in managing the project so that it is completed on time and within budget and satisfies its technical performance objectives.
- The project manager knows how to delegate, demands excellent performance by the functional managers and the members of the project team, and is willing to accept full responsibility for the project.
- The project manager is prompt and judicious in resolving conflicts and disputes that will inevitably arise in the program.
- Project problems are taken to senior management as a last resort, but senior management is informed at all times of the status of the project.
- High performance and quality standards are expected from the functional entities participating on the project.

- The project team does not interfere in the prerogatives of the functional managers and does not permit the functional managers to interfere in the management of those portions of the project that lie within the jurisdiction of the project team.
- The project manager remains focused on the prudent and reasonable management of the project and appreciates that the project is basically a building block in the strategic management of the enterprise.

Clearly, there are many more projects that are successful using the *strong* matrix than those using the weak matrix.

The *bottom line* is to select project managers and other key managers who will be dedicated to their jobs, understand them, and seek unambiguous definition of their roles, and are willing to assume responsibility for the project. Such selection will help ensure that a strong matrix emerges.

8.13 NO ONE BEST ORGANIZATIONAL DESIGN

The best organizational design to use in the management of projects is dependent on the particular circumstances of the project and its organizational and stakeholder environment. Tracey Kidder, in his Pulitzer Prize–winning book *Soul of a New Machine,* describes a product development effort at Data General on the Eagle Team in the development of a new standard in miniframe computers.[19] The book describes the massive effort carried out by a project team of specialists protected from organizational politics and interruptions and engaged in creating something that had not been done before.

There are many alternative ways to organize for the management of a project. One approach is to have a functional organization manage the project using an individual acting as a focal point in the functional entity. A functional organization is simply an organizational unit of work, configured on a hierarchical basis, with each person having one superior. People are grouped by specialty such as marketing, engineering, finance, and construction. Sometimes no single individual is designated as having overall project responsibility. Rather each department and section within the function performs its work needed to ensure input into the project. There is no one person maintaining oversight for the management of the project except the functional manager, who is likely to be busy maintaining oversight over the operation of the total functional organizational unit. Some of the likely problems that such an organizational design can create include (1) interdepartmental politics and territorial battles, (2) avoidance of conflict resolution, (3) overdependence on the existing formal communication networks, (4) having to depend on people to provide schedule and cost control support who lack the proper credentials, (5) dependence on accounting and financial information systems that are based on department needs and are fiscal year oriented rather than project oriented, (6) propensity of department

[19]Tracey Kidder, *Soul of a New Machine* (Boston: Little, Brown, 1981).

personnel to compromise schedule and cost needs in order to meet quality standards, and (7) general lack of concern for what goes on relative to the project.

What can be done to reduce harm to the project's needs when the functional organizational design model is used? Insist on having the project managed on a total systems basis. Insist on having a specific designation of relative authority and responsibility for the project. The use of rigorous established early planning on the project could help. Make sure that representatives from the functional organization help in developing the project plan. Finally, take some time to train the functional representatives that are working to support the project in the basics of project management processes and techniques done within the context of the matrix design.

8.14 GLOBAL PROJECT ORGANIZATIONS

As global competition intensifies, there will be more global projects and strategic alliances among companies and countries. Project managers will no longer be concerned solely with a "domestic" project—each domestic project has a good likelihood of becoming global in nature.[20] Each global project, like a domestic project, is unique—one of the key characteristics of projects. But global projects will be distinctive in that the project team, working across companies and countries, will encounter situations in which boundaries will cause new challenges in customs, cultures, and practices. The traditional matrix structure common to the project-functional interface will take on a global nature. Granted that the matrix structure in a domestic project is complex, in the global project this structure becomes even more complex. It is important that the formal role of the project manager be carefully delineated and that the roles of the team members be specific in terms of their authority, responsibility, and accountability. The chances of project success in the global project depend on many major forces and factors. If care is not taken at the outset of the project to clearly stipulate to all the stakeholders understanding the managerial and leadership role of the project manager and the project team, the opportunity for a successful project is clearly diminished. Organizational design arrangements for "managing" the customer need to be considered.

8.15 PROJECT-CUSTOMER RELATIONSHIPS

The interactions between a customer project office and industry agencies can be appreciated by reviewing Fig. 8.3. The interactions suggested by the figure are only a partial illustration of the number, size, and intensity of the project interrelationships. For example, on a major government project the project manager and

[20]See David I. Cleland and Roland Gareis, *Global Project Management* (New York: McGraw-Hill, 1993), for a comprehensive review of the management of global projects.

office personnel interact with the highest levels of government and industry. Contractors doing business with these organizations tend to develop project offices, which mirror the skills of the government project office. The relationship of the two organizations—the military department and the defense contractor—revolves around the two project managers, as illustrated. Although we use an example drawn from the defense industry, the same basic model could be used to describe any customer–project management situation.

Unifying the parts of the organizational components of the project and its stakeholders is a necessary activity of organizing the project. Unification is particularly important between the project prime contractor and the project owner. A prudent project owner will want to have a sound organizational design through which the owner's needs and the needs of the project contractor can be planned, understood, and met. This organizational design must reflect the reciprocal authority and responsibility tied to the work packages of the project, essentially providing answers to the specificity of individual and collective roles and the level of involvement of each party in the management of the project.

8.16 ORGANIZATIONAL NETWORKING

A project manager is at the focal point of an interconnected network of alliances with members of the project team and with a varied set of people inside and outside the organization, in short, the stakeholders. A network is a set of reciprocal relationships that stabilizes the project work, giving it predictability and synergism. Networks stretch horizontally, vertically, and diagonally to the project's internal and external stakeholders. The strength and viability of these networks depend much on the ability of the project manager to build and maintain alliances with the many people who can help, hinder, or be indifferent to the needs of the project. These networks of relationships with all the project's stakeholders are a

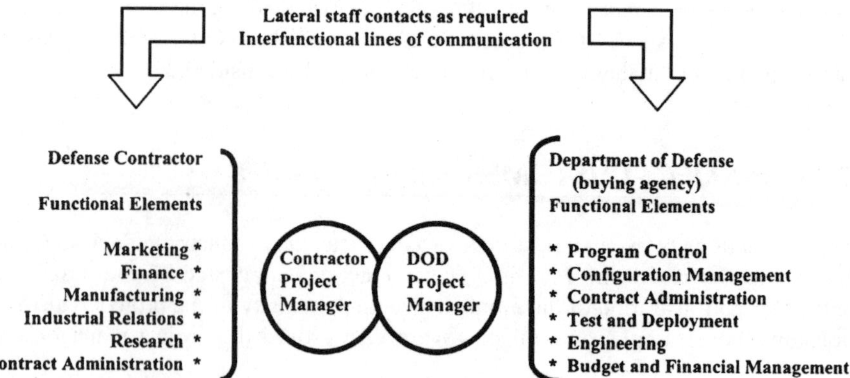

FIGURE 8.3 Customer–project manager relationships. (*Source: Adapted from David I. Cleland, "Project Management—An Innovation in Management Thought and Theory,"* Air University Review, *January–February 1965, p. 19.*)

valuable asset that the project manager possesses in meeting the opportunities and problems of a project.

A project manager must network with the project stakeholders for one compelling reason. The project manager depends on these stakeholders and cannot get the project finished without them. The project manager's ability to build and maintain these networks depends on the project manager's authority and how that authority is perceived by the project stakeholders. The project manager's reputation, alliances, position, favored standing, diplomacy, influence, communication skills, and persuasive skills all help facilitate the building and maintenance of the network. The project network connects like tentacles with diverse project stakeholders, establishing relationships and quid pro quo designed to support the project needs.

The art of networking is one of the most unceasing challenges facing the project manager. Most of the project manager's daily activities deal with the ongoing discovery and creation of relationships directed toward supporting project needs. The project team members need freedom in pursuing their technical expertise, on one hand, and yet must be brought together and unified in supporting the project needs, on the other. A healthy and successful team is marked by healthy relationships. Without this relating that leads to networking, the team weakens and may stumble along as a collection of individuals, but die as a team. Interacting, interfacing, and building networks in harmony with others, the team should become a unity of cooperative effort.

Most successful relationships in a project team are an ongoing process of trial and error, negotiations, resolution of conflict, authority, responsibility, evaluation, planning, execution, commitment, accountability, organizing, control, and communication. The elements are as complex as any scientific formula. Team members must approach these relationships as a creative challenge requiring concentration, innovation, and careful tending and cultivation. Networking requires an open mind and courage and flexibility to compromise when the project team's well-being and the project's outcome are at stake. It demands that the team members seek maximum fulfillment of their technical expertise, yet tolerate disappointment when their technical position is reduced to preserve the overall synergy of the project. Such disappointment and even feelings of rejection require that the team members nourish the attitude that they will try again and again without any guarantee that future disappointments will not happen. The project manager and the team members must continuously work at the skills required for building and maintaining relationships necessary for effective networking.

The experts who are members of a project team can impair the team ambience by always insisting that they are right and by being afraid to reveal their imperfections. Team members all too often see this as something they must do, lest they lose the respect of their contemporaries and their status. So strongly are team members affected by the need to hide their imperfections that they may even run the risk of destroying the valued relationships that make the team effective and a winner. Rigidly adhering to their rightness, the team members (including the project leader) stifle discussion and exasperate others on the team who grow weary of always hearing about "the world according to me." The project team all too often fails to see that nothing has been gained if the final result means being fearful of building relationships, of networking, and of confronting issues.

In the past several years there has been a growing appreciation of the "project office" as an organizational unit in the project-driven organization.

8.17 THE PROJECT MANAGEMENT OFFICE

In the project-driven enterprise, the use of project management concepts, processes, and techniques tends to dominate the culture of the organization. In maintaining an active portfolio of projects, many of the methods and processes of how these projects are dealt with during their various life cycles should be done in an efficient and effective manner so that costs are minimized and maximum profits accrue to the enterprise. During the evolution of the use of project management as a key enterprise strategy, consideration should be given to the establishment of a project management office (PMO) as a strategic initiative of the enterprise. Such an office can be a focal point through which both programs and projects are managed as products, services, and organizational processes change in the enterprise.

The basic organizational design of the project-driven enterprise will consist of (1) the key functional elements that provide support to the portfolio of projects such as design, engineering, finance, marketing, research and development, and production or construction; and (2) the project management office. The organizational units suggested in (1) and (2) would be expected to report to the senior executive, such as the president or general manager. The functional departments would be expected to provide technical support to the portfolio of projects, and the PMO would provide management support to the programs and projects under way in the enterprise. The PMO would further be expected to provide the following support to the enterprise's programs and projects:

- Manage the portfolio of programs and projects strategically as core components in the enterprise's strategies, to include how well this portfolio supports organizational goals, objectives, and mission.
- Provide a focus for the development, publication, and use of enterprise resources, policies, procedures, protocols, and systems to support programs and projects.
- Facilitate the development of modified and new enterprise products, services, and organizational processes.
- Assist in the development of a cultural ambience in the enterprise that supports the use of programs and projects as key organizational initiatives in the strategic management of the enterprise.
- Provide consulting services to enterprise program and project managers for the improvement of the management strategies being used in support of enterprise purposes.

The reader is cautioned that the material presented in this chapter does not reflect the broader context in which organizations—and the organizing process—are found. Figure 8.4 portrays the larger context in which organizing a project is

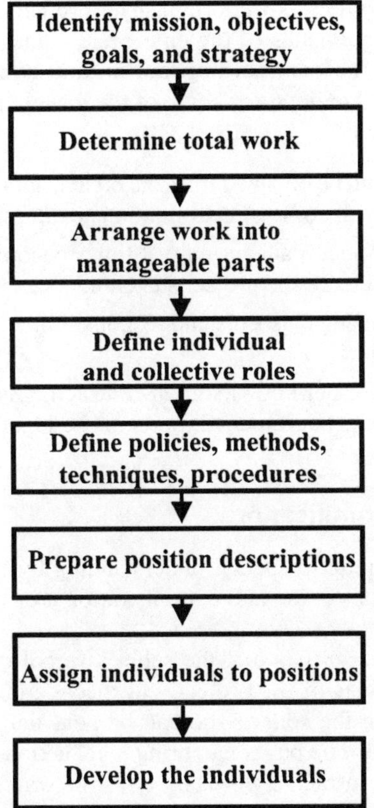

FIGURE 8.4 The organizing process. *[Source: David I. Cleland and Harold Kerzner,* Engineering Team Management *(New York: Van Nostrand Rheinhold, 1986), p. 171.]*

found. Explaining all the elements implied in this figure would be beyond the scope of this chapter.

8.18 PROCUREMENT AND CONTRACT NEGOTIATIONS/ADMINISTRATION

The procurement function is an important area of specialization that supports the management of projects in the enterprise. Sometimes the procurement activities come under the direct oversight of the project manager. In other cases, procurement is considered to be another of the enterprise functions that is linked to the project through some variation of the "matrix" organization. Whatever case is used, the project manager's role in project contract negotiations and administration activities is important. The project manager's role varies, depending on the policy wishes of the enterprise managers. Project managers should be involved in the contract negotiation and administrative processes. Project managers should understand some of the key elements that are involved in such processes, namely,

- General guidance should be obtained from the organization's legal office regarding the best way to deal with contract negotiation and administration matters.
- Recognize that the project manager may at times become the de facto contract manager when dealing with the project stakeholders.
- Learn to appreciate the legal aspects and responsibilities that a project manager has regarding the project.
- Understand when his or her knowledge in contracting is limited—and learn to seek counsel from the procurement experts or the legal office when needed.

Warranties and Indemnification

A couple of areas that require special understanding by the project manager in dealing with contract negotiation and administration are *conrtact warranties* and *indemnification.*

The concept of a *warranty* is that the seller's verbal or written commitment means that the deliverables of the project will meet certain standards. The warranty imposes a duty on the seller, who can be held liable by the buyer if this commitment is breached. The buyer can bring legal action to recover damages or rescind or cancel the contract. Two basic types of warranties exist: First, the *verbal or written warranties,* which pledge a specific commitment to perform on the contract. Second, *implied warranties,* which are assurances or promises that are a matter of law and general usage, rather than a specific promise made in the contracts. Implied warranties arise from specific laws or what is by precedence expected in the product or service. The implied warranty of a product or service holds that such deliverables must be reasonably suited for the ordinary purpose for which they are used.

Indemnification is the act of protection to guard against legal suit or bodily injury to a person or the organization for a loss incurred by that person or organization. There are two types of indemnification: common law and contractual. Indemnity provisions vary considerably from contract to contract as to the extent of the liability transferred. These provisions are generally of three types:

- *A broad form,* which obligates the indemnitor to indemnify and hold harmless the indemnitee against all loss arising out of the contract
- *An intermediate form,* which holds the indemnitor responsible for all claims or suits arising out of the contract except those arising out of the sole negligence of the indemnitee
- *A limited form* where one party agrees to indemnify the other only for the claims arising out of the indemnitor's negligence

Contract Administration

Project contract administration includes several activities: (1) oversight of the work to be done under the terms of the contract; (2) preparing and processing contract changes that come about; (3) providing interpretation of contract language and forms; and (4) approving invoices as the work is performed. Several key standards regarding contraction administration include:

- All required signatures, comments, and approvals have been obtained and documented before the contract is issued.
- No work should be performed before the contract is issued; however, pending the finalization of the contract, a formal letter of authorization to proceed may be provided.
- Contract and performance documentation should be stored in a secure place, be organized in a rational manner, and made available to those people who have a need to know.
- A policy and procedure for contract change should exist, and be followed closely in changing the contract.

If unique circumstances occur, seek the counsel of the legal office or contract specialists before taking action.[21]

8.19 TO SUMMARIZE

The major points that have been expressed in this chapter include:

- In modern organizations, managing across organizational hierarchies and boundaries is as important as managing up and down the hierarchy.
- Enterprises are organizing more and more using teams to manage around the core processes required to create and deliver value to customers.
- Examples were given of situations where a failure to prescribe an appropriate organizational design for projects caused serious problems in the ability of the project team to accomplish project objectives.
- Some organizations operate effectively without any discernible structural hierarchy.
- In terms of being responsive to contemporary challenges, the traditional organizational design has serious shortcomings.
- The project-driven matrix organizational design has a distinctive structure, which at first assessment seems to contradict some basic management principles, such as unity of command.
- Several different kinds of project organizational designs have been studied by researchers in the field.

---

[21]Material on contract administration is paraphrased from David I. Cleland and Lewis R. Ireland, *Project Manager's Portable Handbook* (New York: McGraw-Hill, 2000), pp. 6.36–6.40.

- The matrix design is a compromise between the pure project organization and the traditional functional organizational design.
- In spite of its perceived shortcomings, the matrix design has growing support from contemporary theorists and practitioners.
- Table 8.1 offers a basic prescription of the complementary role of the project manager and the functional manager.
- Project managers have helped show the way for the appointment and use of process managers, who have responsibility for managing an organizational process, such as order entry, across organizational boundaries and extending to process stakeholders.
- In the matrix, organization relationships exist among the project manager, team members, work package managers, functional managers, general managers, senior managers, members of the board of directors, and stakeholders such as suppliers, customers, and regulators.
- In its most elementary form, the interface between the project effort and the function effort constitutes the key focus of the matrix organization carried out through the project work package.
- The characteristics of a weak and a strong matrix organizational design were described.
- The project manager occupies a unique position as the key interface between the project team and the customer organization, working through the customer's project manager.
- "Networking" is an important role—not to be neglected by the project manager.
- Today, there is a growing interest in the project office.
- Procurement is a functional area in which the project manager needs to gain some familiarity.
- The project manager must understand warranty and indemnification issues to avoid problems.

8.20 ADDITIONAL SOURCES OF INFORMATION

The following additional sources of project management information may be used to complement this chapter's topic material. This material complements and expands on various concepts, practices, and theory of project management as it relates to areas covered here.

- Charles J. Teplitz, "Making Optional Use of the Matrix Organization," chap. 14 in David I. Cleland (ed.), *Field Guide to Project Management* (New York: Van Nostrand Reinhold, 1997).
- F. Paul Khuri and H. M. Plevyak, "Implementing Integrated Product Development: A Case Study of Bosma Machine and Tool Corporation"; P. Kayes, "How ICL Used Project Management Techniques to Introduce a New Product

Range"; and Max P. Shrontz, George M. Porter, and Norman L. Scott, "Organization and Management of a Multi-Organizational Single Responsibility Project," in David I. Cleland, Karen M. Bursic, Richard M. Puerzer, and Alberto Y. Vlasak, *Project Management Casebook,* Project Management Institute (PMI). (Originally published in *Project Management Journal,* September 1994, pp. 10–15; *International Journal of Project Management,* October 1995, pp. 321–328; and *Proceedings,* PMI Seminar/Symposium, Chicago, Ill., 1977, pp. 258–264.)

- David I. Cleland, *Matrix Management Systems* (New York: Van Nostrand Reinhold, 1984). This handbook, the first of its kind when it was published, provides managers and professionals with a reference guide for the design and implementation of matrix management systems in their organizations. The book provides a pragmatic explanation of what matrix management is all about. Alternative forms of the matrix organization are presented in this book from practitioners who have been successful in setting up and using effectively the new matrix organization approach.
- Stanley M. Davis, Paul R. Lawrence, and Harvey Kolodny, *Matrix* (Reading, Pa.: Addison-Wesley, 1977). This book, one of the first to appear on the subject of *matrix,* suggests a new alternative to what the authors call the 1-boss command structure that evolved from the industrial revolution. The matrix structure that is described in the book is one that grew out of the unique management problems of the American space effort of the 1960s. In the foreword, the then chairman of Citicorp-Citibank makes a key point that "matrix represents a sharp break with traditional forms of business organizations" and offers us another choice in the selection of organization models.
- C. J. Middleton, "How to Set Up a Project Organization," *Harvard Business Review,* April 1967. This was an early and classic article that describes a strategy for establishing the project organization. The author reviews the need for a balance of power in the matrix organization between the project manager and the functional managers. He describes the relative roles of the project manager vis-à-vis the functional managers, and the responsibility that the sponsoring general managers have for the resolution of conflict between these two managers.
- John F. Mee, "Matrix Organization," *Business Horizons,* Summer 1964. This very short article is truly a "classic." Professor Mee set forth what is believed to be the first basic definition of the matrix organization and what it looks like. According to him, the emerging matrix organization was creating new relationships of established organizational concepts and principles. Professor Mee believes that the matrix organization entails an organizational system designed as a "web of relationships" rather than a line and staff relationship of work performance. There have been many articles and books published which describe the matrix organization. None of these publications has done it any better than Professor Mee did back in 1964.
- Edward J. Morrison, "Defense Systems Management: The 375 Series", *California Management Review,* Summer 1967. This article describes the

United States Air Force (USAF) basic documentation on the concept and process of managing projects in that organization. This article provides an excellent summary of development projects and systems. One key USAF publication is described, namely, *System Program Office Manual, AFSCM 375-3,* which sounds much like some of the recent publications on the Project Management Office that have appeared in the project management community. A perusal of this article will give the reader considerable insight into how project management emerged in the USAF—where the credit must be given for major contributions to the emergence of the theory and practice of project management.

8.21 DISCUSSION QUESTIONS

1. Discuss the importance of an adequate organizational design in the management of a project.
2. For what reasons might an organization need to modify its organizational design?
3. Discuss the range of matrix organizational forms.
4. What factors contribute to the dynamic nature of a matrix organization?
5. Discuss the various forms of traditional departmentalization. In what situations would each of these forms be advantageous?
6. List and discuss the weaknesses of the pure functional organization. What kinds of failures could result from using this form of organizational design on a large project?
7. Discuss the advantages and disadvantages of the pure project organization. In what situations might this form be best?
8. Describe the matrix organizational form. What are its advantages and disadvantages?
9. What are some of the unnecessary characteristics of a successful matrix organization?
10. Discuss the advantages and disadvantages of the alternative forms of the matrix organization. In what situations would each work best?
11. Why is it important for project managers to develop networking skills?
12. It has been stated that the matrix organization is a state of mind in the mature project organization. Why is this so?

8.22 USER CHECKLIST

1. Do the project managers in your organization understand the interrelatedness of organizational forces? Why or why not?
2. Does the current design of your organization leave lines of authority and responsibility clear? Why or why not?

3. Project organizations range from pure functional to pure project. Where does your organization's design fit? Is it appropriate?
4. Are the factors that contribute to a dynamic organization present within your organization? Why or why not?
5. How is your organization departmentalized? Is this the most efficient departmentalization possible? What design might improve organizational efficiency?
6. Does the management of your organization understand the advantages and disadvantages of the various organizational forms? How does it use this knowledge in designing organizational structure?
7. Are the work packages of each project carefully related to the organizational structure? Is the organizational design appropriate for managing the work breakdown structure?
8. Is the use of human resources and skills balanced within the total organization? Explain.
9. Is there an effective means for conflict resolution over organizational roles established within your organization? How are conflicts handled?
10. Is your current organizational design successful and effective? Why or why not? What criteria for success are lacking?
11. Has the management of your organization considered possible alternative forms for structuring the organization? What other forms might be effective?
12. Do the project managers within your organization understand the notion of networking? Are they effective at forming alliances with project stakeholders?

8.23 PRINCIPLES OF PROJECT MANAGEMENT

1. The project-driven organization has distinct characteristics not found in the traditional organization.
2. The matrix organization provides for the organizational design of project teams superimposed on the traditional organizational hierarchy.
3. The project-functional interface in the matrix organization provides for the individual and collective roles in that organizational design.
4. Authority and responsibility are matched pairs in the matrix organization if the roles of the project manager and the functional managers have been accurately designated.
5. The project work package is the organizational unit around which the authority and responsibility between the project manager and the functional manager are designated.
6. Managing across organizational hierarchies and boundaries is as important as managing up and down the hierarchy.

8.24 PROJECT MANAGEMENT SITUATION—UNDERSTANDING THE MATRIX ORGANIZATION

When the matrix organization came into use, many managers, and professionals, were uncomfortable with the "web of authority and responsibility relationships" that emerged when a matrix organizational design was used. Much of this discomfort came from the perceived violation of Henri Fayol's principle of "unity of command," which meant that an individual should receive direction and orders from only one individual. Then too, those people who were disenchanted with the matrix organization claimed that "parity of authority and responsibility," another principle put forth by Henri Fayol, was also violated.

Even today, after we have learned much about the matrix organization and how it operates, some organizations have difficulty understanding and using an effective design of this nature. People seem to long for the more simple authority-responsibility work relationships found in the classical, bureaucratic organizational design. Yet if all organizations would revert to that design, great difficulty would be experienced in using project management as a focus for integrating the work of the many diverse stakeholders that are characteristic of modern organizations.

8.25 STUDENT/READER ASSIGNMENT

To develop a further understanding of the matrix design, the student/reader should try to answer the following questions about this design:

1. What were the reasons that an alternative organizational design like "matrix" emerged in modern organizations? If the matrix design were not used, what would be an effective alternative design?
2. In the organization in which the student/reader works has there been an adequate effort made to define why the matrix design is used—and how that design changes the working relationships in the organization?
3. Has the student/reader ever been in a situation when there was no definition of what his or her specific role was to be? How did the student/reader cope with such a situation?
4. What really helps us influence the people with whom we work—the assigned authority we have over those individuals or our interpersonal skills in working with those people?
5. The matrix organizational design seems to be working today, yet it is still emerging in theory and practice. What might be the key characteristics of the organizational design of the future, which might replace the current matrix design?

CHAPTER 9

PROJECT AUTHORITY[1]

"We trained hard...but it seemed that every time we were beginning to form up into teams, we would be reorganized. I was to learn later in life that we tend to meet any new situation by reorganizing and what a wonderful method it can be for creating the illusion of progress while producing confusion, inefficiency, and demoralization."

PETRONIOUS, ARBITER
GREEK NAVY, 210 B.C.

9.1 INTRODUCTION

In their consulting experiences, the authors have often heard project managers and functional managers lament their lack of sufficient authority to do their jobs. Such laments are not restricted to people in the project environment; others who work in today's complex organizations often have the same complaint. The authors believe that such complaints have as their root causes the lack of understanding of what authority is, how it is defined, and how to develop the ability to exercise more authority in today's organizations.

Accordingly, in this chapter authority is defined as consisting of two elements: (1) the legal defined authority; and (2) the authority that one has as a function of knowledge, skills, and attitudes in working with people associated with projects. Authority's documented definition, how to delegate authority, and the pitfalls of "reverse" delegation are provided in this chapter. In addition, the roles of responsibility and accountability—as key forces in the management of projects—are delineated. The linear responsibility chart (LRC) is suggested as a way to more accurately define individual and collective roles in a matrix organization. The key role that the project work packages play in defining roles in the matrix environment is described, along with a prescription of how to develop and use the LRC as a way to provide a better understanding by people of what is expected of them as they work with and support the project purposes.

[1]Some of the ideas in this chapter have been paraphrased from David I. Cleland, "Understanding Project Authority," *Business Horizons,* Spring 1967.

9.2 AUTHORITY, RESPONSIBILITY, AND ACCOUNTABILITY

In the previous chapter, we described several organizational design alternatives for managing projects. These descriptions dealt with the structural alignment of the matrix organization. In this chapter we will broaden the concepts of authority, responsibility, and accountability.

Authority is essential to any group or project team effort. The legal authority that is exercised by an individual comes from the organizational position occupied by the individual. Such authority is granted or delegated from a higher authority level in the organization. The ultimate source of authority in organizations can be traced to the owners of the organization. In a business organization, the shareholders elect the board of directors of a company. These directors have the authority given to them by the corporate charter and bylaws to manage the corporation on behalf of the shareholders. The authority of the board of directors is broad, is of a fiduciary nature, and is the starting point for the delegation and redelegation of authority within the organizational structure. The board of directors' authority role in project management is to study and approve key strategy proposals, particularly those risky projects that involve a substantial portion of corporate resources, and to maintain surveillance of the project during its life cycle.

Project managers face a unique authority challenge in the management of their projects. Usually project managers have only a few people working directly for them—their small administrative staff. Yet the project manager has to practice a subtle form of delegation in letting others—the functional specialists—become the experts and provide the technical input to the project team.

Sometimes the authority of the project manager is very explicit. For example, at Honda the project team that developed new vehicles had engineers, designers, financial analysts, marketing experts, and manufacturing people all report to a single project leader who had line authority over them and their work. Chrysler, in contrast, was divided by functional disciplines, as departments with their functional agendas competed. The result? The Chrysler system took longer, cost more, and sometimes led to compromises such as in quality.[2]

There is little doubt that the degree of control through using legal grants of authority that can be exercised by the traditional line manager is greater than what can be used by the project manager. In the traditional organization the manager would typically have de jure, or legal, authority to schedule and control work, evaluate performance of subordinates, reward and discipline employees, and hire and fire people. Because project teams typically operate in a complex interdisciplinary setting and possess limited command and control authority, the degree of control managers have is limited. Lacking such traditional line authority, project managers and other members of the project team rely on informal modes of authority through a variety of influence bases.

---

[2]Bradley A. Stertz, "Detroit's New Strategy to Beat Back Japanese Is to Copy Their Ideas," *The Wall Street Journal,* October 1, 1992.

In a survey conducted by two individuals from the project management community involving the polling of 283 project specialists, project managers, and their functional managers in a variety of technology-oriented organizations, those skills and competencies that are required to effectively lead cross-functional multidisciplinary project teams were identified. A key conclusion of this study was that dealing effectively with project team members and subordinates in today's project organization requires high levels of managerial competency. An effective project leader needs to be highly analytical to understand technical subtleties, cope with system inconsistencies, and develop insight to manage technical projects effectively.[3]

A project manager has to watch someone else provide the technical input in which the project manager may have experience and expertise. The project manager must be patient when someone accomplishes a task less proficiently than the project manager might be able to. The project manager must shift from the role of specialist to generalist, a leader in the management functions of planning, organizing, motivating, directing, and controlling. This takes the project manager away from the technical aspect of the project, allowing the project team members to be the experts in the technical work they represent.

9.3 DEFINING AUTHORITY[4]

Authority is a conceptual framework and, at the same time, an enigma in the study of organizations. The authority patterns in an organization, most commentators agree, serve as both a motivating and a tempering influence. This agreement, however, does not extend to the emphasis that the different commentators place on a given authority concept. Early theories of management regarded authority more or less as a gravitational force that flowed from the top down. Recent theories view authority more as a force which is to be accepted voluntarily and which acts both vertically and horizontally.

Although authority is one of the keys to the management process, the term is not always used in the same way. Authority is usually defined as a legal or rightful power to command or act. As applied to the manager, authority is the power to command others to act or not to act. The manager's authority provides the cohesive force for any group. In the traditional theory of management, authority is a right granted from a superior to a subordinate.

There are two types of project authority. One, de jure project authority, is the legal or rightful power to command or act in the management of a project. Inherent in this authority is the legal right to commit or withdraw resources supporting the project. The legal authority of a project manager usually is contained in some form of documentation; such documentation of necessity must contain, in addition, the complementary roles of other managers (e.g., functional managers, work package managers, general managers) associated with the project.

[3]Richard G. Donnelly and Deborah S. Kezsbom, "Overcoming the Responsibility-Authority Gap: An Investigation of Effective Project Team Leadership for a New Decade," *Cost Engineering,* May 1994, pp. 33–41.

[4]Portions of this material have been taken from David I. Cleland and William R. King, *Systems Analysis and Project Management,* 3d ed. (New York: McGraw-Hill, 1983), chap. 12.

Having legal authority is a start. However, to be a successful manager, an individual must develop capabilities in the de facto aspects of authority.

The second type of authority, de facto project authority, is that influence brought to the management of a project by reason of a particular person's knowledge, expertise, interpersonal skills, or personal effectiveness. De facto project authority may be exercised by any of the project clientele, managers, or team members. In another study it was found that project managers and project personnel believe that expertise and reputation are the most helpful sources of influence in the management of technical projects. It was further determined that technical expertise and organizational expertise are two sources of influence that are available to project managers. Expert power comes to the project manager through background and experience, technical achievement, participation in past projects, and longevity.[5]

Ford and McLaughlin in their research remind us that classical management theory holds that parity of authority and responsibility should exist. In project management there may not be such parity across the various stages of the life cycle. They note that few empirical data have been collected to test the hypothesis that parity does not exist and that this lack of parity is the cause of many management problems. In their research report collected from 462 information system managers, the data indicated that in the majority of cases parity did not exist.[6]

A major part of de facto authority is the ability of the project manager to influence others whose cooperation and support are needed to provide timely resources to support the project. Part of the ability to influence is the competence to work effectively with project team members, functional managers, general managers, and project stakeholders. A project manager must have some technical skill in the technology embodied in the project, not only to participate in the rendering of technical judgments but also to gain the respect of team members who have in-depth technical knowledge and skills. Interpersonal skills provide power to the project manager in influencing the many professionals and managers with whom the project manager works. Developing and maintaining a successful track record that gets people to work with the project manager are, in themselves, a form of power in influencing. The ability to influence is directly related to how others perceive one's expertise.

Another source of power is to pay attention to and recognize the performance of other people who work with you, such as team members, managers, and stakeholders. In other words, acknowledge the performance of other people just as you would like to have your own good performance recognized. This recognition can take many forms, such as letters of appreciation, phone calls to thank the person, a public thanks in a meeting, comments to a person's manager, a citation in the person's personnel file, stopping by the person's desk to say, Thanks for your help, a personal note of thanks, or some token of appreciation such as a lunch, a book, flowers, or pen and pencil set. Sometimes praising a person's work to members of the peer group works well; inevitably that praise will be reported to the person.

---

[5]Christopher G. Worley and Charles J. Teplitz, "The Use of 'Expert' Power as an Emerging Influence Style within Successful U.S. Matrix Organizations," *Project Management Journal,* March 1993, pp. 31–34.

[6]Robert C. Ford and Frank S. McLaughlin, "Using Project Teams to Create MIS Products: A Life Cycle Analysis," *Project Management Journal,* March 1993, pp. 43–47.

The ability to exercise de facto authority is dependent on the competency of the individual. This competency is essentially a combination of the knowledge, skills, and attitudes that an individual possesses. Figure 9.1 portrays these elements of competency within the context of project management. The reader should note that the elements of knowledge, skills, and attitudes reflected in this figure are described in various chapters in this book.

9.4 POWER

The theory of power can be traced back to sociologist Max Weber. He described three kinds of authority: charismatic, traditional, and bureaucratic. Charismatic authority is where people follow the leader because of his or her inspiration, exemplary character, or behavior, for example, Jesus Christ, Martin Luther King, or Thomas Edison. Often those people who represent this kind of authority are change leaders such as Trotsky. Traditional authority is where obedience is given to an individual who occupies a traditional or inherited position such as in theocracies, patriarchies, and family businesses like the House of Windsor or Anheuser-Busch. Bureaucratic authority (or the role of law) is where power is vested in a hierarchical position and where the authority comes from, say, an elected person who holds office, such as a person holding military rank, or one who occupies an organizational position in the enterprise. All of these sources of power do not provide enough clout to get the job done in today's complex organizations, particularly in those organizations that use alternative teams in their organizational design. Modern organizations depend on the personal power that an individual is able to wield using sources of knowledge, skills, expertise, track record, interpersonal skills, attractiveness, dedication, networks, alliances, and tenacity, to name a few.

Knowledge + skills + attitude = competency

Knowledge	Skills	Attitude
(Familiarity, awareness, or comprehension acquired by study or experience)	(The ability to apply knowledge)	(A state of mind or feeling)
• Project "technology" • Strategic management • Project management theory and practice • Project management processes • Project management systems model	• Interpersonal skills • Communication skills • "Systems" application • Political sensitivity • Style • Building conceptual models	• Maslow's hierarchy of needs • McGregor's theory X and theory Y • Authority and responsibility • Emotional intelligence • Trust

FIGURE 9.1 Individual competency model.

Power coming from the position that one occupies is not enough to get the job done. Hierarchy confers less power because there is less of it in modern organizations. The collective authority that comes from the team and the knowledge and skills of its cross-functional and cross-organizational networks and working arrangements is effective power. In such situations the team depends on many people over whom there is no formal authority, and peer and stakeholder networks are more important. Indeed, in today's complex organizations, power is all about empowerment of people to the lowest possible level, which will enable them to carry out their responsibilities without having to check with the boss. The act of empowering people through the delegation process actually results in an increase of power for the one who delegates. In such environments politics, networking, and listening are the core of people skills matter.

Empowerment is like a coin—it has two sides. On the one side is the official authority or legal power that is given to an individual who is occupying an organizational position, such as a project manager and the positions that the members of the project team hold. On the other side of empowerment is the influence that an individual has with regard to the organization's stakeholders. The first side of empowerment stated above is granted through documentation such as a position description, letters of appointment, a project charter, a policy, and procedure documents. The second side cannot be delegated. It depends on the knowledge, skills, and attitudes of the individuals and the competency that they are able to develop and sustain in the management of the project and in their dealings with project stakeholders.

9.5 MATRIX IMPLICATIONS

The matrix organizational design to support the management of projects has been given much attention in the project management literature. Whatever controversy and disenchantment that the matrix design has caused, it cannot be forgotten that the different alternative uses of the matrix that have been tried have been a search for how authority and responsibility could be shared by those organizational entities cooperating in bringing about a focal point to manage the sharing of resources to support organizational projects. Most failures in the use of the matrix have been caused by one or more of the following relative authority-responsibility factors:

- Failure to define the specificity of authority and responsibility of the project and functional people relative to the work packages for which each is solely and jointly responsible.
- Negative attitudes on the part of project, functional, and general managers and team members who support a sharing of authority and responsibility over the resources to be used to support organizational projects.
- Lack of familiarity with the theoretical construction of the matrix and the context in which that organizational design is applied.

- Failure on the part of senior managers to bring about the development of some basic documentation in the organization that prescribes the formal and relative authority of managers and team members associated with a project team.
- Failure to do adequate project team development to include how the team will operate in a cultural ambience of the enterprise where project resources, results, and rewards are shared.
- Existence of an organizational culture that believes and reinforces the traditional command and control notions of authority and responsibility being primarily vertical in their flow downward through the organizational hierarchy.
- Failure on the part of organizational leaders to recognize that the traditional organizational model in the vertical flow of authority and responsibility is rapidly being eroded by the increasing use of computer and communication technology, the increasing pace of change, and the success which alternative organizational designs are enjoying such as found in the use of self-directed teams, quality teams, task forces, and the growing use of participative management to include employee empowerment.
- Failure to modify the traditional pyramid to a design that has fewer levels, with more options for personal movement and flexibility among and within organizational levels. This modification includes the reduction in the number of middle managers and the changes in their roles from one of approval and control to problem solving and facilitation of the means for people to work together to accomplish organizational ends.
- And finally, the failures of managers to promote synergy and unity within and between organizational levels and with outside stakeholders so that resources, results, and rewards can be shared. This type of promotion requires true teamwork, discussion, cooperation of all organizational members, education, and the opening and maintenance of many lines of communication.

When project management is introduced in an organization, it is essential that these authority roles be understood and accepted by general managers, project managers, and functional managers. This understanding can be facilitated if all the managers concerned jointly participate in the development and publication of a policy document containing a description of the intended authority and responsibility relationships characterized by Fig. 9.2.

During the early days of the matrix organization, it was not uncommon to hear people express their dissatisfaction with the matrix because it was against their religion, and they would quote the biblical phrase about not serving two masters. There was some basis for their concern.

Conceptual guidance for the relationship of the project team member to the project manager and the functional manager can be found in the Bible. Verse 24, Chapter 6, Matthew, states:

> No man can serve two masters: for either he will hate the one, and love the other; or else he will hold to the one, and despise the other. Ye cannot serve God and mammon.

This verse probably provides the basis for the evolution of the principle expressed by Henri Fayol as unity of command in which one is expected to receive orders from only one individual. The unity of command principle has provided a key basis in the design of the traditional organizational structure in which authority, responsibility, and accountability flow from the senior person through an organizational hierarchy to the worker who is doing the work of the organizational entity. Violation of this key management principle of unity of command, along with the key principle of parity of authority and responsibility, was considered to be serious, potentially laying down the basis for impairment of the efficiency and effectiveness of the enterprise.

As one reads further in the Bible, another insight is gained in how to deal with this apparent violation of a couple of key management principles. In Verse 21, Chapter 22, Matthew, the bible states, "Render therefore unto Caesar the things which are Caesar's, and unto God the things that are God's." By taking license in paraphrasing this verse related to the matrix organization, one could say that the project team members should render unto the project manager the things that are the project manager's, and unto the functional manager the things that are the functional manager's. How to do this is explained in Fig. 9.2, in which the relative roles and authorities of the principal players in the matrix organization are portrayed.

A significant measure of the authority of project managers springs from their function and the style with which they perform it. Project managers' authority is neither all de jure (having special legal foundations) nor all de facto (actual influence exercised and accepted in the environment). Rather, their authority is a combination of de jure and de facto elements in the total project environment. Taken in this context, the authority of project managers has no organizational or functional constraints but rather diffuses from their offices throughout and beyond the organization, seeking out the things and the project stakeholders to influence and control.

9.6 THE POWER TO REWARD

Not only do teams change the culture and the modus operandi of the organization, but also they change the manner in which organizational rewards are provided to people. As people serve on teams and rotate from team to team, performance evaluations are more difficult. In most organizations the team does not yet assume a major part in appraising team performance.[7] Of the organizations surveyed by Development Dimensions International, the Association for Quality and Participation, and *Industry Week,* 46 percent indicated that leaders outside the team handle appraisals, 17 percent said that the responsibility is shared, and 37 percent responded that the team takes the lead in appraising performance.[8] On the basis of these surveys, team performance appraisal is changing—teams are accepting such appraisal responsibility—and at the same time management is moving slowly in relinquishing appraisal prerogatives.

[7]Richard S. Wellins, William C. Byam, and Jeanne M. Wilson, *Empowered Teams* (San Francisco: Jossey Bass, 1991), p. 3.

[8]Ibid.

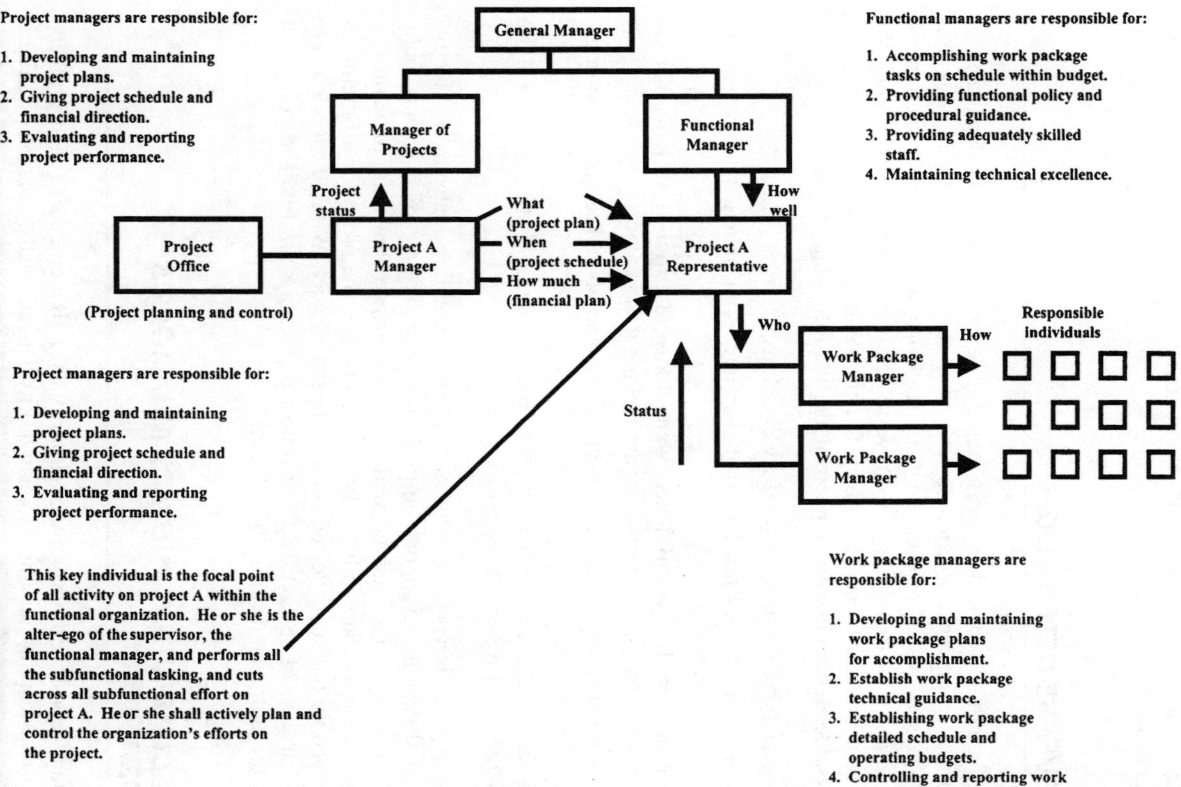

FIGURE 9.2 Project-functional organizational interface. [*Source: David I. Cleland and William R. King,* Systems Analysis and Project Management, *3d ed. (New York: McGraw-Hill, 1983), p. 353.*]

In the project-driven organization, people and teams have considerable mobility. It becomes a challenge to keep abreast of where people are on teams. Cypress Conductor, a San Jose, California, maker of specialty computer chips, developed a computer system that tracks its 1500 employees as they crisscross functions, teams, and projects.[9]

9.7 REVERSE DELEGATION

The effectiveness with which project managers exercise authority depends to a large degree on their legal position as well as on their personal capabilities. But there are ways in which project managers can operate to enhance their basic authority. One way is to guard against reverse delegation, which occurs when the person to whom authority has been delegated gives authority back to the delegator. This reverse delegation usually happens under the following conditions:

- The team member wants to avoid risky decisions.
- The team member does not feel that the functional manager is adequately supporting the project.
- The team member lacks confidence, wants to avoid criticism, or feels that the necessary information and resources are lacking to do the job.
- The team member feels that the project manager wants to keep involved in the details of the project.
- The project manager has not been explicit in establishing what is expected of the team member in supporting the project.

Effective delegation is a necessary but not sufficient condition to ensure an effective organizational design to support the project. Organizing a project means many things, one of which is the establishment and maintenance of meaningful authority, responsibility, and accountability relationships among the project team members and other people having a vested interest in the project. Without an adequate, committed process of delegation, there is no effective organization and things can easily fall through the cracks in the project.

9.8 DOCUMENTING PROJECT MANAGER'S AUTHORITY

Project managers should have broad authority over all elements of their projects. Although a considerable amount of their authority depends on their personal abilities, they can strengthen their position by publishing documentation to establish their modus operandi and their legal authority. At a minimum, the documentation

[9]Brian Dumaine, "The Bureaucracy Busters," *Fortune,* June 17, 1991, pp. 36–50.

(expressed in a policy manual, policy letters, and standard operating procedures) should delineate the project manager's role and prerogatives in regard to:

- The project manager's focal position in the project activities
- The need for a defined authority-responsibility relationship among the project manager, functional managers, work package managers, and general managers
- The need for influence to cut across functional and organizational lines to achieve unanimity of the project objectives
- Active participation in major management and technical decisions to complete the project
- Collaborating (with the personnel office and the functional supervisors) in staffing the project
- Control over the allocation and expenditure of funds, and active participation in major budgeting and scheduling deliberations
- Selection of subcontractors to support the project and the negotiation of contracts
- Rights in resolving conflicts that jeopardize the project goals
- Having a voice in maintaining the integrity of the project team during the complete life of the project
- Establishing project plans through the coordinated efforts of the organizations involved in the project
- Providing an information system for the project with sufficient data for the control of the project within allowable cost, schedule, and technical parameters
- Providing leadership in the preparation of operational requirements, specifications, justifications, and the bid package
- Maintaining prime customer liaison and contact on project matters
- Promoting technological and managerial improvements throughout the life of the project
- Establishing a project organization (a matrix organization) for the duration of the project
- Participation in the merit evaluation of key project personnel assigned to the project
- Allocating and controlling the use of the funds on the project
- Managing the cost, schedule, and technical performance parameters of the project[10]

The publication of suitable policy media describing the project manager's modus operandi and legal authority will do much to strengthen his or her position in the client environment. In practice, we find many types of de jure authority documentation. A sample of a project/program management charter appears in Table 9.1.

As in the example, care should be taken to delineate the legal position of the project manager. This constitutes an obvious source of power in the project

[10]David I. Cleland and William R. King, *Systems Analysis and Project Management,* 3d ed. (New York: McGraw-Hill, 1983), pp. 337–338.

TABLE 9.1 Typical Charter of Program Project Manager (Matrix Organization)

Position Title: Program/Project Manager

Authority

The program/project manager has the delegated authority from general management to direct all program activities. He or she represents the company in contacts with the customer and all internal and external negotiations. Project personnel have the typical dual-reporting relationship: to functional management for technical performance and to the program manager for contractual performance in accordance with specifications, schedules, and budgets. The program/project manager approves all project personnel assignments and influences their salary and promotional status via formal performance reports to their functional managers. Travel and customer contact activities must be coordinated and approved by the program/project manager.

Any conflict with functional management or company policy shall be resolved by the general manager or his or her staff.

Responsibility

The program/project manager's responsibilities are to the general manager for overall program/project direction according to established business objectives and contractual requirements regarding technical specifications, schedules, and budgets.

More specifically, the program/project manager is responsible for (1) establishing and maintaining the program/project plan, (2) establishing the program organization, (3) managing and controlling the program/project, and (4) communicating the program/project status.

1. *Establishing and maintaining the program/project plan.* Prior to authorizing the work, the program/project manager develops the program plan in concert with all key members of the program/project team. This includes master schedules, budgets, performance specifications, statements of work, work breakdown structures, and task and work authorizations. All of these documents must be negotiated and agreed upon with both the customer and the performing organizations before they become management tools for controlling the program/project. The program/project manager is further responsible for updating and maintaining the plan during the life cycle of the program/project, including the issuance of work authorizations and budgets for each work package in accordance with the master plan.
2. *Establishing the program/project organization.* In accordance with company policy, the program/project manager establishes the necessary program/project organization by defining the type of each functional group needed, including their charters, specific roles, and authority relationships.
3. *Managing the program/project.* The program/project manager is responsible for the effective management and control of the program/project according to established customer requirements and business objectives. He or she directs the coordination and integration of the various disciplines for all program/project phases through the functional organizations and subcontractors. He or she monitors and controls the work in progress according to the program/project plan. Potential deficiencies regarding the quality of work, specifications, cost, or schedule must be assessed immediately. It is the responsibility of the program/project manager to rectify any performance deficiencies.

TABLE 9.1 Typical Charter of Program/Project Project Manager (Matrix Organization) *(Continued)*

Position Title: Program/Project Manager
4. *Communicating the program/project status.* The program/project manager is responsible for building and maintaining the necessary communication channels among project team members to the customer community and to the firm's management. The type and extent of management tools employed for facilitating communications must be carefully chosen by the program/project manager. They include status meetings, design reviews, periodic program/project reviews, schedules, budgets, data banks, progress reports, and team collocation.

Source: Harold Kerzner and Hans J. Thamhain, *Project Management Operating Guidelines* (New York: Van Nostrand Reinhold, 1986), p. 68.

environment. Although this gives project managers the right to exercise that power, the significance of authority under the project-functional interface cannot be understated. Even though project managers may have the final, unilateral right to make decisions in the project, it would be foolhardy for them to substitute their views without fully considering the crystallization of thinking of the other stakeholders in their project. Project managers rarely hope to gain and build alliances in their environments by arbitrarily overruling the team members who contribute to a project. They may not have the control for such arbitrary action. Even if they did, they should be most judicious in using authority in such a manner that the culture in which the project team is operating is not adversely affected.

Authority operates in the context of responsibility and accountability. These concepts are presented in the following material.

9.9 WHAT IS RESPONSIBILITY?

Responsibility, a corollary of authority, is a state, quality, or fact of being responsible. A responsible person is one who is legally and ethically answerable for the care or welfare of people and organizations. A person who is responsible is expected to act without specific guidance or being told to do so by a superior authority. To be responsible is to be able to make rational decisions on one's own, to be trusted to make such decisions, and to be held liable for one's decisions. Archibald, a noted author in the field of project management, notes the following concerning the project manager's responsibility:

> If the project manager's responsibilities are divided among several persons (such as one man for engineering, another for scheduling, a third for cost, a fourth for contract administration, etc.) such division is the most common cause of projects not achieving their objectives. Unless one person integrates the efforts of the project engineer, the project contract administrator, and so on, it is not possible to evaluate the project effectively to identify current or future problems and initiate corrective action in time to assure that the project objectives will be met.

> The project manager cannot actually perform all the planning, controlling and evaluation activities needed, any more than he can perform all the technical specialty activities required. Project management support services must be provided to him, and he must direct and control these support activities. The hazard is that the support activities may exist, but in the absence of an assigned project manager, they are not properly used.[11]

Some companies are very explicit about their project manager's responsibilities. For example, within the Fluor Corporation, a major engineering/construction company, project managers have total responsibility for the execution of the project from its earliest stages right through to completion.[12]

9.10 WHAT IS ACCOUNTABILITY?

Accountability is the state of assuming liability for something of value, whether through a contract or because of one's position of responsibility. A professional is held accountable for excellence in the quality of the service rendered to the organization. Project managers have dual accountability: They are held answerable for their own performances and for the performance of people who comprise the project team. One of the basic characteristics of managers is that they are held accountable for the effectiveness and efficiency of the people who report to them.

Authority, responsibility, and accountability can rest with a single person or with a group of people. An example of pluralism in this sense is found in the use of a plural executive at the top-management level of organizations such as a management council or the board of directors. The plural executive serves as an integrator of top-management decision making and implementation. The increasing complexity and size of many large organizations have created managerial responsibilities beyond the capabilities of one individual. The plural executive that has been created by organizations usually acts in an advisory capacity to the chief executive by providing stewardship for the strategic management of the company. The specific authority of such plural executives depends on the character establishing such a body. Authority, responsibility, and accountability within the matrix context are the cohesive forces that hold the organization together and make possible the attainment of the organization's cost, schedule, and technical performance objectives. Figure 9.3 is one way of portraying these forces. The existence of cost, schedule, and technical performance objectives in this figure means that the degree of completeness of authority, responsibility, and accountability at each level in the model can influence any or all of the parameters.

---

[11] R. D. Archibald, *Managing High-Technology Programs and Projects* (New York: Wiley, 1976), p. 39.

[12] Robert M. Duke, "Project Management at Fluor Utah Company, Inc.," *Project Management Quarterly*, vol. 8, no. 3, September 1977.

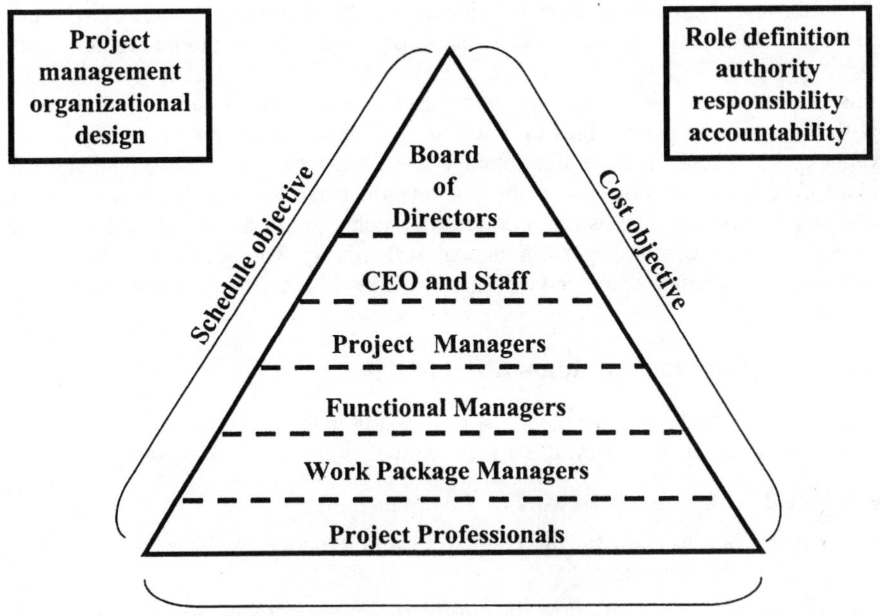

FIGURE 9.3 Project management organizational design.

9.11 PROJECT ORGANIZATION CHARTING[13]

The organizational model that is commonly called the organizational chart is derided in the satirical literature and in the day-to-day discussions among organizational participants. However, organizational charts can be of great help in both the planning and implementation phases of project management.

9.12 TRADITIONAL ORGANIZATIONAL CHART

The traditional organizational chart is of the pyramidal variety; it represents or models the organization as it is supposed to exist at a given time. At best, such a chart is an oversimplification of the organization and its underlying concepts, which may be used as an aid in grasping the concept of the organization.

[13]Some of the ideas in this section have been paraphrased from the following articles: David I. Cleland and Wallace Munsey, "Who Works with Whom?" *Harvard Business Review,* September–October 1967; Dundar F. Kocaoglu and David I. Cleland, "A Participative Approach to the Development of Organizational Roles and Interactions," *Management Review,* October 1983, pp. 57–64.

Unfortunately, too often the policy documentation describing the role of a project manager will describe this manager's relationship with the functional organizations as a "dotted line" relationship, which can mean anything one wishes it to mean. In this respect, Davis and Lawrence note that for generations managers have lived with the fiction of dotted lines to describe secondary reporting relationship in the organization.[14] One suspects that managers use a dotted line on an organizational chart because at the time the chart was developed the relationship had not been completely defined. The use of a dotted-line technique in depicting authority and responsibility gives a manager a great deal of flexibility. The price of this flexibility is confusion and unclear understandings of reciprocal authority and responsibility.

Usefulness of the Traditional Chart

The organizational chart is a means of visualizing many of the abstract features of an organization. In summary, the organizational chart is useful in that:

- It provides a general framework of the organization.
- It can be used to acquaint the employees and outsiders with the nature of the organizational structure.
- It can be used to identify how the people tie into the organization; it shows the skeleton of the organization, depicting the basic relationships and the groupings of positions and functions.
- It shows formal lines of authority and responsibility, and it outlines the hierarchy that fills each formal position, who reports to whom, and so on.

Limitations of the Traditional Chart

The organizational chart is something like a photograph. It shows what the subjects look like, but tells little about how individuals function or relate to others in their environment. The organizational chart is limited as follows:

- It fails to show the nature and limits of the activities required to attain the objectives.
- It does not reflect the myriad reciprocal relationships between peers, associates, and many others with a common interest in some purpose.
- It is a static, formal portrayal of the organizational structure; most charts are out of date by the time they are published.
- It shows the relationships that are supposed to exist but neglects the informal, dynamic relationships that are constantly at play in the environment.
- It may confuse organizational position with status and prestige; it overemphasizes the vertical role of managers and causes parochialism—a result of the blocks and lines of the chart and the neat, orderly flow they imply.

---

[14]Stanley M. Davis and Paul R. Lawrence, "Problems of Matrix Organizations," *Harvard Business Review*, May–June 1978, p. 142.

Role definition within the project team is a key consideration in developing the team. When a new team is formed, or when new objectives and goals are developed for the team, or when any key circumstance about the team or its mission changes, such as additional responsibilities, then the definition and understanding of individual and collective roles become important. If the team is intended to be interactive and synergistic, role understanding is critical. Allocating authority and responsibility to the team is an important first step. But the team must understand the authority and responsibility associated with both individual and collective roles, must be committed to those roles, and must be proactive in developing the personal influence that gives added power to the execution of these roles.

How can the individual and collective roles of the project team be established, particularly as team members work with the project stakeholders? Two organizational charts are needed: the traditional chart, which portrays the general framework of the organization, and the linear responsibility chart, which is useful to determine the specificity of individual and collective roles in the organization.

9.13 LINEAR RESPONSIBILITY CHART

The *linear responsibility chart* (LRC) is an innovation in management theory that goes beyond the simple display of formal lines of communication, gradations, or organizational level, departmentalization, and line-staff relationships. In addition to the simple display, the LRC reveals the work package position couplings in the organization. The LRC has been called the *linear organization chart,* the *responsibility interface matrix,* the *matrix responsibility chart,* the *linear chart,* and the *functional chart.*

Six key elements make up the form and process of an LRC:

- An organizational position
- An element of work—a work package—to be accomplished to support organizational objectives, goals, and strategies
- An organizational interface point—a common boundary of action between an organizational position and a work package
- A legend for describing the specificity of the organizational interface
- A procedure for designing, developing, and operating LRCs for an organization
- A commitment and dedication on the part of the members of the organization to make the LRC process work

The LRC shows who participates, and to what degree, when an activity is performed or a decision made. It shows the extent or type of authority exercised by each position in performing an activity in which two or more positions have overlapping involvement. It clarifies the authority relationships that arise when people share common work.

Figure 9.4 shows the basic structure of an LRC, in terms of an organizational position and a work package, in this case "conduct design review." The symbol "P" indicates that the director of systems engineering has the primary responsibility for conducting the system design review.

9.14 WORK PACKAGES

The work elements of the hierarchical levels of the work breakdown structure are called work packages. They are used to identify and control work flows in the organization, and they have the following characteristics:

- A work package represents a discrete unit of work at the appropriate level of the organization where work is assigned.
- Each work package is clearly distinguished from all other work packages.
- The primary responsibility of completing the work package on schedule and within budget can always be assigned to an organizational unit, and never to more than one unit.
- A work package can be integrated with other work packages at the same level of the work breakdown structure to support the work packages at a higher level

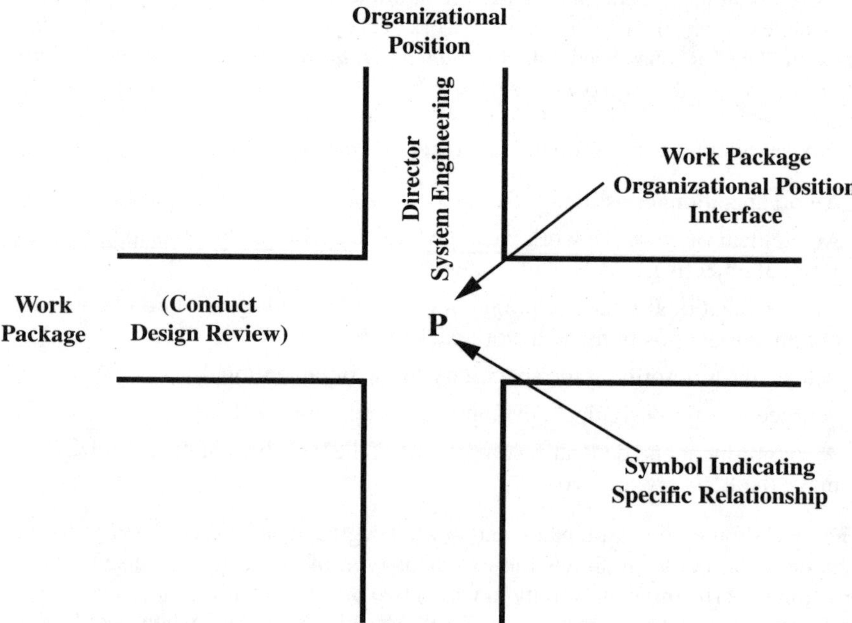

FIGURE 9.4 Essential structure of a linear responsibility chart.

of the hierarchy.

Work packages are level-dependent and become increasingly more general at each higher level and increasingly more specific at each lower level.

9.15 WORK PACKAGE–ORGANIZATIONAL POSITION INTERFACES

The organizational positions and the responsibilities assigned to them in carrying out the work package requirements constitute the basis for the LRC. It is developed by specifically identifying responsibilities on each of the work packages. The responsibilities are defined at the work package–organizational position interfaces, by using symbols or letters to depict relationships.

The LRC is a valuable tool as a succinct description of organizational interfaces. It conveys more information than several pages of job descriptions and policy documents by delineating the authority-responsibility relationships and specifying the accountability of each organizational position. However, by far the most important aspect of the LRC is the process by which the people in the organization prepare it. If the LRC is developed in an autocratic fashion, it simply becomes a document portraying the organizational relationships. But if it is prepared through a participative process, the final output becomes secondary to the impacts of the process itself. The open communications, broad discussions, resolution of conflicts, and achievement of consensus through participation provide a solid basis for organizational development and managerial harmony. By the time the LRC is developed in this way, the organization goes through such an "education" that the chart becomes secondary.

9.16 A PROJECT MANAGEMENT LRC

The LRC can be very useful for project managers to use to understand their authority relationships with their project team members. For a simple project, these relationships may be easy to depict; for more complex projects, a series of descending charts from the macrolevel of the project to successively lower levels may be necessary.

Table 9.2 shows an LRC for project-functional management relationships within a matrix organization. The development of such a chart, combined with the discussions that usually accompany such a development, can help greatly to facilitate an understanding of project management and how it will affect the day-to-day lives and activities of the team members.

In the table the legend depicts the appropriate relationships among the listed positions. Note that there will typically be more than one project manager and more than one functional manager. This table will serve as a guide to developing specific relationship, based on the organization and the projects.

TABLE 9.2 Linear Responsibility Chart of Project Management Relationships*

Activity	General manager	Manager of projects	Project manager	Functional manager
Establishment of department policies and objectives	1	3	3	3
Integration of projects	2	1	3	3
Project direction	4	2	1	3
Project charter	6	2	1	5
Project planning	4	2	1	3
Project-functional conflict resolution	1	3	3	3
Functional planning	2	4	3	1
Functional direction	2	4	4	1
Project budget	4	6	1	3
Project WBS	4	6	1	3
Project control	4	2	1	3
Functional control	2	4	3	1
Overhead management	2	4	3	1
Strategic projects	6	3	4	1

*Legend: 1: actual responsibility; 2: general supervision; 3: must be consulted; 4: may be consulted; 5: must be notified; 6: approval authority.

9.17 DEVELOPING THE LRC

The development of the project LRC is inherently a group activity—getting together with the key people who have a vested interest in the work to be done. The following plan for the development of an LRC has proved useful:

- Distribute copies of the current traditional organizational chart and position descriptions of the key people.
- Develop and distribute blank copies of the LRC.
- At the first opportunity, get the people together to discuss
  - The advantages and shortcomings of the traditional organization chart.
  - The concept of a project work breakdown structure (WBS) and the resulting work packages.
  - The nature of the linear responsibility chart, how it developed, and how it is used.
  - A simple way of establishing a code to show the work package–organizational position relationship (getting a meeting of the minds on this code is very important because individuals who believe the code to be either too fine or too coarse will find it difficult to accept).

  - The makeup of the actual work breakdown structure with accompanying work packages.
  - The fitting of the symbols into the proper relationship in the LRC.
- Encourage an intensive dialogue during the actual making of the LRC. In such a meeting, people will tend to be protective of their organizational "territory." The LRC by its nature requires a commitment to support and share the allocation of organizational resources applied to work packages. This commitment requires the ability to communicate and decide. This process takes time, but when the LRC is completed, the people are much more knowledgeable about what is expected of them.[15]

Much of the success of project management depends on how effectively people work together to accomplish project objectives and gain personal satisfaction. The development of a project LRC can greatly contribute to achieving this.

Once assembled, the LRCs can become a "living document" to

- Portray formal authority, responsibility, and accountability relationships.
- Acquaint newcomers with how things are done in the organization.
- Get people committed and motivated so they know specifically what is expected of them.
- Bring out real or potential conflict over territorial prerogatives in the organization.
- Permit people to see the "big picture"—how they fit into the larger whole.
- Facilitate teamwork so people have greater opportunity to see their specific/individual roles on the project in the enterprise.
- Provide a standard against which the project managers and other managers can monitor what people are doing.

9.18 TO SUMMARIZE

The major points that have been expressed in this chapter include:

- Authority is a force that is essential to the functioning of any organization.
- Authority is like a coin. On one side is the legal or de jure authority that is delegated to the organizational position that a person occupies. The other side of the coin is the de facto authority that an individual has by reason of influence in the organization in which he or she works.
- Great care should be taken to prescribe in appropriate documentation the legal authority that attaches to an organizational position.
- The de facto authority that an individual has comes from knowledge, expertise, interpersonal skills, experience, and ability to work cooperatively with the people associated with the project team to include stakeholders.

[15]David I. Cleland and Dundar F. Kacaoglu, *Engineering Management* (New York: McGraw-Hill, 1981), pp. 47–50.

- Power is an element of authority and can come from several sources.
- A conceptual framework for the proper delineation of roles in the matrix organization can be found in the Bible.
- Project managers—and all managers—should be aware of some of the techniques that people can use to avoid responsibility through the reverse delegation process.
- Some examples of how to document de jure authority in the matrix organization were provided in the chapter.
- Responsibility, a corollary of authority, is a state, quality, or fact of being responsible.
- Accountability is the state of assuming liability for something of value.
- Definitions of the project management role of all principals in the enterprise were given.
- There are serious limitations to the traditional organizational chart—particularly in understanding how individual and collective roles are carried out in the enterprise.
- The material in this chapter includes an identification of the six key elements that make up the form and process of the linear responsibility chart (LRC).
- Individual and collective roles in the project team are key considerations in understanding how the project team is expected to operate in the matrix organizational context.
- An appropriate legend must be developed to use in the LRC to describe the relationships between the work package responsibilities and the organizational position.
- A methodology for how to develop the LRC was provided in the chapter.

9.19 ADDITIONAL SOURCES OF INFORMATION

The following additional sources of project management information may be used to complement this chapter's topic material. This material complements and expands on various concepts, practices, and theory of project management as it relates to areas covered here.

- Robert J. Yourzak, "Motivation in the Project Environment," and Stephen D. Owens and Francis M. Webster, Jr., "Negotiating Skills for Project Managers," chaps. 19 and 21 in David I. Cleland (ed.), *Field Guide to Project Management* (New York: Van Nostrand Reinhold, 1997).
- Tony Yep, "Quality Management Works"; E. A. O'Connor, "Real World Challenges to a Multinational Project Team Building a Manufacturing Facility in India"; and D. H. Stamatis, "Total Quality Management and Project Management," in David I. Cleland, Karen M. Bursic, Richard J. Puerzer, and Alberto Y. Vlasak, *Project Managment Casebook,* Project Managment Institute (PMI). (Originally

published in PMI Canada 1996 Symposium, pp. 40–45; *Proceedings,* PMI Seminar/Symposium, Vancouver, Canada, October 1994, pp. 377–380; and *Project Management Journal,* September 1994, pp. 48–54.)

- David I. Cleland and Lewis R. Ireland, *Project Manager's Portable Handbook* (New York: McGraw-Hill, 2000). This portable handbook describes project management, including the major topic areas of the discipline with summaries, figures, and tables. Appropriate additional references that support the summaries in a topic area of project management are provided for the reader who wishes to learn more. The book presents the concepts and processes, as well as the strategic context of projects, and how they fit into the overall vision of the enterprise. Today's abundance of literature on project *management* makes it difficult to find a single source to keep abreast of the knowledge, skills, and attitudes required to manage projects. This book provides that single source.
- Kimball Fisher, *Leading Self-Directed Work Teams* (New York: McGraw-Hill, 1993). There are many similarities between a project team and a "self-directed" team. This book explains how team leadership skills such as coaching, facilitating, group dynamics, and many more of the issues likely to confront the project manager—and the members of the project team—can be managed. The author profiles the most innovative team leader practices from known and successful industrial organizations. The reader should be able to recognize how the knowledge expressed in this book can be applied to the management of a project team.
- David I. Cleland, "Understanding Project Authority," *Business Horizons,* Spring 1967. This is believed to be the first article that describes management authority and its use in the matrix organization. The focus of the article is a description of the means for the determining how and to whom the "legal right to act" is delegated in those organizations that use project management. Cleland describes "project authority" as applied in the horizontal sense to accommodate the means for empowering the project managers, functional managers, and members of the project team.
- Christopher G. Worley and Charles J. Teplitz, "The Use of 'Expert' Power as an Emerging Influence Style within Successful U.S. Matrix Organizations," *Project Management Journal,* February 1993, pp. 31–34. The authors briefly review the theory behind the matrix structure and the necessary requirements for its successful implementation. They then review the matter of power and influence in the matrix organization, including a description of prior research on the subject. Then the authors report on the results of a survey of project managers and teams within U.S. matrix organizations.
- Valerie Lynne Herzog, "Trust Building on Corporate Collaborative Project Teams," *Project Management Journal,* March 2001, pp. 28–35. This article looks at collaborative team trust building. The article recommends strategies for building corporate team trust. A specific model for trust building in project management is suggested. The author makes the strong point that integrating trust-building strategies into a team environment will help teams and their respective companies become more competitive.

9.20 DISCUSSION QUESTIONS

1. Describe a project management situation from your work or school experience. What role did project authority play in the management of the project? Did authority ambiguities exist?
2. Discuss the importance of clear definitions of project authority.
3. Define authority. Discuss the changing view of traditional authority. Discuss the difference between de jure and de facto authority. What is power?
4. What difficulties do project managers often face in exercising project authority?
5. Discuss the project-functional interface. How can clear lines of authority help in managing this interface?
6. What is meant by reverse delegation? Under what conditions might it be present? How can it be avoided?
7. Discuss the importance of negotiation between project and functional managers.
8. What is the purpose of documenting project authority?
9. What is the difference between authority, responsibility, and accountability?
10. What role does power play in project management? List and discuss some power sources.
11. What are some of the advantages of the traditional organizational chart? What are its limitations?
12. Define the linear responsibility chart in terms of its structure.
13. Define each of the symbols used to describe the responsibilities at the work package–organizational position interface.
14. List the steps involved in the development of the LRC.
15. Why is it important for this development to be a group effort?

9.21 USER CHECKLIST

1. Do the managers in your organization understand the limitations of the traditional chart for managing projects? How do they address these limitations?
2. Are the responsibilities and roles of project team members clear to the project manager and other managers? Are they clear to the team members themselves?
3. Are discussions held between the project managers, team members, and other project stakeholders to clarify authority, responsibility, and accountability? Why or why not? How can these discussions contribute to the success of the project?
4. Think about the various projects within your organization. How is project authority managed? Are there authority ambiguities?
5. Do you think that the authority of the project managers in your organization is clearly defined? Why or why not?

6. Do the managers of your organization understand the need for definition of authority relationships? Explain.
7. Do the managers of your organization use both de jure and de facto authority? How?
8. Is the project-functional interface effectively managed within your organization? Why or why not? How can clearer lines of authority assist in this management?
9. How is project authority granted within your organization?
10. What barriers to delegation exist on the projects within your organization? How can these barriers be better managed?
11. Is project authority documented? How?
12. What power tactics are used by managers in your organization? Is the use of power tactics productive or destructive toward achievement of organizational and project goals?

9.22 PRINCIPLES OF PROJECT MANAGEMENT

1. There are two types of project authority: de jure and de facto.
2. De jure authority is the legal or rightful power to command or act in the management of a project.
3. De facto authority is the influence brought to the management of a project by reason of a person's knowledge, skills, and interpersonal skills.
4. Authority and responsibility are shared in the matrix organization among the project manager, functional manager, the general manager, the work package manager, and the professionals on the project team.
5. The focus of authority and responsibility is at the project-functional interface, and centers around the project work package.
6. Project managers should have broad authority over all elements of their projects.
7. The authority and responsibility that is shared in the matrix organization should be documented.
8. Responsibility, a corollary of authority, is a state, quality, or fact of being responsible.
9. Accountability is the state of assuming liability for something of value.
10. The linear responsibility chart is an effective way to determine and assign the authority and responsibility for the management of the project.
11. Authority is a force that is essential to the functioning of any organization.
12. The project manager should have a balance between assigned de jure authority and the capability to exercise de facto authority.

9.23 PROJECT MANAGEMENT SITUATION—PRESCRIBING PROJECT MANAGEMENT AUTHORITY

Authority is the legal right to make decisions that affect the organization and the people in that organization. Responsibility is the obligation to make decisions that will impact the organization and the people in that organization. Authority and responsibility are matched pairs in the management of any organization. Authority and responsibility are provided to other people through the process of delegation. When a manager assumes a new position in an organization, one of the first things the manager should be concerned about is what authority and responsibility he or she has for making and implementing decisions on the project.

There are two basic types of authority defined and discussed in this chapter: (1) de jure, or the legal right to make decisions; and (2) de facto, or the influence that an individual brings by reason of her or his competency. Indicated below are some of the sources of de jure and de facto authority portrayed in the context of project management:

De jure:

- Policy/procedure manuals
- Project charter
- Letter of appointment as a project manager
- Contractual provisions
- Project plan
- Position dscription

De facto:

- Interpersonal skills
- Ability to communicate
- Expertise
- Team-building skills
- Negotiating skills
- Political skills
- Attitude
- Image with project stakeholders
- Ability to resolve conflict
- Coaching abilities

9.24 STUDENT/READER ASSIGNMENT

Assume that you have just been assigned as a project manager of a large product development project. One of the first questions that you have is what authority and responsibility you will have in making and implementing decisions on this project. In order to gain insight into this question, you decide to design a "model" that will describe the specific strategies/actions you would take to capitalize on both the de jure and de facto sources of authority. Describe such a model, and be as specific as you can.

Note that the authors make the strong point that integrating trust-building strategies into a team environment will help teams and their respective companies become more competitive.

CHAPTER 10
PROJECT MANAGEMENT MATURITY

"It takes a long time to bring excellence to maturity."

PUBLILIUS SYRUS, CIRCA 42 B.C.

10.1 INTRODUCTION

The project management discipline continues to ripen in its progression as a building block in general management theory and practice. Theorists and practitioners in the field have offered various project management maturity models. These models are preliminary—but show promise of becoming an effective means of describing where project management has been and where it is today, and give insight into where it is likely to go in the future.

In this chapter a total organizational capability maturity model is presented. Then, instructions are provided for how to assess contemporary project management maturity. Insight into how to build a mature capability in the context of project management is suggested. The roles of benchmarking and competitive analysis are also offered as a means of providing insight into project maturity. In addition, benchmarking and competitive intelligence are described as strategies for improving organizational effectiveness through collection of business information about competitors.

10.2 ORGANIZATIONAL PRODUCTIVITY IMPROVEMENTS

Organizations continue to search for better means of improving their productivity and profit through changes to the manner in which work is done and the type of work being done. Project management is being embraced, to some extent, by most companies as the best way to develop and deliver new or improoved products, services, and organizational process changes. Enhanced capability to conduct project management is often sought through adopting new, innovative ways to perform project management and through the use of new tools and techniques.

Recent initiatives to improving an organization's project management capability are through capability maturity models. These models provide a structure for an organization to follow that will lead to more efficient and effective operations. Several models have been described in literature, each containing similar features and approaches to implementing the process.

Other efforts preceded the maturity models for improving organizational capability. Using different names and having different focuses, each of the initiatives made significant contributions to advancing productivity and profit for businesses.

- In the early 1980s, there was a concerted effort within organizations to improve quality. This movement had several different titles, but total quality management (TQM) seems to be the most enduring and lasting label. This TQM movement consists of the right approach to how we do our work and how we achieve good products by doing things right. The quality movement stimulated other activities, directly or indirectly, that improved American productivity.
- Philip Crosby, a quality guru who is instrumental in spreading the word about quality, made a significant contribution to improving products, services and processes in all forms of industry. His book *Quality Is Still Free* has provided thousands of organizations a framework for pursuing and attaining businesses that changed their quality focus.
- Another quality guru who changed the face of quality in American is Dr. W. Edwards Deming—a statistician by training and education. Dr. Deming, from his initial quality work in Japan during the 1950s to his lectures around the United States in the 1980s, consistently emphasized the use of data to support decisions—management by fact—and continually improve products and processes—remove the variances from the processes.
- In 1988, the National Institute of Standards and Technology (NIST) partnered with the American Society for Quality (ASQ) to establish The Malcolm Baldrige National Quality Award Program. The purpose of the program is to recognize U.S. organizations that meet rigorous criteria in seven core areas:

  1. Leadership
  2. Strategic planning
  3. Customer and market focus
  4. Information and analysis
  5. Human resource focus
  6. Process management
  7. Business results[1]

The Software Engineering Institute–Capability Maturity Model (SEI CMM) is a derivative from Philip Crosby's "The Quality Management Maturity Grid," as outlined in *Quality Is Still Free.* This grid provides the basis for five stages of

[1]*Baldrige National Quality Program 2001: Criteria for Performance Excellence, 2000,* National Institute of Standards and Technology, Bethesda, Md., 67 pages.

maturity and understanding the advances made at each stage. Crosby's quality grid is summarized in Table 10.1.[2]

These landmark quality efforts have made a major difference in how some companies look at quality and what has changed for the better within the organizations. The successful efforts of Crosby, Deming, and NIST and ASQ—and others—have changed how quality is perceived and improved through dedicated effort.

All these efforts through quality improvement programs focus on what it takes to improve parts of the business, and all of these parts are important. It may be more effective to keep the total system in focus and improve those parts that are holding the business back or keeping the business from achieving its best. A total systems perspective should be used to "fix" those parts that are operating at less than full efficiency and effectiveness.

In Chap. 1, choice elements are described as being integral components of an organization's approach to using project management as the process for conducting business. Figure 10.1 (also Fig. 1.2) depicts the choice elements of an organization using project management as its primary approach to building products, services, and organizational change processes.

Figure 10.1 consists of several interlocking nodes that must work in harmony to provide a process solution for any business that represents the optimal structure. It is intuitively obvious that weaknesses in any node will adversely affect the other nodes. For example, if the mission statement is flawed, it would be nearly impossible to develop relevant and supportive objectives and goals. Any implementing strategy would also be difficult to implement for achieving the desired results that move the organization ahead in the best manner.

This diagram and brief explanation demonstrate the need to address the entire organizational concept though a maturity model approach rather than selecting random weaknesses to repair. The randomly identified weaknesses can yield some dramatic results, but have greater value if the entire organization is fine-tuned to support projects as the building blocks.

TABLE 10.1 Quality Management Maturity Grid

Stage	Title
I	*Uncertainty*—unknown status of quality in the organization.
II	*Awakening*—realization that quality is important and there are quality issues.
III	*Enlightenment*—program initiated and progress made in correcting quality issues.
IV	*Wisdom*—quality culture in the organization and everyone works toward quality.
V	*Certainty*—quality initiatives working and organization producing products and services for clients

Source: Paraphrased from Philip Crosby's *Quality Is Still Free: Making Quality Certain in Uncertain Times* (New York: McGraw-Hill, 1996), pp. 32–33.

[2]Paraphrased from Philip Crosby's *Quality Is Still Free: Making Quality Certain in Uncertain Times* (New York: McGraw-Hill, 1996), pp. 32–33.

FIGURE 10.1 Choice elements of strategic management.

10.3 PROJECT MANAGEMENT MATURITY MODELS

The continual evolution and advancement of maturity models provide examples of the thinking and the direction that maturity models are taking. These efforts to define and implement a project management maturity model emphasize the challenges and successes of developing such a model.

Carnegie Mellon University's Software Engineering Institute (CMU/SEI) defined a capability maturity model for software in an effort to improve the success rate for software projects. This initiative resulted in a five-level model to characterize the behavior of maturity as it progressed from an ad hoc stage to one of continuous improvements. The resultant levels are defined as follows:

Level 1—initial level. The software process is characterized as ad hoc, and occasionally even chaotic. Few processes are defined, and success depends on individual effort.

Level 2—repeatable level. Basic project management processes are established to track cost, schedule, and functionality. The necessary process discipline is in place to repeat earlier successes on projects with similar applications.

Level 3—defined level. The software process for both management and engineering activities is documented, standardized, and integrated into a standard software process for the organization. All projects use an approved, tailored version of the organization's standard software process for developing and maintaining software.

Level 4—managed level. Detailed measures of the software process and product quality are collected. Both the software process and products are quantitatively understood and controlled.

Level 5—optimizing level. Continuous process improvement is enabled by quantitative feedback from the process and from piloting innovative ideas and technologies.

CMU/SEI also developed key process areas (KPAs) for the different levels of maturity. Key practice areas follow:

Level 1 key process areas—none developed because this is the initial stage of entry for an organization. There are no standard process requirements. An audit is required to identify those processes in use and how well they are being accomplished.

Level 2 key processes:

- Requirements management
- Software project planning
- Software project tracking and oversight
- Software subcontract management
- Software quality assurance
- Software configuration management

Level 3 key processes:

- Organizational process focus
- Organizational process definition
- Training program
- Integrated software management
- Software product engineering
- Intergroup coordination
- Peer reviews

Level 4 key processes:

- Quantitative process management
- Software quality management

Level 5 key processes:

- Defect prevention
- Technology change management
- Process change management[3,4]

The CMU/SEI capability maturity model only supports software development although the model provides a general framework for any capability maturity model. This same model has been translated to a project management capability

[3]Mark C. Paulk, Bill Curtis, Mary Beth Chrissis, and Charles V. Weber, *Capability Maturity Model*(sm) *for Software,* Version 1.1, CMU/SEI-93-TR-024, February 1993.

[4]Mark C. Paulk, Charles V. Weber, Susanne M. Garcia, Mary Beth Chrissis, and Marilyn Bush, *Key Practices of the Capability Maturity Model*(sm) Version 1.1, CMU/SEI-93-TR-025, February 1993.

maturity model with different criteria for achieving levels of maturity. Typically, the new models use the nine functional areas of the Project Management Institute's *Guide to the Project Management Body of Knowledge.*

The Federal Aviation Administration (FAA) put in place a Software Process Improvement Program, which follows the CMU/SEI capability maturity model. In a December 2, 1999, presentation, Dr. Linda Ibrahim of the FAA described the what, why, and how of software improvements.

1. *What is the meaning of integrated process improvement?* Both improving integrated processes and improving process integration are essential. An integrated process was further defined by Ibrahim: "An integrated process requires the participation of experts from more than one functional area or discipline." The integrated process involves multidiscipline—not just software. FAA's examples of processes that needed improvement are (1) requirements, (2) acquisition management, and (3) outsourcing.
2. *Why improve integrated processes?* The objectives are to improve the effectiveness and efficiency of the FAA systems.

   a. Effectiveness:

   (1) Avoid suboptimization because it may be a system problem, not a software problem.
   (2) Do not focus too much on development measures because that may adversely impact maintenance.
   (3) Address interfaces and interrelationships of the parts of the system.
   (4) Mark importance of hardware-software interfaces.
   (5) Obtain integrated product teams that perform integrated processes.
   (6) Align processes with business objectives before integration across the business.
   (7) Enhance the corporate culture of continuous improvement.
   (8) Conduct teamwork across lines of business and from executive to practitioner.

   b. Efficiency:

   (1) Conduct training and appraisal, resulting in fewer processes to improve.
   (2) Reduce complexity of processes and systems.
   (3) Pursue common goals of vision, focus, and clarity.
   (4) Get everyone involved to pursue the same goals.

3. *How to achieve integrated process improvements.* Be prepared to change when an opportunity emerges for integrated process improvement.

   a. Prepare process improvement guidance that relates to actions across the enterprise.
   b. Have enterprisewide vision and goals.
   c. Have a strategy and plan to take advantage of opportunities.
   d. Have the right resources to implement the change.

The FAA-iCMM^{sm} is the model for FAA activities to improve its processes throughout facilities conducting such vital services as air traffic control. Derived from the CMU/SEI capability maturity model, FAA-iCCM has reduced the 53 separate process areas of the CMM to 23 integrated process areas. There is only one appraisal compared to the three of the CMM.

The FAA went beyond software improvements to the more robust thinking of *system integration.* Software improvements are one step forward in an uncoordinated way, whereas system integration considers the enterprisewide positive and negative impact of change. It is the goal of the FAA to achieve greater improvements through a systems approach rather than just focus on software improvements.[5]

The Project Management Institute has been researching and developing an organizational project management maturity model that includes both a staged and continuous approach. This work, conducted under the direction of John Schlichter, is progressing toward a fully capable model as defined by the Project Management Institute. Reports of progress have been made by Margaret Combe in 1998 on preliminary results.[6] In 1999, Schlichter and Duncan (PMI Standards Committee) reported on their interpretation of what project management maturity means to them.[7] The outcome of the Project Management Institute's efforts will certainly have a major influence on the future of project management maturity models and will provide a baseline for further evolution of the concept of maturity measurements for project management capability.

10.4 TOTAL ORGANIZATIONAL CAPABILITY MATURITY MODEL

Building a capability maturity model by leveraging the work that has been done for software (SEI CMM) and integrated processes (FAA-iCMM) permits structuring to engage all elements of a business. The SEI CMM and FAA-iCMM's strengths in selected areas demonstrate the need for improving business for better productivity and profit. Business in general needs a model from which to leverage its strengths and to advance in a competitive marketplace.

Figure 10.1 provides the framework for balancing business initiatives by first establishing the mission with its objectives and goals within an overall vision of where the company wants to be. Any capability maturity model must consider that these choice elements flow down the requirements of what the business must accomplish and when the goals will be met.

The company's strategies, or the means by which the business will be achieved, are critical to success. Unrealistic strategies to fulfill the objectives and goals, such as use of resources that are beyond the capacity of the organization, will materially

[5]Paraphrased from Dr. Linda Ibrahim, "Integrated Process Improvement: Is Software Improvement Enough?" SPI 99 Conference, December 2, 1999, Barcelona, Spain, 52 slides. (No copyright.)

[6]Margaret W. Combe, "Standards Committee Tackles Project Management Maturity Models," *PM Network,* August 1998, p. 21.

[7]John Schlichter and William R. Duncan, "An Organizational PM Maturity Model," *PM Network,* February 1999, p. 18.

affect implementation through projects and programs. The first order of any maturity model must be an examination of (1) vision, (2) mission, (3) objectives and goals, and (4) strategies. These top-level business elements cannot have any significant flaws or the implementation will be impacted.

A total organizational capability maturity model could be structured as follows:

- *Vision.* Establishes an image of a future position for the enterprise that is favorable to its business.
- *Mission.* A broad, umbrella statement of the business that the company is pursuing that establishes the guidelines, in general terms, that focus the organization's use of resources.
- *Objectives and goals.* Establish narrow areas of work to be achieved and, by doing so, represents a detailed elaboration on the enterprise's mission.
- *Strategies.* Establish the means through which the objectives and goals will be pursued.
- *Projects and programs.* Sharply focus objectives and goals that give the organization the methodology, techniques, and tools to achieve selected products, services, or processes.
- *Operational plans and organizational design.* Align the organization and its resources to accomplish work through project management or other appropriate means.
- *Policies, procedures, protocols, and systems.* Align practices through published guidance and establishment of a systems approach to performing work.

A total organizational capability maturity model would examine the top-level guidance to determine the adequacy for the business being pursued. Table 10.2 outlines some of the questions that might be asked.

The questions posed in Table 10.2 should all be answered yes to have a solid foundation from which to conduct further examination of an enterprise. If the answer is no, can the projects be expected to be successful? Any capability maturity model must have a foundation that provides the stable basis for project actions.

There are typically two types of maturity models being described in current literature: the continuous project management maturity model and the staged project management maturity model. These two models form the overall framework for improving project management through a structured approach.

- *Continuous project management maturity model.* A model that establishes a baseline for an organization through an assessment; specific elements are then used as criteria to establish what will be improved and at what rate. Figure 10.2 depicts the continuous project management maturity model.
- *Staged project management maturity model.* A five-step model that establishes criteria for each step. This permits incremental improvements in all areas being addressed for projects. All areas are considered essential to project maturity and given similar weight for improvements. Figure 10.3 depicts the staged project management maturity model.

TABLE 10.2 Strategic Element Evaluation

Vision
- Is there a vision for the organization?
- Does that vision convey the future business in a clear fashion?
- Do the people in the organization understand and agree with the vision?

Mission
- Does the mission statement clearly define what business the enterprise is in?
- Does the mission statement match what the enterprise is pursuing for business?
- Is mission a broad statement of what business the enterprise is in?

Objectives
- Do the objectives provide guidance on what will be achieved?
- Do the objectives tell what is to be achieved?
- Are the objectives an elaboration of the details of the mission?

Goals
- Are the goals milestones to be achieved in executing the mission?
- Are the goals time-focused and specific as to the expected results?
- Are the goals consistent with the business the enterprise is pursuing?

Strategies
- Are the strategies the means by which the enterprise will achieve the objectives and goals?
- Are the strategies realistic within the resource constraints of the enterprise?
- Are the strategies success-oriented?

Source: Adapted from the discussion of the term "strategic choice," which was previously used in William R. King and David I. Cleland, *Strategic Planning and Policy* (New York: Van Nostrand Reinhold, 1978), chap. 6.

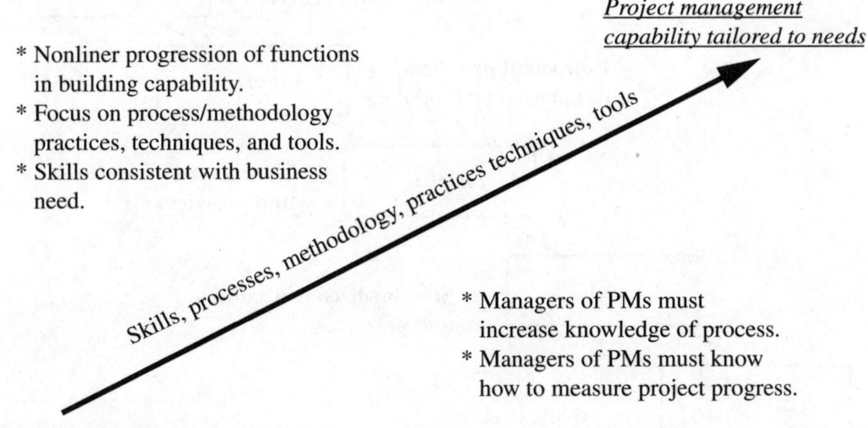

FIGURE 10.2 Continuous project management maturity model. (*Source: Lewis R. Ireland, "Executive Project Management Training Course," prepared for PMI® Pikes Peak chapter training course, Colorado Springs, Colo., 1998.)*

The continuous improvement model has to be tailored to fit the organization's needs. For example, some organizations do not include procurement in their model because it is not a critical part of their projects. There would be little or no gain for the organization to divert resources to improving an element not essential to the business. The development of an improved project management capability also needs to be consistent with the advancement of capability. Training in project management must be consistent with the organization's needs.

At the lower end of Figure 10.2 is the label "Haphazard project management." This is identified as such because the work is done in a random manner and each project manager has a different methodology for projects. The inconsistent approach does not maximize the effort to achieve the best results. Following an assessment of the current practices and the organization's project management needs, a plan for a continuous improvement model can be established and implemented.

The staged project management maturity model has criteria for levels 2 through 4; level 1 is the unaudited current situation in an organization. The capability at level 1 can have significant strengths and weaknesses in the capability to implement project management. Bringing the strengths and weaknesses into balance for a full capability will require leveraging the strengths and eliminating the weaknesses.

Both models have used the Project Management Institute's division of project management into nine functional areas—integration, scope, time, cost, human resources, risk, communications, quality, and procurement. These elements are widely recognized as the standard for project management areas and defined in the *Guide* to *Project Management Body of Knowledge,* published by the Project Management Institute.

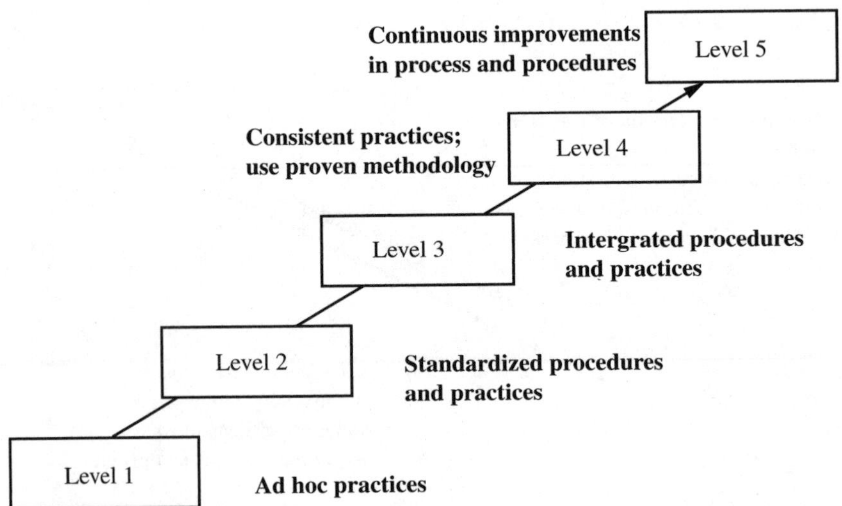

FIGURE 10.3 Staged project management maturity model. (*Source: Lewis R. Ireland, "Executive Project Management Training Course," prepared for PMI® Pikes Peak chapter training course, Colorado Springs, Colo.,1998.)*

Each area must be assessed against a standard practice and criteria to provide consistent results for sequential audits. The criteria need to be developed on the basis of business needs to achieve a tailored solution specifically for an enterprise.

Table 10.3 gives a layout for the responses to a questionnaire that may be used to collect information regarding project element maturity. The questions are developed to evaluate the element and to determine the degree to which certain practices are employed. Subelements are questions that specifically relate to the element, for example, Scope: Do you consistently develop a scope management plan?

General questionnaires would be supplemented by interviews with project stakeholders to obtain information on the project management practices and their effectiveness. Organizational guidance and project documents would be assessed for their completeness and effectiveness. Any assessment should also determine whether the published guidance is used.

Interviews are typically conducted to follow up on items in the general questionnaire that are either skewed to one side or the other, lack answers or have insufficient responses to determine what is the true situation, or have answers uniformly distributed across a range of solutions. Interviews should focus on resolving the data that give no clear indication of the situation. Interviews may also probe into areas that appear to be important following the general questionnaire analysis. Interviews may be the only means of resolving apparently conflicting results from the questionnaires.

Document review should be conducted in order to assess the highest level of guidance to the project level. Conducting a document review may find the following weaknesses in the system.

- Organizational documents are nonexistent, provide inadequate guidance, are not being used by the project team, are not available to the project team, or are overly detailed and cumbersome to use.
- Project charter may either be nonexistent, inadequate in detailing responsibilities, or overly restrictive for the project manager.
- Project plan may have too little or too much detail. It should include:
    - Scope statement.
    - Specification for product.
    - Statement of work.
    - Schedule.
    - Budget.
    - Risk plan.
    - Communication plan.
    - Procurement plan, if needed.
    - Human resource plan.
    - Quality assurance and quality control plan.
    - Other plans tailored to the project.

TABLE 10.3 Evaluation Elements and Subelements (Example for Questionnaire)

Element and questions	Score* N/A	1	2	3	4	5
Integration						
1. Do you consistently prepare an integration plan?						
2. Are integration activities included in the WBS?						
3. Are integration activities included in the schedule?						
4. …						
Scope						
1. Do you consistently develop a scope management plan?						
2. Is the scope management plan used on the project?						
3. Is the scope management plan updated to reflect changes to current practices?						
4. …						
Time						
1. Is the schedule based on a WBS?						
2. Is a schedule developed for each project to reflect the tasks/activities and their durations?						
3. Are schedules realistic and based on sound estimates of the work to be done?						
4. …						
Cost						
1. Is a cost estimate done for each project?						
2. Does each project have a time-phased budget?						
3. Are expenditures controlled through a positive process?						
4. …						
Risk						
1. Is a risk management plan developed for each project?						
2. Has a risk assessment been conducted for each project?						
3. Have contingency and management reserves been established for risk?						
4. …						

TABLE 10.3 Evaluation Elements and Subelements (Example for Questionnaire) (*Continued*)

Element and questions	Score* N/A	1	2	3	4	5
Quality						
1. Have quality control and quality assurance plans been developed for each project?						
2. Is there a quality assurance process in place for each project?						
3. Are quality validation procedures in place for each project (tests, demonstrations, audits)?						
4. …						
Communications						
1. Is there a communication plan prepared for each project?						
2. Have meeting procedures been established to most effectively manage the meeting time?						
3. Have standard reporting procedures been established for each project?						
4. …						
Procurement						
1. Is there a project procurement plan in place for each project?						
2. Is there an established procedure for requesting goods and services?						
3. Is there a procurement expert identified to support your needs?						
4. …						
Human resources						
1. Is there a human resource plan developed for each project?						
2. Are all the skill requirements identified for the project during planning?						
3. Do the skill requirements typically match the available resources?						
4. …						

*Legend: N/A: not applicable in my organization; 1: never done in my organization; 2: infrequently done in my organization; 3: about half the time done in my organization; 4: done most of the time in my organization; 5: always done in my organization.

Source: Adapted from presentations to PMI® chapters (Colorado Springs, Colo.; Denver, Colo.; Tacoma, Wash.; and Seattle, Wash.), Lewis R. Ireland, "Assessing Project Management Maturity," November 2000.

- Training records reflect the wrong training for the project team or inadequate knowledge and skills training.
- Active issues logs, action item logs, configuration change logs, and reports may not be current and prosecuted.
- Contractor and vendor agreements may be inaccurate or out of date.
- Time cards or time reports may be inaccurate and poorly administered.

10.5 ASSESSING PROJECT MANAGEMENT MATURITY

Assessing project management levels of maturity is accomplished through a series of audits. The first audit establishes where the organization is currently in terms of maturity and identifies areas for improvement. Audits are typically conducted in the following sequence:

- Identify and adopt the capability maturity model for the organization.
- Identify the areas to be audited and the criteria for each area of audit.
- Select an audit team either from internal resources or external sources.
- Conduct a briefing for the audit team to set expectations regarding how and when the audit will be accomplished.
- Conduct the audit.
  - Administer questionnaires to selected personnel.
  - Interview selected personnel.
  - Review project documents.
  - Collect artifacts from projects.
- Prepare audit report.
- Brief audit results.

Senior management sets the tone for any audit—it is a fact-finding effort, not a fault-finding effort. To obtain full cooperation from those persons being audited, the questionnaires and interviews should be confidential and reports should not make direct reference to anyone or any group. The only exception to any audit confidentiality is that unsafe or unlawful acts must be acted upon; all persons being interviewed should understand this.

Conduct of the audit must be professional and made without comment to the persons being interviewed. No “results” type of information should be released to individuals because emerging results do not consider all aspects of the audit and may be misleading. The final report should be prepared and briefed to senior executives. Those individuals participating in the audit should also be provided feedback on the results.

10.6 BUILDING A MATURE CAPABILITY

Recognizing opportunities for improvement is the first step in building a mature organization. Any weakness at the top of the organization will limit the ability of elements at lower levels. Take, for example, outdated or erroneous guidance in the organization's policies and procedures that are being used by project managers. This guidance can cause a wide variance from the most favorable path and consume resources for no benefit.

Selecting the right model and assuring that it is a fit with the strategic focus of the enterprise initiates the process. Using the model's criteria and developing new criteria will align the model to the organization's vision and mission as well as validate the objectives and goals.

Starting at the top of the organization to correct weaknesses will yield better results than correcting the small items within projects. For example, training seems the obvious and immediate need in order to enhance knowledge and skills within projects. If the methodology is not in place, training would not be appropriate for any particular method of performing project work. Further, once trained individuals use a specific method, it takes more effort to change them to a new method. Vision, mission, objectives, goals, and strategies need to be validated first.

An audit will identify many of the opportunities for improvement and point out strengths and weaknesses. There needs to be a plan for when, how, and why the changes will be made. Some changes may eliminate weaknesses in other areas. For example, a change to the scheduling software may improve resource management on projects when resource allocation has been identified in the resource management area.

Table 10.4 summarizes actions needed to build an improved project management capability.

10.7 BENCHMARKING

Benchmarking may be a tool in developing maturity criteria for any organization that is tailoring a capability maturity model for its use. The dynamic nature of business often dictates that an organization find new and better practices that meet the competition's approach. Benchmarking is also presented in Chap. 21.

Benchmarking is a strategy for measuring organizational products, services, and organizational processes against top-of-the-line competitors and industry leaders. This measurement is accomplished to determine whether an organization is using best-in-class practices for business operations and for development of new performance standards against which to evaluate the enterprise.

The use of benchmarking varies among industries. Some reports claim that benchmarking is as low as 5 percent within an industry, whereas others claim that it is a widespread practice. One claim is that there was a savings of $7.3 million

TABLE 10.4 Actions to Build on Project Management Capability

Item	Action
Select maturity model.	Evaluate project management maturity model that meets the needs of the organization's business.
Review vision, mission, objectives, goals, and strategies for alignment and focus.	Senior management conducts a review or has an outside consultant review the organization for strategic alignment with business.
Audit project management functions.	Identify strengths and weaknesses to determine opportunities for improvement within the overall strategic focus of the business.
Audit project management support functions.	Identify strengths and weaknesses of support to projects to determine opportunities for improvement.
Plan for improvements.	Prepare a business improvement plan that establishes priority and criteria for implementing changes.
Perform improvements.	Initiate changes and assess impact of changes.
Conduct follow-on audits.	Validate effect of changes and effect of improvements on the business.
Position for continuous improvement.	Change the culture over time to encourage continuous improvements in both product and process.

in oil exploration operations because benchmarking found a more efficient method for drilling. Another claim by a university is that benchmarking was used to develop the curriculum for an MBA program.

Some examples of benchmarking results demonstrate the effects of programs within major industries:

- General Motors has pushed benchmarking in a big way. Every new operation must be benchmarked against the best in the class, including companies not in the car-making business. A project team of 10 people from GM helped coordinate GM's worldwide benchmarking strategies.[8]
- When Digital Equipment benchmarked its manufacturing operations, it discovered that its costs were 30 to 40 percent too high.[9]
- Ma Bell's Global Information Solutions (GIS) constantly benchmarks itself against rivals and surveys customers constantly, turning the results into a measure

[8]Alex Taylor, III,"GM's $11,000,000,000 Turnaround, *Fortune,* October 17, 1994, pp. 54–74.
[9]Stratford Sherman, "Are You as Good as the Best in the World?" *Fortune,* December 13, 1993, pp. 95–96.

of "customer delight" on a scale of one to seven. Key characteristics of the culture at GIS include vision, trust, rewards, and compassion—all the tools of enlightened leadership.[10]

Benchmarking is accomplished through teams and is used in several different yet complementary contexts:

- *Competitive benchmarking.* The five or six most threatening competitors are studied and evaluated to gain insight into their strengths, weaknesses, strategies, and performance capabilities. This insight is then compared to the strategies and performance of the enterprise.
- *Best in industry.* The strategies, practices, and performance of the best performers in selected industries are studied and evaluated. Sometimes those enterprises selected for further study may or may not be in the same industry.
- *Best in class.* When information on a process is not typically available within an industry, the best source may be another organization that is not considered either within the industry or a competitor. The search is then on the process and who performs that process most efficiently and most effectively. For example, a medical manufacturing company may find the best process for delivery services in a food company.

Any benchmarking effort should consider the following questions in developing sources of information:

- Who are the most relevant competitors to examine?
- Who are the best performers in the industry to examine?
- Who does the best job of a certain type of service?
- What processes should be examined and to what depth?
- How will the benchmarking information be used to improve the enterprise's operations?

Every endeavor needs a plan to lay out the procedures for conducting the work effort. Develop a benchmarking plan that identifies what is being sought and how the information will be used. This plan should include:

- Purpose of benchmarking effort
- Anticipated results of benchmarking examination of other industries/businesses
- Anticipated organizations to be examined
- Sources of information about organizations identified
- Benchmarking team to collect information
- Relevant information
- Information analysis

[10]Thomas A. Stewart, "How to Lead a Revolution," *Fortune,* November 28, 1994, pp. 60–61.

- *Best competitor* practices, *best-of-industry* practices, and *best-of-class* practices
- Dissemination of results to organizational units

Information useful for benchmarking can come from a wide variety of sources, both in the private and public domains. Company records, site visits, interviews, customers, suppliers, regulatory agencies, periodical literature, seminars and symposia, investment bankers, and brokerage firms are a few key sources for information. Use of the Internet has facilitated the ability to do effective benchmarking. There is usually more information available than can be reasonably collected and analyzed; therefore, information needs and the collection processes must be studied carefully before launching the benchmarking.

Many of the websites are transitory and new ones will emerge. Finding information on the Internet is best accomplished by using the "search" function for the category that a person is seeking. A "benchmarking" search will identify several websites and then a person can select the sites providing the appropriate information.

One Internet site lists more than 600 best practices of industry available for accessing information. Major categories of this website are:

- Customer service
- Human resources
- Internet and e-business
- Knowledge management
- Sales and marketing
- Business operations

Many companies have websites that list the basic information on their business. These websites may only reveal top-level information, but it is a good place to start identifying organizations, what they do, what products they sell, and their focus.

10.8 COMPETITIVE INTELLIGENCE

Competitive intelligence (CI) is the process of monitoring the competitive environment. CI enables senior managers in companies of all sizes to make informed decisions about everything from marketing, R&D, and investing tactics to long-term business strategies. Effective CI is a continuous process involving the legal and ethical collection of information, analysis that doesn't avoid unwelcome conclusions, and controlled dissemination of actionable intelligence to decision makers.[11]

Competitive intelligence refers to the collection of information on other businesses. It has been called business intelligence, but the current usage is competitive intelligence. This field has grown to the extent that a professional association has

[11]Parafrased from Society for Competitive Intelligence Professionals. 1700 Diagonal Road, Suite 600, Alexandria, VA. 22314. See www.scip.org.

been formed—Society of Competitive Intelligence Professionals (SCIP)—and publishes a bimonthly magazine—*Competitive Intelligence Magazine.*

Competitive intelligence, when properly collected, analyzed, and utilized, provides early warnings or alerts to shifts in trends. This intelligence aids senior managers in making the right decisions in a timely manner to maintain a competitive edge and for strategic positioning of the enterprise.

Intelligence refers to competitive and environmental data that have been evaluated to be useful in a specific situation, project, or class of situations found today in the global marketplace. The types of information about competitors that can be considered to be *fair game* for collection include a wide range of information on market pricing, discounts, terms, specifications, market volume, historical performance trends, estimates of competitor's share, reverse engineering of competitive products, marketing policies, major strengths and weaknesses, to name a few. Marketing information, production and product information, organizational and financial information, are additional areas on which competitive intelligence is needed.

To systematize the competitive intelligence process means that questions such as those that follow need to be answered and the answers melded into the competitive strategies of the enterprise:

- What needs to be known?
- Where can the data be obtained?
- Who will gather the data?
- How will the data be gathered?
- Who will analyze and interpret the data?
- How will the data collected be stored and disseminated?
- Who will use the data?
- How will the information collected be protected from "leakage" and sabotage?

The enterprise competitive data collection should be organized on a program basis. Interdisciplinary and interorganizational project teams can serve as the organizational design for the collection of competitive intelligence. The operation of these teams should follow the conceptual framework for the development of a project management process, and the execution of those processes should be similar to what would be done for any project.

An important part in the analysis of data and the formulation of the intelligence results is an analysis of the credibility and reliability of information. This means that experts from the different disciplines and functions of the enterprise should be able to accomplish two key appraisals of information: (1) appraisal of the source and (2) appraisal of the content.

Competitive intelligence is vital information in the strategic management of the enterprise. All intelligence data gathering, analysis, and use should be done to ensure the maximum value and credibility of the decisions being made and executed by the strategic managers.

10.9 TO SUMMARIZE

The major points that have been expressed in this chapter include:

- Several capability maturity models have been developed, most notably the SEI CMM, FAA-iCMM, and a variety of similar models.
- Capability maturity models, like the quality movement, improve productivity and competitiveness in the marketplace.
- The SEI CMM is a software improvement model developed through the Department of Defense by Carnegie Mellon University.
- The FAA-iCMM started as a software improvement model but evolved to a system model for the FAA.
- All maturity model implementation requires that assessments be made of the situations and compared to maturity criteria in order to advance within the model.
- Benchmarking is a means of identifying the best practices of other organizations and applying them to the enterprise's situation to achieve better results.
- Benchmarking provides the best practices for industry, class, and competitive analysis.
- Competitive intelligence is analyzed information about competitors and potential competitors that contributes to the decision-making process.
- Competitive intelligence was formerly called business intelligence and focuses on collection, analysis, dissemination, and use of intelligence.

10.10 ADDITIONAL SOURCES OF INFORMATION

The following additional sources of project management information may be used to complement this chapter's topic material. This material complements and expands on various concepts, practices, and theory of project management as it relates to areas covered here.

- John E. Martin and Pierre-Francois Healume, "Risk Management: Techniques for Managing Project Risk," chap. 12 in David I. Cleland (ed.), *Field Guide to Project Management* (New York: Van Nostrand Reinhold, 1997).
- Francois Lacasse, "Goal Definition and Performance Indicators in Soft Projects: Building a Competitive Intelligence System," chap. 1 in David I. Cleland, Karen M. Bursic, Richard J. Puerzer, and Alberto Y. Vlasak, *Project Management Casebook,* Project Management Institute (PMI).
- Michael O. Tingey, *Comparing ISO 9000, Malcolm Baldrige, and SEI-CMM for Software* (Englewood Cliffs, N.J.: Prentice Hall, 1997). This book's comparison of three leading programs for organizational improvement brings out the best practices to provide a framework to apply to any capability maturity

model. The author highlights the strengths of each program, such as ISO 9003 having the best process orientation and Malcolm Baldrige for having the best approach to leadership. This book serves as a guide through the competing frameworks and allows the reader to select the optimum approach for an organization. As an added bonus, the author discusses the "whats" and "whys" behind each program to clarify and promote understanding.

- Christopher E. Bogan and Michael J. English, *Benchmarking for Best Practices: Winning through Innovative Adaptation* (New York: McGraw-Hill, 1997). The authors, experienced in benchmarking practices and quality, describe their nine-step benchmarking model that leads through the fundamentals to world-class quality. This book focuses on successful benchmarking to change corporate attitudes to bring an end to "not invented here." The book contains many examples and suggestions for how to select benchmarking targets, organize for rapid learning, and implement strategies. One theme in the book is "learn by borrowing and adapting approaches" to improve practices.
- Charles Halliman, *Business Intelligence Using Smart Techniques,* Information Uncover, Houston, Tex., 2001. This book explores environmental scanning and business analysis with ideas that can potentially improve an organization's strategic outlook. It shows how to examine the business environment for threats and opportunities. It explains the environmental forces: regulatory, marketing, foreign, management, and competitor. There are examples of business concepts and competitor activities presented.
- Scott Campbell, "Benchmarking Centers Pushed by Support Net, IBM," *The Newsweekly for Builders of Technology Solutions,* May 21, 2001, p. 92. This article describes how benchmarking centers are being encouraged by Support Net and IBM. These facilities are expected to cost $1 million each and will store information to be accessed for business purposes. The reported heavy investment demonstrates the value assigned to benchmarking and its anticipated payoff through the sale of information to various consumers.
- Sandra Bolan, "Competitive Calibration," *Computing Canada,* May 4, 2001, p. 24. This article describes benchmarking as a way of calibrating an organization against competitors to discover and adopt best practices. It reports that only 5 percent of companies have active benchmarking programs and that companies must "do it right" with up-to-date internal analysis and external metrics. An assertion is made that companies that perform annual benchmarks typically have a 20 percent higher improvement ratio than other companies.
- Stephen H. Miller, "Competitive Intelligence—An Overview," *Competitive Intelligence Magazine,* Society for Competitive Intelligence Professionals, Alexandria, Va. (available through *www.scip.org*). This article provides an overview of what competitive intelligence is and isn't. It demonstrates the value of competitive intelligence and identifies some of the people involved in competitive intelligence activities such as financial planners, business development people, strategic planners, and marketing planners. This is a quick look at the role of competitive intelligence in an organization and SCIP.

10.11 DISCUSSION QUESTIONS

1. Discuss the concept of a capability maturity model for a company that you are familiar with and have knowledge of its project management system.
2. Discuss the requirements for a basic repeatable process within an organization using project management as its primary approach to developing and delivery of products.
3. Compare the continuous improvement model to the staged project management maturity model and identify the benefits of both in a given organization.
4. Identify some criteria for attaining each level of maturity in the five-stage model.
5. Create five criteria for project management maturity at the repeatable level in the SEI CMM system.
6. Discuss the functions that are pertinent to improving one's project management capability. Are the nine functions from PMI appropriate for your situation?
7. Discuss the efficacy of using only a software capability maturity model for an organization and what effect it may have on other project management functions not related to software?
8. Benchmarking requires that sources of information be available. Identify six sources of information that are quick to access.
9. Benchmarking for better practices can be done in any industry. Discuss why a practice in one industry may be applicable to a business in another industry.
10. Benchmarking plans have specific goals. Discuss some goals that may be appropriate for benchmarking a product distribution system.
11. Competitive intelligence relates to the business information collected from other companies to aid in assessing current information for strategic uses. Discuss what type of information might be appropriate for strategic planning.
12. Competitive intelligence is collected from what another company may consider its trade secrets or proprietary information. Discuss the legal and ethical merits of collecting such information.

10.12 USER CHECKLIST

1. What efforts are being made in your organization to formalize improvements to project management capability and why?
2. What maturity model has your organization selected for improving its capability? If there is not one being used, does your company plan to adopt a capability maturity model in the near future?
3. Why and when would you use the continuous project management maturity model over the staged model? What advantages are there to the continuous model over the staged model?

4. For your organization, what elements would you select to improve on a continuous improvement model? Why those elements?
5. What strategic choice elements would be assessed in your selection of a capability maturity model and its implementation? Why?
6. Who in your organization is aware of the need for an improved or more mature project management capability? What authority or influence does that person have on implementing such a program?
7. If you are not currently implementing a capability maturity model, where do you think your organization is with respect to the five-level staged model? Why do you believe that (what evidence)?
8. Where is the strategic linkage of the project management capability maturity model within your organization? Who is managing this linkage?
9. What recent uses have been made of benchmarking to collect practices for adaptation to your environment? If none, why hasn't benchmarking been done?
10. What areas within your organization would you determine need a benchmarking program to improve practices, methods, or procedures? What would be the results of your benchmarking program?
11. What items would you include in a benchmarking plan if you were going to improve your administrative procedures for hiring technical people? Why would your hiring be different from what it is today?
12. Does your organization collect competitive intelligence to support planning and marketing efforts? How does this help your organization improve decision making?

10.13 PRINCIPLES OF PROJECT MANAGEMENT

1. A capability maturity model provides the means to enhance the organization's project management capability through top to bottom assessment and alignment of practices.
2. Capability maturity models are evolving to meet the needs of the total organization, not just software or just projects.
3. Assessing an organization identifies strengths and weaknesses for improvement opportunities.
4. Benchmarking supports project management initiatives through gathering and applying better practices of an industry, other organizations of a like type, or organizations unrelated to the operating industry.
5. Competitive intelligence is valuable in developing strategies for the organization or the project.

10.14 PROJECT MANAGEMENT SITUATION—GAINING PROJECT MANAGEMENT MATURITY

Semtac Corporation has been struggling with its project management capability. About half the completed projects are considered successful and provide the anticipated benefits. The best results seem to be obtained by a set of project managers who are "movers and shakers" in the organization. The not so successful project managers are quiet and appear to be efficient.

Senior management immediately concluded that the weakness was in the knowledge and skills of the project managers. Everyone agreed that training was the answer to the problem. The HR department was instructed to coordinate a training program for project managers to raise their competency in leading projects to success.

Everett Smith was tasked with developing the training program. Everett decided to first identify the shortfall in knowledge and skills by administering a comprehensive questionnaire on project management as a means of determining who would receive the training and what knowledge or skills were needed.

When the results of the testing was completed, there was no discernable difference between the "movers and shakers" and the other project managers. Everett then had the problem of sorting through the task that he was given. It was specific: "Train those individuals in project leadership who need to upgrade their knowledge and skills." This appeared to be everyone tested.

The problem changed for Everett from training several project managers to convincing senior management that the problem was not in a shortfall of knowledge or skills on project management. There was another underlying cause of the failures on projects. Training for all project managers would have made some minor improvement to the situation, but not the dramatic shift in success rates desired by senior management.

10.15 STUDENT/READER ASSIGNMENT

1. In the project management situation above, it was assumed that a shortfall in skills and knowledge could be corrected by training. What could be the contributing causes of project failures?
2. If the problem with project success is not the project manager, but the system that project managers must operate within, what would you recommend to senior management?
3. If the project management process is a random one that depends upon the behavioral style of the project manager, what is the solution to correct the situation and increase success rates?
4. Describe how you would conduct an assessment of the Semtac Corporation to determine their level of project management maturity capability.
5. What model for capability maturity would you select and what modifications would you make to apply to Semtac's situation?

P · A · R · T · 4

PROJECT OPERATIONS

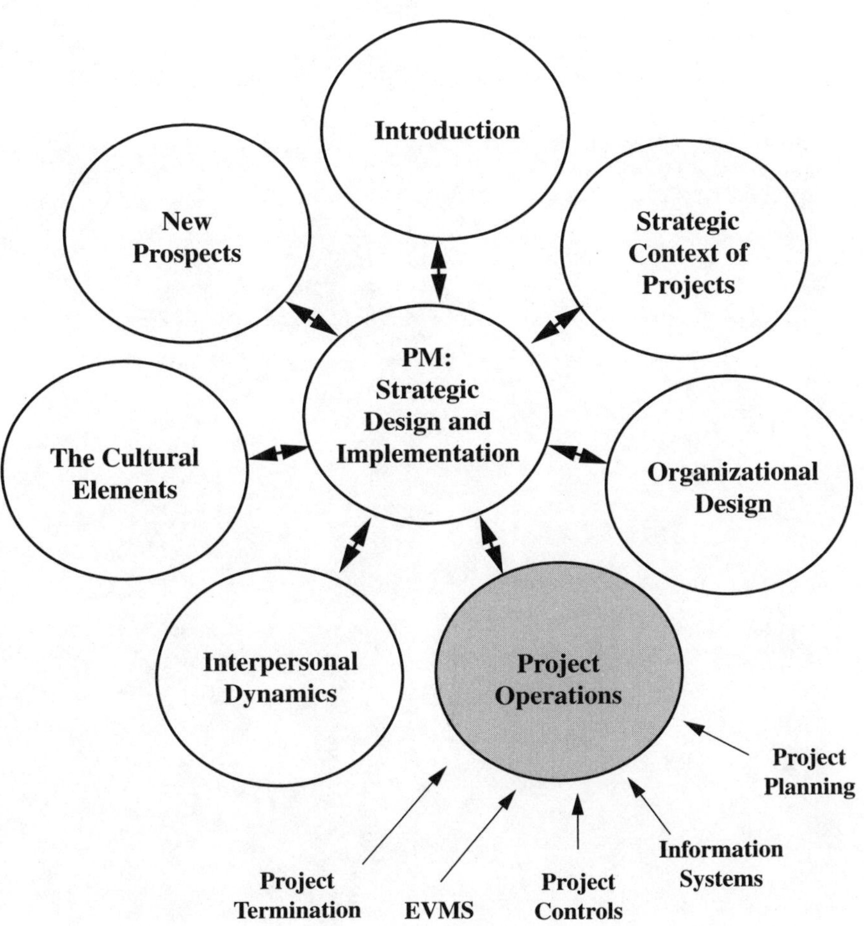

CHAPTER 11
PROJECT PLANNING

"Amid a multitude of projects, no plan is devised."

PUBLILIUS SYRUS, CIRCA 42 B.C.

11.1 INTRODUCTION

The most important responsibility of the project team is to develop the project plan in consort with other supportive stakeholders. Project planning is reflective thinking about the project's future in relationship to its present role in the design and execution of enterprise strategies. The project plan must be harmonious with the strategic plan of the enterprise, the functional plans, and, where appropriate, with the plans of the relevant stakeholders. If adequate project plans are developed, then an important standard for monitoring, evaluating, and controlling the application of project resources is available. If the project plans are inadequate, then the review of the project during its life cycle is greatly impaired.

In this chapter, a conceptual model of project planning is offered, along with a description of planning processes, considerations, and results that can be expected from adequate planning. The work breakdown structure is put forth as an absolute preliminary initiative to build on for the development of other project plans—such as schedules, scheduling techniques, planning charts, and networking techniques. A summary is given of the project planning elements, along with life-cycle planning, cost estimating, statement of work, project specification, and supporting functional plans. A citation of the generic work packages of project planning is offered as a template to guide the development of more specific planning work packages for individual projects. Included in the chapter content is a summary of the planning for project partners and outsourcing of project management, two growing areas of interest to project-driven organizations today.

11.2 THE IMPORTANCE OF PLANNING

Project planning is an important part of the "deciding" aspect of the project team's job to think about the project's future in relationship to its present in such a way that organizational resources can be allocated in a manner which best suits the project's purposes. More explicitly, project planning is the process of thinking through and making explicit the objectives, goals, and strategies necessary to bring the project through its life cycle to a successful termination when the project's product, service, or process takes its rightful place in the execution of project owner strategies. This chapter offers an overview of the project planning function placed in the context of enterprise planning.

In Chap. 1, strategic and organizational planning is covered. The context or planning within the organization is depicted in Fig. 1.2. This figure shows the hierarchical relationship for the enterprise, with projects subordinate to the organization's mission, objectives, goals, and strategies. The operational plans and organizational design influence how project planning will be accomplished. For purposes of this chapter, project and program planning are considered the same.

Three types of plans are interrelated in an enterprise: the strategic plan, the functional plans, and the project plans. Project planning involves the development of a strategy for the commitment of resources to support the project objectives and goals. The project plan reflects the strategic plan of the enterprise in providing guidance in the likely forthcoming "strategic fit" of the "stream of projects" in the enterprise. The functional plans provide detailed guidance on how resources will be used to support the project purposes. All three plans are essential guides for the use of resources, as well as providing reciprocal guidance on how the three plans work together during their execution in creating value for the enterprise.

Project planning is a rational determination of how to initiate, sustain, and terminate a project. McNeil and Hartley define the basic concepts of project planning as developing the plan in the required level of detail with accompanying milestones, and the use of available tools for preparing and monitoring the plan.[1]

Project planning and controls are interrelated. Planning prescribes the path to be followed in executing the project, whereas the controls are the means to collect, analyze, compare, and correct. Project controls are an integral part of project planning. A useful flowchart model (Fig. 11.1) shows this interrelationship and shows the sequence in which planning is required.

FIGURE 11.1 A flowchart of typical planning and control functions.

[1]Harold J. McNeil and Kenneth O. Hartley, "Project Planning and Performance," *Project Management Journal,* March 1986, p. 36.

Project planning has played a key role in the outcome of successful nuclear power plant projects in an industry where many projects have had grave difficulties. For example, at the Erie Nuclear Power Plant project, an overall plan was prepared at the project beginning. This plan provided the basis for controlling and coordinating the activities of the participating parties.[2]

At Florida Power and Light's St. Lucie Unit 2 plant, adequate project planning contributed to project's success by calling for the appointment of a project management organization in the early stages of the project, an early total project schedule, and the planning, scheduling, and implementation of an effective start-up program.[3]

11.3 PLANNING REALITIES

The planning ethos of the 1970s, rooted in the extrapolation of history, has been discredited by the "bends in the trends" exemplified by the oil crises and the political and social upheavals of the late 1980s and early 1990s. It is giving way to a new approach to strategic planning. This new approach is based on a visionary view of the future gained by a growing awareness of the possibility of long-term strategic alliance building, project partnering, sharing of risk in exploiting new technologies and processes leading to earlier commercialization, and continuous improvement of product, service, and process development and implementation in maintaining a competitive edge.

Planning is the most challenging activity for a leader or manager. Planning starts with the development of a vision—*the ability to see something that is invisible to others.* People in general find that it is more comfortable to do the work than to plan. All too often people equate activity with progress. Taking time to think through a plan of action for the future is not considered active management or leadership. Planning involves thinking through the possibilities and the probabilities of the future, and then developing a strategy for how the organizational resources will be positioned to take advantage of future competitive conditions.

Planning is a responsibility of the project leader. Finding ways to get the full-hearted cooperation of team members and other stakeholders will facilitate the planning process and improve the chances of the development of a project plan to which members of the project team are committed.

Planning for the use of resources precedes the monitoring, evaluation, and control of resources. Insufficient front-end planning, unrealistic project plans, failing to estimate the degree of complexity, and lack of consideration for the project's objectives will lead to reduced accomplishment of project objectives. When planning is done by an active, participating project team, the interactions and communications give greater insight to the project work. Interactions among the

---

[2]Barry M. Miller and Charles D. Williams, "Management Action through Effective Project Controls: A Case Study of a Nuclear Power Plant Project," *Proceedings,* PMI Seminar/Symposium, Los Angeles, October 1978, vol. 2, pp. G.1–G.5.

[3]Paraphrased from W. B. Derrickson, "St. Lucie Unit 2—A Nuclear Plant Built on Schedule," *Proceedings,* PMI Seminar/Symposium, Houston, October 1983, vol. 5, pp. E.1–E.14.

team members help develop the team and give the team members greater ease in dealing with each other. This guides the future use of organizational resources.

11.4 A CONCEPTUAL MODEL OF PLANNING

Project planning is preceded by comprehensive organizational strategic planning, because projects are integral elements of organizational strategies. A conceptual model, depicting the strategic context of organizational planning that includes both strategic planning and strategic implementation, appears in Fig. 11.2.[4]

Strategic considerations are addressed in Chap. 1 and will not be restated here. This section will address the project and how planning is accomplished. It is well to remember that projects are building blocks of the enterprise and that all projects must contribute to the organization's mission, as connected through organizational objectives and goals that are implemented through organizational strategies.

11.5 PROJECT PLANNING MODEL

Project planning begins within the framework of strategic planning in the organization. For example, the strategic planning phase at a steel corporation led to the approval of a comprehensive facility feasibility study for the location and configuration of the steel plant. As a result of this feasibility study, which evaluated seven alternative sites, the plant location was fixed at Cleveland, Ohio. During the planning for this facility, several options were considered, ranging from turnkey contract to construction management consultant to the owner acting as its own general contractor with subcontractors and/or in-house personnel. These options were considered in detail before final project planning was carried out with approved cost estimates and milestone schedules.

The strategic context of organizational planning, depicted in Fig. 11.2, sets the stage for project planning. Projects must adhere to the strategic "umbrella" to assure flow down of the enterprise's mission, vision, objectives, goals, and strategies. Using these concepts, project planning becomes an elaboration of the overall approach to business and provides for consistently building on the enterprise's capability to perform work through projects.

Understanding the strategic approach to business and linking that to the project planning will provide the means to pursue work that supports the organization's objectives and goals by using strategies that are adopted and used by the organization. Ensuring that projects, as building blocks for the business, contribute to the organization's growth and improvement is critical to future well-being.

Table 11.1 depicts the hierarchical approach at the project level and establishes an overall framework for project planning. This is an extension of the concepts in Fig. 11.2, which provides the umbrella for project planning. The definitions provide an understanding for each key element.

---

[4] Adapted in part from David I. Cleland and William R. King, *Systems Analysis and Project Management,* 3d ed. (New York: McGraw-Hill, 1983), p. 63.

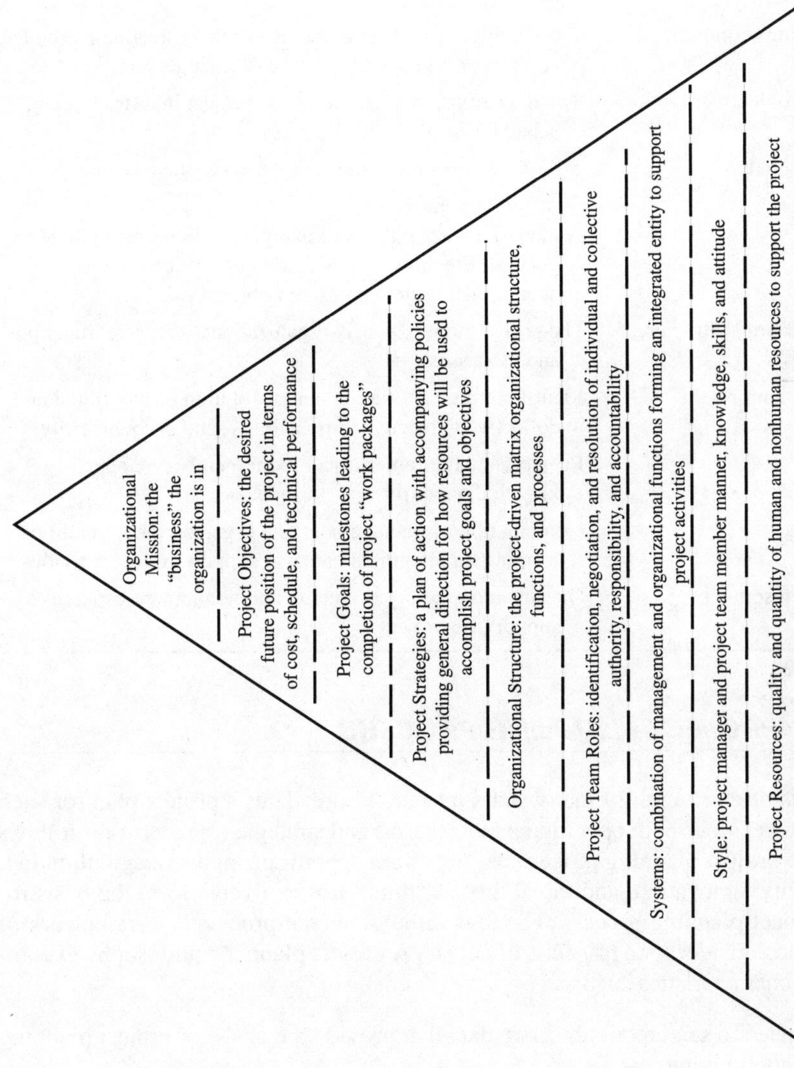

FIGURE 11.2 Elements in the project-driven matrix management philosophy.

TABLE 11.1 Project Planning in a Strategic Context

Strategic vision, mission, objectives, and goals for the enterprise that drives project planning	
Project planning element	Definition
Project mission/purpose	The central reason for the project, such as creating a product, service, or organizational process change.
Project objectives	The desired future position of the project in terms of cost, schedule, and technical performance.
Project goals	Milestones leading to the completion of the project's "work packages."
Project strategy	A plan of action with accompanying policies providing general direction of how resources will be used to accomplish project goals and objectives.
Organizational structure	The project-driven, matrix organizational structure, functions, and processes.
Project team roles	Identification, negotiation, and resolution of individual and collective authority, responsibility, and accountability.
Style	Project manager and project team member manner, knowledge, skills, and attitudes.
Systems	Combination of management and organizational functions forming an integrated entity to support project activities.
Project resources	Quality and quantity of human and nonhuman resources to support the project.

11.6 PROJECT PLANNING PROCESS

Projects often extend for many years into the future. Thus a project plan for such projects becomes both operational (short term) and strategic (long term). It follows that the project planning process requires both operational and strategic thinking. Creativity, innovation, and the ability to "think prospectively" form the basis for the project planning process. The real value of such a process is a framework of things to consider for a project's life cycle. A project planner's philosophy encompasses characteristics such as:

- The need to search out objective data that provide the basis for project planning decision making
- The value of questioning assumptions, databases, and emerging project strategies to test their validity and relevance
- An ongoing obsession with where the project should go and how it is going to get there

- A demonstrated ability to view project opportunities in the largest possible context and to constantly seek an understanding of how everything fits together during the project's life cycle
- A faith that, given ample opportunity, a *bisociation* will occur: the fitting together of separate events or forces on the project.[5]

Individuals making key project planning decisions today will have a long-term strategic impact on the organization. Generally, the strategic roles of key individuals involved in project planning are as follows:

- The *board of directors* reviews and approves (or redirects for further study) key project plans and maintains surveillance over the implementation of the plans.
- *Senior management* directs the design, development, and implementation of a strategic planning system and a project planning philosophy and process for the corporation.
- *Functional managers* are responsible for the integration of state-of-the-art functional technology into the project plans.
- The *project manager* is responsible for integrating and coordinating the project planning activity.
- The *work package manager* is responsible for providing input to the project plans.
- Professionals participate as required in contributing to the project planning processes.

By involving these individuals in the roles as described, key people are afforded the opportunity to participate in project planning. Of course, such participation requires relevant knowledge, skill, and insight into both the theory and the practice of project planning. By maximizing the participation of key individuals in project planning, the overall value of the project plan should be improved. One large project-driven organization recognized the value of project planning like this:

- During the early 1960s, after hundreds of projects had been completed, it became apparent that many projects successfully achieved their basic project objectives, whereas some failed to achieve budget, schedule, and performance objectives originally established.
- The history of many of these projects was carefully reviewed to identify conditions and events common to successful projects, vis-à-vis those conditions and events that occurred frequently on less successful projects. A common identifiable element on most successful projects was the quality and depth of early planning by the project management group. Execution of the plan, bolstered by strong project management control over identifiable phases of the project, was another major reason why the project was successful.[6]

---

[5]Arthur Koestler, *The Act of Creation* (London: Hutchinson, 1964). He explains creativeness as the result of bisociation, of putting together unconnected facts or ideas to form a single new idea.

[6]Robert K. Duke, H. Frederick Wohlsen, and Douglas R. Mitchell, "Project Management at Fluor Utah, Inc.," *Project Management Quarterly,* vol. 3, no. 3, September 1977, p. 33.

Thus, project planning is the "business" of many individuals in the organization. Understanding planning concepts and how to develop realistic plans adds value to the organization and its ability to reach into the future to lay out a path for success.

Project planning may be considered a form of information development and communications. As the project team develops the project plan, the project team should learn more about the project goals, strategies, and team member roles. The project objectives then can be decided in terms of cost, schedule, and technical performance. Satisfaction of project goals is accomplished through the completion of the project work packages. The project strategy is a plan of action with accompanying policies, procedures, and resource allocation schemes, providing general direction of how the organizational effort will be used to accomplish project goals and project objectives.

Simultaneous project planning is the process of having the project team consider all aspects, issues, and resources required for the project plan on a concurrent basis. Concurrent planning means that everything that can or might impact the project is reviewed during the planning phase to ensure that an explicit decision is made concerning the role that all resources, however modest, might have on the project. Project start-up workshops can be useful in the planning phase of a project to help identify and get people committed to the notion of thinking through all probable and possible aspects of the project to be reflected in the project plan.[7]

11.7 PROJECT PLANNING CONSIDERATIONS

Many projects are started before the requirements are fully defined and understood. The lack of requirements, which may be the specification for a product, can allow the planners to drift away from the customer's needs. Once project planning starts in the wrong direction, it is difficult to correct. Like many other phenomena, the first 20 or 30 percent of the planning effort establishes a direction for the project.

The customer's requirements define what the project should be. Planners who understand the customer's requirements can collect the information to plan the project and apply appropriate project management practices and techniques to build a "road map" to project implementation, control, and closeout.

All too often, when people think of project planning they perceive the use of only techniques and concepts such as PERT or CPM networking. These concepts are briefly discussed in this chapter. The footnote references can serve as useful guidelines in using PERT and CPM networking.[8] These techniques are important to use in the development of a project schedule; however, project planning includes a much wider scope of activity. Such concerns as objective and goal setting,

[7]For a meaningful description of the role that planning workshops can play in project planning, see Alexander Walton, "Concurrent Planning Workshops: The Best Way to Communicate during Project Planning," *Proceedings*, Project Management Institute, 26th Annual Seminar/Symposium, New Orleans, October 1996, pp. 357–366.

[8]Joseph J. Moder, "Network Techniques in Project Management," in D. I. Cleland and W. R. King (eds.), *Project Management Handbook* (New York: Van Nostrand Reinhold, 1983), chap. 16, pp. 303–309; and James J. O'Brien, *CPM in Construction Management*, 3d ed. (New York: McGraw-Hill, 1984).

cost estimating and budgeting, scheduling, resource usage estimating, and specification of deliverables are key concerns. Project planning also involves a delineation of the organizational design to support the project as well as the information system and the control system, which are used to model, evaluate, and reallocate resources as required during the execution of the project plan.

Project planning deals with the determination of the activities and resources that have to be utilized to ensure that the project is adequately executed. Authority, responsibility, and accountability have to be planned so that members of the project team know what their specific roles are and how they relate to other members of the project team who are involved in executing work package activity. The following key questions need to be answered:

- When is activity due?
- What is the time duration of each activity?
- What human and nonhuman resources are needed to execute each activity on the project?
- What are the estimated costs?
- How are the budget and financial plans to be established to support the cost considerations of the budget?

One of the changes under way in contemporary organizations is that more people are involved in and carry out the management functions.

Participative planning has been used effectively by AT&T. Participation is obtained through the use of workshops that include the entire project team and even customers in joint planning sessions. A planning process facilitator helps guide the activities and keeps the project planning moving forward. The purpose of these workshops is to have the participants agree on high-level project plans, schedules, and project monitoring and evaluation strategies. Held at the beginning of a project, the workshops achieve the benefits of early planning, including overcoming planning problems and getting the team members involved early in the planning, which leads to more commitment and dedication to their role on the project. In addition, team members are given an early exposure to their individual and collective roles in the project and an opportunity to identify any interpersonal anxieties that might hinder team development and operation at a later date. These start-up workshops have been successful in producing planning deliverables, developing planning skills, and building team interaction and cohesiveness.[9]

The project manager is responsible for initiating action to bring about the development of a plan. In discharging the project leadership role, the project leader has the final responsibility for ensuring that "the right things are done" about the project plan. The complexities of deciding what the details of the project should be and doing things right rest with the specialists, who are members of the project planning team. Planning becomes a method for coordinating and synchronizing the forthcoming project activities. Project planning should be undertaken

---

[9]Dan Ono and Russell D. Archibald, "Project Start-up Workshops: Gateway to Project Success," *Proceedings*, PMI Seminar/Symposium, San Francisco, September 17–21, 1988, pp. 500–554.

after the project has been positioned in the overall strategy for the enterprise; then the detailed planning can be carried out with a high degree of assurance that the project planning team is working on the right areas.

Because planning involves thinking through the probabilities and possibilities of the project's future, a detailed cookbook recipe for planning cannot be provided. However, certain key work packages and planning tools have to be addressed in the development of the project plan of action. These planning considerations are described in the next section.

11.8 WORK BREAKDOWN STRUCTURE[10]

The most basic consideration in project planning is the work breakdown structure (WBS). The WBS divides the overall project into work elements that represent singular work units, assigned either to the organization or to an outside agency, such as a contractor.

The WBS process is carried out in the following manner: Each project must be subdivided into tasks that can be assigned and accomplished by some organizational unit or individual. These tasks are then performed by specialized functional organizational components. The map of the project represents the collection of these units and shows the project manager the many organizational and subsystem interfaces to be managed.

The underlying philosophy of the work breakdown structure is to divide the project into work packages that are assignable and for which accountability can be expected. Each work package is a performance-control element; it is negotiated and assigned to a specific organizational manager, usually called a work package manager. The work package manager is responsible for a specific measurable objective, detailed task descriptions, specifications, scheduled task milestones, and a time-phased budget in dollars and work force. Each work package manager is held responsible by both the project and the functional managers for the completion of the work package in terms of technical objectives, schedules, and costs.

The work breakdown structure is a means for dividing a project into easily managed increments, helping ensure the completeness, compatibility, and continuity of all work that is required for successful completion of the project. The WBS provides the basis for a fundamental understanding of the scope of the project and helps ensure that the project supports organizational objectives and goals.

The process of developing the WBS is to establish a scheme for dividing the project into major groups, and then dividing the major groups into tasks, subdividing the tasks into subtasks, and so forth. Projects are planned, organized, and controlled around the lowest level of the WBS. The organization of the WBS should follow some orderly identification scheme; each WBS element is given a distinct identifier. With an aircraft, for example, the WBS might look like the information shown in Fig. 11.3.

---

[10]Paraphrased from D. I. Cleland and W. R. King, *Systems Analysis and Project Management,* 3d ed. (New York: McGraw-Hill, 1983), pp. 255–258.

X-33 Aircraft

1.0 System
- 1.1 Airframe
- 1.2 Tail section
- 1.3 Wings
- 1.4 Engines
- 1.5 Avionics

2.0 Documentation
- 2.1 Operator's manual
- 2.2 Repair manual

3.0 Test and Demonstration
- 3.1 Static system test on the ground
- 3.2 Dynamic air test
  - 3.2.1 Initial flight for aerodynamics
  - 3.2.2 Initial flight maneuver test
  - 3.2.3 Endurance flight test

4.0 Logistics
- 4.1 Maintenance tools
- 4.2 Repair parts (spares)

FIGURE 11.3 Work breakdown structure coding scheme for an aircraft (example).

In the WBS, the decomposition of the aircraft includes both system components and the functions to support the project. The indenture at different levels indicates the components of that item. Decomposition is only done to the level at which the item will be managed.

A graphic representation of the WBS can facilitate its understanding. Using numerical listings with deeper indentation for successively lower levels can aid in communications and in developing understanding of the total project and its integral subsystems, sub-subsystems, and so forth. Figure 11.4 demonstrates the graphic approach to displaying the information.

Work packages follow from a WBS analysis on the project. When the WBS analysis is completed and the work packages are identified, a WBS comes into existence. As shown in Fig. 11.4 a WBS can be represented similar to traditional organizational structure.

In the context of a project, the WBS and the resulting work packages provide a model of the products (hardware, software, services, and other elements) that completely define the project. Such a model enables project engineers, project

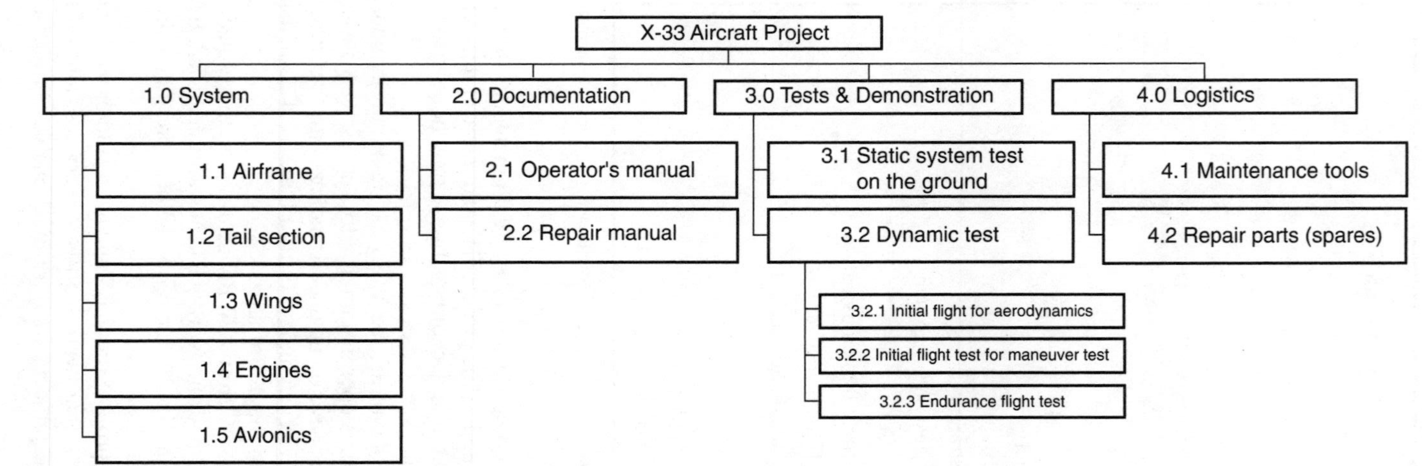

FIGURE 11.4 WBS in a graphic diagram (example).

managers, functional managers, and general managers to think of the totality of all products and services comprising the project as well as its component subsystems. The model is the focus around which the project is managed. More particularly, the development of a WBS provides the means for:

- Summarizing all products and services comprising the project, including support and other tasks
- Displaying the interrelationships of the work packages to each other, to the total project, and to other engineering activities in the organization
- Establishing the authority-responsibility matrix organization
- Estimating project cost
- Performing risk analysis
- Scheduling work packages
- Developing information for managing the project
- Providing a basis for controlling the application of resources on the project
- Providing reference points for getting people committed to support the project

Work packages are the goals to be accomplished on the project. There are certain criteria that should be applied to the project goals:

- Are the goals clear?
- Are they specific?
- Are they time based?
- Are they measurable?
- Can they be communicated easily to the project team?
- Can they be clearly assigned to the work package managers/professionals?

The WBS provides a natural framework or skeleton for identifying the work elements of the project: hardware, software, documentation, and miscellaneous work to be accomplished to bring the project to completion. The WBS provides an identifier and a management thread to manage myriad aspects of the project. In some projects unique work packages are found. For example, in some global projects a cultural planning work package is included in the work breakdown structure. From this work package, successful cultural training and orientation can be carried out.

11.9 PROJECT SCHEDULES

A key output of project planning is the project master schedule, along with supporting schedules, which is a graphic time representation of all necessary project-related activities. The project schedule establishes the time parameters of the project and helps the managers effectively coordinate and facilitate the efforts of the entire project team during the life of the project. A schedule becomes an effective part of the project control system. For a project schedule to be effective, it must be:

- Understandable by the project team
- Capable of identifying and highlighting critical work packages and tasks
- Updated, modified as necessary, and flexible in its application
- Substantially detailed to provide a basis for committing, monitoring, and evaluating the use of project resources
- Based upon credible time estimates that conform to available resources
- Compatible with other organizational plans that share common resources

Several steps are required to develop the project master schedule. These steps should be undertaken in the proper sequence.

- Define the project objectives, goals, and overall strategies.
- Develop the project work breakdown schedule with associated work packages.
- Sequence the project work packages and tasks.
- Estimate the time and cost elements.
- Review the master schedule with project time constraints.
- Reconcile the schedule with organizational resource constraints.
- Review the schedule for its consistency with project costs and with technical performance objectives.
- Senior managers approve the schedule.

11.10 SCHEDULING TECHNIQUES

Several scheduling techniques are useful in dealing with the timing aspect of the project resources. The typical graphic illustration is the bar chart, whereas the networks for PERT (program evaluation and review technique) and CPM (critical path method) show the connectivity between work packages or activities.

Project Planning Bar Charts

A technique for simple project planning and scheduling is based on the bar chart (or Gantt chart, after Henry Gantt, one of the early users of a bar chart). This chart consists of a scale divided into units of time (e.g., hours, days, weeks, or months) across the top and a listing of the project work packages or elements down the left-hand side. Bars or lines are used to indicate the schedule and status of each work package in relation to the time scale. Figure 11.5 is an example of a project planning bar chart for the development of an electronic device.

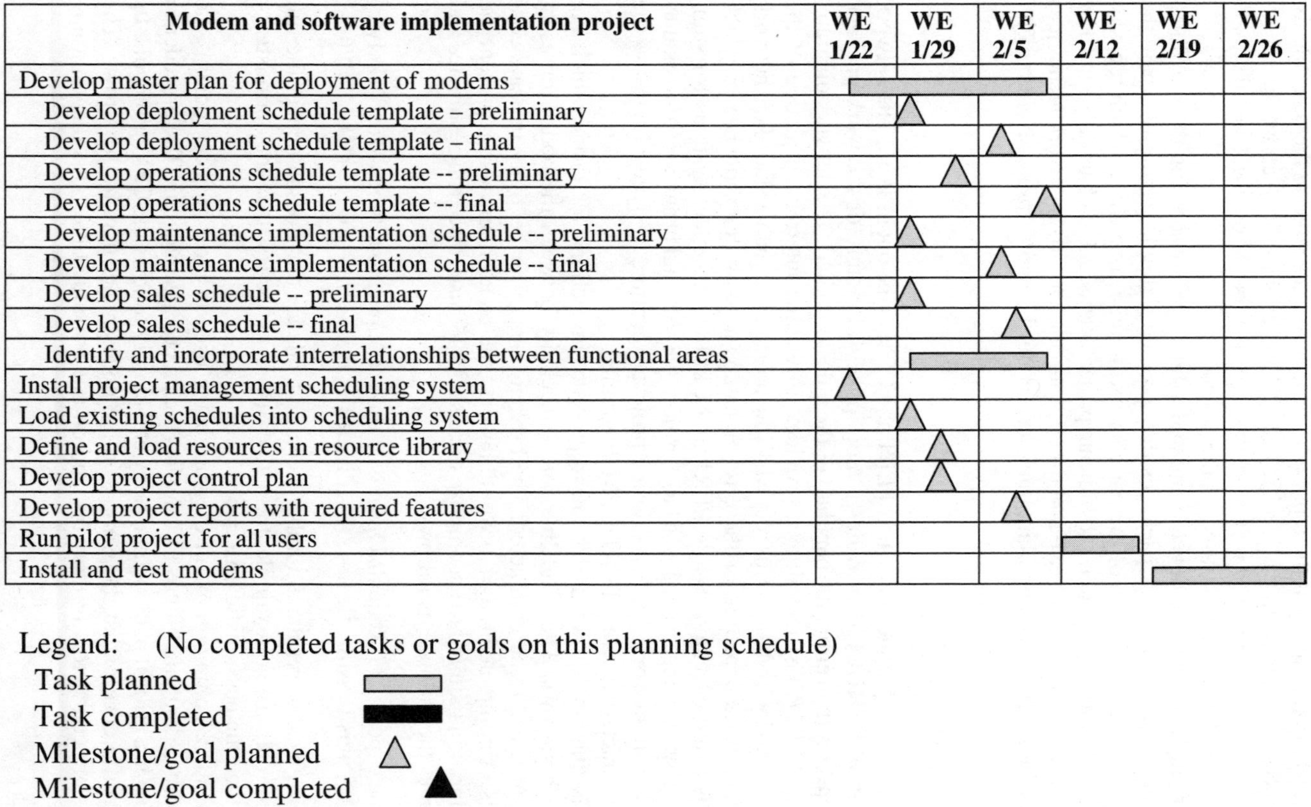

Modem and software implementation project	WE 1/22	WE 1/29	WE 2/5	WE 2/12	WE 2/19	WE 2/26
Develop master plan for deployment of modems						
Develop deployment schedule template – preliminary						
Develop deployment schedule template – final						
Develop operations schedule template -- preliminary						
Develop operations schedule template -- final						
Develop maintenance implementation schedule -- preliminary						
Develop maintenance implementation schedule -- final						
Develop sales schedule -- preliminary						
Develop sales schedule -- final						
Identify and incorporate interrelationships between functional areas						
Install project management scheduling system						
Load existing schedules into scheduling system						
Define and load resources in resource library						
Develop project control plan						
Develop project reports with required features						
Run pilot project for all users						
Install and test modems						

Legend: (No completed tasks or goals on this planning schedule)

Task planned

Task completed

Milestone/goal planned

Milestone/goal completed

FIGURE 11.5 Project planning bar chart.

The work packages of the project are listed on the left-hand side, and the units of time in weeks are shown at the top. Light horizontal lines indicate the schedule for the project elements, with the specific tasks or operations written above the schedule line. Work accomplished is indicated by a heavy line below the schedule line.[11] The triangles represent milestones in the project.

Bar graphs are easy to develop and understand, and by showing the scheduled start and finish of the work packages they provide a simple picture of where the project stands. A variation of the bar chart is the milestone chart, which replaces the bar with lines and triangles to indicate project status. A bar chart does not show work package interdependence and time-resource trade-offs. Network techniques used on larger projects help plan, track, and control complex projects effectively.

Network Techniques

The network techniques known as PERT and CPM came out in 1956. The network diagram of PERT/CPM provides a more powerful picture of the work package relationships than either the bar chart or the milestone chart. The network diagram, basic to PERT/CPM techniques, provides a dynamic interrelated picture of the activities and interrelationships relative to the project. The main value of the network technique is its depiction of these relationships to show the sequence in which work is being executed. While PERT and CPM are excellent systems for summarizing work being tracked on a large project, they also aid in analysis of work during delays in one or more tasks during project execution.[12] Figure 11.6 shows the basic characteristics of a PERT/CPM diagram.

This diagram shows nodes that represent connections for tasks, which are shown by arrows. The solid lines of connectivity between the nodes represent work, or tasks, to be completed. The broken lines of connectivity indicate constraints. A completed diagram would have a time on each arrow to represent the time duration required to accomplish the task. PERT uses three time estimates and computes the estimate as follows:

Time expected (T_e) equals Optimistic time (O), plus four times Most Likely (ML), plus Pessimistic time (P), divided by six, or $T_e = (O + 4\text{ML} + P)/6$.

This gives the resultant number a central tendency that does not cover the full range of options. For example, the actual time could be less than the optimistic or more than the pessimistic estimates.

The more common type of diagram being used in most projects today is the precedence diagram. Figure 11.7 depicts the precedence diagram. The precedence diagram is a critical path methodology and closely follows the CPM protocols. Its major difference is that the work is in the nodes rather than in the arrows. The arrows are connectors to the work and do not consume time. This diagramming

---

[11]Ibid.

[12]For a summary of network techniques see Hans J. Thamhain, *Engineering Program Management* (New York: Wiley, 1984), pp. 109–140.

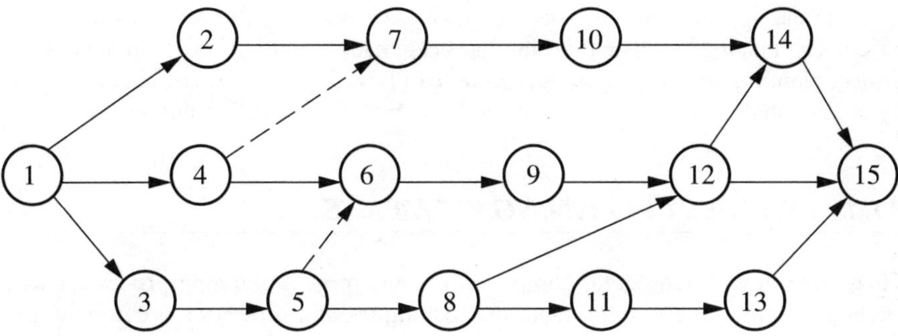

FIGURE 11.6 PERT/CPM diagram.

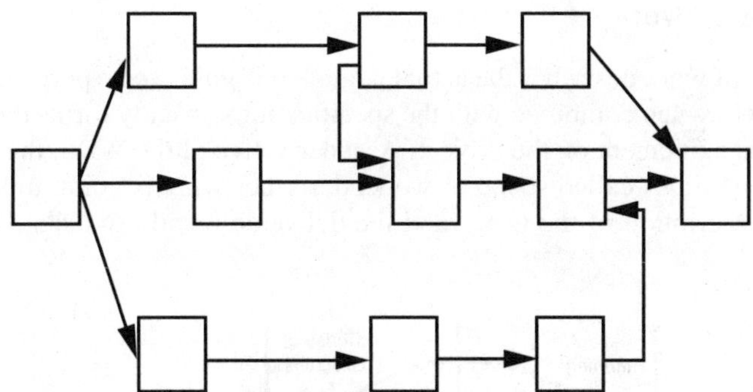
FIGURE 11.7 Precedence diagram.

method is more flexible than CPM/PERT in that no lines constrain the work like the dashed lines of CPM.

Another advantage of the precedence diagram is that the connectivity can be arranged in the schedule through four relationships—finish-to-start, finish-to-finish, start-to-start, and start-to-finish. Each connectivity relationship has its advantages. However, the most common is the finish-to-start, or finish one activity and start another.

11.11 PROJECT LIFE-CYCLE PLANNING

The project life cycle is a key consideration in project planning. Project life cycles contain phases, or control mechanisms, to divide the project into different work efforts. The typical project life cycle has four phases: initiation, planning, execution and control, and closeout. These phase can be different to meet the needs of a specific industry and the names may differ. However, the concept is the same—divide the life of the project into manageable parts, each one representing a control point at the end.

Once the appropriate work packages for each phase of the project's life cycle have been depicted, a substantial start has been made toward the development of the project plan. Figure 11.8 shows an example of how the work is to be accomplished by project phase. This model is also shown in Fig. 2.3 in a different context.

11.12 PROJECT PLANNING ELEMENTS

There are a few fundamental components in the project planning process. These elements include the outputs from the techniques and processes previously discussed as well as the elements discussed in the material that follows.

Statement of Work

A statement of work describes the actual work that is going to be performed on the project which, when combined with the specifications, usually forms the basis for a contractual agreement on the project. As a derivative of the WBS, the statement of work (sometimes called scope of work) describes what is going to be accomplished, a description of the tasks, and the deliverable end products that will be

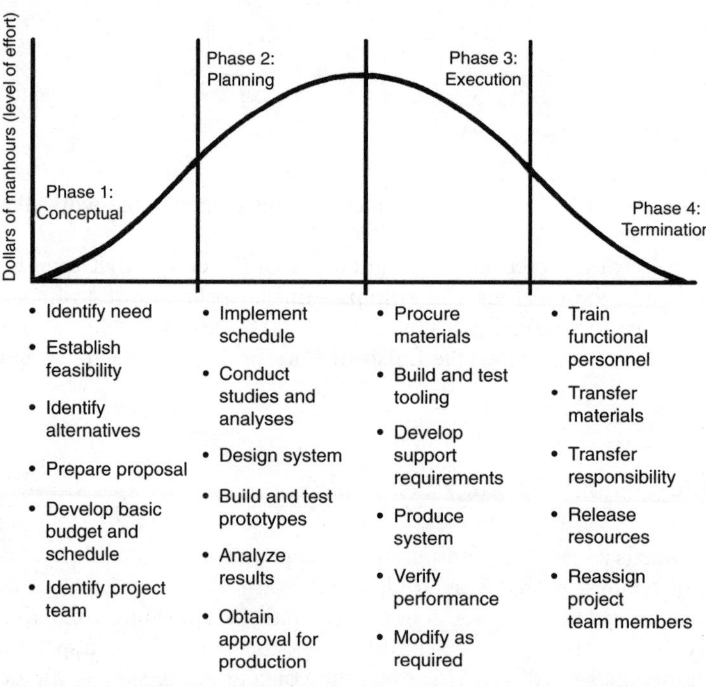

FIGURE 11.8 Tasks accomplished by project phase. [*Source: John R. Adams and Stephen E. Brandt, "Behavioral Implications of the Project Life Cycle," in David I. Cleland and William R. King (eds.),* Project Management Handbook *(New York: Van Nostrand Reinhold, 1983), p. 227.*]

produced, such as hardware, software, tests, documentation, and training. The statement of work may also make reference to specifications, directives, or standards, that is, the guidance to be followed in the project work. The statement of work includes input required from other tasks involving the project and a key element of the customer's request for a proposal.

Project Specification

Specifications are the descriptions of the technical content of the project. These specifications typically describe the product of the project and the requirements that the product must meet. Further, they contain the characteristics of the various subsystems in the project's product or service, in order to include an overall system specification, hardware, software, test specifications, and logistics support.

Cost Estimate

The cost estimate forms the baseline budget from which all actual expenditures will be measured. Typically, the cost estimate follows the WBS during development and implementation. The type of cost estimate determines the precision associated with individual cost items.

Detailed cost estimates are expensive because of the time and labor effort required to perform the estimate. The less precise the estimate, typically, the less the cost that is associated with the estimate. The three types of cost estimates are as follows:

- *Order of magnitude (−25%, +75%).* Used to make an initial estimate of a project for purposes of getting a rough idea of the cost. Note that the project may have a 100 percent variance, that is, 25 percent less or 75 percent more.
- *Budget (−10%, +25%).* Used to perform a "tighter" estimate than can be used for decision making concerning whether the project is within the tolerances of the organization.
- *Definitive (−5%, +10%).* Used to start a project based on solid planning and a good estimate of all parts of the project.

During implementation, the cost estimate forms the baseline for project expenditures and provides a means of comparing actual costs to the estimate. Projects may require weekly or monthly cost reports to reflect the actual expenditure as compared to the baseline estimate. An element of the cost system is a report, which provides weekly actual cost with estimated cost as well as a comparison of actual worker-hours with target worker-hours in manufacturing or construction.

The *cost account* usually is considered the basic level at which project performance is measured and reported. This account represents a specific work package identified by the WBS, usually tracked by information on a daily or weekly time card, which ties in with the organizational cost accounting system.

Financial Plan

Assuming that project budget, work package budget, and budgets for the appropriate cost accounts have been developed, financial planning involves the development of action plans for obtaining and managing the organizational funds to support the project through the use of the work authorization process.

The project manager usually authorizes the expenditure of resources on the project for work to be accomplished within the organization as well as on work subcontracted to vendors. The work authorization process is an orderly way to delegate authority to expend resources for the project. The work authorization document usually includes:

- The responsible individual and/or organization
- A work package WBS
- A schedule
- Cost estimate and funding citation
- A statement of work

Usually the work authorization document is in a one-sheet format that is considered a written contract between the project manager and the performing organization and/or person.

Functional Plan

Each functional manager should prepare a functional operations plan that establishes the nature and timing of functional resources necessary to support the project plan. For example, the accounting/financial organization that supports the project manager should establish a plan for how the project budget can be monitored. Such a plan would be an information system for monitoring actual project costs and comparing them with budgeted costs.

11.13 PLAN FORMAT

The organization and arrangement of the project plan depend on the nature of the project. The bare essentials of a project plan include:

- A summary of the project that states briefly what is to be done and the methods and techniques to be used. It lists the deliverable end products in such a way that when they are produced, they can be identified easily and compared with the plan.
- A list of tangible and discrete goals, identified in such a way that there can be no ambiguity about whether a goal has been achieved.
- A WBS that is detailed enough to provide meaningful identification of all tasks associated with job numbers, plus all higher-level groupings such as work units or work packages.

- A strategy outlining how organizational resources will be used to accomplish project objectives and goals.
- An activity network that shows the sequence of elements of the project and how they are related (which can be done in parallel, which can start only when another is finished, etc.).
- Separate budgets and schedules for all the elements of the project for which some individual is responsible.
- An interface plan that shows how the project relates to the rest of the world, most particularly to the customer.
- An indication of the review process: who reviews the project, when, and for what purpose.
- A list of key project personnel and their assignments in relation to the WBS.

11.14 PROJECT MANAGEMENT MANUAL

An important part of project planning is the development of organizational policies and procedures that support the project plan. Many organizations use a project management manual that tells all project participants what they have to do and how they have to do it. Bitner suggested the contents for a representative project procedure manual.[13]

The project management manual is the document that establishes standing procedures for project planning and implementation. When properly constructed, the project management manual covers aspects of the organization's project management that are consistent across projects and programs. For example, it may state that all projects will have a formal review by senior management no less than once each quarter and once a month for projects that are experiencing difficulties.

11.15 PROJECT PLANNING WORK PACKAGES

Even the task of planning a project should be broken down into work packages. Here is a general guide to these work packages, not necessarily done in the sequence indicated:

- *Establish the strategic fit of the project.* Ensure that the project is truly a building block in the design and execution of organizational strategies and provides the project owner with an operational capability not currently existing or improves an existing capability. Identify strategic issues likely to affect the project.

[13] L. M. Bitner, "Project Management: Theory versus Application," *Project Management Journal,* June 1985, p. 67.

- *Develop the project technical performance objective.* Describe the project deliverable end product(s) that satisfies a customer's needs in terms of capability, capacity, quality, quantity, reliability, efficiency, and so on.
- *Describe the project through the development of the project WBS.* Develop a product-oriented family tree division of hardware, software, services, and other tasks to organize, define, and graphically display the product to be produced, as well as the work to be accomplished to achieve the specified product.
- *Identify and make provisions for the assignment of the functional work packages.* Decide which work packages will be done in-house, obtain the commitment of the responsible functional work managers, and plan for the allocation of appropriate funds through the organizational work authorization system.
- *Identify project work packages that will be subcontracted.* Develop procurement specifications and other desired contractual terms for the delivery of the goods and services to be provided by outside vendors.
- *Develop the master and work package schedules.* Use the appropriate scheduling techniques to determine the time dimension of the project through a collaborative effort of the project team.
- *Develop the logic networks and relationships of the project work packages.* Determine how the project parts can fit together in a logical relationship.
- *Identify the strategic issues that the project is likely to face.* Develop a strategy to deal with these issues.
- *Estimate the project costs.* Determine what it will cost to design, develop, and manufacture (construct) the project, including an assessment of the probability of staying within the estimated costs.
- *Perform risk analysis.* Establish the degree or probability of suffering a setback in the project's schedule, cost, or technical performance parameters.
- *Develop the project budgets, funding plans, and other resource plans.* Establish how the project funds should be utilized, and develop the necessary information to monitor and control the use of funds on the project.
- *Ensure the development of organizational cost accounting system interfaces.* Because the project management information system is tied in closely with cost accounting, establish the appropriate interfaces with that function.
- *Select the organizational design.* Provide the basis for getting the project team organized, including delineation of authority, responsibility, and accountability. At a minimum, establish the legal authority of the organizational board of directors, senior management, and project and functional managers, as well as the work package managers and project professionals. Use the LRC (linear responsibility chart) process to determine individual and collective roles on the project team.
- *Provide for the project management information system.* An information system is essential to monitor, evaluate, and control the use of resources on the project. Accordingly, develop such a system as part of the project plan.

- *Assess the organizational cultural ambience.* Project management works best where a supportive culture exists. Project documentation, management style, training, and attitudes all work together to make up the culture in which project management is found. Determine the project management training that would be required. What cultural fine-tuning is required?
- *Develop project control concepts, processes, and techniques.* How will the project's status be judged through a review process? On what basis? How often? By whom? How? Ask and answer these questions prospectively during the planning phase.
- *Develop the project team.* Establish a strategy for creating and maintaining effective project team operations.
- *Integrate contemporaneous state-of-the-art project management philosophies, concepts, and techniques.* The art and science of project management continue to evolve. Take care to keep project management approaches up to date.
- *Design project administration policies, procedures, and methodologies.* Administrative considerations often are overlooked. Take care of them during early project planning, and do not leave them to chance.
- *Plan for the nature and timing of the project audits.* Determine the type of audit best suited to get an independent evaluation of where the project stands at critical junctures.
- *Determine who the project stakeholders are and plan for the management of these stakeholders.* Think through how these stakeholders might change through the life cycle of the project.

If the project team follows the suggested development of the project planning work packages suggested in this chapter, it is probable that all the needed project planning will be carried out. As these project planning work packages are prepared, it is likely that planning initiatives not reflected in these work packages will be identified and can be planned for accordingly.

11.16 MANAGEMENT REALITIES

Complete plans are but a "best effort" of the planning team based on information and guidance received. The larger the project, the greater the chance that the plan will change through either new guidance or updated information being received. Plans are "living documents" that change as competitive and environmental systems change. Continual update of information and additional planning are typically required throughout the life of the project.

Implementation of plans requires that communications and coordination take on added emphasis. Teams, rather than fixed organizational structures, are the order of the day. Integration becomes supreme as organizational processes are crossed as needed, and when needed, pulls together synchronized and quality efforts to produce customer value.

For many managers, who grew up in a "command and control" culture, the new paradigm of "consensus and consent" management is disquieting. They will need to adapt to the new way of getting things done—that is, by sharing the decision making and power in the project. Managers have become the servants of those they choose to rule. This new world is concerned with flexibility in strategies, markets, projects, resources, and people.

In the past, we managed as if the optimization of the parts of the organization—research, manufacturing, engineering, marketing, product development, and so forth—would lead to the optimization of the whole of the organization. Today, we optimize the integration of the organizational processes by using project teams as focal points to pull together the human and nonhuman resources needed to do the job. The major breakthroughs in improving organizational efficiency and effectiveness have come about because of the management of the organizational processes rather than the functional entities of the organization. The management of the organizational processes through a self-directed team has finally captured the essence of the interdependencies of the organizational functions.

All planning efforts need to consider the new world and dynamic changes that occur, some on a daily basis. Planning does provide the framework and the thought process to visualize the work required to build a product or service that brings benefits to the customers. Planning also gives a foundation from which to initiate change, when required to meet new situations. No plan is perfect to carry one through an entire project, but a good plan does provide a path from which one can adjust to meet the changes.

11.17 PROJECT PARTNERING[14]

Partnering in projects has emerged in recent years as a means of sharing equally in managing large projects by two or more organizations. One organization may initiate partnering to gain additional capability for a project as well as share in the risks of large complex projects. Partnering is also a means of combining information, such as in research and development projects, to improve the chance of success.

Partnering may be accomplished between public, private, and for-profit and not-for-profit organizations. A government agency may, for example, partner with a private organization to conduct research. Another example is for a private for-profit professional association to partner with a not-for-profit company to develop a commercial product from intellectual property.

There is no limit to project partnering by organizations. A common goal and complementing capabilities are the basis for a project partnership. Finding the talent needed to perform specific project work and developing cooperative arrangements for the partnering are needed for a working partnership.

[14]Paraphrased from David I. Cleland and Lewis R. Ireland, *Project Manager's Portable Handbook* (New York: McGraw-Hill, 2000) pp. 4.8–4.15.

11.18 TYPES OF PROJECT PARTNERING ARRANGEMENTS

Project partnerships may take many forms of cooperative agreements. The formality and contractual relationship are determined by the needs of all partners. Some examples of these relationships are depicted in Table 11.2.

The number of arrangements is left to the organizations desiring to partner and work together. One important aspect is for the visibility that the customer has of the accomplishment of different work packages. In some partnering, the customer wants to be aware of who is doing the work, and in other situations, the customer is only concerned with the quality of the work.

TABLE 11.2 Types of Project Partnering Arrangements

Relationship and binding documentation	Description of possible working arrangements	Remarks
Formal and contract	Two companies obligate themselves to perform parts of a project. The allocation of work is based on expertise in the type of work to be performed.	One or both companies must commit to the customer for delivery. One or both may sign the contract with the customer.
Formal and consortium	Two or more companies obligate themselves to perform project work through a single contract that joins them in a separate legal entity.	The consortium represents itself to the customer as the contracting entity. The individual identities of the companies may not be visible to the customer.
Formal and contract	One company may bid on a project and use another company's resources. The resources are excess to the "loaning" company, but become a part of the project under the direction of the host company.	The arrangement is invisible to the customer. The "loaning" company has an obligation to provide qualified resources to the project and may or may not provide supervision.
Informal and agreement	One company bids and wins work on a project. One or more other companies agree that they will contribute to the success of the project through selected work.	Second-tier companies are invisible to the customer. Work performed by second-tier companies is accomplished as needed, but the customer does not know other companies are involved.

11.19 EXAMPLES OF PROJECT PARTNERING ARRANGEMENTS

There are many examples of project partnering that demonstrate the concept and the future of this type of business relationship. Figure 11.9 shows several examples of project partnering.

- Engineering and construction firms partner to obtain the best mix of talent and capability for projects. Partnering to obtain excellence in project control for major projects is often found in major projects such as the Super Collider Project in Dallas, Texas.
- Aerospace companies partnered to build a stealth bomber on a $4 billion project. Several companies worked together to develop the best in stealth technology.
- Three small companies combined their capabilities to bid and win a project requiring expertise in computer technology, computer network operations, and procurement knowledge.

Examples of Project Partnering

E&C firms mix capabilities

Aerospace uses best capabilities to build airplanes

Small companies partner to expand capability

Licensing to obtain technology

FIGURE 11.9 Examples of project partnering.

- Licensing of intellectual property of a professional association to a company to build software products is a current project. The association's standards were licensed to a company for the specific purpose of expanding the distribution of knowledge in the standards and to generate a modest profit for the association.

11.20 MANAGING PARTNERED PROJECTS

Customers are concerned with the management of the project work and who will be responsible for such items as reports, corrective actions, changes to the project, and overall project direction. Strong management capability builds confidence with the customer, whereas a weak or vague project management structure erodes confidence. Figure 11.10 shows typical types of arrangements for managing partnered projects. A discussion follows.

Some management structures for partnered projects are:

- *Steering committee.* Senior managers from all partnering organizations. This committee reviews progress and sets direction. A project manager or co-project manager is designated for all partnered work and reports to the steering committee.
- *Project manager and deputy project manager (or co-project manager).* Two individuals are designated from the partnering companies to lead the project. These individuals may report to their respective company executives or to a steering committee.
- *Project manager.* A single individual is appointed as the project manager for all the project work. Project team members report to the project manager for performance issues.

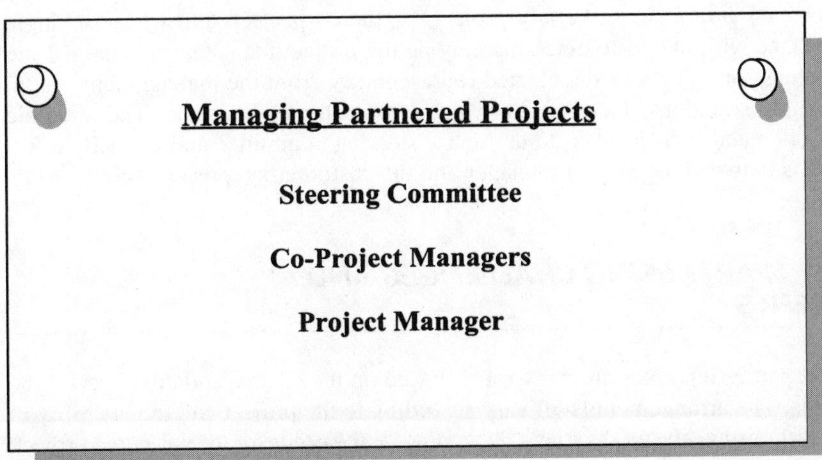

FIGURE 11.10 Managing partnered projects.

These three management structures can be modified to meet any situation. Large projects need senior guidance from a group such as a steering committee, whereas small projects would only require a single project manager.

11.21 TECHNICAL ASPECTS OF PARTNERED PROJECTS

One of the primary reasons for partnering is to gain additional technical capability. A customer wants assurances that the project will be technically successful and that the performing organization has the capability to accomplish the work. Partnering gives that assurance by bringing the best talent from more than one company.

Strategies of partnering will guide companies to better work solutions and better products for the customers. The following is a list of sample strategies:

- *Work allocation.* Divide work so the work packages are performed with an integral team when possible.
- *Project control.* A team could be assigned responsibility for controlling the project work. This team may be individuals from different companies, but should have clear work responsibilities.
- *Project management.* There needs to be a single person responsible for managing the work. Some work may be performed under the supervision of a manager, but that manager is responsible to the project manager for delivering a product component.
- *Company participation.* Company managers must not, as individuals, become involved in the direction of the project. Direction must be through a consolidated body, such as a steering committee or senior management representative group.
- *Customer interface.* Like any project, partnered projects must have a single interface with the customer. This may be the project manager, the chair of the steering committee, or the elected representative from the management group. In some situations, there are two levels of customer interface. The strategic direction and liaison are from the senior steering committee and the daily interface is between the project manager and the customer's representative.

11.22 PARTNERING CHALLENGES AND BENEFITS

Project partnering takes on many forms based on the desires and creativity of the partners. The arrangements will vary according to the project and the visibility of organizations may be more or less, depending on the needs for the partnered project. Organizations, public and private, for-profit and not-for-profit, can create partnerships to meet the needs of a small, medium, or large project.

Managing a partnered project is typically driven by the desires of the customer. Customer confidence and visibility into the project dictate the management approach as well as the size of the project. It is not uncommon to have a two-tier management structure, that is, a project manager and a steering committee to which the project manager reports.

Many benefits are derived from partnered projects. These include learning from other organizations' management and technical processes, obtaining technical knowledge about project management methodologies and processes, acquiring improved work methods from others and large profit shares through improved productivity.

11.23 OUTSOURCING PROJECT MANAGEMENT[15]

Outsourcing, or out-tasking as some organizations call it, is a growing field. Historically, the soft areas such as custodial services, food services, and landscaping have dominated the outsourced work; nearly 50 percent of recently surveyed organizations identify these areas. Maintenance of fleet automobiles and aircraft are also examples of areas for outsourcing.

Outsourcing is contracting for services that could be provided by the organization, if the organization had the capability and desire to perform those functions. Outsourcing has many advantages in that it is typically more economical to buy the products and services than provide them in-house. There is no investment in the function in terms of people, equipment, or maintenance for the outsourcing organization.

At GE, under the banner of "no back offices," CEO Jack Welch told managers to digitize or outsource the pants off their business that didn't touch the customer.[16] Outsourcing of core competencies is not recommended because an organization loses its capability to function effectively. Core competencies must be nurtured and grown rather than contracted for through another organization. A providing organization has little incentive to improve on another organization's core competencies as long as the products and services are being purchased.

11.24 PROJECT MANAGEMENT AS AN OUTSOURCED SERVICE

Project management services have been outsourced in several instances and the future looks promising for more outsourcing. The benefits of outsourcing project management include improved and more economical operations.

---

[15]Paraphrased from David I. Cleland and Lewis R. Ireland, *Project Manager's Portable Handbook* (New York: McGraw-Hill, 2000), pp. 7.52–7.59.

[16]"It's All Yours, Jeff. Now What?" *Fortune,* September 17, 2000, p. 66.

Project management services can be improved by transferring the functions to an organization specializing in project management. This specialized company hires the right skills and uses the best of practices because it is their core competency. They have resources that focus on providing these services and have the in-depth expertise to perform at high-performance levels.

Outsourcing relieves the parent company of the burden of managing project management services. Through contractual relationships, the parent company states its needs and then manages the contract and delivery of services. There is less effort required to receive the products and services from outsourced project management than to manage it in house.

Outsourcing companies may be concerned with loss in the project management function. Organization outsourcing will lose visibility into the details of work, but there is no need to see detail when the end product meets the requirements. Outsourcing companies should only be concerned with the delivered product and less concerned over the details of production.

11.25 OUTSOURCING TRENDS

Project management outsourcing has been primarily one of providing services to organizations by one person at a time. It is referred to as "body shop" or "hired gun" because the project management professional is typically working on-site with the organization's staff. This type of arrangement is profitable for the organization doing the work, but is not outsourcing.

The "body shop" or "hired gun" approach mixes different levels of proficiency and competency. The lower level of project management proficiency and competency is typically the resident employee of the organization and the outside person brings the expertise. This places the resident employee as the driver and the expert as the follower. There is considerable waste in talent, time, and money with this approach.

Outsourcing of project management services will, like other professional services, continue to be used to fulfill the needs of organizations. This has a high-growth potential because of the ability of project management service providers to deliver better products in a timely manner at lower cost. Project management service providers will have the expertise and skills to build better products than a part-time effort in-house.

11.26 SELECTING AN OUTSOURCE PROVIDER

Successful selection of an outsource provider is accomplished in four sequential steps. These steps give a high degree of assurance that the best provider is selected and that the outsourcing relationship is satisfactory. Figure 11.11 graphically displays the steps.

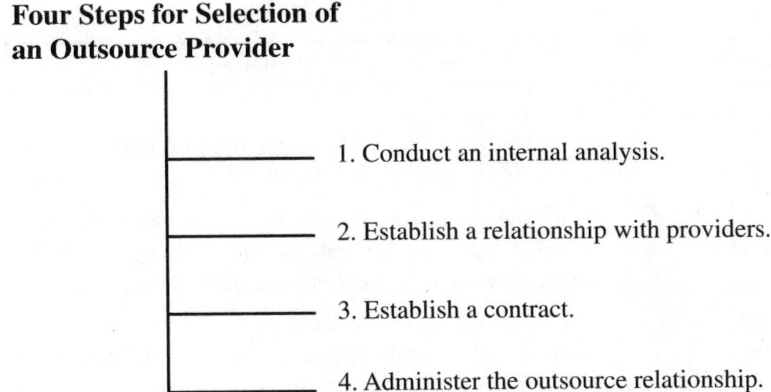

FIGURE 11.11 Selecting an outsource provider.

- *Conduct an internal analysis.* Identify those functions that can be outsourced, assess the tactical and strategic impact of outsourcing each identified function, evaluate the total cost of each function selected for possible outsourcing, and determine the advantages and disadvantages of outsourcing. Classify each function as "no outsourcing," "possible outsourcing," and "definite outsourcing." Use criteria for each category such as listed in Table 11.3.
- *Establish a relationship with providers.* Issue a request for information (RFI) that solicits interest in providing products/services, select two or three respondents, and ask for due diligence. Conduct a due diligence survey to validate the capabilities of the potential providers.
- *Establish a contract.* From the three respondents meeting the due diligence survey criteria, negotiate a contract with one. Establish scope and boundaries for the contract and describe the following:
    - Resources to be used to produce the products and services
    - Key deliverables and the schedule for delivery
    - Performance measures and other quality metrics
    - Invoice and payment schedule to include provisions for timing of payments
    - Change and termination provisions of the contract
- *Administer the outsourcing relationship.* Establish the contract management process, establish the technical review process, establish a change order process, and establish a steering committee or oversight committee. Involve the users or consumers of the products and services in the steering/oversight committee.

Identifying providers who can meet the organization's needs involves more than establishing a contract and working to enforce the provisions and clauses. A good contract is the basis for promoting understanding with the provider of products and services. It is also essential that the provider have a record of meeting contract provisions.

TABLE 11.3 Criteria for Outsource Provider Selection

Category	Criteria
No outsourcing of function	• Part of core competency. • Negative benefit or value added to outsourcing. • Impacts strategic goals.
Possible outsourcing of function	• Cost and other benefits show no advantage or disadvantage. • Neither contributes to strategic goals nor impacts strategic goals. • All factors equal for in house or outsourcing.
Definite outsourcing of function	• Definite cost savings to outsource. • Reduces complexity of management for in-house work. • Better product/service results from outsourcing. • More responsiveness to needs of organization. • Outsourcing uses better practices and technology than available in-house.

References should be checked during due diligence and references should be asked the following questions:

- Did the contractor deliver the products and services called for in the contract?
- Were the products and services usable by the consumer as delivered?
- Did the contractor demonstrate flexibility in minor changes to the contract or was each minor change an issue?
- Were change orders to the contract performed on a fair and equitable basis?
- Do you recommend the contractor for these products and services?
- Would you award a contract to this contractor again?

11.27 OUTSOURCING PROJECT MANAGEMENT SERVICES AND PRODUCTS

From an organization's perspective, there is a need to determine what may be outsourced and to which project management service provider. First, the areas of project management that may be outsourced need to be identified. Second, identify the best provider.

Project management products and services that may be outsourced depend upon the organization's structure for project management, the degree of maturity in project management, the number of projects, and the management style. The organization must know what is needed in terms of project management products and services. An organization with a loosely structured project management capability may not know what is needed.

The entire project may be outsourced. This frequently happens within such fields as information technology or information systems where the organization only wants to be concerned with software solutions and not the challenges of designing, developing, testing, and delivering software releases. When only components of project management are outsourced, Table 11.4 may be helpful.

11.28 PROJECT MANAGEMENT OUTSOURCING GUIDELINES

Outsourcing for any products and services needs to consider all aspects of the relationship and how it is managed. Transition of the responsibility for products and

TABLE 11.4 Outsourcing by Project Management Component

Project management component	Outsourcing potential
Project planning	*High.* Project planning expertise is typically not resident in companies. A centralized planning effort with the project team can provide many benefits by getting the projects started from a solid basis.
Project scheduling and maintenance	*High.* Scheduling and schedule maintenance skills are often resident in house, but the work is time-consuming for highly qualified technical people.
Project cost estimating	*High.* Cost estimating is an art that requires specific skill sets that are typically not resident in a company. It can be time-consuming for technical people who are better employed on other tasks.
Project progress reporting	*High.* Status and progress reporting are part of the schedule maintenance. Consistent reporting to senior leaders will be achieved as well as standard reports across projects.
Change control • Product • Project	*High for both categories.* This is the administration of the change control process. Initiation of changes and decisions on changes are typically performed in house.
Issue tracking	*High.* Tracking issues by a standard procedure and collecting information on the status is a routine matter that should not burden the project manager.
Problem tracking	*High.* Tracking actions by a standard procedure and collecting information on the status is a routine matter that should not burden the project manager.
Risk assessment and risk tracking	*Medium.* Conducting risk assessments and mitigation actions can be performed by an outsource provider. There must be good coordination with the project manager for any project being assessed.
Other components	This open item allows tailoring to meet individual organizational requirements.

services must be coordinated and performed according to an agreed-upon plan. The following items are helpful in guiding an organization through outsourcing of project management services and products.

- *Products and services.* Identify the products and services to be transitioned (or initiated if there are no current comparable in house). Describe the products and services in detail and conduct a mutual understanding meeting to ensure both parties have full agreement on the products and services.
- *Standards and formats.* Identify the standards and formats that are considered a requirement. Standards and formats provide a consistency to the delivered products, but will change the current in-house products.
- *Frequency of delivery.* Determine the frequency of delivery for products and time required for preparing products. Typically, reports are provided on a weekly basis, but there may be requirements for rapid turnaround for some products.
- *Method of delivery.* Products may be delivered in hard copy, electronic copy, or a combination of these. Determine the most appropriate and effective means of delivery for all products.

11.29 OUTSOURCING POTENTIAL

Outsourcing project management products and services has the potential for significant gains for an organization. Outsourcing relieves the outsourcing organization of a technical area for which they may not have the trained resources to perform and to perform at a lower cost. Outsourcing can place many of the project management functions in the hands of project management professionals.

Entering into an outsourcing arrangement requires some background work to identify and select the provider with the best qualifications and reputation as a contractor. The time and effort spent on selection of the provider will save effort in managing the relationship.

11.30 TO SUMMARIZE

The major points that have been expressed in this chapter include:

- Project planning involves the development of a strategy for the commitment of resources to support the project objectives and goals.
- There is a high degree of interdependence among the strategic plan, functional plans, and project plans of the enterprise.
- Project planning is a rational determination of how to initiate, sustain, and terminate a project.
- Planning starts with the ability to sense and develop a vision involving the ability to see something that is invisible to others.

- Without a vision, the project may very well deteriorate and fail.
- An organizational mission is a summary statement of the "business that the enterprise is in," and projects are the way the business is conducted.
- Organizational objectives are ongoing end purposes of the enterprise.
- A project goal is a milestone for the project.
- A project strategy is the design of the means, through the use of resources, to accomplish end purposes.
- A project is more than a schedule and a spending plan; it is an integrated effort to produce a product, service, or organizational process change.
- The quality and quantity of the organizational resources constitute the common denominator of planning, as well as the competencies of the enterprise and the projects involved in that enterprise.
- The roles of people regarding their project planning responsibilities were indicated.
- There are several specific project planning techniques that were briefly described in the chapter with references to more extensive literature on these techniques.
- The work breakdown structure (WBS) is the most basic consideration in project planning. The key reasons for using a WBS were described in the chapter.
- The general nature of a project schedule was noted in the chapter along with some general guidelines on the processes to be followed in developing a project schedule.
- A guide for project planning was presented in the chapter with examples of techniques.
- An important part of project planning is a consideration for partnering and outsourcing.

11.31 ADDITIONAL SOURCES OF INFORMATION

The following additional sources of project management information may be used to complement this chapter's topic material. This material complements and expands on various concepts, practices, and theory of project management as it relates to areas covered here.

- Thomas C. Belanger, "Choosing a Project Life Cycle"; Paul Warner, "How to Use the Work Breakdown Structure"; and Hans J. Thamhain, "Developing Winning Proposals," chaps. 6, 7, and 11 in David I. Cleland (ed.), *Field Guide to Project Management* (New York: Van Nostrand Reinhold, 1997).
- Mehdi Adib, "Managing Kuwait Oil Fields Reconstruction Project"; Ian Boggan, "The Benfield Column Repair Project"; and Larry Johnson, "Managing Resources and Communicating Results of Sydney's $7 Billion Clean Waterways

Programme," in David I. Cleland, Karen M. Bursic, Richard J. Puerzer, and Alberto Y. Vlasak, *Project Management Casebook,* Project Management Institute (PMI). (Originally published in *Proceedings,* PMI Seminar/Symposium, Vancouver, 1994, pp. 184–190; *PM Network,* February 1996, pp. 25–30; and *Proceedings,* PMI Seminar/Symposium, Pittsburgh, PA, 1992.)

- Stephen A. Devaux, *Total Project Control: A Manager's Guide to Integrated Project Planning, Measuring, and Tracking,* Wiley Operations Management Series for Professionals (New York: Wiley, 1999). This book focuses on managing projects in their totality, which includes all resources and responsibilities for managers. It includes new metrics for assessing project performance and profitability. Individual, multiple, and portfolios of project controls are described to give the reader a full range of information.
- Jim Fuller, *Managing Performance Improvement Projects: Preparing, Planning, and Implementing* (San Francisco: Jossey-Bass, 1997). This book focuses on improving the performance of project participants through better planning of projects. It describes an approach for better management of projects through such areas as project sponsorship, managing performance change, and accelerating a project for rapid completion. It is filled with examples that convey concepts and practices that lead to project results.
- William H. Roetzheim, "Creating the Project Plan," *Software Development,* January 2001, pp. 61–64. This article describes the planning process for a software project. It includes the fundamental processes and techniques in building the plan and extends to the maintenance requirements of the project following delivery of the product.
- Al Picardi, Mark Hodges, and Nancy Tarr, "Fast Track Development Strategies," *Electric Perspectives,* March–April 2001, pp. 12–21. This article concentrates on the initiation process for a project and the complexity of long-term projects when there are several individuals and agencies involved in project selection. Collection of information and coordination are prerequisites for good project planning.
- Christian Mitchell, "Planning New Facilities," *Broadcast Engineering,* May 2001, pp. 68–73. This article addresses the life cycle of planning facilities and the issues associated with taking shortcuts. The author emphasizes that cost and time are driving factors in virtually every TV station facility installation or construction. Good planning is therefore required to know in advance the critical parameters of the work.

11.32 DISCUSSION QUESTIONS

1. What are some of the basic concepts involved in project planning? How can a complete project plan contribute to project success?
2. Define each of the key elements of project planning.

3. Discuss the relationship of project planning to strategic planning and functional planning. Why is this relationship important?
4. List and discuss some of the traditional and modern planning practices.
5. Nearly all project stakeholders have some role in the project planning process. How are each of the key people afforded the opportunity to participate in project planning?
6. Describe a simple project management situation from your work or school experience. Develop a work breakdown structure for the situation.
7. What are some of the characteristics of an effective project schedule? Describe some of the scheduling techniques used in project management.
8. What are the relationships between the statement of work, project specification, cost estimate, financial plan, functional plan, and implementation plan? How is each developed and used?
9. List and describe the project planning work packages.
10. What is the purpose of a project management manual? How does it support the project plan?
11. Discuss the nature of the relationship between effective planning and effective control. What are some of the elements of each?
12. Discuss the following statement: Project planning is the most important work carried out by a project team.

11.33 USER CHECKLIST

1. Are the key elements of organizational planning mission, objectives, goals, strategies, and so forth evident in your organization? Is project planning linked to strategic planning? Explain.
2. Which project planning practices are used by your organization? How?
3. What role does each of the project stakeholders on projects within your organization play in the project planning process?
4. Do the project managers in your organization understand the concept of a work breakdown structure? Are work breakdown structures developed for every project? Why or why not?
5. How are project schedules developed? What scheduling techniques are used? Are these techniques effective? What other techniques might be more effective for project planning in your organization?
6. Is the project life cycle understood by the project managers in your organization? How is this concept integrated into the project planning process?
7. Which project planning elements (e.g., statement of work, project specification, cost estimate, financial plan, functional plan) are used by your organization? Which are not? Explain.

8. Are organizational policies and procedures developed to support the project planning? Would the development of a project management manual increase the effectiveness of the project planning process in your organization?
9. Do managers consider the strategic fit of projects to ensure that the project is truly a building block in executing organizational strategies? Explain.
10. Consider the project planning work packages. How do these work packages fit in your project planning?
11. How are the work packages assigned on major projects? Does management consider subcontracting work packages? Are logic networks and relationships developed for the work packages?
12. Is the cultural ambience of the organization supportive of rigorous project planning?

11.34 PRINCIPLES OF PROJECT MANAGEMENT

1. Planning projects is a critical function of project managers or leaders.
2. Planning projects will result in a plan that typically will require some change because of new information or a change to the basic requirements.
3. Planning projects is a top-down activity that pulls information from the organization as well as the project's goals and objectives directly related to the expected benefits.
4. The quality of project planning relates directly to the chance of success for the project.
5. All plans will require revision to meet emerging changes and new information updates.

11.35 PROJECT MANAGEMENT SITUATION—DEVELOPING A PROJECT PLAN

An organization that builds aircraft controls has received a request for proposal (RFP) to bid on the manufacture and sale of 10,000 aircraft cabin temperature controls. The RFP has a statement of work and a technical specification that details what is required and to what standards. There is no drawing or design that details the temperature control mechanism, but the specification gives the space allocated for the mechanism, power requirements, durability requirements, and reliability requirements.

A decision has been made that a proposal will be submitted. However, the competition is expected to have a control that requires only a small amount of modification to meet the client's requirements. Therefore, this job will be a project that must be planned to optimize all cost, schedule, and technical performance parameters.

Management has the following questions and guidance:

- How much will it cost for the project in total and what will be the average cost per unit?
- How long will it take to establish a design that meets or exceeds the specifications?
- How long will it take to build and deliver 10,000 units?
- What are the chances that the client will change the requirements after reviewing our design?
- What is our capability to start the project within 30 days following award of the contract?
- Give me a summary project plan before we submit our bid.
- Give me your answers and the plan in 2 weeks.

The prospective project manager is tasked with developing the summary project plan and the questions are to be answered by the functional managers. Coordination among the managers is essential to ensure that senior management gets the best single report.

11.36 STUDENT/READER ASSIGNMENT

1. As the project manager, what elements would you include in a summary project plan?
2. What type of cost estimate will be appropriate for the senior decision maker and why?
3. What type of schedule would be included in the summary project plan, if any?
4. What resource requirements are needed for the summary project plan and why?
5. What type of project organization should be considered for the summary project plan?

CHAPTER 12
PROJECT MANAGEMENT INFORMATION SYSTEM

"Knowledge is of two kinds: We know a subject ourselves, or we know where we can find information about it."

SAMUEL JOHNSON, 1709–1784

12.1 INTRODUCTION

Accurate and timely information is essential for the management of a project. Project planning, organizational design, motivation of project stakeholders, and meaningful project reviews simply cannot be carried out without information on the project—and how it relates to the larger organizational context in which the project is found. An accurate and complete project management information system must exist to provide the basis for how the project is doing. The project manager—or any other manager for that part—simply cannot make and execute meaningful decisions without relevant and timely information.

In this chapter, a project management information system (PMIS) is presented. Project failures attributed to lack of information are offered. The value of the PMIS, a description of a PMIS, and the uses to which a PMIS can be put are offered. How to use the PMIS in the management of a project is described, along with how project information can be shared. The role of technology vis-à-vis the PMIS is provided. A summary description of PMIS hardware and software is suggested, along with descriptions on how to plan for the PMIS. A description of the essential elements of a PMIS closes the chapter.

12.2 THE PROJECT MANAGEMENT INFORMATION SYSTEM

In Chap. 4, the project management system and its subsystems are described. Figure 12.1 shows the project management system and its subsystems. The project management information system (PMIS) is intended to store information essential to the effective planning, organizing, directing, and controlling of the project, as well as provide a repository of information to be used to keep stakeholders informed about the project's status. The essential elements of a PMIS are covered in this chapter.

All too often projects are characterized by too much data and not enough relevant information on where the project stands relative to its schedule, cost, and technical performance objectives as well as the project's strategic fit and function in the parent organization's strategies. The 80-20 rule tells us that typically there will be *the vital few and trivial many,* or 20 percent will be relevant and the remaining 80 percent will be of significantly less importance.

Information is essential to the design and execution of management decisions allocating resources in a project. Decisions coming from project planning, organizing, direction, motivation, and control must be based on timely and relevant information. Motivation of the project team and discharge of leadership responsibilities by all managers associated with the project require information by which informed decisions can be made and executed.

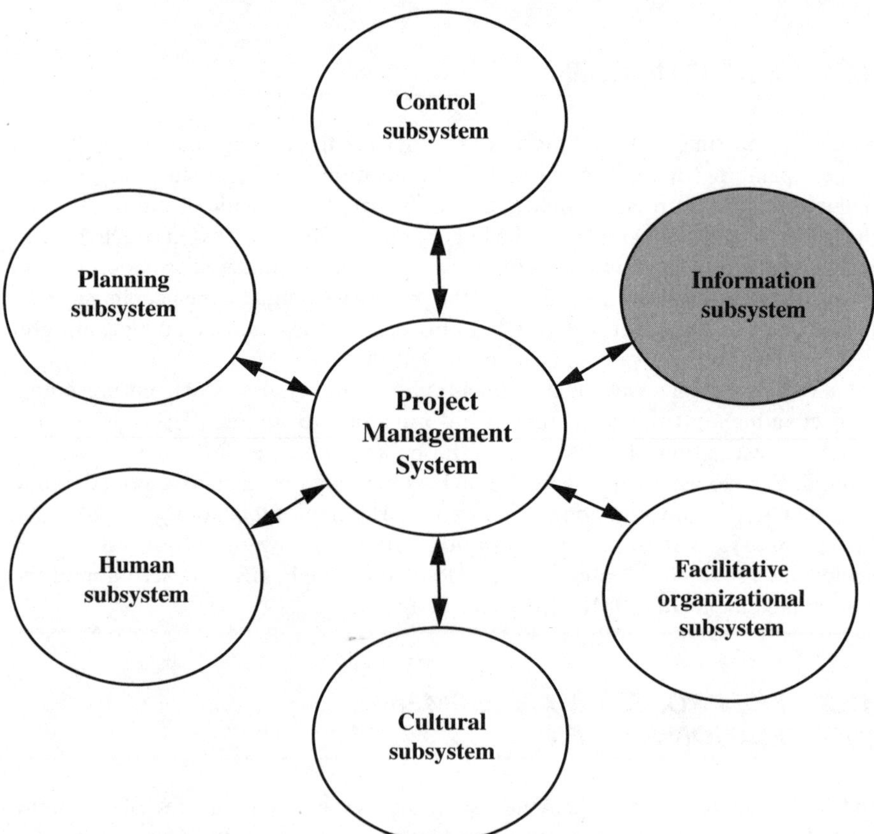

FIGURE 12.1 Project management system—information subsystem.

Information is required for the operation of any enterprise. In organizations, making and implementing decisions depend on the character of the information available to the decision makers. Information availability and flow are critical considerations in the speed and eloquence with which the efficient and effective use of resources is carried out in meeting the purposes of the enterprise.

Organizations of all sizes need information to design, produce, market, and provide after-sales support to the products and services that are offered to customers. In large organizations the flow of information can be incomplete and sequential, often not getting to the people who need the information for their work in time to make the best decisions. Information may be found lying around in organizations waiting for someone who has the authority to make a decision. The best information loses its value if it is not available to people who need it to make decisions and direct actions.

A system for collecting, formatting, and distributing information is needed for the organization and each project. The organization's management information system will contain some information that is needed for the projects, but there is a need for additional project-related information as well as that information generated as a result of the project's activities.

An important part of the management of any project is a well-developed strategy for understanding and managing the set of procedures and documents that establish information used in the management of the project. One author has suggested a strategy for the development of such documentation.[1]

Sometimes the initiation of a project for the development of an information system for one element of the enterprise results in the broadening of information usage. For example, at 3M during the development of a computer-integrated manufacturing (CIM) approach for the company, a total integration of all the information technology for one of the company's plants was initiated. The name given to this effort became *integrated manufacturing system* (IMS). Tying the administrative systems into their CIM structures provided for further broadening the notion of concurrency in the management systems of the organization.[2]

In addition to the immediate participants to a project, there is a need to consider all stakeholders. A project manager might characterize the PMIS as being able to provide information that he or she needs to do the job and information that the bosses need. Typically, stakeholders have various information needs that can often be satisfied through the information stored in the PMIS. Table 12.1 describes some of the stakeholders' information needs on a routine basis.

Those individuals with real or perceived information needs about the project soon become disenchanted when inadequate or inaccurate information is provided. No stakeholder likes surprises that reflect a change to the project plan or anticipated progress. Surprises quickly erode confidence in the project manager's capability to manage the work and keep key stakeholders fully informed on progress. One corporate vice president in Rochester, N.Y., stated to her managers, "Surprises on projects are not career-enhancing moves."

[1]Henry J. McCabe, "Assuring Excellence in Execution in Construction Project Management," *PM Network*, October 1995, pp. 18–21

[2]Tom Waldoch, "From CIM to IMS Spelled Success at 3M," *Industrial Engineering*, February 1990, pp. 31–35.

TABLE 12.1 Stakeholder Information Needs

Stakeholder	Type of information needed (examples)
Customer	• Status and progress of project • Significant changes to cost, schedule, or anticipated technical performance • Any difficulty in converging on the project's objectives and goals
Senior management	• Status and progress of project • Significant changes to cost, schedule, or anticipated technical performance • Changes to resource requirements • Any difficulty in converging on the project's objectives and goals
Project manager	• Status and progress of project • Significant changes to cost, schedule, or anticipated performance • Changes to resource requirements • New project requirements or changes to specification or statement of work • Issue resolution or delay in critical decision
Functional manager	• Status and progress for their respective project elements • Changes to design or specification for their respective area of responsibility • Requirement for additional resources from their respective area of responsibility
Project team member	• Status and progress of project • Changes to project goals or objectives • New requirements for the project • Issue resolution • Change to work assignment

12.3 INFORMATION FAILURES

Not all projects are managed by using a relevant and reliable information system. For example, on the Shoreham project, the administrative law judges found that the Long Island Lighting Company (LILCO) nuclear power plant's measurement and reporting systems continually and repeatedly failed to accurately depict cost and schedule status at Shoreham. LILCO managers were unable to use LILCO's measurement system to gain an accurate picture of what was happening on-site and complained that LILCO's reporting systems were confused and cluttered.[3]

[3]Recommended Decision by Administrative Law Judges William C. Levey and Thomas R. Matias, Long Island Lighting Company–Shoreham Prudence Investigation, Case no. 27563, State of New York Public Service Commission, March 13, 1995.

The judges left no doubt as to the overall responsibility of the LILCO board of directors for the Shoreham project:

> We conclude that the limited information presented to the Board was inadequate for it to determine project status or the reasonableness of key management decisions or to provide requisite guidance and direction to LILCO management.[4]

Inadequate information systems on the Trans-Alaskan Pipeline System (TAPS) project contributed to the lack of adequate controls. Crandall testified:

> [T]here is little question that the control of TAPS required an adequate and well designed formal control environment to provide control information for senior managers. The volume of data to be processed indicated the need for computers in at least parts of this control environment. Thus, had cost controls been in place in early 1974 at the very start of the project, the controls would have allowed management to minimize costs while still attaining realistic schedule goals. Thus, it is my opinion that if prudent cost controls, as part of a comprehensive control environment, had been installed at the start of construction, they would have helped assure completion of the project on or even before the schedule date.[5]

These major projects were materially affected by the lack of adequate information with which to make informed decisions. There is no doubt that if a functioning information system had been in place, the outcome would have been significantly different and perhaps would have avoided the external reviews and criticism.

These two examples demonstrate the need for an information system for projects to collect, format, and distribute information to the decision makers. Without information, decisions are made through "best effort" based on something other than the facts. Managers at all levels must manage by facts if the enterprise is to achieve the best results from projects.

12.4 VALUE OF THE PMIS

The PMIS is a vital part of the communications for the project. As a store of knowledge, the plans, practices, procedures, standards, guidelines, and methodologies are readily available to consult prior to making a decision or taking an action. A single store of information facilitates the collection and recovery of key data at any time—during planning, project implementation, and postproject activities.

Figure 12.2 shows a conceptual arrangement of the project's information. This diagram depicts the organizational information being loaded into the PMIS from the computer on the left. Organizational information would be all background

[4]Ibid.

[5]Keith C. Crandall, prepared direct rebuttal testimony, Alaska Public Utilities Commission, Trans-Alaska Pipeline system, Federal Energy Regulatory Commission, Washington, D.C., January 10, 1984, pp. 8–9.

information such as contracts, strategies, operational plans, policies, procedures, and other documents influencing how and when the project will be implemented.

The three computers on the right depict the interaction between the project team and the store of knowledge. The project team would be populating the PMIS with such information as the project plan, including all its subordinate documents, schedules, budget, correspondence, specifications, statements of work, and drawings. Once the initial data are loaded, the project team would maintain the system through updates.

A fully populated PMIS would be accessed anytime there was a need for information. It would be the first source of information for managing the project with the relevant information from both the enterprise's information system and the project-generated information.

During postproject assessments, the PMIS can provide a wealth of information on what was accomplished, what should have been accomplished, and how it was accomplished. The actual performance data for the project provide a record of how well the project accomplished its purpose. This written record is more reliable than the memory of individuals. When individuals typically transition through the project to complete their work, they may not be available for postproject questioning.

One project may generate significant information that has value for future projects. The PMIS, as the store of knowledge, can be made available at any time to support the enterprise's work on another project. Although the project may be ongoing, there is still valuable information that can support planning and initiation of new projects.

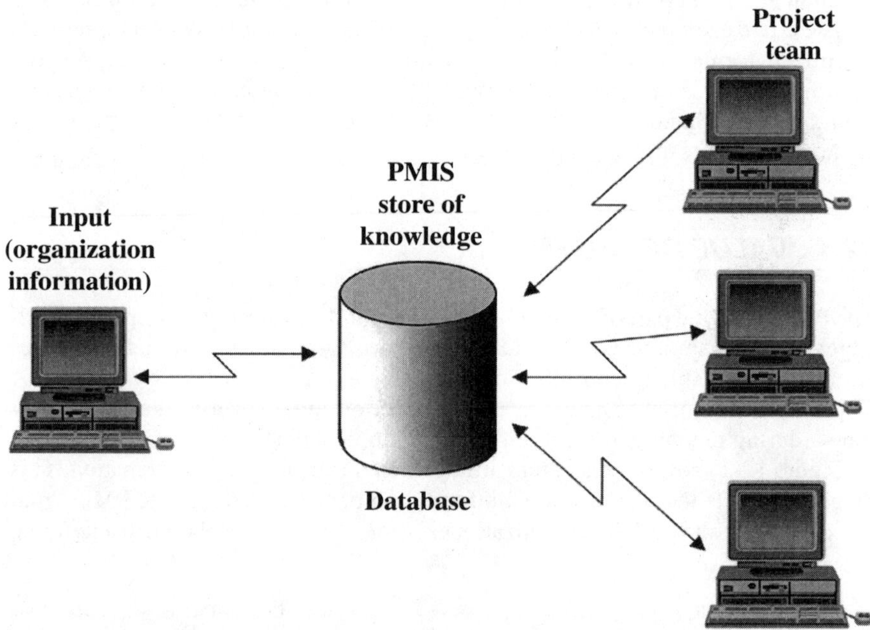

FIGURE 12.2 Project store of knowledge.

One may give a value to the PMIS by comparing a current method of managing a project's knowledge store with a model of what it could be. The comparison should consider the project's needs for information and the benefits derived for the enterprise's other projects, whether they are ongoing, being planned, or being assessed in a postproject audit. Some questions that could be asked:

- How are the present project management information needs being met and are they adequate?
- What improvements are needed to support projects in the future?
- What information is provided by completed projects to support planning and implementation of other projects?
- How are best practices captured and passed on to others?
- How is project information distributed to functional departments?
- What benefits could be derived from an improved PMIS?
- What are the cost and benefits of a PMIS?

12.5 DESCRIBING A PMIS

There are many descriptions of a PMIS. The authors believe that a fully capable PMIS consists of all information needed by the project team to conduct its business. This includes information from the organization that guides the project as well as background information on the project.

In Fig. 12.3, Tuman presents a model of an information system that is the minimum type of PMIS for a project. Tuman's model concentrates on schedule, cost, and technical performance information related to the project's objectives and goals—and does not present the interface with the strategic management of the enterprises. In the context of information and control, Tuman's model serves as a very effective means of describing the process. In describing this systems model, he states:

> With this brief view of the *system,* we can define the project management information and control system as the people, policies, procedures, and systems (computerized and manual) which provide the means for planning, scheduling, budgeting, organizing, directing, and controlling the cost, schedule, and performance accomplishment of a project. Implicit in this definition is the idea that people plan and control projects, and systems serve people by producing information. The design and implementation of the procedures and methodologies, which integrate people and systems into a unified whole, are both an art and a science. Some of the more pragmatic aspects of these procedures and methodologies are considered.[6]

[6]John Tuman, Jr., "Development and Implementation of Effective Project Management Information and Control System," in D. I. Cleland and W. R. King (eds.), *Project Management Handbook* (New York: Van Nostrand Reinhold, New York, 1983), p. 500.

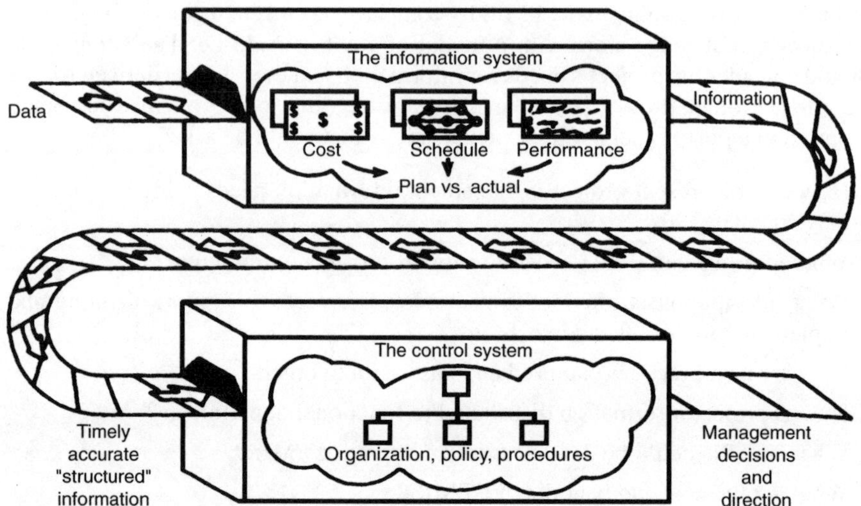

FIGURE 12.3 Information control system.[*Source: John Tuman, Jr., "Development and Implementation of Effective Project Management Information and Control Systems," in David I. Cleland and William R. King (eds.),* Project Management Handbook *(New York: Van Nostrand Reinhold, 1983), p. 499.*]

Another view of a PMIS is that project teams need information to support their efforts in the project. Information should be readily available and easy to obtain, preferably by computer. The PMIS should be a store of knowledge for the project and the first source for information about the project. It should include background information on the project, current information on project activities, and information that reflects organizational guidelines. The PMIS is a critical area that supports the project and allows it to be managed by fact.

The PMIS store of knowledge should be an enabling tool for the project manager and project team. It does not replace leadership or project methodologies, but will provide the means to make the projects more successful. The PMIS may be divided in to four categories of information.

1. Organizational guidance or support information in the PMIS could be:
   - Project management manual
   - Project management methodologies
   - Organizational polices for projects
   - Organizational procedures for projects
   - Organizational briefings on project capabilities and implementation
2. Historical information in the PMIS could be:
   - Files from other projects that contain performance data and best practices
   - Proposal, quotes, and bids on this project
   - Budgets, schedule, and technical performance measures from prior projects
   - Project plans from prior projects
   - Marketing presentation for this project

3. Current project information in the PMIS could be:
   - Contracts for easy access by the project manager
   - Project charter
   - Specifications on the project's product
   - Statements of work
   - Drawings, schematics, and illustrations related to the project
   - Schedules
   - Budgets
   - Risk assessments
   - Risk plans
   - Communication plans
   - Project correspondence
   - Project internal policies and procedures
   - Resources lists (human and nonhuman)
   - Approved vendor list
   - Names and addresses of key organization people
   - Stakeholder management plan
   - Functional or operational plans prepared by the functional departments
   - Project diary
   - Product standards
   - Time cards for project team
   - Briefings
   - Issue log
   - Action item log
   - Lessons learned
4. Old files from the current project that are no longer needed for the project's ongoing work could contain:
   - Old or superseded materials (schedules, briefings, expenditures, plans)
   - Records of former project team participants
   - Closed out contracts or closed invoices
   - Inactive files for correspondence
   - Superseded policies, procedures, standards, and decision papers

12.6 USES OF THE PMIS

The objective of an information system is to provide the basis to plan, monitor, do integrated project evaluation, and show the interrelationships among cost, schedule, and technical performance for the entire project and for the strategic direction of the organization. In addition, information should provide a prospective view in order to identify project problems before they occur, so they can be avoided or their results minimized.

Information is required so that the project team can continuously monitor, evaluate, and control the resources used on the project. Also, higher management must be kept informed of the status of the project to satisfy its strategic responsibility. There will be times when the project status will require the active involvement of senior management and/or the project owner. Thus, when project status is reported to higher-level management, the report should contain the key data stating the problem; the circumstances surrounding the problem; the cause; the potential impact on project cost, schedule, and other pertinent areas; a recommendation for the action to be taken; the expected outcome of the action; and the assistance sought from senior management.

Several additional methods can be used to keep abreast of the project status, assuming that an effective project management information system is in place. An easy and important method is to go down and "kick the tires," to observe what is going on in the project. Informal discussions with project team members during these visits can also provide insight into the status of the project. Analysis and interpretation of formal written and oral reports are useful, as are graphic displays of information. An essential way of keeping informed is to have formal project evaluation and control meetings.

The hardest part of any management job is not having all the right information yet having the responsibility of making the right decisions. Some companies find the project evaluation and control process so important that they have set up a project war room to facilitate the review process. A war room or information center has significant implications for improving project management. At an aerospace company, an information center provides information as well as information services such as analysis of user information requirements, specialized assistance, technology support, education, and training. As a clearinghouse for information, the facility provides database searches and assists users in deciding which products and services to use.

12.7 INFORMATION CHARACTERISTICS AND ATTRIBUTES

These characteristics and attributes are the following:

- *Accurate.* Information in the PMIS should be accurate and represent the situation. Erroneous information can lead to wrong decisions and failed projects. Accurate information provides the best chance for managing by fact.
- *Precise.* The precision of the information needs to be only to the level of granularity dictated by the project decisions. For example, there is typically no need to estimate project labor-hours to less than an hour. It is a special case where labor estimates are to the nearest minute or nearest 10 minutes.
- *Reliable.* The information must be derived from a source that gives confidence that it is real and representative of the situation. Information from an unknown source or stated in terms that permit more than one interpretation should be labeled "questionable."

- *Level of detail.* The information should be at a level of detail that permits easy translation to the current project. Too much detail masks the purpose and too little detail is not supportive of the project team.
- *Graphics, pictures, and illustrations.* The use of graphics, pictures, and illustrations can convey information more quickly than narrative text. These items can be supplemented with textual descriptions or highlights.
- *Mathematics and numbers.* Mathematics and numbers are a precise means of providing information. These are especially good to use for performance measures or product performance requirements.

The PMIS is an essential part of the project and critical to making the project successful. It takes an initial effort to provide the organizational and historical information as well as the project planning data. Once the PMIS is activated for a project, that project assumes responsibility for sustaining the system. It soon becomes outdated and loses its effectiveness if new information is not entered on a timely basis.

Files and documents that establish the project baseline should not be deleted, but placed in an archive when new files or documents supersede them. For example, the organization may issue a new project management manual that significantly changes the project methodology. It is important to maintain the superseded copy for reference when questions arise as to why something was accomplished a certain way.

12.8 SHARING INFORMATION

It is becoming more common for project information to be shared with the project stakeholders. When the project management information system provides information to stakeholders, the conditions for getting the stakeholders working together are facilitated. When project problems, successes, failures, challenges, and other issues are brought to the attention of the project stakeholders, there will likely be closer identification of the people with the project. If the stakeholders sense that the project manager is withholding information, there is the risk that stakeholders will perceive that the project manager does not trust them, because the information is not being shared. The sharing of information can promote trust, empathy, and more mature relationships among project stakeholders. Then, too, as the project stakeholders review information on the project, such as the problems that the project faces, they may have suggestions that can contribute to the solution of the problems.

Sharing of project information is one of the more important dimensions of keeping the team members working together cohesively and concurrently in the utilization of the project resources. Such sharing also facilitates the building of networks with the stakeholders through continuous interpersonal contact and dialogue. By using technology and a willingness to communicate, information systems can be designed for the project team that help everyone do a better job of making and implementing decisions in the utilization of project resources.

Every project manager has to ask key questions about the quality and quantity of information available to manage the project:

- What information do I need to do my job as project manager?
- What information must I share with the project stakeholders to keep them informed on the status of the project?
- What information do I need about other projects in the organization that interface with my project?
- What information do I require about the enterprise that provides me with insight into how the project fits into the overall strategy of the organization?
- What information do I require to coordinate my project's activities with other initiatives in the organization?
- What is the cost of my not having adequate information about my project and how that project interfaces with other projects in the overall organizational strategy of the enterprise?
- What information about the project do I not need to do my project management job? Remember, too, that I can be overloaded with untimely and irrelevant information.

Following are two examples of sharing information, or making it available to anyone who requested it. The outcome of each project was materially affected by the two instances.

- One program manager would not share information on the performance of his major contractor. Information was exchanged between the program manager and the contractor's project manager on a *confidential* basis and the contractor committed through his project manager to a certain course of action. The project team was aware of the *confidential* relationship, but did not have access to the information. When the contractor's performance was questioned, the program manager insisted that there was a personal commitment by the contractor (through the project manager) to correct the identified deficiencies. The contractor's senior management denied that there was any commitment and discharged its project manager. Partly as a result of the program manager and contractor's project manager withholding information on performance deficiencies, there was a cost overrun of nearly five times the original cost estimate and the project was delayed 3 years before it failed for technical reasons.
- Another program manager established an open system of sharing information and allowing anyone to communicate with any other person in the project. The only stipulation was that anyone sharing information had to be certain that the facts were correct. The sharing of information was between the project office and several contractors. Electronic mail was the primary means of communicating information between locations around the United States, which permitted easy exchange of information in written form without regard to time zones or duty hours. This free exchange of information was viewed as one of the major contributing factors in the success of the early completion of the project within budget, and the product's performance exceeded expectations.

12.9 INFORMATION VALUE

Information provides the intelligence for managing the project. Information must be processed so that decisions can be made and executed with a high degree of assurance so that the results will contribute to the project's success. In the project planning role, information provides the basis for generating project action plans, schedules, network diagrams, projections, and other elements of planning. Information is essential to promote understanding; establish project objectives, goals, and strategies; develop mechanisms for controls; communicate status; forecast future performance and resources; recognize changes; and reinforce project strategies. The project planning function establishes a structure and a methodology for managing the information resources, which encompass defining, structuring, and organizing project information, anticipating its flow, reviewing information quality, controlling its use and source, and providing a focal point for the project's information policies.[7]

Information is a valuable resource to be developed, refined, and managed by the project principals: project managers, functional managers, work package managers, project professionals, and the project owner. Project information is as much an essential resource as people, materials, and equipment. Information is also a key tool which facilitates the project management process. Information is needed to prepare and use the project plans, develop and use budgets, create and use schedules, and lead the project team to a successful conclusion of the project. Information, then, becomes both a key resource to the project stakeholders and a tool for all concerned to do their job. Information is important, but its role is limited. As Gilbreath states:

> It does not take the place of management skill, planning, project controls, experience, well directed intentions, or other project essentials. It will not shore up inherent inadequacies in organizations, approaches, or individuals.[8]

Gilbreath differentiates data and information. He states:

> A common misconception is that data equals information. Nothing could be further from the truth. Data is merely the raw material of information. It means virtually nothing without refinement. By refinement we mean the structuring of data into meaningful elements, the analysis of its content and the comparisons we make among data and preexisting standards, such as cost, schedule, and technical performance baselines. Only then does data become transformed into information. Data has no value unless it is transformed into structured, meaningful, and pertinent information, and information has no value unless it leads to needed management action or precludes unnecessary action.[9]

Information's real value is when it is used effectively in the management of the project. Information does not automatically lead to effective management of project,

---

[7] M. D. Matthews, "Networking and Information Management: Its Use by the Project Planning Function," *Information and Management,* vol. 10, no. 1, January 1986, pp. 1–9.

[8] R. D. Gilbreath, *Winning at Project Management—What Works, What Fails, and Why* (New York: Wiley, 1986), p. 147.

[9] Ibid., pp. 146–147.

but lack of information can contribute to project failure. Information may be in varying degrees of completeness when the PMIS is not properly populated—both on a timely basis and an accuracy basis. Partial information can be misleading and inaccurate information can lead to the wrong decision.

Information is perishable. Managing a project requires planning, organizing, and controlling of resources on a moving target as the project evolves through its life cycle. Information on a project at a particular point in that life cycle can change quickly as new project problems and opportunities emerge. Aged information will provide a distorted picture for the decision maker as well as give undue confidence in the decision. Only current information gives the best picture of the situation and allows decisions based on facts.

Gilbreath believes that information with a detail structure adds to the project's value as well as that of the parent organization.[10] Analyzed and structured project management data become information that is summarized for ease of reading and understanding. This analyzed information is disseminated up to senior managers of the organization and used within the project for measuring results. Reliable information has an audit trail from its source through the analysis process to the dissemination points.

Analyzed information provides the project team with the knowledge of where it has been in preceding periods, where it is today, and the direction the project is heading in. The proper amount of project information will support these goals, whereas too little information will not give the clear picture. Too much information has the tendency to overload the project team with information that must be filtered to properly view the past, present, and future situations. Senior managers also need the proper amount of pertinent information with which to make sound decisions on the project's future.

Information provides the basis for continuation of the project in the absence of the project manager. The project team can monitor the progress of the project and compare it to the project plan to assure that work is progressing satisfactorily. An effective PMIS provides the information that demonstrates when the project is on track or when it has exceeded the allowable limits of performance.

An important purpose served by a PMIS is that it can track at the work package level for early identification of schedule slippage or significant cost overruns on detailed work areas. Early identification of small problems permits the attention to detail before there are major impacts on higher-order work. This is especially important on large projects or projects that have a very rigorous schedule to meet the enterprise's or customer's goals.

12.10 TECHNOLOGY AND THE PMIS

Technological innovations support the development of a sophisticated PMIS that incorporates early warning measures to highlight variances from standard practices. The design of the PMIS should be general to meet the needs of projects across the

[10] Ibid., p. 148.

board in an enterprise, but may incorporate unique items such as the means to sort information by date, by source, by originator, and by type of information.

Although most information may be stored in the PMIS for a project, there are other sources that may be more effectively used. For example, the Internet, or World Wide Web, has a wealth of information for background and ongoing information needs. Some examples of what is available through the Internet include building code ordinances, information on stakeholders like chambers of commerce, conference proceedings, databases on topics ranging from accounting to zoology, exchange rates and other money matters, design and engineering data, document delivery, government information, industry information, and patent data. The information available on the Internet is extensive and reasonably accurate, depending upon the source. The Internet allows the project manager the ability to search, ask questions, and find an incredible range of data that can be integrated into the management of the project. The Internet provides the project team more than 56 million information sources in over 150 countries—primarily in the English language.[11]

Companies today have in many instances separated their operational units by thousands of miles to achieve advantages associated with locales. Warehouses are built in Ohio because of its heartland location and data processing offices are set up where skilled farm-belt workers are available. This decentralization is supported by the growing sophistication of telecommunications. Facsimile machines, cellular phones, and toll-free telephone numbers have offset the disadvantage of distance. Computer and telecommunication technologies developed over the past two decades facilitate remote decentralized locations providing instantaneous communications between distant posts. Advanced technology for communications around the world minimizes the number of on-site meetings and hours spent traveling across country. Time that was previously spent traveling can now be devoted to productive tasks that are coordinated electronically over thousands of miles, across multiple time zones.[12]

Telecommunications is one of the industries that is growing rapidly throughout the global community. The $600 billion per year global telecommunications industry is changing from a cartel of monopolies and entrenched suppliers to a global free-for-all. Corporate customers want global telephone networks; the companies that build them want global profits. Developed countries are trying to encourage competition, whereas developing countries see the clear need for improved communications to attract business investments. Some of the developing countries are demanding stringent conditions. For example, the Indonesian government put out project proposals for bids for 350,000 telephone lines of digital switching capacity. The government will not consider a bid unless it could offer a 25-year grace period before any payment became due. These conditions are beyond the capability of any corporation, especially for the financial considerations of a project of this magnitude.[13]

One aircraft manufacturer believes that air travel for business purposes will become less important as more sophisticated communication devices and techniques

---

[11]For a summary description of how the Internet can benefit project management, see "The Internet and the Project Management," *PM Network,* October 1996, pp. 17–26.

[12]Brent Bowers, "Technology Allows Small Concerns to Exploit Distances," *The Wall Street Journal,* October 28, 1991.

[13]Andrew Kupfer, "Ma Bell and Seven Babies Go Global," *Fortune,* November 4, 1991, pp. 118–128.

become available. This company believes that the people on business travel from one location to another carry within themselves a bundle of information that will be transferred to other person(s) at the end of the journey. Once the information is transferred, the traveler gets on the returning airplane and returns to the home office. Through modern telecommunications the bundle of information can be transmitted quickly without time-consuming air travel for humans.

12.11 CHALLENGES TO AN EFFECTIVE PMIS

There are real challenges in developing and implementing a PMIS. These problems must be avoided or overcome if an effective PMIS is to be developed. Gilbreath cites uses and abuses of information. He opines that misuse of information is common—often sophisticated, and limited only by our imagination.[14] He delineates the acceptable uses and common misuses of information. When correctly used, information helps to:

- Promote understanding among the project team
- Target controls (by quantifying risks, testing proposed controls, and initiating corrective action)
- Dispel project phantoms (artificial failure factors)
- Allow project transactions (such as progress payments)
- Communicate status
- Predict the future
- Satisfy outside inquiries
- Enhance resource usage (efficiencies)
- Validate plans
- Comprehend change
- Sharpen and reinforce perspectives
- Test expectations
- Recognize failure

Information is often misused, in order to:

- Deceive or confuse
- Postpone a decision or action
- Create errors in the information department
- Justify errors
- Slow or divert processes
- Support the status quo
- Mask failure (or dress it up)[15]

---

[14] Gilbreath, op. cit., p. 152.
[15] Ibid., pp. 152–153.

A PMIS can fail to support the project through a host of reasons. Perhaps the most important is design of the system for the required information. The system must be capable of receiving the right information for the project and can be easily used to retrieve information in a timely manner. Too complex an arrangement for either inputting information into the system or retrieving information will frustrate many users—who will not take the time or make the effort to use the PMIS.

Second, there is a need to have confidence in the information stored in the PMIS. Users of the information will avoid the PMIS if the data are aged or inaccurate. Information in the system can easily fall into disuse when the project team loses confidence in the system to provide usable information in support of their work. Either poor initial population of the system or failure to maintain the system through timely updates with new project information can create this situation.

More information in a PMIS is not necessarily good. Overpopulating the system with irrelevant information or information that masks the *vital few* can lead to errors in decisions or avoidance of the system. When there is too much information to be sorted and reported, this can overwhelm the decision maker. Decision makers need the right amount of information to analyze and distill into a decision that supports the project. Populating the system with erroneous information can create conflicts between sources that waste time and efforts to sort though opposing information.

Some studies and observations of decision-making theory have concluded that one should have about 80 percent of the information to make the best decision. In some instances when there is a need for immediate action, there may be only about 40 percent of the information available—still the decision must be made without regard to the shortfall of information if a disaster is to be averted. It is accepted that a person will never have 100 percent of the pertinent information and many times will not have even 80 percent. Delays in decision making for lack of information may be worse than making a decision on inadequate information. To offset this situation, it may be well to anticipate the type of decisions that will be required and the supporting information needed. The PMIS should be designed to receive, store, sort, and retrieve the required information.

Project managers may not want to have a fully capable PMIS. Because of their personality and decision-making style, there is often reluctance by some project managers to be *burdened* with facts that do not permit the use of subjective, intuitive information. The management style exhibited by these individuals may include the following:

- Only my experience counts in decisions.
- Don't tie me to decisions based on partial information; let me use my judgment.
- A PMIS would only confirm my decisions—if it had the right information.
- The facts in the PMIS are wrong; I can see what is happening.

Often, we forget that a system includes people with all their good intentions and differences. A computer cannot think and recognize situations as easily as a skilled person observing the parameters of the system. Yet people function in various modes—some good and some not so good. For example, who will give the bad news to the boss when *the boss shoots the messenger?* In this situation, people represent weakness in the system—the person who fails to report the bad news and the boss who blames the messenger for the bad news. It is up to the project manager to ensure that the people side of the system works—technology will not replace those people in the system.

Gilbreath believes that management reports are only as good as the information they contain that promotes analysis and evaluation of the project. The best reports, according to Gilbreath, manage to:

- Isolate significant variances and identify the reasons they occurred
- Emphasize the quantitative and specific rather than the subjective and general
- Describe specific cost, schedule, and technical performance impacts on other project elements (other contracts, areas, schedules, organizations, plans)
- Indicate effects on project baselines (what revisions are needed, when, why)
- Describe specific corrective actions taken and planned
- Assign responsibility for action and give expected dates for improvement
- Reference corrective action plans in previous reports (what happened)[16]

12.12 PMIS HARDWARE AND SOFTWARE

Computer-based information systems have become valuable for project managers to use in managing projects. In the 1970s and into the 1990s, project managers could use a computing capability through a substantial infrastructure. Today, the desktop computer has led to a flood of project management software packages. Archibald separates computer-based software for project management into three categories: scheduling, cost and resource control, and cost/schedule/integration and reporting.[17]

Although much of the computer software is used for project scheduling and tracking, with capability to load resources and generate budgets, there is still a need for databases that can accommodate large volumes of data. These scheduling software programs must be supplemented with a means of electronically storing information in a structure that can be rapidly retrieved for use in the project.

Desktop computers currently have storage capacities of several gigabytes that can accept large volumes of data. Organizations are using an intranet—the internal network linking desktop computers with a mainframe computer—for distribution of information. The data may be stored on the mainframe computer and accessed at any time. Project-specific data are most often maintained in the desktop computer and a copy filed on the mainframe computer.

Figure 12.4 shows a typical configuration for linking desktop computers with a central computer that stores pertinent project and organizational data. Information is available to all project team members any time of day and the electronic linkage

---

[16] Ibid., p. 160.

[17] Russell D. Archibald, *Managing High Technology Programs and Projects* (New York: Wiley, 1976), pp. 204–210.

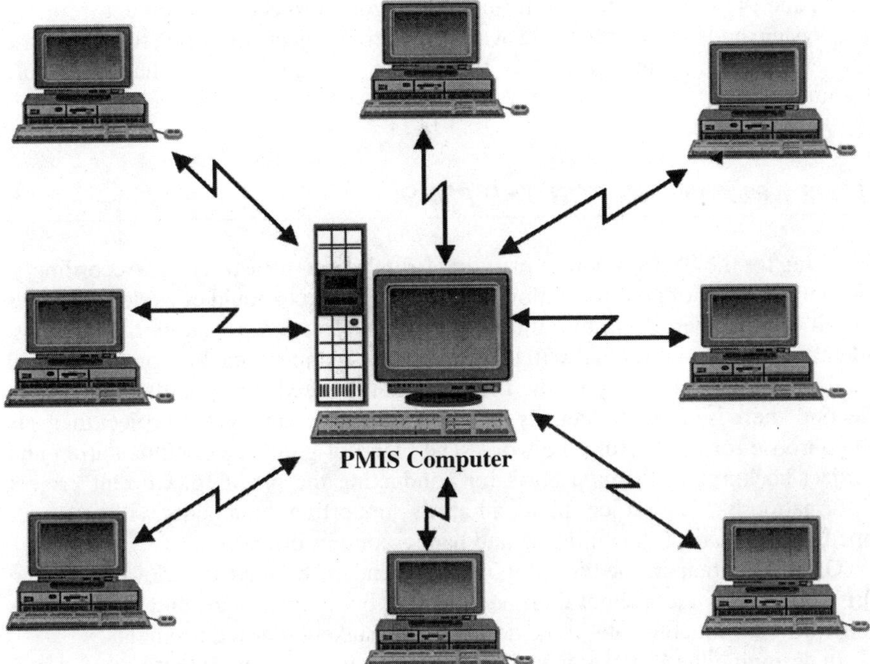

FIGURE 12.4 PMIS with project team member linkages.

permits rapid access to any part of the database. Moreover, it permits any team member to contribute to the store of knowledge around the clock without regard to that person's location.

The power of the computer has made information sharing easy and rapid. Many laptop computers are used by project participants to store data while moving between locations. Typically, the laptop computers can be linked to the central mainframe computer for an interchange of information.

The power of the computer is harnessed only when there is a single person or group involved in sustaining the database. This may be a person on the project or it may be a project office. The individuals responsible for the mainframe computer do not populate data or manipulate the data, but will most often provide the framework for storing, accessing, backing up, transferring, and deleting the data.

Gilbreath offers advice on how to consider the use of project management software by noting that we need to understand our information needs and how we intend to organize and use the information. He notes that software failures often occur because the software does not match needs, organization, and intended uses. He concludes that poorly planned or performed projects are not helped by software.[18]

A good information system provides key input to the project decision makers. Projects that get into trouble often are found to lack information, or they have too

[18]Gilbreath, op. cit., p. 160.

much and the wrong kind of information. A good project management information system adds value to the data available on the project, and when those data are properly organized and structured, the project management team has a valuable resource to use.

12.13 PLANNING FOR THE PMIS

Planning for the PMIS is part of planning for a project infrastructure. Accordingly, the development of an information system for a project should consider the needs of all stakeholders and the timeliness of information to support decisions. Identifying information that will support project planning, such as organizational guidance documents, may be the first information needed to populate the PMIS. Second, there is a need to identify background information on the project that sets the purpose for conducting the work. The blend of guidance documentation and project background forms a basis for conducting the planning. Current project information like the project plan and all its supporting documents will be developed on the basis of the guidance and background information.

One may anticipate the type of information and the timing for retrieval by asking different stakeholders about their needs. Also, ongoing projects and past projects may provide valuable data in the design of the stakeholder requirements.

In designing the PMIS, it is well to remember that all required information will not be stored in the PMIS. Some information is readily available through other systems such as the enterprise's management information system. Other information may be needed, but the PMIS hardware and software may limit its storage and retrieval in an orderly and efficient fashion. Knowing where there are limitations in the PMIS will cause the project team to identify other sources—if the information is needed.

How much information is needed in the PMIS? To paraphrase an Army general in 1974, the project manager might say, "I need that amount of information to do my job and to keep my stakeholders informed." Using this criterion, it is easy to see that each project will have a unique PMIS. Many parts and functions will have common ground with similar types of information—take the guidance information from the enterprise, for example. Background information on a project should, for example, be uniquely fitted to that project.

In the planning context of a PMIS, a number of factors are essential to the establishment of an information system. The parent organization of the project should have in place the following:

- An information clearinghouse function, particularly in the design and execution of projects to support corporate strategy
- An established organizational design with supporting policies, procedures, techniques, and methodologies to manage the organizational information bases
- Appropriate people who can work at the interfaces between information technology and the project needs

By forcing a systematic delineation of the project work packages, the project work activities, and the integrated information components, the project manager can analyze the best way to translate the information to a format that produces useful information. Within an information center, project information networks are a medium for organizing the structure and continuity of information over the life of the project. These networks provide an integrated perspective of the project work packages and their interrelationships. The networks provide methodology to identify work packages, information requirements and sources, information flows, and decision parameters.[19]

A project plan may be considered to be an information system, which provides a time-phased array of work packages, appropriately sequenced to the WBS with resource estimates, to accomplish the project plan's scope of work within an appropriate time frame. This baseline project plan eases monitoring and analysis by showing the information needed to measure against the proper control points. As the project is worked and the actual status data become available, the project plan and the networks are updated to provide for the tracking and monitoring process.

12.14 ESSENTIAL ELEMENTS OF A PMIS

In the design, development, and operation of a project management information system, a few essential elements can be applied:

- Information is needed to manage the project—to plan, organize, evaluate, and control the use of resources on the project.
- Information is needed to satisfy stakeholder queries about the project's status and progress.
- The quality of management decisions in the project is related to the accuracy, currency, and reliability of the information on the project.
- Enterprise guidance and project background information form the basis for planning the project. This information should be a part of the PMIS.
- The information requirements for all stakeholders drive the design and development of the PMIS's contents. The project manager and project team will be the primary users of the PMIS, but will need to consider stakeholders such as senior management, customers, and functional managers.
- The PMIS supports the full range of the project life cycle to include preproject analysis and postproject reviews.
- Information establishes the basis for all project decisions and commitment of resources. The PMIS is the repository of much of this information.
- Information to manage a project comes from a wide variety of sources, including formal reports, informal sources, observation, project review meetings, and

[19]Paraphrased from M. D. Matthews, "Networking and Information Management: Its Use by the Project Planning Function," *Information and Management,* vol. 10, no. 1, January 1986, pp. 1–9.

questioning—aided by formal evaluation and analysis as to what the information says about the status of the project.

- Information systems reflect the user's needs for making and executing decisions in the managing of project resources.
- The PMIS should interface with larger organizational information systems to permit smooth, efficient interchange of information in support of organizational and project objectives and goals.
- Planning for a PMIS requires that information be selectively included and irrelevant information omitted to preclude an overabundance of data and little relevant information.
- The PMIS should be prospective and capable of providing intelligence on both the current and probable future progress and status of the project.

Each PMIS is tailored to project situations to meet specific requirements for managing the project. General characteristics that should be in all PMISs would include the following:

- Be adaptable to differing customer requirements.
- Be consistent with organizational and project policies, procedures, and guidelines.
- Minimize the chances that managers will be surprised by project developments.
- Provide essential information on the cost-time-performance parameters of a project and on the interrelationships of these parameters, as well as the strategic fit of the project.
- Provide information in standardized form to enhance its usefulness to all managers.
- Be decision oriented, in that information reported should be focused toward the decisions required of the managers.
- Be exception oriented, in that it focuses the manager's attention on those critical areas requiring attention rather than simply reporting on all areas and requiring the managers to devote attention to each.
- Be a collaborative effort between users and analysts.
- Be executed by a multidisciplinary team that views the design, development, and implementation of the information system as a project itself, amenable to project management approaches.

12.15 TO SUMMARIZE

The major points that were presented in this chapter include:

- The project management information subsystem is an important subsystem of a project management system.
- Relevant and timely information is essential to the management of an enterprise and to the management of a project.

- Plans, policies, and procedures are really a repository of information providing guidance on how the enterprise or project will be managed.
- Several examples were given of where the failure on large projects was caused in part by inadequate and inappropriate information systems.
- There is a direct relationship between quality and quantity of information and project planning and control results.
- When an adequate information system exists for a project, management of that project is made much easier with a higher probability of success.
- In this chapter, an important number of key questions were suggested that a project manager could ask about the quality and quantity of information available for the management of the project.
- Project and project-related information should be considered a valuable resource—and easily accessible to the project manager.
- Computers, telecommunications technology, and the Internet are revolutionizing the availability and use of information as a key element in the management of an enterprise and projects.
- Information available through the Internet is a major information source for the project team and for other stakeholders.
- Hardware and software for a project management information system permit round-the-clock access to the project's store of knowledge.
- A comprehensive set of principles for a project management information system was presented. Following these principles should enhance the design and use of suitable project management information.
- The availability of timely and relevant information is critical to carrying out the monitoring, evaluation, and control of projects—and is particularly valuable during project progress review meetings. Availability of information permits decisions based on facts.

12.16 ADDITIONAL SOURCES OF INFORMATION

The following additional sources of project management information may be used to complement this chapter's topic material. This material complements and expands on various concepts, practices, and theory of project management as it relates to areas covered here.

- Daniel F. Green, "What to Include in Project Management Information Systems," Harvey A. Levine, "Selecting the Best Project Management Software," and James J. O'Brien, "Calculating Costs and Keeping Records for Project Contracts," chaps. 24, 25, and 10 in David I. Cleland (ed.), *Field Guide to Project Management* (New York: Van Nostrand Reinhold, 1997).

- Brant Rogers, "Food Waste Composting at Larry's Markets," and Julie M. Wilson, "R&D in the Insurance Industry: PM Makes the Difference," in David I. Cleland, Karen M. Bursic, Richard J. Puerzer, and Alberto Y. Vlasak, *Project Management Casebook,* Project Management Institute (PMI). (Originally published in *PM Network,* February 1995, pp. 32–33; and *Proceedings,* PMI Seminar/Symposium, Pittsburgh, 1992, pp. 223–231.)
- Amrit Tiwana, *The Knowledge Management Toolkit: Practical Techniques for Building a Knowledge Management System* (Upper Saddle River, N.J.: Prentice Hall PTR, 2000). This book is a guide to building knowledge (information) management systems that are aligned with an organization's strategic goals. The author addresses the value of knowledge, how to categorize knowledge, and how to manage it for the most effective organizational results. The book contains practical tips and techniques for building a knowledge management system that easily translate to building a project management information system.
- Stephen A. Devaux, *Total Project Control: A Manager's Guide to Integrated Project Planning, Measuring, and Tracking* (New York: Wiley, 1999). This book addresses the need to manage all aspects of a project, to include data. The author deviates from traditional project management methods while focusing on the three parameters of project: cost, schedule, and work scope. Using "value concepts," the author objectively balances complex, multiproject resources.
- Samuel W. McDowell, "Just-In-Time Project Management, *IIE Solutions,* Norcross, Ga., April 2001, pp. 30–33. This article describes projects that have a need for immediate, accurate, and full information in several knowledge areas. Project planning, control, and management are viewed with a critical eye to define the information needs. The author uses specific examples of projects to place the issues in context.
- Rebecca Somers, "Can I get That In Writing?" *Geospatial Solutions,* Duluth, Minn., April 2001, pp. 20–23. This article addresses the challenges of developing policies and procedures for management and administrative matters. These polices and procedures communication documents prescribe the means to get things done. A PMIS may host many project management documents that describe practices. This article highlights the role of data management and data distribution—a critical part of any PMIS.
- Laurent Dubernais, "Yesterday's Lessons, Today's Advanced Tools, Tomorrow's Business Success," *Buildings,* Cedar Rapids Iowa, June 2001, pp. 96–97. This article examines practices in the construction industry where data management has moved from manual, hand-written schedules to the powerful computer. The author asserts that this move to computers streamlines the project management process for faster project completion. Examples are given for the type of data managed through the use of computers and for the use of the Internet.

12.17 DISCUSSION QUESTIONS

1. What is the importance of the information subsystem to the project management system (PMS)? What is the relationship of this subsystem to the other subsystems of the PMS?
2. What is the difference between data and information?
3. List and discuss some of the essential elements of a project management information system (PMIS). What is the purpose of a PMIS?
4. What gives information value? How can information be used as a resource for project stakeholders?
5. What are some of the challenges in developing a PMIS?
6. Discuss some of the uses and misuses of project information.
7. Describe a project management situation from your work or school experience. How was information managed? Was it an effective resource in project control?
8. PMIS software can be essential to managing projects. What characteristics of software must project managers assess in order to determine appropriateness?
9. Discuss the essential factors in the establishment of an information system.
10. What is the difference in a PMIS for a small project, say 3 months' duration, and a large project, say 2 years' duration? What design/content difference might there be between the two PMISs?
11. List and discuss some of the principles of PMIS.
12. Why would you want to have background documents, such as the contract, the project management manual, and organizational polices and procedures, in the PMIS? What is the significance of having these documents in electronic form on a PMIS network?

12.18 USER CHECKLIST

1. Does your organization have an effective PMIS for each of its projects? Explain.
2. Do project team members understand the difference between data and information? Do the measurement and reporting systems that they use generate data or information? Explain.
3. Describe the flow of information of projects within your organization. What information is effectively communicated? What information is often lacking?
4. Does the information system used in projects within your organization contain all the essential elements of a PMIS as described in the chapter? Why or why not?

5. Think about some of the recent projects completed by your organization. Was information effectively managed? Did the information system contribute to the success of the project?
6. Do the project managers in your organization understand the purpose of a PMIS? Explain.
7. Do the project managers in your organization understand the value of information? Is information used effectively as a tool in controlling projects?
8. What problems has your organization had with its information systems? How can these problems be managed on future projects?
9. Do you believe that the information is used for legitimate purposes within your organization, that is, to promote the enterprise versus protect incompetent or wrong behavior? Explain.
10. What PMIS hardware and software comprise the system in your organization? How? Is the system capable of meeting your requirements? What alternatives exist?
11. Have the managers of your organization taken the time to assess the effectiveness of the information systems within your organization? Explain.
12. Is there sufficient information stored within your organization's PMIS to facilitate the smooth and efficient support of projects, that is, provide accurate, timely information to facilitate answering most questions? What is missing?

12.19 PRINCIPLES OF PROJECT MANAGEMENT

1. Project management requires good information management to give the greatest chance of success.
2. Project data are unanalyzed information; information is analyzed, formatted, and distilled data.
3. A PMIS must be tailored to fit the projects being served for the best results.
4. A PMIS is an essential part of the communication process for a project and must serve all stakeholder needs.
5. Electronic database PMISs are the trend of the future.

12.20 PROJECT MANAGEMENT SITUATION—PRESCRIBING A PMIS

Betac Corporation has just been awarded a contract to build 550 metal detection devices for airports in a foreign country. Betac has experience in metal detection devices, but has not kept up with the technology over the past 5 years. It is known that many advances have been made in detecting various quantities of metal that pass through airport entrances. Betac is an experienced user of project management to meet its business commitments.

The strategy for building the 550 devices is to construct and test three prototypes and then build the production models. Production models will be shipped from the United States directly to the airports, where they are to be installed. All production will be accomplished in the United States and installation will be accomplished by a joint team of U.S. personnel and indigenous personnel at each site.

Major concerns follow:

- What is the latest technology and digital processing for metal detection devices?
- How long will it take to build and successfully test the three prototype models?
- When can production be started on the final design models?
- What is the sequence of installation for the various sites?
- What training is required for operation by indigenous personnel?
- What are the contractual requirements for scheduled installation, training of operators, and training for diagnostic procedures?
- Who is available to travel to the installation sites and what skills are required?

This project requires a lot of coordination and keeping people informed of progress. A PMIS is needed that hosts all the information and includes daily updates on schedules. The project team will have access to the mainframe computer through their respective desktop computers. The installation team will have three laptop computers that connect to the host computer by telephone line and modem.

Senior management has committed to building a PMIS specifically for this project and stated that the project manager could construct any database of information required to assure success with this project.

12.21 STUDENT/READER ASSIGNMENT

You are a part of the project planning effort and have been specifically tasked with determining what information should be in the PMIS host computer.

1. What technology information will be in the PMIS and why do you believe there should be technology information, if any?
2. What cost information should be in the PMIS and what would it be used for during the project?
3. What schedule information should be in the PMIS and who should be responsible for populating the system and maintaining the information?
4. What type of feedback would you expect from the installation team and what should be entered into the PMIS—for immediate use, for post-project use?
5. How much of the PMIS information will be of interest to individuals external to the project team? What uses would these external people have for PMIS information?

CHAPTER 13

PROJECT MONITORING, EVALUATION, AND CONTROL

"I claim not to have controlled events, but confess plainly that events have controlled me."
ABRAHAM LINCOLN, 1809–1865

13.1 INTRODUCTION

The unexamined project is not worth much. No matter how perfect the plan, without regular reviews during the life of the project neither the project progress nor the realty of the plan can be assessed. Control is one of the key functions of the management process. To ask the question, How are we doing today on this project? cannot be answered without an effective and efficient control system in place. How often should a project be reviewed? This depends on many variables; the size of the project, its complexity and priority, and how it fits into the strategic intent of the enterprise are all-important considerations.

This chapter presents a model of the generic steps in the control cycle, and describes these steps within a project context. How to evaluate the adequacy of the project progression and of how the project management functions are being carried out helps give insight into which actual or potential results that the project is likely to provide are worth examining. The evaluation of the project from a perspective of project planning, project organization, project accomplishments, and project information is suggested. What should be examined about the project and on what time intervals? The chapter closes with a recommendation on who should monitor the project evaluation processes during the life of the project.

13.2 PROJECT CONTROL CYCLE

The final management function carried out by the project team is control, which is discussed in this chapter. Control is the process of monitoring, evaluating, and comparing planned results with actual results to determine the progress toward the project cost, schedule, and technical performance objectives, as well as the project's "strategic fit" with enterprise purposes. The management function of control may be visualized as distinct steps in a control cycle model, as portrayed in Fig. 13.1. Monitoring and control are universal activities indispensable to effective and efficient operation of the control cycle.

Fayol noted that "to control means seeing that everything occurs in conformity with established rule and expressed command."[1] Control is a fact-finding and remedial action process to facilitate meeting the project purposes. Its primary purpose is not to determine what has happened (although this is important information), but rather to predict what may happen in the future if present conditions continue and if there are no changes in the management of the project. This enables the project manager to manage the project in compliance with the plan. The basis of effective project monitoring, evaluation, and control is an explicit statement of the project objectives, goals, and strategies which provide performance standards against which project progress can be evaluated. The following concepts and philosophies are essential elements for the assessment of project results:

- The objective is to develop measurements of project trends and results through information arising out of the management of the project work breakdown structure.
- Performance measurements are always to be tempered by the judgment of the managers and professionals doing the measurement.
- The use of common measurement factors arises out of the status of project work packages consistent with the organizational decentralization of the project.
- Measurements should be kept to a minimum relevant to each work package in the project work breakdown structure.
- Measurements of work packages must be integrated into measurement of the project as a whole.

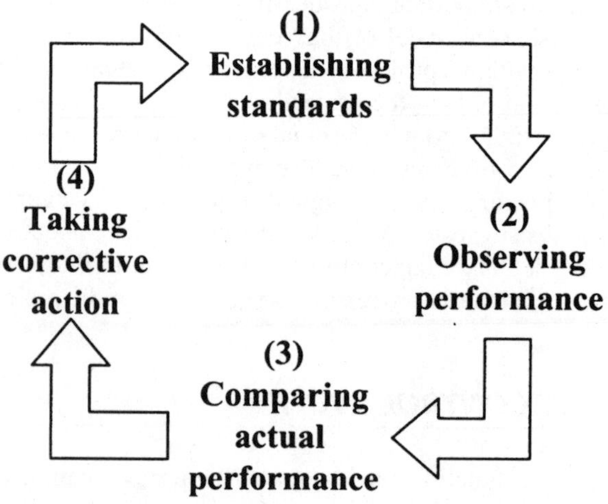

FIGURE 13.1 The control system.

[1] Henry Fayol, *General and Industrial Management* (London: Pitman, 1949), p. 6.

- Measurements should be developed that are applicable to both current project results and future projections to project completion.
- Measurement should be conducted around previously planned key result areas.

In the material that follows, more information will be given about each of the steps in the control cycle.

13.3 STEPS IN THE CONTROL CYCLE

There are several distinct steps in the control cycle, depicted in Fig. 13.1. These steps in one sense are independent. In another, they are interdependent in the execution of project control. We'll consider performance standards first.

Performance Standards

Project performance standards are based on the project plan, including at minimum the expectations for the project, established in the project objectives, goals, and strategies, relative to project cost, schedule, technical specifications, and strategic fit. Some key standards in project control include the following:

- Scope of work
- Project specification
- Work breakdown structure
- Work packages
- Cost estimates and budgets
- Master and supporting schedules
- Financial forecasts and funding plans
- Quality
- Project owner satisfaction (strategic fit)
- Project team satisfaction
- Senior management satisfaction
- Stakeholder satisfaction
- Reliability
- Physical quantities of work
- Vendor/contractor performance
- Project management
- Innovation
- Resource utilization
- Productivity

A project should be evaluated by using several additional key standards:

- Effectiveness and efficiency in the use of the enterprise resources supporting the project. Were the right resources used in the most productive fashion to support the project?
- Expected technical performance quality of the product or service resulting from the project. Does the project promise to provide value to the customer?
- Development cycle time. Is the project being developed in sufficient time to meet or preempt competition?
- Strategic fit in organizational purposes. Do the project results complement existing products and services being provided in the marketplace?
- It is important to recognize that performance standards are derivatives of project planning, as well as organizational planning, keynoting again the basic (but often forgotten) principle that proper planning facilitates proper control.

Performance Observation

Project performance must be sensed and that is where performance observation comes into play. Performance observation is the receipt of sufficient information about the project to make an intelligent comparison of planned and actual performance. Information on project performance can come from many sources, both formal and informal. Formal sources include reports, briefings, participation in review meetings, letters, memoranda, and audit reports. Informal sources include casual conversations, observations, and listening to the inevitable rumors and gossip that exist within the project team and in other parts of the organization. Talking and listening to the project stakeholders can be a useful source of information on the project's status. Informal meetings for lunch or coffee breaks can help provide the total information "system" the project manager needs to have to know fully what is going on. Both formal and informal information sources are needed to keep up with the project's status. Feedback during performance observation consists of relevant data on the result of the project management process and provides the basis for making a judgment on performance through doing comparative analysis.

The process of performance observation can take many forms, such as the following:

- Regular receipt of formal information reports on the project's performance, obtained through formal project reviews, along with other supporting formal information.
- Walking the project, sometimes called "kicking the tires," and observing what people are doing as they work on the project in their various capacities.
- Conversations with people, especially with the project manager taking the lead, to ask the project stakeholders on a regular basis about "how things are going on the project."

- From regular formal project meetings where progress is discussed on the project to the identification of remedial action that is required and who will take the responsibility for following up on the correction of the perceived deficiencies.
- Informal sources talking and listening to project team and other stakeholders on a regular basis to seek their assessment of "how the project is going."
- Briefings about the project concerning specific problems and opportunities that have arisen or are expected to arise concerning the status of the evolving strategy of the project.
- Maintaining a concentrated effort to listen and listen and listen to what project stakeholders are saying about the project.

Comparing Planned and Actual Performance

Comparing planned and actual performance based on the desired project standards gives the opportunity to get answers to three key questions about the project:

- How is the project going?
- If there are deviations from the project plan, what caused these deviations?
- What should be done about these deviations?

Assessment of the project's status is an ongoing responsibility of the project team and the senior managers. Information obtained by performance observation is compared with the performance standards laid down in the project plan and when analyzed, forms the basis for reaching a judgment about the project's status and whether corrective action is required.

Corrective Action

Corrective action can take the form of replanning, reprogramming, or reallocating resources, or changing the way the project is managed and organized. The corrective actions that are available to the project manager center on the cost, schedule, and technical performance parameters of the project. The project owner may have finalized one or more of these parameters. Correcting a problem with one of the parameters of the project may have reverberations on one or both of the other parameters. Such potential reverberations should be considered by the project team when the alternatives for corrective action are being studied.

13.4 MONITORING AND EVALUATION

Monitoring and evaluation to collect information are integral to control, as depicted in Table 13.1, and are key companions of the control function. The questions listed in Table 13.1 are intended to stimulate discussions and generate information about the project. Monitoring means to keep track of and to check systematically all project activities. This enables the evaluation, an examination and appraisal of how things are going on the project. As a direct link between planning and control, the monitoring and evaluating functions provide the intelligence for the members of the project team to make informed decisions about the project performance. Monitoring should be designed so that it addresses every level of management requiring information about project performance and reflects the work breakdown structure of the project. Each level of management should receive the information it needs to make decisions about the project. In addition, monitoring should be consistent with the logic of the planning, organizing, directing, and motivating systems on the project.

Monitoring means to make sure sufficient intelligence is gained on the status of the project so that an accurate and timely evaluation of the project can be conducted. Several issues have to be addressed by the project team in considering their monitoring and evaluation responsibilities:

- What should be monitored and evaluated.
- What monitoring tools should be used.
- When the project should be monitored and evaluated.
- Who should monitor and evaluate.
- Where the monitoring and evaluation should be carried out.

TABLE 13.1 Key Project Control Questions

1. Where is the project with respect to schedule, cost, and technical performance objectives and goals?
2. What is going right (according to plan) on the project?
3. What is going wrong on the project?
4. What problems are emerging?
5. What opportunities are emerging?
6. Does the project continue to have a strategic fit with the enterprise's mission?
7. Is there anything that should be done that is not being done?
8. Are the project stakeholders comfortable with the progress of the project?
9. How is the project customer image—is the customer happy with the project's progress?
10. Has an independent project evaluation been conducted?
11. Is the project being managed on a total "project management systems" basis?
12. Is the project team an effective organization for the project's work?
13. Does the project take advantage of the enterprise's strength?
14. Does the project avoid a dependence on the weakness of the organization?
15. Is the project making money for the organization?

All activities of the project and its stakeholder environment should be monitored and evaluated, of course, done on an exception basis through the delegated responsibilities of the project team. A framework for doing the evaluation can consist of a series of key questions about the project, which must be answered on an ongoing basis. If the project team can ask questions and get timely, credible answers, then the chances of knowing the project's true status are enhanced considerably.

Questions of the type shown in Table 13.1 can be used during regularly scheduled project review meetings to motivate discussions among the project team members and to encourage them to think retrospectively as well as prospectively about the project. Such thinking will prompt the team members to evaluate the project. Project review meetings should be held regularly by the project owner, senior managers of the project organization, the project team, the work package managers, and the project professionals.

A key question in reviewing any project is the degree of success that the project management team has had in the development of an integrated project management system for the project. Project evaluation, to be effective, must look both at the efficacy of the parts of the project (the subsystems) and at the project totality, expressed in such factors as attainment of the project's technical performance objective, completion schedules, and final cost. Monitoring and evaluation of the project require that the project team look inward to the project and the sponsoring organization as well as outward to the stakeholders and the general "system" environment in which the project is found.

The avoidance of cost and schedule overruns should be one of the key outcomes of any project control system. Following are examples of how projects were able to stay within budget or exceed budget:

- The $630 million upper-atmosphere research satellite is one of two large U.S. space projects that was kept on cost and schedule in part because the project team combined political savvy with technical conservatism to guard the project from controversy and keep it moving in the right direction. To cut costs and improve reliability, the spacecraft was designed by using technology that had been used before, such as plug-in modules for propulsion, communications, and navigation. By keeping a low profile, the project proceeded without controversy. In addition, much of the success of this project can be attributed to a good plan, which is always an important factor in controlling the use of resources on a project and in determining the success of a project.[2]
- Another project, the Earth Observing System (EOS), an environmental satellite project, was, by some estimates, $13 billion above its original cost projections and 5 years behind schedule. Its managers overestimated their political support and underestimated the technical challenge of the project. This project became mired in controversy from the start; the space agency proposed to build six of the largest, most complex satellites ever conceived for EOS and to back them up with one of the world's most sophisticated computer systems. The project was

[2]Bob Davis, "A NASA Satellite Project Accomplishes Incredible Feat: Staying within Budget," *The Wall Street Journal,* September 9, 1991.

taking so much money that lawmakers and scientists feared it would take away funds from other projects considered to be more worthy. Although the White House and Congress approved the start of the project, both backed off as the project's risk level became apparent.[3]

13.5 MANAGEMENT FUNCTIONS EVALUATION

You can use management-related activities to address representative key questions to evaluate the project. Assuming that a project management functions viewpoint is used as a baseline for evaluating a project, what should be measured?

Answers to questions such as the following can give insight into what should be measured and how well the project is doing.

Project Planning

- Are the original objectives and goals realistic?
- Is the plan for the availability of project resources adequate?
- Are the original project schedule and budget realistic?
- Is the plan for the organization of the project resources adequate?
- Are there adequate project control systems?
- Is there an information system for the project?
- Were key project stakeholders brought into project planning?
- Was facility planning adequate?
- Was planning completed before the project was initiated?
- Were potential users involved early in the planning process?
- Was there adequate planning for the use of such management tools as project control networks (CPM/PERT), project or study selection techniques, and information systems?

Project Organization

- How effective is the current organizational structure in meeting the project objective?
- Does the project manager have adequate authority?
- Is the organization of the project office staff suitable?

[3] Ibid.

- Have the interfaces in the matrix organization been adequately defined?
- Do key project stakeholders understand the organization of the project office?
- Have key roles been defined in the project?

Project Management Process

- Does the project manager adequately control project funds?
- Are the project team personnel innovative and creative by suggesting project management improvements?
- Does the project manager maintain adequate management of the project team?
- Do the project team people get together on a regular basis to see how things are going?
- Does the project office have an efficient method for handling engineering change requests?
- Does the project staff seek the advice of stakeholders on matters of mutual concern?
- Have the project review meetings been useful?

Project Accomplishments

- To what extent have the original project goals been achieved?
- How valuable are the technical achievements?
- How useful are the organizational and/or management achievements?
- Are the project results useful in accomplishing organizational objectives?
- Are the results being implemented?
- Are the users being notified properly?
- Is the customer happy with the project results to date?

Effective project control can be carried out only if there is adequate information about the project that can be used for monitoring and control.

Project Information

The project team requires a project control system that provides key information on the status of the project. It needs several key systems to provide such information:

- An equipment, labor, and material information system that provides the basis for the effective and efficient utilization of the work force on the project. These cost factors are usually the large contributors to the overall project cost. Their status should be known to the project team and the owner.

- A cost control system so that the project team can determine whether the costs are in line with the project plan and to help understand deviations that may occur.
- A schedule control system to identify schedule problems so that cost-effective trade-offs can be carried out as needed.
- A budget/financial planning/commitment approval system so that the data on the commitment, expenses, and cash flow of the project can be collected and analyzed and appropriate remedial action taken.
- A work authorization system that provides for the allocation of project funds to the functional organizations and outside vendors.
- A method of using the collective judgment of team members to judge the progress being made to satisfy the project's technical performance objective. To reach this judgment, the progress of the individual work packages must be assessed along with the progress on the integration of all work packages. This judgment can best be made by the project team in a group session by reviewing and assessing all the information the team has assembled and then reaching an informed judgment of where the project stands, for example, on project costs.

13.6 WHEN TO MONITOR AND EVALUATE

When should you monitor and evaluate the project? The answer to this question is simple and straightforward. Monitor and evaluate the project during its entire life cycle.

For example, the James Bay project had management controls that tied together all the project efforts from conceptual design through contract closeouts. Furthermore, the engineering department of the James Bay Energy Corporation (the project management organization for the James Bay project) conducted quarterly board of consultants meetings to review engineering designs.[4]

Project evaluation is a process that extends throughout the project life cycle through to a "postmortem" that assesses the capability of the project to support organizational strategy in terms of a useful product or service, which supports the organizational mission. There are four major types of project evaluation:

- Preproject evaluation for the selection of a project to determine if it shows promise to the objectives and overall strategy of the organization or enterprise.
- Ongoing project evaluation for measuring the status of a project during its life cycle.
- Project completion evaluation for an immediate assessment of success upon project completion.
- Postproject evaluation for a down-the-road assessment of project success after the dust and confusion have settled.

[4]Peter G. Behr, "James Bay Design and Construction Management," *Proceedings of the American Society of Civil Engineers,* vol. 104, no E12, April 1978, p. 146.

13.7 PLANNING FOR MONITORING AND EVALUATION

Clearly, part of project planning should include the development of a strategy on how the project will be evaluated during its life cycle. This planning is just as important as the planning for any other aspect of the project. In approaching the development of project evaluation strategy, there are several key requirements: the inclusion of an evaluation policy and the process, the commitment of all key managers involved in the project to an evaluation strategy for evaluation methodologies, and the use of both inside and independent evaluators who have the professional credentials to do a credible job of the evaluation process.

Not only does a periodic evaluation of the project as a normal and expected responsibility of the project team make good sense, but also evaluations on a periodic basis to examine the rationale and mandate for the project provide benefits. Such evaluations show that the principal managers have a concern for the degree to which the project objectives and goals are being achieved and the identification of any shortcomings in the management of the project. By having the principal managers insist on periodic and special evaluations, an important message is sent throughout the organization. A project owner who has contracted for the engineering, design, and development of the project's product would be foolhardy not to insist on regular and special evaluations of the project's progress. Indeed, a prudent project owner would actively participate in such evaluations.

13.8 WHO MONITORS AND EVALUATES?

The principal responsibility for project monitoring and evaluation rests with the project team and the project owner. The manager who has "general management" or "project owner" jurisdiction also shares in the residual responsibility for keeping informed on what is happening on the project. Where are the monitoring and evaluation carried out? Simply put, as close as possible to the action on the project at the individual professional's level where the work is being done, as well as at:

- The work package level
- The functional manager's level
- The project team level
- The general manager's level
- The project owner's level

Each successive level's monitoring and evaluation are more integrated, dealing with the project's total schedule, cost, and technical performance objectives. Finally, the project's strategic fit in supporting the owner's mission is evaluated.

13.12 PROJECT AUDITS

An important part of project evaluation can be done through the conduct of a project audit. See Section 13.16 for additional discussions on project evaluations. Project audits provide the opportunity to assess how work is being managed. Periodic audits have to be planned. An audit should have as its purposes:

- Determining what is going right, and why
- Determining what is going wrong, and why
- Identifying forces and factors that have prevented or may prevent achievement of cost, schedule, and technical performance goals
- Evaluating the efficacy of existing project management strategy, including organizational support, policies, procedures, practices, techniques, guidelines, action plans, funding patterns, and human and nonhuman resource utilization
- Providing for an exchange of ideas, information, problems, solutions, and strategies with the project team members

A project audit should cover key functions, depending on the nature of the project, in both the technical and nontechnical areas, such as:

- Engineering
- Manufacturing
- Finance and accounting
- Contracts
- Purchasing
- Marketing
- Human resources
- Organization and management
- Quality
- Reliability
- Test and deployment
- Logistics
- Construction

How often should an audit be conducted, given that a thorough audit takes time and money? Generally, audits should be carried out at key points in the life cycle of a project, and at times in those phases of the life cycle that represent go/no-go trigger points, such as preliminary design, final design, first prototype, commitment to production, first use, warranty, and maintenance and service contracts.

An unplanned audit may be called for during the project life cycle if there is a sense by the principal managers that the project is in trouble or is heading for trouble. If there is uncertainty concerning the project's status, or if there has been a change in the strategic direction of the organization, an audit of the project may be in order. When a new project manager takes over a project, she or he should order an audit, both to become familiar with an unbiased view of the project and to come up to speed with the key issues and problems that are facing the project.

A project audit is a formal independent evaluation of the effectiveness with which the project is being managed. A typical audit evaluates the adequacy of the project management system being used, the effectiveness of project planning and implementation, and the adequacy of project guidelines, policies, and procedures. Its purpose is to provide an objective and impartial assessment of the manner in which the project is being managed and the results that are likely to be accomplished through the use of the project resources. The audit participants should ensure that adequate documentation is provided to the audit team and that suitable presentations are given to the team to acquaint them with the status of the project; they should participate in interviews with members of the audit team, and they should work with audit team to develop remedial plans for the full exploitation of the audit team's recommendations. The principal responsibilities of the audit team include:

- Critical review of the project documentation.
- Interview of the project team and other project stakeholders to gain insight into their perceptions of the project affairs.
- Participation in enough of the project activities to gain an appreciation of what is going on regarding the project and insight into the project problems and opportunities.
- Preparation and submission of a final audit report and the debriefing of the project stakeholders on the results of the audit.

13.10 POST-PROJECT REVIEWS

Much can be learned about the efficiency and effectiveness with which projects are managed in the organization through a postproject review (PPR). Postproject reviews are gaining more favor in the management of projects. In the nuclear power plant industry, such reviews have become commonplace to determine which project costs have been incurred reasonably, so the public utility commission can decide which costs can be passed on to the consumers of the nuclear plant's electricity. Other industries are conducting PPRs in the capital budgeting process. One view of this process is depicted in Fig. 13.2.

PPRs take a large view to examine the rationale for the project in the first place. The PPR also examines the strategic fit of the project into the overall organizational strategy. PPRs offer insight into the success or failure of a particular project as well as a composite of lessons learned from a review of all the projects in the organization's portfolio of capital projects. At British Petroleum, PPRs have become an integral part of the corporate planning and control process. British

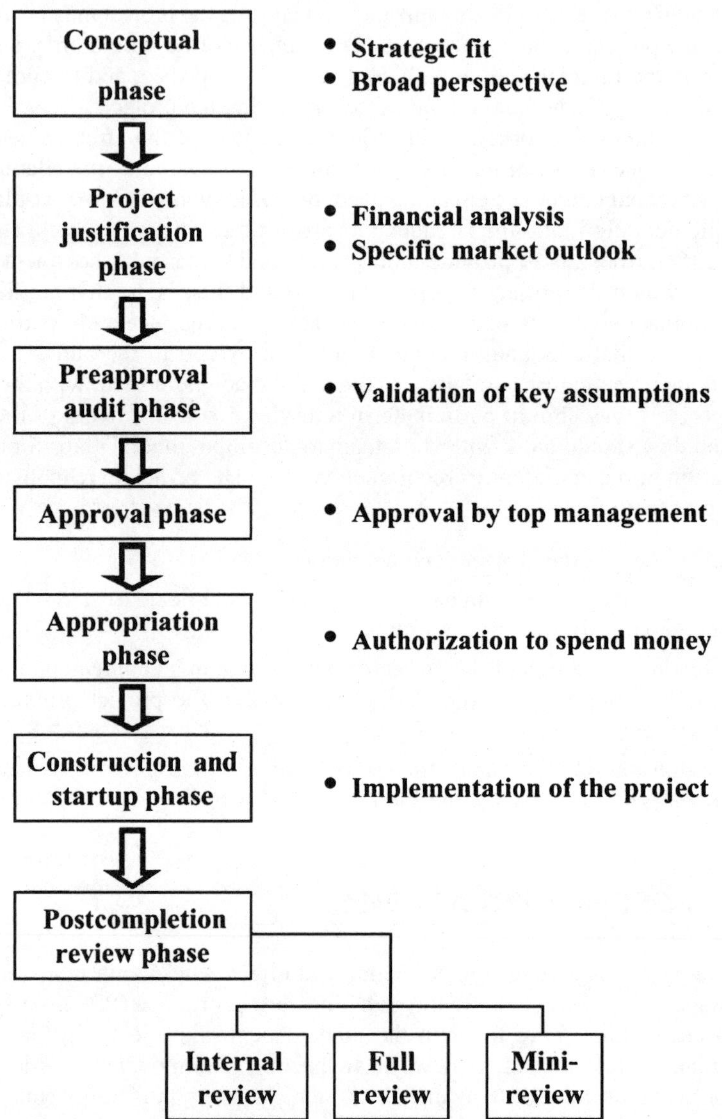

FIGURE 13.2 PPRs in the capital budgeting process. (*Source: Surendra S. Singhvi, "Post-Completion Review for Capital Projects,"* Planning Review, May *1986, p. 37.*)

Petroleum has learned valuable lessons on each capital project, and general principles about project management have emerged over the 10 years that British Petroleum has been conducting such reviews. These principles are as follows:

- Determine costs accurately.
- Anticipate and minimize risk.

- Evaluate contractors more thoroughly.
- Improve project management.[5]

If the project plan contains a specific strategy for the conduct of PPRs, there will be subtle benefits through the discipline and team effort needed to do an adequate review. If the project team members know that the success (or failure) of the project will be evaluated at the project's completion, they should be motivated to do a better job of managing the project during its life cycle. These attitudes will permeate the culture of the organization and improve decision making in the planning process for new projects to support the organizational strategies. This should result in better capital investment decisions and should improve the organization's competitive chances.

13.11 CONFIGURATION MANAGEMENT AND CONTROL

Most people recognize the need to plan and control budgets, schedules, and even performance specifications for a system that is under development. However, in complex projects, the elements that must be controlled are much more varied and detailed than is generally understood.

Configuration control (or engineering change management) represents one such level of specificity. In the development of complex projects, changes in the configuration of hardware and software reverberate through the system, causing problems with budgets, schedules, and so forth. Thus, such changes must be directly addressed and controlled. It is inadequate to attempt to effect control solely at a higher level.

Configuration management is the discipline that integrates the technical and administrative actions of identifying the functional and physical characteristics of a system (or product) during its life cycle. It is directly related to the project technical performance objective. Configuration management provides for control of changes to these characteristics and provides information on the status of engineering or contract change actions.

The essence of configuration management comprises three major areas of effort: configuration identification, configuration status accounting, and configuration control.[6]

Configuration Identification

Configuration identification is the process of establishing and describing an initial system baseline. This baseline is described in technical documentation (proposal

[5]Frank R. Gulliver, "Post-Project Appraisals Pay," *Harvard Business Review,* March–April 1987, pp. 128–132.

[6]Paraphrased from David I. Cleland and William R. King, *Systems Analysis and Project Management,* 3d ed. (New York: McGraw-Hill, 1983), pp. 376–377.

terms, specifications, drawings, etc.). The identification function provides for a systematic determination of all the technical documentation needed to describe the functional and physical characteristics of items designated for configuration management. Configuration identification also ensures that these documents are current, approved, and available for use by the time needed.

The concept of a baseline system requires that the total system requirements and the requirements for each item of the system be defined and documented at designated points in the evolution of the system. An evolutionary life cycle of the system from paper study to inventory items is prepared to plan for development and production status and to permit changes in the scope of the system.

There must be a recognized and documented initial statement of requirements. Once stated, any change in the system's requirements must be documented so that the current status may be fairly judged for performance requirements. A baseline is established when it is necessary to define a formal departure point for control of future changes in performance and design. Configuration at any later time is defined by a baseline model, plus all subsequent changes that have been incorporated. This baseline model provides a point of departure to manage future engineering and/or contract changes.

Configuration Status Accounting

Configuration status accounting is the process of recording and documenting changes to an approved baseline to maintain a continuous record of the status of individual items that make up the system. Configuration status accounting also shows which actions are required and which engineering changes are complete. Status accounting will identify all items of the initially approved configuration and then continually track authorized changes to the baseline.

Configuration Control

Configuration control is the process of maintaining the baseline identification and regulating all changes to that baseline. Configuration control prevents unnecessary or marginal changes while expediting the approval and implementation of those changes that are necessary or that offer significant benefits to the system. Necessary or beneficial changes are typically those that correct inefficiencies, satisfy a change in operation or logistics support requirements, effect substantial life cycle cost savings, or prevent or eliminate slippage in approved scheduling.

A configuration control board is a joint agency of the project clientele that acts on all proposed changes. The configuration control board recommends final decisions on engineering changes and installs good engineering change discipline in the system. This board can provide a single-point authority for coordinating and approving engineering change proposals.

There are two basic potential costs of a contract change such as an engineering change. The first is the direct cost of the change itself, expressed in performing the substantive work of the change, for example, redesign, engineering, and construction/manufacturing. The second is related indirectly to the change order, or the "ripple effect," for example, additional supervision, consequential damages, decrease in productivity during the implementation of the substantive work, and so on.

The approach to configuration management in the context of engineering change management is undergoing change as concurrent engineering techniques continue to be used. In a team charged with the responsibility of planning and executing a concurrent engineering initiative, an ongoing configuration management process is carried out. As the interdisciplinary concurrent engineering team does its work, an ongoing design review continues on a regular basis. Also, because there is a much closer coordination and working relationship between the design and manufacturing disciplines, there are likely to be fewer engineering changes on the project. Most if not all compromises between design engineering and manufacturing engineering have been studied and resolved by the team before the product or service is committed to production.

13.12 PLANNING AND CONTROL IMPLICATIONS FOR PROJECT SUCCESS OR FAILURE

The success or failure of a project is directly related to is objectives and goals. These objectives and goals establish the baseline by which one can measure degrees of success and degrees of failure. Typically, there is not a simple success or failure of a project because successes do not meet all stakeholders' expectations and failures provide some benefits, but perhaps at a cost that is more than expected.

In the initial planning stage, cost, schedule, and technical performance goals and objectives are established. In a stable project, these goals and objectives remain the same through completion. A stable project where there is no migration of the goals and objectives, which includes the initial requirements, is a rare situation. Projects that "grow" in requirements and "discovered" work may be the norm.

From an organizational point of view, a project brings benefits that are determined through the business process for identifying and quantifying jobs. The goals and objectives for the project relate directly to the "business" and its strategic objectives. Changes in the business direction or changes by a client may materially alter the need for a project, its benefits to both the organization and the client, or the number of benefits delivered. Business objectives and project objectives should be in alignment, although they are not necessarily the same.

Criteria for project success are established from the business requirement and relate to the three areas of the project's scope—cost, schedule, and technical performance. An example of this could be a new venture to build an electric car for commuters. Objectives could be:

- *Schedule.* Build and test a prototype electric car in 12 months.
- *Cost.* The budget for the project is not to exceed $1.4 million.
- *Technical performance.* The prototype car will operate for 3 hours between charges, recharge the batteries fully in less than 6 hours from 110-volt AC current, transport 400 pounds of weight to include the operator, accommodate two people (driver and passenger), have a speed of at least 65 mph on the highway, incorporate all safety features for highway operation, and meet all U.S. DOT requirements.

Criteria for success for the business requirement could be:

- Build and sell 2000 electric automobiles for sale in the year 2004.
- Generate total revenues of $20 million in 2004.
- Establish dealership relations for sales and maintenance of the electric cars.
- Double sales for the year 2005.

Success for the project is stable and the project manager can plan and work toward meeting the objectives. Any change to the business requirement may affect the project's objectives. Some examples could be:

- The sales force underestimated the demand for an electric car. The car is needed in 2003 to meet competitors that are building a similar model. The project schedule is negatively impacted and the prototype must be built and tested in 6 months.
- The competition is building a car that will operate on the average for 4.5 hours. The technical specification changes to 5 hours of operation with the same weight capacity and recharge time. This would drive the technology for the project and perhaps look at means of reducing use of electrical energy and improving on the batteries' storage capacity.
- The prototype car does not include the aesthetic features that give the car an image and the marketing department wants the prototype to look like a production model. This will permit smooth transition from the prototype to production with only a little engineering and design change. The project's technical objectives may be significantly changed to require more design work, more engineering, and greater expenses for the "production prototype."

13.13 RESULTS OF PROJECTS—SUCCESS OR FAILURE

Project success or failure appears to be a simple equation: The project met its cost, schedule, and technical performance goals. However, success or failure is relative to the interests of the party viewing the project. Each stakeholder can have interests that do not necessarily agree with another stakeholder. These potentially conflicting views can result in one person calling a project a success and another calling it a failure.

A brief look at criteria that different stakeholders may use in assessing the success of project is helpful to understand the outcome of projects. Some of the stakeholder criteria for success could be:

- *Customer or user.* The product of the project meets my needs and has all the characteristics, features, attributes, and functions required. Success is receiving the full capable product on time and at a reasonable price.
- *Senior corporate management (sponsoring organization).* The project generates a reasonable monetary contribution to the organization (i.e., profit or fee), the product performs to the customer's satisfaction, the project and product add to the organization's capability through improved processes, and the organization's image is enhanced to promote future sales. Success is with project results that create value in the marketplace.
- *Project manager.* The project is completed within budget and schedule while meeting the technical performance criteria, the customer is pleased, and senior management is pleased with the results. Success is measured against the project's objectives of cost, schedule, and technical performance.
- *Project team.* The project is completed and members are rewarded for their contribution to the effort. Success is that each member receives a reasonable compensation for his or her work and grows professionally through added knowledge, skills, and experiences.
- *Second-party stakeholders.* The project and its product do no harm to their interests. Success in this instance is when there is no disruption of these stakeholders' perceived interests.

Generally, project success is measured by the cost, schedule, and technical performance objectives. Success of the project may also be measured through the by-products that contribute value to stakeholders. R&D projects, for example, will often have by-products that can have strategic value to the sponsoring organization. The strategic by-products may be such items as a database of collected information on a phenomenon of nature—take the space program experiments as an example.

Some projects labeled complete failures may have by-products that can be "harvested" for future project work and future business. These benefits may not be stated in a project report, but can have a direct influence on an organization's future capability within a given field or discipline. These by-products may indirectly deliver benefits greater than the project, whether it is considered a success or a failure.

Some by-products that may result from projects are as follows:

- New technical or business purposes that improve project management and operations
- Trade secrets and practices that improve work capacity

- Patents and copyrights (intellectual property) for the developer or customer
- New uses for developed products and services
- Enhanced corporate image for building products
- Individual and team experiences in project management processes
- Corporate knowledge of new technology and manufacturing processes
- Best practices in project management
- Trained work force (project teams, individuals, senior managers)

13.14 EXAMPLES OF PROJECT SUCCESS/FAILURE

Discussion of project success or failure gives a better understanding of project results and the relative nature of defining the outcome. Some examples of project success/failure from literature are helpful toward understanding the current method of classifying projects in discrete *success* or *fail* categories. There are seldom grades or degrees of success when following the project's objectives of cost, schedule, and technical performance.

Thomas and colleagues state at a research conference in June of 2000 that a survey revealed the following:

- More than 30 percent of projects surveyed are canceled before completion.
- More than one-half the completed projects have cost overruns nearly double the original budget.
- More than one-half the completed projects exceed the schedule by more than twice the planned time.
- Lack of senior management commitment in more than 1400 surveyed organizations was a key factor for project failure.[7]

Thomas and her colleagues identify one problem that challenges successful projects: Senior managers do not connect project management's value to the organization's goals and objectives. The use of project management appears to be a response to internal crises rather than a formal strategy to promote business.

A review of the record concerning project success and failure reveals that most projects have problems that result in delays in schedules, cost overruns, and/or inability to attain the desired technical objectives. For example:

- In an assessment of approximately 8400 projects it was determined that about one-third were outright failures, 50 percent were undergoing remedial strategy and would likely overrun their initial cost estimates by almost 200 percent, and

[7]Cited in Janice Thomas, Connie L. Delisle, Kam Jugdev, and Pamela Buckle, "Selling Project Management to Senior Executives: What's the Hook?" *Proceedings of PMI Research Conference,* Houston, June 2000, pp. 431–442.

only 16 percent were done on time, within budget, and meeting their technical performance specifications.[8]

- Another assessment states that a series of studies conducted over the years indicated the presence of an ever-increasing phenomenon of project "failure." The study presented what seemed to be the "big four reasons of project failure," namely, inadequate project definition, lack of general information, poor scheduling and allocation of resources, and loss of control of the project.[9]
- There have been many causes of failure in projects. Perhaps the most general cause—and the most probable reason—is argued to be that "most project failures occur because basic and obvious principles of management are ignored."[10]

The past editor of the *Project Management Journal* provides insight into the general state of performance on a variety of projects drawn from diverse environments. For example:

- The General Accounting Office of the U.S. government claims that there has been a consistent pattern of overruns beyond the initial baseline standards in federal government–funded projects.
- The Rand Corporation studies show a consistent pattern of overruns in technically complex projects, whether in aerospace or construction.
- U.S. nuclear power plants, tunneling, highway, water, building, defense, and other projects experience overruns.[11]

Project *success* and *failure* are relative terms. Failure is the condition or fact of not achieving the desired end or ends; success is the achievement of something desired, planned, or attempted. *Whether a project is a success or a failure is in the eye of the beholders,* that is, those individuals, enterprises, agencies, institutions, who are the stakeholders. For example:

- A project that overruns its cost and schedule and experiences delays in attaining its technical performance objectives may be considered a success if the project's outcomes ultimately fit into the choice elements of the organization.
- A project that experiences problems in attaining its desired objectives in costs, schedules, and technical objectives—and is not available to fit into the strategic management purposes of the enterprise—would likely be considered to be a failure.
- A project that is terminated prematurely because its final outcome is expected to be inconsistent with the choice elements of the enterprise could be considered to be a success by the strategic managers of the enterprise.

---

[8]Gopal Kaput, "IT Project management Can Succeed!" *Managing Office Technology,* 1997, pp. 22, 42.

[9]Cited in William G. Wells, Jr., "From the Editor," *Project Management Journal,* December 1998, vol. 29, no.4, pp. 4–6.

[10]Paraphrased from Michael William Hughes, "Why Projects Fail: The Effects of Ignoring the Obvious," *Industrial Engineering,* April 1968, pp. 14–18.

[11]Paraphrased from William G. Wells, Jr., "From the Editor," *Project Management Journal,* June 1999, vol. 30, no. 2, pp. 4–5.

- An individual who gains valuable experience in serving on a project team could view the project as a success—regardless of its ultimate outcome.
- A key project stakeholder, such as an environmentalist concerned with the pending construction of a nuclear power plant, could view the project as a failure even though the plant finally comes on line.
- The manager of a financial institution who sees that a project's financial management is inadequate could view the project as a failure.
- A project team could feel strongly that the project they are working on is a success if extraordinary problems on the project had been finally solved. Yet, if the project were canceled because it no longer promised to make a suitable contribution to the strategic choice elements of the enterprise, the project team might well feel that the project was a failure because of its early termination.

13.15 THE CAUSES OF SUCCESS OR FAILURE

A project's success or failure should be attributed to the philosophy of management that is carried out during the project life cycle:

- By those *strategic managers* who have responsibility for the management of the enterprise as if its future mattered—to include continuous oversight of the project in its journey toward becoming a potential contributor to the choice elements of the enterprise.
- By the *project manager* and the *team* who have responsibility for the attainment of the project's cost, schedule, and technical performance objectives.

If we take these two types of managers and investigate their responsibilities further, we find that the strategic managers are responsible for:

- Providing adequate resources to support the project during its life cycle
- Defining how the project results will contribute to the choice elements of the enterprise
- Establishing and defining how the individual and collective roles of the project manager, the project team members, and the functional managers are to be carried out
- Providing for a regular review of the emerging project cost, schedule, and technical performance results as well as the likely fit of these results into the future of the enterprise
- Maintaining a general perspective of the satisfaction that the stakeholders have with the project.
- Promoting a culture that supports the management of projects in the enterprise.

The project managers are responsible for:

- Use of project resources in an effective manner that enhances the probability of the project achieving its schedule, cost, and technical performance objectives
- Keeping the project stakeholders informed of the progress of the project as needed, and to the extent possible, managing the project stakeholders to gain and hold their support for the project
- Maintaining ongoing interfaces with organizational units to develop and continue support of the project
- Enhancing the knowledge, skills, and attitudes of the project team members to increase the effectiveness of their support of the project
- Keeping the enterprise managers informed of the status of the project at all times
- Developing and using appropriate information systems to keep project stakeholders informed of the status of the project

An empirical investigation of the sources of major project problems found that project managers felt a lack of ability to influence external stakeholders. Comments suggesting how external stakeholders could be influenced in even minor ways were rare.[12]

There have been many attempts to provide a description of the success factors and failure factors for projects. A few rigorous research initiatives have been designed and executed to identify the critical success factors.

The U.S. Air Force's acquisition of the T-3A "Firefly " trainer was a troubled project. Rather than develop a new aircraft, the Air Force decided to save time and money by buying a commercial off-the-shelf (COTS) trainer. But significant aircraft modifications undermined the integrity of the COTS strategy.[13] Choices made early in a project determine future success. Missteps in early phases will cause trouble later in the project's life cycle.

There are four lessons learned from this project:

- Any project must be managed as a system of interrelated parts.
- A project strategy must be flexible to accommodate changing circumstances.
- Testing must be done in realistic environments.
- Concurrency carries with it benefits and dangers.

Evaluating the success of projects is rarely a precise science. Examples of project ambiguity abound: The Hubbell Telescope started its operational life as a sort of national joke, a case study of project failure. Yet today it continues to reveal never-before-seen views of the heavens, views unobtainable from any other source. At its completion, the Sydney Opera House was seen by most as a stupendous failure:

---

[12]Paraphrased from Alan L. Brumagim, "An Empirical Investigation of the Sources of Major Project Problems: A Project Manager's Perspective," *Proceedings of PM Research Conference,* 2000, June 2000, pp. 223–237.

[13]Paraphrased from U.S. Air Force Report, *The Fall of the Firefly: An Assessment of a Failed Project Strategy,* March 2001.

a music hall with poor acoustics, stunningly over cost and behind schedule. Decades later, that same structure is a unique national treasure, its massive cost and schedule overruns long ago forgotten.

Andersen and Jessen provide a scheme for project evaluations that can be used as a checklist or guide from the start of a project through its postproject reviews. Sixty items provide critical statements that could be used to validate the planning, execution, and control phases of projects. Their information is interpreted and summarized here to provide the essence of the assessment checkpoints[14]:

- *Project scope.* Mission and goals are clearly stated and unambiguous in their wording. The mission and goals cannot be misunderstood because the originator uses concrete terms that lead to a clearly defined, deliverable product or service.
- *Terms of reference.* Economic and time parameters are clearly stated for the project and the product is sufficiently defined. All stakeholders are apprised of the project's parameters and agree to the key goals and objectives. There is a well-documented business strategy for the project.
- *Project planning.* The plan is simple and easily understood with milestones. The strategy is embedded. It has sufficient detail to delineate responsibilities and make clear the direction is understood. Adequate tools are incorporated to support the implementation process.
- *Project organization.* A project organization chart depicts the hierarchy for responsibility as well as linkages to other departments or parties involved in the project. Show coordination lines to other organizations. Ensure the organization does not inhibit communication with key decision makers or supporting organizations.
- *Project execution.* Implementation of the project plan by designated individuals is in accordance with the procedures prescribed. Changes to the plan are made in a timely manner and communicated to all stakeholders. Senior management stays involved in the decision making and stays apprised of the project's progress. The project team maintains awareness of progress and status.
- *Project control.* Active tracking of progress, including the corrective action necessary to sustain project convergence on the end goals and objectives. Keeping stakeholders informed of the progress and status as well as any challenges to completing the project. Obtaining stakeholder concurrence for changes to the project plan, if a change is necessary.

13.16 PROJECT EVALUATION

Project evaluations are performed either on a routine basis or for special reasons. The project team, for example, should routinely conduct assessments of their project to determine whether it is progressing within the scope as expressed in the

[14]Paraphrased from Erling S. Andersen and Sven Arne Jessen, "Project Evaluation Scheme: A Tool for Evaluating Project Status and Predicting Project Results," *International Project Management Journal,* vol. 6, no. 1, 2000, app. A, pp. 68–69.

project's goals and objectives. Senior enterprise management may require that a project be evaluated for progress on a routine basis—such as during stages of project execution—to determine how the project is progressing and to build confidence that the project will meet its goals and objectives. Senior enterprise management or the customer may direct a project evaluation and review because of perceived problems or to validate the status of the project. These ad hoc reviews typically are triggered by major challenges to completing the project.

Major problems that have the potential for materially affecting the outcome of the project require unanticipated, immediate action to define the problems, identify corrective actions needed, and change the direction of the project. Special evaluations and reviews typically have limited scope objectives: Identify and fix problems that block the project from progressing as originally planned.

Historically, these special project reviews have had dramatic outcomes such as major replanning efforts, reduction of the scope of the project, and replacement of the project manager. It has been observed that new project managers assigned as a result of a failing project have greater authority and senior management support than previous project managers. Furthermore, the failing project will often have a higher priority for resources than before the project review.

Scheduled project evaluations have two primary purposes. First, an audit is conducted to determine how well the project is progressing toward satisfaction of its cost, schedule, and technical performance objectives and in what manner. Second, an audit is conducted to determine whether the project promises to contribute to the strategic management choice elements of the enterprise, primarily to the goals of the enterprise, as envisioned during the planning phase.

As indicated in Table 13.2, an audit may be conducted on a project for any number of reasons. Perhaps the only criterion for conducting an audit is that the benefits exceed the cost of the audit.

The project team is best suited to maintain an ongoing evaluation of the cost, schedule, and technical performance goals of the project. Although the project team should maintain an ongoing audit of the project's status, formal evaluation of the project's progress should be done on a regular basis, probably at least monthly for medium- and large-size projects. The project sponsor (or owner) should be present, along with other select stakeholders, during this formal review. As often as needed, senior enterprise managers should be present at these formal reviews to gain insight into how well, and in what manner, the project is likely to make a contribution to the strategic objectives of the enterprise.

An ongoing overall assessment of a project by the project's team is not the same as a regular project review. A project review is an event to examine the current status of the project's progress in meeting its cost, schedule, and technical performance objectives. The project review should be held for the benefit of the project team, the sponsor(s), which includes the owners, and other interested and appropriate stakeholders. A project review should be held as often as needed so that the key stakeholders can determine how well the project is doing—and how likely it is that the project will contribute to the goals of those stakeholders.

TABLE 13.2 Types and Purposes of Audits

Type of audit	Purpose of audit
Routine project progress and status validation by the project team	Conducted by the project team to self-assess the project's progress and determine whether the project is converging on its goals and objectives—cost, schedule, technical performance. Results of the audit will be adjustments to work practices and perhaps some minor replanning.
Routine scheduled project progress and status validation by senior management	Conducted by an audit team that may be completely outside the project team or may be a mixture of project team and independent experts. Typically, the audit will be conducted by individuals from the organization's functional staff. Expertise may be requested on a temporary basis from an outside consultant. Results of the audit will usually be general guidance to the project team that may include emphasis on specific project work. It may include addition of resources to meet specific skill requirements or to shorten the project's schedule.
Specific-purpose audits of the project as a result of perceived problems with convergence	The audit is specifically designed to determine what is adversely affecting the project's advancement and what needs to be fixed. The expertise to conduct the audit may be either from internal enterprise resources or from an external audit team. The project team will not participate as part of the audit team. Results will be a statement of the problems, a recommended procedure of fixing the problems, and a recommendation for a follow-up audit to determine whether the corrective actions fix the problems.
Other ad hoc audits conducted on projects	Audits may be conducted during any phase of the project. Sometimes is it helpful to assess only the project plan at the end of the planning phase to validate its efficacy and the probability that it is adequate to guide a project team to the successful conclusion of a project. An audit may be conducted on the control mechanisms only of a project to determine if they are adequate for the size project and the type of project. An audit may be made of a completed project to identify what went right and what went wrong. This may be significant if there is potential litigation pending.

A project office is the likely organizational unit to conduct the project review. The project office can prepare an inventory of topics that should be considered in the reviews; each stakeholder should be available to suggest topic areas that can be reviewed. The purpose of the formal project review is to determine if the project is likely to contribute to the "stake" that the stakeholders have regarding the project and to identify problems with the project that might adversely impact the stakeholders' goals.

Each project reviewed should provide written summary information of the project's status that can be disseminated to those attending the review. As a courtesy, the project team should have the privilege of looking at this summary information to determine its validity and reliability—and what corrective action, if any, should be carried out.

Formal project reviews constitute a major effort to evaluate a project's status. A few summary considerations should be acceptable to those individuals who conduct a review—and those individuals for whom the review is carried out. The project plan should stipulate how often and under what considerations a project review should be carried out—and contain as a minimum the following strategies:

- A description of when, how, and what project review methodology and administrative processes will be carried out.
- Commitment of those organizational managers who use projects as building blocks in the design and implementation of choice elements in the strategic management of the enterprise.
- The propagation of an organizational culture that views project control in a positive way—both from the project's standpoint and the strategic management context of the enterprise.
- Use of project performance standards as a guide to reviews—carried out in a professional manner.
- Project reviews done in the spirit of improving the management of the project as needed, and certainly not to find and place fault if there are problems in how the project is being carried out.

Those people who participate in, or will benefit from, project reviews should not forget that most project failures happen because basic project management principles and processes have been inadequately applied.

One large, and ultimately successful, project was evaluated using the following philosophy: The use of a postproject review can provide valuable information to be used in the management of future projects. In order to do such postproject reviews, several requirements must be met:

- A standard methodology and philosophy must exist as a template for managing projects in the enterprise.
- A project management review board, which will act as a sponsor, must be created, as well as a project management review team, which should be organized and used to carry out the review.

- The review must be carried out in the spirit of "lessons learned" and not as a subterfuge for finding fault.
- During the review, the project team is to be asked questions concerning how the project was managed during its entire life cycle. The planning phase of the project should be given particular scrutiny
- The findings of the review process should be summarized in the spirit of the strengths and weaknesses with which the project was carried out.
- Particular attention should be given to the conditions—and restraints—under which the project team had to operate.
- The results of the postproject reviews are provided to existing project teams. Each time a team is formed for a new project, the existing database of "lessons learned" from previous projects should be reviewed by the new project team.

A project audit is another way to gain in-depth understanding of what is going on in a project—and in the management of that project. Project audits take time and money to do properly. Because of the cost in time and money, project audits are not often used. But, an audit can provide much value for ensuring that project controls are working and that stakeholders really understand what is going on in the project. What can a project audit do for you?

- Provide complete and objective information on the progress of the project and include the likely contributions that the project makes to the choice elements of the enterprise.
- Provide insight into the effectiveness and efficiency of the support that the enterprise functional elements and other organizational units are providing to the project.
- Ensure consistency across the enterprise on how projects are managed and measured in terms of quality, reliability, contribution of organizational choice elements, and other performance standards for the project and the enterprise.
- Provide for standards in following up on collective and preventive action.
- Identify areas in which a project's management can be improved—and how such improvements can serve as valuable lessons learned for ongoing and future projects. Project audits can be general, in which the overall project is evaluated. In some cases a part of the project may need to be audited, such as technical assessment, financial prudence, systems support, customer satisfaction, planning or control systems adequacy, or probable contribution to the enterprise's choice elements.

There are several steps to be carried out in the audit of a project. The following list is a general guide to this process:

- Selection and preparation of a team to carry out the audit.
- Selection or development of performance standards that will be used for comparison purposes during the audit.

- An initial meeting of the audit team and selected project stakeholders to be briefed—and to reach consensus on what will be audited and how the audit will be carried out.
- Collection of relevant information through questionnaires and interviews with appropriate stakeholders, study of project documentation, and briefings on the project's progress regarding its cost, schedule, and technical performance objectives.
- Analysis of the differences between what the project should have accomplished vis-à-vis what has been done—expressed in the form of gap analysis.
- A closing meeting of project sponsors, project team, and other stakeholders to communicate the results of the audit.
- Feedback to appropriate senior enterprise managers, the project team, and other project stakeholders regarding development and delineation of appropriate strategies to correct any shortcomings of the project.
- Follow-up on recommended corrective actions for the audited project and implementation of preventive measures to avoid weaknesses in the enterprise's project management methodology, policies, procedures, best practices, and techniques.
- Integration of audit results into an inventory of "lessons learned" for dissemination to existing project teams and future project teams.

The benefits of project audits, mindful of the costs, include the improvement of the enterprise culture for project management, opportunity for closed-loop corrective action, evidence that project monitoring, evaluation, and control are important, and the indoctrination of present and future project managers in the wisdom of exercising reasonable and prudent management of assigned projects. A key benefit is that the carrying out of such audits sends a clear message to the enterprise *that projects are important in the management of this organization.*

13.17 TO SUMMARIZE

Some of the major points that were expressed in this chapter include the following:

- Henri Fayol, often called the father of modern management, stated that control involves whether everything occurs in conformity with the plan adopted, the instructions issued, and the principles established.
- The control cycle presented in Fig. 13.1 is the key model to use in designing and executing control systems for an enterprise and for a project.
- A series of concepts and philosophies were suggested as essential elements for the assessment of project results.

- Performance standards are reflected in the project plan and are the key targets against which to judge project results.
- Performance observations take many forms, both informal and formal, and should be expanded as needed to garner feedback on project performance.
- When planned and actual performance is carried out, the comparison module of the control cycle is being used.
- Corrective action to modify in some manner the use of resources on the project has as its primary purpose the reallocation by some means of the resources committed to the project.
- Systems do not control people and the project team and other key stakeholders have an inherent responsibility to implement the monitoring, evaluation, and control functions on a project.
- In Table 13.1 some key questions regarding carrying out the monitoring, evaluation, and control functions on a project were suggested.
- A series of management function evaluation questions were suggested to provide additional importance to the monitoring of project progress and results.
- Project monitoring, evaluation, and control cannot be done without relevant and timely project information.
- A project should be evaluated at least during the key points in the project's life cycle.
- A project review should be done at the work package level, the functional manager's level, the project level, the general manager's level, and the project owner's level.
- Project audits can be carried out at any time during the project's life cycle when the efficacy with which resources are being used on the project comes into question.
- A general paradigm for what to review during a project audit was suggested.
- The use of postproject reviews can provide valuable lessons for how to better manage future projects. It is important that a postproject review not be used to "find fault" in the management of a project.
- Although the concept and process of configuration management and control are important, their use will continue to be modified as the process of concurrent engineering continues to grow in usage.
- The control function should never be considered in the context of "command and control" or dominion or dominance. Rather, monitoring, evaluation, and control are the organized means used to determine progress in the management of a project.
- Project success or failure is relative to the interests of the stakeholder.
- Projects may fail in the classic sense of not meeting cost, schedule, and technical performance criteria or not meeting the customers' needs, but may provide benefits through by-products that far exceed the cost.

- Projects succeed most often when properly planned, implemented, and controlled while receiving senior management support.
- Project audits and evaluations identify opportunities for improvement.

13.18 ADDITIONAL SOURCES OF INFORMATION

The following additional sources of project management information may be used to complement this chapter's topic material. This material complements and expands on various concepts, practices, and theory of project management as it relates to areas covered here.

- Pierre-Francois Heaulme and John E. Martin, "Risk Management: Techniques for Managing Project Risk"; Dan Ono, "Project Evaluation at Lucent Technologies"; and James R. Snyder, "How to Monitor and Evaluate Projects," chaps. 12, 27, and 23 in David I. Cleland (ed.), *Field Guide to Project Management* (New York: Van Nostrand Reinhold, 1997).
- Edward W. Ionata, "Managing Environmental Regulatory Durations," and Robert H. Kohrs and Gordon C. Weingarten, "Measuring Successful Technical Performance: A Cost/Schedule/Technical Control System," in David I. Cleland, Karen M. Bursic, Richard J. Puerzer, and Alberto Y. Vlasek, *Project Management Casebook,* Project Management Institute (PMI). (Originally published in *Proceedings,* PMI Seminar/Symposium, San Diego, 1993, pp. 152–156; and *Proceedings,* PMI Seminar/Symposium, Montreal, Canada, 1986, pp. 158–164.)
- Al and Jackie DeLucia, *Recipes for Project Success,* Project Management Institute, 2001. This book offers 10 key tips for project success. The authors give tips on planning, running a project, and controlling the work through analogy to cooking. Basic project management information is provided for the newly appointed project manager, who needs a quick, simple explanation of techniques and principles. This book gives good tips on what to do and what issues to avoid that place the project at risk. The principles given are proven through experience and good, sound logic.
- John P. Kotter, *Leading Change,* Harvard Business School. This book addresses change and altering behavior. Project corrections often require that human behavior be changed to something other than the current approach to work to provide a better product or service while meeting the established project goals. The author describes an eight-step process that establishes a sense of urgency to meet competitive forces. Further, the author identifies the need for leadership and the use of teams to assist in change.
- Scott Berinato, "Recipes for Disaster," *CIO,* Framingham, Mass., July 1, 2001, pp. 82–83. This article describes how and why two information technology projects went wrong and what could have been done to save them. Further, it

identifies three key reasons for the collapse of Federal Express's business-to-business parcel delivery project. Reasons for another company's software project problems include sponsor apathy, lack of focus, no deadlines, and erratic executives.

- Pimm Fox, "Project a Mess? A Charter Can Help," *Computerworld,* Framingham, Mass., July 16, 2001, p. 26. This article demonstrates the need for a project charter that has all stakeholders approving prior to project initiation. Success is started with the agreement by the stakeholder as to what the project is about and what can be the expected results. Once a project has been started, there is little incentive for stakeholders to agree on the benefits of the project or support the project if they disagree.
- Janette Simpson, "The Measure of Success," *Intelligent Enterprise,* San Mateo, Calif., June 29, 2001, pp. 22–23. This article addresses the need for metrics in a project to be able to measure success. A project needs some agreed-upon method of determining whether a project is successful and metrics that stakeholders commit to. These metrics need to be in place prior to the start of a project and used to gauge the progress. Through these metrics, one can assess the project's degree of success.

13.19 DISCUSSION QUESTIONS

1. List and briefly define the elements of the project control cycle. Why is control an important project management function?
2. What performance standards must be set in order to control a project? How can these standards be observed throughout the life of the project?
3. In what ways is corrective action carried out? What are some of the potential impacts of these corrective actions?
4. Discuss the personal nature of control. How can managers foster an environment that supports the control system?
5. Monitoring and evaluating are integral to control. Discuss what is meant by each. How do they relate? What kinds of questions must project team members ask in order to continually monitor and evaluate the project?
6. Discuss the importance of monitoring and evaluating with respect to all project stakeholders.
7. When should the project be monitored and evaluated? Discuss the ability to make changes over the project life cycle.
8. Who should be responsible for monitoring and evaluating projects? Where should monitoring and evaluating take place?
9. What is required in a project audit? What purposes do project audits serve?
10. List and discuss some of the factors that lead to success of a project.
11. List some of the by-product benefits of projects for an organization, whether they are successes or failures.

12. List the criteria that a customer might use to define success. List those criteria for failure.

13.20 USER CHECKLIST

1. Do project managers establish appropriate performance standards based on the project plan so that the control cycle can be effectively carried out? What performance standards are overlooked?
2. How is project performance observed in your organization? What comparisons are made to standards? What corrective actions are taken?
3. Is the cultural ambience of your organization supportive of the control systems? Why or why not?
4. What questions do project managers ask in order to monitor and evaluate project performance? Do managers consider the requirements of all project stakeholders in monitoring and evaluating project performance? Explain.
5. Are monitoring and evaluating done at the appropriate points in the project life cycle? Do project managers take advantage of the early opportunities to influence project success? Explain.
6. Are project monitoring and evaluating a part of the early project plans? Do evaluation policies and procedures exist? Are key managers committed to the evaluation strategy? Explain.
7. Are project audits performed on projects within your organization? Who is responsible for project audits? Are the audits effective? Why or why not?
8. What types of postproject reviews are carried out by your organization? How do these reviews contribute to the success of other ongoing projects?
9. What factors (with respect to project control) have contributed to the success of projects within your organization? Explain.
10. Does your organization have an evaluation system for projects after they are completed? Describe it.
11. How does your organization define a successful project? What objective criteria are used?
12. Who in your organization decides whether a project is a success or a failure?

13.21 PRINCIPLES OF PROJECT MANAGEMENT

1. Project success or failure are relative terms, defined by the project stakeholder and based on his or her interests.
2. Project audits are required to measure the degree of success of projects.
3. Projects have by-products that benefit stakeholders in other ways than direct return on the projects.

4. Project controls measure progress and guide required corrective actions to meet the project's objectives.
5. Project control is exercised against the project plan and an organization's project performance standards.

13.22 PROJECT MANAGEMENT SITUATION—ESTABLISHING A PROJECT CONTROL SYSTEM

An organization has been managing projects for several years through a group of project managers specifically hired for their successes with this type of work. They have consistently performed well in other organizations, demonstrating an exceptional understanding of the technology and the project management discipline. They are excellent communicators and keep all parties informed of the progress of projects.

Recently, the company finance director stated that many of the projects were losing money. This was determined when the new accounting system was implemented and costs for projects were collected, collated, and analyzed. Nearly one-third of the projects were not performing well in the cost area. Only 18 percent of the projects were producing revenues that met the company's fee structure.

A review of project documents and plans as well as discussions with some project participants revealed the following:

- There was no consistency in the project plans. Some were missing vital information and others were inconsistent with the company's policies for project operations.
- Each project manager followed a different life-cycle model and methodology for similar projects. There was no single methodology being used for projects.
- Project managers relied on their experience, methodologies, and styles from their prior organizations.
- Project managers considered all projects to be successful if they met the cost, schedule, and technical performance objectives. Because cost data were not provided by the accounting department, it was assumed that projects were making money.
- No project audit had been conducted in at least 3 years and there were no procedures for conducting an audit. No trained people were available to conduct audits. The project managers were hired because they were "experts" in managing projects.

Senior management was in a quandary—experienced project managers were not providing the desired cost performance results and there may be other areas that were in need of fixing. Further, the company relied heavily on projects to generate most of its revenue. There was a need for some immediate action to move the company into a better cost performance position.

13.23 STUDENT/READER ASSIGNMENT

1. From the information given in the project management situation, what is the most likely problem and how would one fix it?
2. Why do highly qualified project managers use different methodologies in the same company? Is this a problem?
3. What are some of the reasons for lack of meeting cost performance (profit) goals? What can be done to improve the situation?
4. Does the company's project control system work? What, if any, changes should be made?
5. What type of audit would you perform on the projects to determine critical success factors? What would be your initial focus?

CHAPTER 14
THE PROJECT EARNED VALUE MANAGEMENT SYSTEM

"Value is in the eye of the beholder."

ANONYMOUS

14.1 INTRODUCTION

Earned value is a concept whose use is spreading today. In essence the concept means that any use of resources should return something of value to those who provide the resources. It is a way of measuring both efficiency and effectiveness in the use of resources for some particular purpose. The concept and process of earned value measurement came about through the desires of managers to more accurately predict future project returns while working toward organizational objectives and goals.

In this chapter, the application of earned value is applied to a project—as a means of gathering additional intelligence on how well the project resources are being utilized. The background of EVMS (earned value management system) is given, along with the capability that EVMS can offer, and some of the more important EVMS implementation considerations are presented. How to plan for a project-related EVMS is suggested, along with a description of how to measure progress on the project and how the project's performance achievement can be assessed. The chapter closes with some general considerations regarding EVMS that should be of interest to the reader.

14.2 BACKGROUND

The earned value management system is a project performance measurement system that integrates cost and schedule aspects. It is indirectly related to technical performance in that satisfactorily completed work packages represent technical performance achievement. Completed work packages are the basic building blocks for the cost-schedule measurement and performance reporting.

EVMS was first developed as cost/schedule control system criteria (CSCSC) in 1963 by the U.S. Department of Defense (DOD) to control major projects. The focus is on "criteria" in this system, which requires users to implement adequate planning, tracking, and control systems. Each qualified user must have a planning and tracking system that meets the criteria for a reliable capability to collect and report information.

EVMS has also been called earned value analysis, which implies that there is an emphasis on being able to analyze the data derived from a system. "Analysis" is only one part of the entire process. It is the entire system that provides the planning, analysis, reporting, and corrective action for projects. EVMS is the more commonly accepted term for the process of measuring and tracking cost-schedule data.

EVMS has been slow to gain acceptance in the commercial community although the government has relied heavily on its use in major projects. More specifically, the Department of Defense and its military components, as well as the Federal Aviation Administration, have required that EVM or its predecessor be invoked for the management of contracts. Contractors were required to implement EVM in order to be awarded the contract and for incremental payments based on performance.

It is perhaps fair to say there was not a good implementation of EVM by both the government and contractors in a cooperative mode. Contractors, in some instances, say EVM was used to punish them. The government, on the other hand, was trying to implement a system that gave some degree of control of the expenditures over time.

14.3 EVM CAPABILITY

Major projects are difficult to control when there is an accordion affect on schedules that contract and expand on a continual basis and when there are changes to the product's scope and resultant cost. EVM provides a means of representing cost-schedule progress through schedule achievement and expenditures. Information derived from EVM is only as accurate as the planning, data collection, and timeliness of collection.

Some criticism has been made of EVM in that it always looks at past data to predict future performance. "It is like trying to drive 100 miles per hour while looking out your rearview mirror," is one such criticism. These criticisms are valid when the data are not collected and formatted in a timely manner for analysis and trend development. No other known system has the capability to provide the cost-schedule trends and performance measurement for major projects.

On the other hand, some project managers have had great success in using EVM as a tool to support their project management efforts. When there is cooperation between the project manager and the performing contractor, success is more likely to occur. EVM supports both the project manager and the performing contractor because it:

- Provides early identification of adverse trends and potential problems.
- Provides an accurate picture of contract status with regard to cost, schedule, and technical performance.
- Establishes the baseline for corrective actions, as needed.

- Supports the cost and schedule goals of the customer, project manager, and performing contractor.

While the government agencies do not require performing contractors to establish a specific management control system, they will typically demand that they meet 32 separate criteria for an effective system. These 32 criteria are in the following areas:

- Five in organization
- Ten in planning and budgeting
- Six in accounting
- Six in analysis
- Five in revisions and access to data[1]

The EVM process is defined at the top level as a management system that meets 32 criteria. It must consist of the following:

- The performing contractor must establish a management system and demonstrate that the system meets all 32 criteria.
- An integrated baseline is established that defines the work, schedule for performing that work, and the resources allocated to perform the work.
- Work packages and resources must be allocated at the lowest level for execution.
- Value is earned as work is accomplished according to the plan.
- Progress and status are always reported against the baseline plan with separate schedule and cost variances maintained.
- Reported trends give early insight into final estimated cost for projects.[2]

Two government program managers successfully used EVMS to meet their management responsibilities. Both programs were exceptionally good examples of what EVMS can do to assist in managing to success. Quotes from each express their view of EVMS:

> Are we looking good, or are we in trouble? And, how do we know? (V. Adm. (then Capt.) Joe Dyer, USN, F/A-18 E/F Program Manager)
>
> It forces you to plan, and then to manage to the plan. (Lt. Col. Paul Vancheri, USAF, JSTARS Production Program Manager)

In the fall of 1998, David Christensen estimated that the marginal benefits of EVM were greater than zero. He goes on to list benefits of EVMS as follows:

- The criteria concept allows contractors to use their internal management systems.
- All authorized work must be defined using a WBS (work breakdown structure).
- Consistent reporting is established.

---

[1]Paraphrased from Eleanor Haupt, "Basic Earned Value Management for Program Managers," a training presentation from Air Force System Command, slide #11, c. 1998.

[2]Ibid., slide #9.

- Proven early cost management is essential.
- All work must be scheduled and traceable to a master program level.
- Management attention is directed to critical problems.[3]

In 1998 at the 10th Annual International Integrated Program Management Conference, Michael M. Sears, McDonnell Douglas's Program Manager for the F/A-18E/F aircraft, presented his views on EVM. Some of the more important points made are:

- No single management tool is more useful, more effective, or more important to successful program management.
- To ensure that we "get it right" [program management], things had to be done in new and better ways.
- We set out to develop a disciplined, credible set of program processes. We made a commitment to use them all for our programs.
- Past practices allowed program managers to apply different program management approaches, which resulted in a lot of inconsistency in our program performance.
- We committed to ourselves and our customers to manage all our programs with the same basic set of tools and a common overall program management system.
- We organized ourselves around integrated product teams that included our customers and our supplier-partners. We established teams with clear lines of responsibility, authority, and accountability for the work they were doing.
- EVM is now one of the handful of companywide program management best practices. We even migrated it to the commercial side of our business.[4]

14.4 EVM IMPLEMENTATION CONSIDERATIONS

Effective application of EVM principles requires that the project planning team review some critical aspects of the project to determine whether EVM is appropriate for a project. EVM may be inappropriate for a project or the organization may not have sufficient expertise in EVM to effectively use its full capability. Some questions that must be asked include:

- Is the project of sufficient size to warrant the expense of planning for and implementing an earned value management system?
- Will the planning be in sufficient detail to develop logical work packages?
- Is the scope of the work sufficiently defined to allow development and use of a work breakdown structure?

[3]Paraphrased from David S. Christensen, Ph.D., "The Costs and Benefits of the Earned Value Management Process," *Acquisition Review Quarterly,* Fall 1998, pp. 373–385.

[4]Paraphrased from Michael M. Sears' opening speech, 10th Annual International Integrated Program Management Conference, October 19, 1998.

- Will the project team be able to collect the right data on a timely basis to permit accurate, timely measurement of progress?
- Is there a system in place to measure technical performance achievement?
- Do the operators of the earned value management system have the expertise to prepare reports for decision makers?
- Do the decision makers understand the meaning of trends, variances, and issues reported?
- Is there a simpler system that will give similar results for less effort and cost for this project?

Project planning required for effective EVM purposes may be more rigorous than an organization typically is capable of. The appropriate level of detail and discipline in conducting the planning to remove uncertainty in the scope of the project is mandatory. Poorly defined projects will materially impact the ability of the project team to collect, analyze, and report the cost and schedule data.

A sequence for the planning of the project to meet EVMS criteria is:

- Establish the project's requirements through defining goals to be met in terms of technical performance, project completion, and project cost.
- Define the work to be accomplished by dividing the project into work packages using a work breakdown structure.
- Define the control process for the project and all associated costs, if not included in the work breakdown structure.
- Identify the overhead costs and the organization responsible for controlling those indirect costs to the project.
- Identify the project's organizational structure, including subcontractors performing work for the project. Vendors providing major components may also need to be identified.
- Define the cost and schedule data to be collected, the method of collection, and the frequency of collection. Identify who will provide the data to the project control section.
- Define the reports that will be produced and to whom they will be distributed.[5]

In the material that follows, some general planning considerations are given. General planning considerations are helpful in developing an integrated coherent project plan for accomplishment of the work. The planning sequence is critical because one part builds on information for a following part. The planning sequence should be:

- Define the requirement and the product to be delivered. The product's technical performance criteria are crucial to project success. The product features, functions, and characteristics are what define customer satisfaction. Describing

[5]Paraphrased from "Industry Standard Guidelines for Earned Value Management Systems," National Security Industrial Association Subcommittee, EVMS Work Team, August 8, 1996.

the product with all its attributes provides the basis for the time and cost of the project.

- Define the schedule for building and delivering the project's product. This time line describes the logical sequence for buying, fabricating, constructing, assembling, testing, and delivery of the product's components. Sequencing the schedule's activities establishes the best estimate of the project duration. The schedule dictates when the technical activities will be accomplished on the time line.
- Define the cost of performing activities and the human and nonhuman resources required to accomplish the work. Costs include the direct cost of resources, any subcontracted work, and indirect costs such as senior supervision and special training. Costs may be burdened (include overhead, fringe payments, and fees) or unburdened (exclude overhead, fringe payments, and fees). The cost account practice must be determined during the planning and agreed to with the customer.
- Planning for schedule and cost is typically accomplished as an integrated effort. Modern project management software tools facilitate simultaneous planning for both functions. For example, labor-intensive projects often estimate the number of labor hours required to perform an activity and insert the number of people planned to work on that activity. The tool calculates the activity duration. An activity requiring 40 labor hours may be assigned four people to generate a 10-hour-duration activity.

14.5 PLANNING FOR EVMS

EVMS starts with planning the project through a work breakdown structure to build work packages that have estimated costs and estimated completion durations. These work packages form the building blocks for the project and for the EVMS. Cost and schedule performance are measured on the basis of work packages completed.

The work breakdown structure provides a rigorous means of dividing the work into manageable elements called work packages. Figure 14.1 is an example that

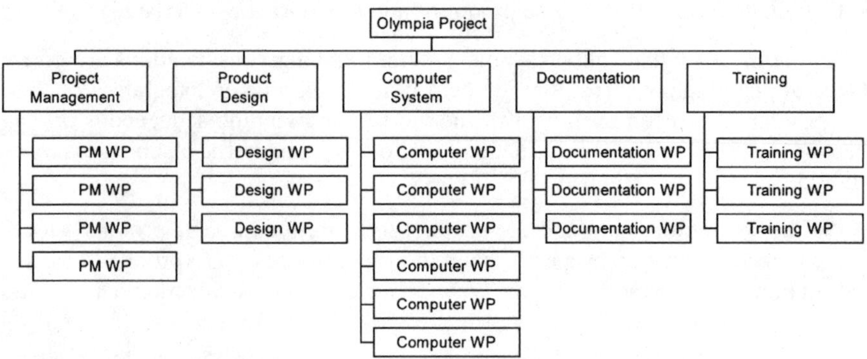

FIGURE 14.1 Example of a work breakdown structure.

shows both the product and the functions being divided into work packages. (The reader will note that a fuller description of the development of the WBS is discussed in Chap. 11.) These work packages are the lowest level of the work breakdown structure.

For illustration purposes, the example consists of only three levels of decomposition of the project. It is often the case that decomposition may be five or more levels for large, complex projects. Divide the project to the work package level at which it can be best managed.

Much like the WBS, an organizational breakdown structure (OBS) is often used to define the organization to the lowest operating level. The OBS is often numbered like the WBS and provides a unique identifier for all elements. Integrating the OBS and WBS, as shown in Fig. 14.2, provides an overlay of work assignments and responsibilities for completing work packages.

This overlay of the WBS and OBS gives the capability to also roll up expenditures for both the work to be done and the performing department. The intersections, or cost accounts, provide an easy way to track work responsibility as well as the expenditures against those work packages.

Using work packages, the EVM's basic integration of cost and schedule is depicted in Fig. 14.3. This simplified diagram shows the cumulative effect on cost and schedule when work packages, defined in terms of cost and time duration, are scheduled for the project.

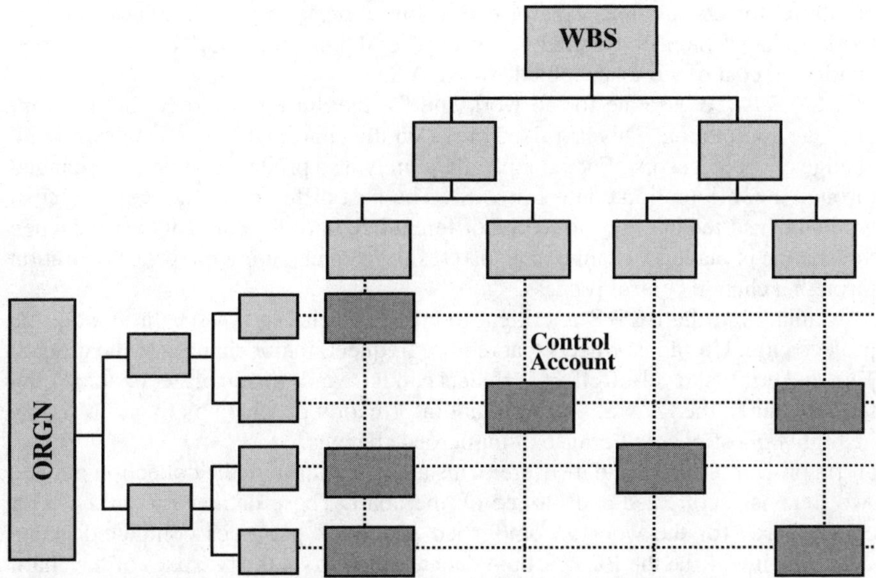

FIGURE 14.2 WBS-OBS integration.[6]

[6]Adapted from Eleanor Haupt, "Basic Earned Value Management for Program Managers," a training presentation from Air Force System Command, slide #43, c. 1998.

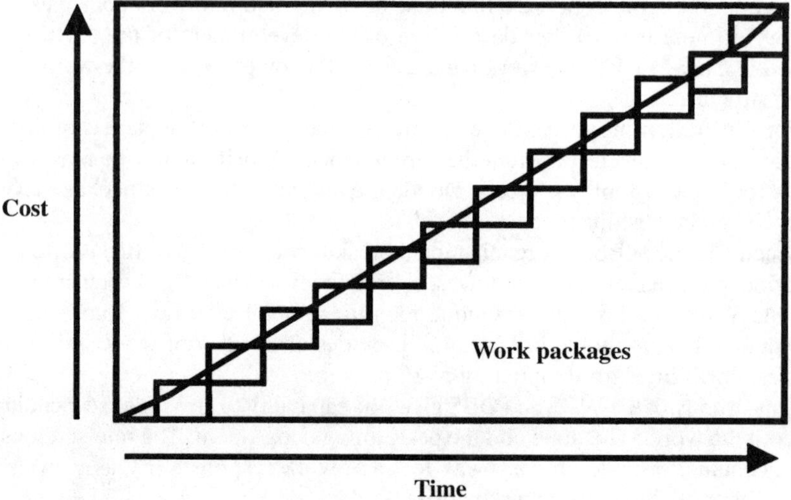

FIGURE 14.3 Construction of the basic EVMS chart.

Cost is always depicted on the vertical axis, whereas time is depicted on the horizontal axis. The line bisecting the chart is the result of laying out work packages over time, where work packages are defined in terms of both "cost" and "time duration" for completion. This bisecting line represents the cumulative cost of work packages planned for completion at several points in time. This line is called "budgeted cost of work scheduled" or BCWS.

BCWS is the baseline for all work and for measuring progress, that is, work package completion. This is a fixed curve on the chart and only changes through change of project scope. For example, there may be a product component changed through a configuration control process. The cost differential, more or less cost, would be inserted in the system for a different height of the curve at the time when the change is made. A change may also result from changing the project duration through a change control process.

Maintaining the BCWS is critical to ensure a reliable system throughout the project's life. Unstable projects that require frequent, major changes to the original plan and erratic "level-of-effort" projects can have a dramatic effect on managing the baseline in the EVMS. The resultant information provided by the EVMS may be highly questionable because of numerous changes.

As project progress and measurements are taken through the collection of data, two items are collected and plotted on the chart. These data are actual costs or expenditures for the work accomplished and work packages completed in the schedule time. Like the BCWS, both data are plotted on the two axes of the chart.

Cost data are plotted to form a line called actual cost of work performed (ACWP). This collection and plotting of data for cost over time represents the real cost of completing the work packages rather than the estimated cost in the BCWS. Schedule data, similarly, show the real time consumed rather than estimated time in the BCWS, or the budgeted cost of work performed (BCWP). Typically, the three cumulative curved lines will not overlay each other—the perfect project solution—but will closely parallel each other. Figure 14.4 shows an example of the chart depicting the three cumulative curved lines.

The EVMS graph that depicts the trends for a project, shown in Fig. 14.4, is an example of a typical project. The "Now" line represents the time for the data collected, which may lag the current date by 1 or 2 weeks. Any delay between the "data date" and the current date gives a less accurate picture of the project's status.

BCWS is the plan for expenditures over time and does not show any changes to the baseline. Changes would be depicted as a step function, up or down, on the cumulative curve. The dashed line representing the ACWP, or actual expenditures, is above the BCWS line. This variance shows that project cost is exceeding the estimated (budgeted) cost. Similarly, the BCWP line is below the BCWS, and the variance shows the work is not proceeding as planned.

Variances to the budget and schedule need to be investigated to see what is causing the adverse situation. The amount of variation and a recovery plan would give the decision makers additional information on required corrective actions. Trends and variances alert decision makers, but do not necessarily dictate corrective action.

14.6 MEASURING PROGRESS

Work packages represent the basic building blocks for measuring progress. Work packages result from planning the work effort in short-span jobs or material items

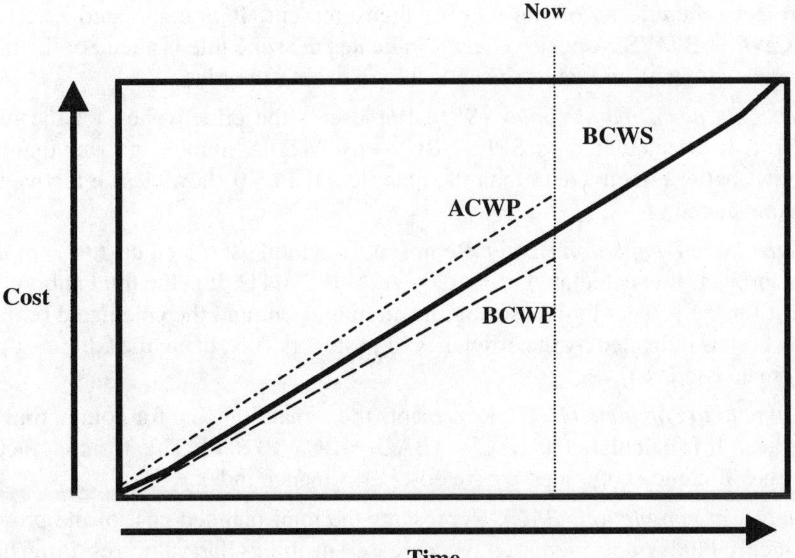

FIGURE 14.4 EVMS graph with plotted data.

that are required to complete the project. A work package may be described by the following characteristics.

- Clearly distinguishable from other work packages and typically identified by a unique number such as a work breakdown structure code
- Work accomplishable by a single responsible team, individual, or element
- Start and completion dates as well as a distinguishable output
- An estimated value assigned, typically expressed in dollars but may be expressed in labor-hours
- A measurable component of the project represented

Some of the formulas used in the EVMS are shown below. The definitions give an indication of their use and utility. These formulas, although typical of those used in EVMS, will be supplemented by other more sophisticated approaches to measuring the progress of projects.

- *Cost variance (CV).* Represents an over- or underexpenditure from the planned project cost as of the date of measurement. It is calculated as CV = BCWP − ACWP. A positive number indicates expenditures less than planned; a negative number indicates greater cost than planned.
- *Cost performance index (CPI).* Represents the efficiency of the work effort. It is calculated as CPI = BCWP/ACWP. A number greater than 1.0 shows greater efficiency than the plan; less than 1.0 shows less efficiency than planned.
- *Schedule variance (SV).* Represents an over- or undertime from the planned project schedule as of the date of measurement. It is calculated as SV = BCWP − BCWS. A positive result indicates the schedule is ahead of the plan; a negative result indicates the schedule is behind the plan.
- *Schedule performance index (SPI).* Represents the effectiveness of the work effort. It is calculated as SPI = BCWP/BCWS. A number greater than 1.0 shows better effectiveness than the plan; less than 1.0 shows less effectiveness than planned.
- *Estimate at completion (EAC).* Represents the total estimated cost to complete the project. It is calculated as EAC = ACWP + ETC. It is the total anticipated cost for the project by measuring the amount spent and the calculated remaining cost as indicated by the efficiency of past work [i.e., using the CPI (cost performance index.)]
- *Estimate to complete (ETC).* Represents the remaining cost for completing the project. It is calculated as ETC = (BAC − BCWP) /CPI. This formula factors in the efficiency computed by the cost performance index.
- *Budget at completion (BAC).* Represents the total planned cost of the project, which includes any increases or decreases in the dollar value resulting from changing the scope of work.

14.7 PERFORMANCE ACHIEVEMENT

The EVMS can best be explained through an example to demonstrate how the system works. As stated earlier, detailed planning to develop work packages with the appropriate cost and time estimates is required to establish the basis for measuring earned value. Technical performance is not included in the earned value concept, but is a function of tracking the work packages and their output.

First, a project is selected as having the right characteristics for using earned value. The project is large and it is unmanageable to try to establish cost and schedule performance metrics that give reasonable control over the work progress. EVMS is selected for implementation as the system for controlling the business aspects of the project.

Detailed planning to the work package level is accomplished by using a work breakdown structure. A schedule and time-phased budget are prepared that reflect the work packages. Subcontractors have prepared project plans to a similar level of detail. Reporting procedures for weekly data collection have been instituted to reflect which work packages have been started, which are works in progress, and which have been completed.

Rules have been established for reporting work accomplishment. During planning, work packages could not exceed more than 40 hours in duration, that is, no more than 1 week. Credit for work accomplishment would be granted only for work packages completed. Once a work package was completed, it could not be opened again to perform more work. Additional work would require a new work package.

These rules brought discipline to the process of earned value. Results of the rules were anticipated to be the following:

- Forty-hour work packages gave tight control in that work started would typically bridge no more than 2 weeks.
- Giving credit for earned value only when the work packages were completed provides incentive to complete the work package rather than open it or complete a percentage of it.
- Not permitting a work package to be opened again after it received credit provides incentive to actually complete the work before reporting completion.
- New work packages for additional work on a closed work package highlights the shortfall in performance on the original work package. New work packages do not receive earned value credit for completion.

EVMS does not control project cost and schedule overruns, but should signal the trend in sufficient time to permit some action by decision makers.

Figure 14.5 shows variances between the planned activity and the actual results. If performance was the same as the actual performance, ACWP and BCWP would meet at point B. In this example, cost variance is the difference between point A and point C. Subtract the amount of BCWP from ACWP. Even without a scale on the left axis, it is apparent that there is an excess cost for the work accomplished. Schedule variance is the difference between point C and point B, or BCWS from BCWP. Again, it is apparent that the schedule performance is poor.

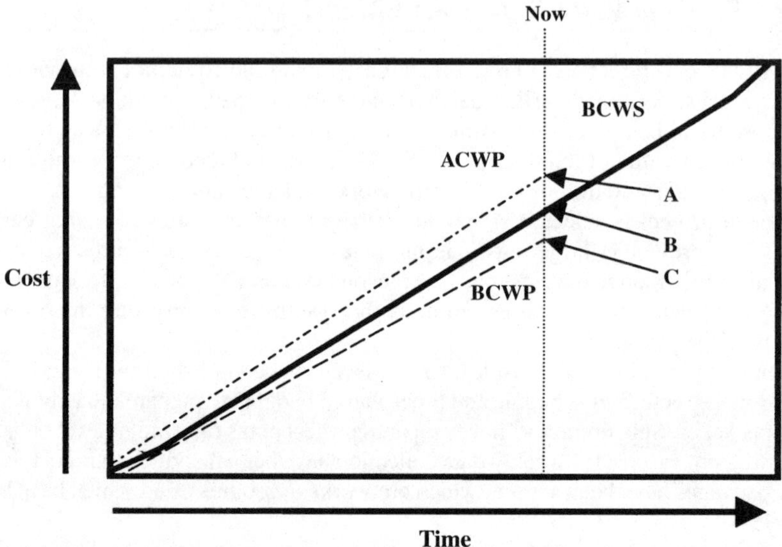

FIGURE 14.5 EVMS graph showing variances.

Graphically, this displays the status and the trend. Both are adverse and both show continual deviation from the baseline (BCWS) at a steady rate. Table 14.1 describes the cost performance index (CPI) that is used to show variance in a decimal value. For illustration purposes, let us say the CPI is 1.04, or the performance is 4 percent better than expected. If this is a trend, the end cost of a project valued at $100,000 would have a final cost of $96,154 (100,000 divided by 1.04).

Because projects frequently change in scope (technical need, cost, and schedule), the EVMS needs to reflect the current situation. If near the end of the project, for example, there is a technical difficulty that dictates some rework, this would increase the costs as well as extend the duration of the project. If senior management approves the change to the plan and the plan is adjusted, the EVMS is then corrected.

Figure 14.6 shows the change to the baseline (i.e., BCWS) by a vertical shift. This shift upward on the chart reflects the additional funds for the project. The extension of the BCWS line indicates the additional time required to complete the project with its new tasks. The change authorized new work packages after a technical failure, which did not replace work planned but added work to the project.

While new work was authorized on the project and that work is in the future, the cost and schedule performance for past work remains the same. In this graph, the less-than-planned performance stays the same. Efficiency indicators can be continued as a trend for the new work. If schedule efficiency is 0.94, one can expect that same efficiency to continue as a trend regardless of the work added or removed.

When projects have growth problems with the cost and schedule, typically both items cause EVMS graphs to change. In Fig. 14.7, for example, it is easy to see a change could extend the BCWS baseline from X, or the budget at completion (BAC), to Y, estimated cost to complete (ETC). The graph also gives a rough

TABLE 14.1 Formula and Definitions for EVMS Measures

Measurement	Formula	Definition
Cost variance	BCWP − ACWP	A variance from the planned expenditure in terms of dollars or labor-hours.
Schedule variance	BCWP − BCWS	A variance from the planned time in terms of dollars or labor-hours
Cost performance index (CPI)	BCWP/ACWP	A cost performance measure expressed in a decimal figure. A value of 1.0 indicates the expenditures equal the planned expenditures, more than 1.0 indicates the better than expected performance, and less than 1.0 indicates lower performance than expected.
Schedule performance index (SPI)	BCWP − BCWS	A schedule performance measure expressed in a decimal figure. A value of 1.0 indicates the schedule equals the planned progress, more than 1.0 indicates the better than expected performance, and less than 1.0 indicates lower performance than expected.

approximation of the changes. It would add 5 months to the schedule, a 22 percent increase in time, and approximately $200,000 in cost, or a 20 percent increase in cost.

Although the graph provides a rapid visual indication of problems, more precise figures may be derived from calculating the additional cost and time required to complete the work. Research into project overruns has shown that variances when the project is 30 percent complete will remain throughout the life of the project. Simplified, if there is a 10 percent cost variance at the 30 percent completion time, the entire project will have a minimum of 10 percent variance at completion. A $5 million project would cost at least another $500,000.

14.8 EVMS CONSIDERATIONS

EVMS is not the perfect tool for measuring cost and schedule performance, but it is the best tool currently available in terms of reliability and accuracy. It is best suited for medium to large projects as a means of keeping control over the authorization of work packages to start, measuring the completion of work packages, and analyzing the level of cost-schedule performance. It also provides a means of conducting trend analysis to project future efficiencies.

EVMS does not measure technical performance achievement in a project, but tracking the quality of work at the work package level can provide assurances that the project's product is being designed, built, and tested to appropriate grades or

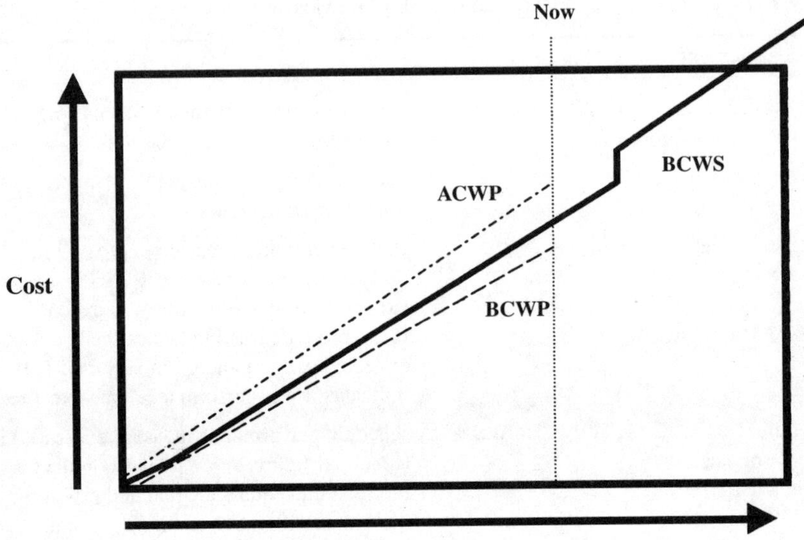

FIGURE 14.6 EVMS graph reflecting increase in project scope.

specifications. A technical performance program can be implemented separately from EVMS without affecting the cost-schedule tracking.

Essential elements of an EVMS consist of detailed planning, a means to collect relevant data on cost and schedule progress, a capability to format and display information on a periodic basis, and an understanding of what the data mean. Weaknesses in any area can materially affect the reliability of the system and minimize the effectiveness of this tool.

Graphic displays of the information help communicate the status and trends of the project. These graphics highlight issues for further prosecution. The actual figures can provide more precise information and the trends. Actual numbers are typically used by the EVMS specialist or analyst to quantify the data to determine whether any operating criteria have been exceeded.

Issues identified during analysis need to be investigated and solutions developed. These solutions may range from doing nothing to fully funding overruns. When there are variances that may indicate the project was not adequately planned and the work plans are unstable, it may require complete replanning or cancellation of the project.

EVMS provides a capability at a cost to deliver trend information against an established baseline. It is useful for many large projects and has other uses such as computing "earned value" for payment to contractors. Smaller projects are in need of a less costly and time-consuming method, but a method that is effective.

Typically, projects are independently assessed for the technical performance, schedule, and cost. These three parameters form the basis for meeting project objectives and goals. Through flaws in planning or guidance from seniors, the project's goals may not be achievable—they may be "stretch goals," or goals that

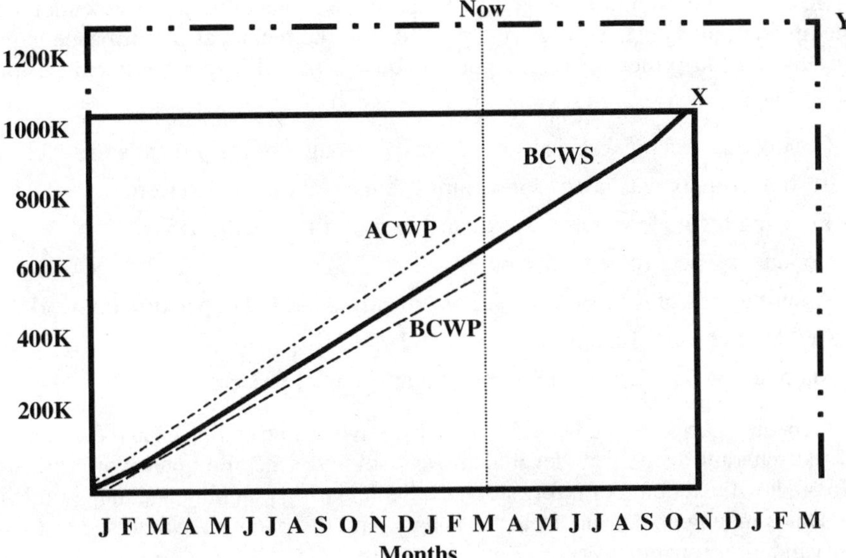

FIGURE 14.7 EVMS graph for cost-schedule growth.

exceed the normal limits with the expectation that greater productivity will be achieved through more distant targets. The following paragraphs address the three parameters of project management and the goals.

First, one must develop the technical performance goals. These goals are the basis and reason for the project. These goals provide the focal point for what will be delivered and what the customer needs. Schedule and cost must be subordinate to technical performance.

Second, the schedule goals must follow the technical performance goals to provide a path that is unconstrained by cost considerations at this time. These goals, when the schedule is complete, should result in the "best schedule" for everyone based on information available.

Third, the cost goals follow both technical performance and schedule goals. These goals are derived from computing the cost of the individual parts of the project and any changes in prices resulting from a time-driven cost change.

Given these three parameters and the associated goals, there must be a reconciliation to assure that all goals are compatible. Building a product with many more features and functions than are needed will affect the schedule and cost. Therefore, the technical performance goals need to be linked to what the customer wants and is willing to pay. Schedule goals are often driven by the availability of resources: human, nonhuman, and financial. Ideally, the schedule describes the time path for delivery of the product when the customer needs it. Customer need for the product is a major consideration in schedule development. Cost, of course, is a function of the price of developing and constructing the product with the associated overhead costs and any change of price over time.

In consideration of the foregoing discussion of project goals, it is essential that the project manager have a means of validating the technical performance gains throughout the project. Some suggested means of validating technical performance are:

- Conducting technical reviews periodically throughout the project's life cycle
- Testing components, assemblies, subsystems, and the final system
- Ensuring the materials and workmanship meet the specifications
- Conducting peer review of work
- Assuring the human resources are technically qualified to perform the work
- Conducting form, fit, and function audits
- Maintaining a rigorous configuration management process

Schedule performance is based on realism in the time estimates and the quality of the schedule when it is developed. Realism and schedule quality are directly affected by the amount of information available during planning and the optimism of senior management, the project manager, and the planners. Some suggestions for validating schedules are:

- Conduct an independent assessment of the schedule before project implementation to test for realism and level of detail.
- Check for the number of assumptions used and the validity of the assumptions. Too many assumptions or optimistic assumptions materially affect the schedule.
- Check schedule data against actual data frequently early in the project to validate the ability to implement the schedule as planned.
- Watch technical challenges that can drive the schedule if technical performance is not met or there is rework.
- Informally talk to the performing individuals to obtain their insight on schedule viability.
- Catch problems early and correct them before they become big problems.

Good cost performance results from the efficiency of the technical work and the schedule stability. Some suggestions for validating the cost are:

- Check the degree of precision of all major cost items and cost drivers for the project to ensure valid estimates.
- Check expenditures to ensure they are required and directly related to the project.
- Maintain an informal system to collect costs as they occur if the organization's accounting system does not provide timely cost data.
- Ensure there is a work authorization plan in place and working to release work that is in the project.
- Update cost estimates for major items to ensure there have not been major changes to prices.

- Check to ensure that the price of labor being used matches that in the plan; don't use expensive labor for jobs that can be accomplished by less expensive resources.
- Maintain tight control over contractors and any anticipated change of contract scope.
- Establish a contingency reserve for discovered new work and a management reserve for inefficiencies in the system.

Perhaps the most important aspect to any project, large or small, is to have a valid plan in place before the work is started. There is an old saying, "Plan your work and work your plan." Poor planning will typically result in substandard performance. Plans may not always work, but they provide the best chance to meet the goals.

The project plan includes all the metrics for the project. These "benchmarks" over time will tell whether the project is converging on the right solution. The situation may change, but a good project plan can be easily adapted to meet the new requirements. A weak or flawed plan causes major replanning efforts as well as disruption in the flow of work.

14.9 TO SUMMARIZE

Some of the key messages that were presented in this chapter include:

- EVMS is a mature system that has been used since 1963 in the Department of Defense, but called cost/schedule control systems criteria (CSCSC).
- EVMS demands that project planning be thorough and detailed to be effective in measuring performance.
- EVMS is a cost-schedule integration that links work package cost and schedule duration.
- EVMS provides trend data to predict positive and negative trends for the project's cost and schedule.
- There are 32 criteria in five categories: organization, planning and budget, accounting, analysis, and revisions and access to data.
- Major projects are the best candidates for EVMS implementation because of the cost associated with establishing and maintaining the process.
- Use of EVMS dictates that a disciplined process be used for planning, data collection, data analysis, and data reporting.
- Budgeted cost of work scheduled (BCWS) is the planned baseline for the EVM implementation.
- Actual cost of work performed (ACWP) is actual cost data for completed work based on the real cost for the work.
- Budgeted cost of work performed (BCWP) is the estimated cost of the work completed.

- Changes to the project's scope are integrated into EVMS through a step function shown on the graphic chart.
- Through trend analysis, EVMS can predict the estimated cost to complete the remaining work and the total estimated cost at completion.
- EVMS only indirectly measures the technical performance of a project and that is through assessing the technical aspects of completed work packages.

14.10 ADDITIONAL SOURCES OF INFORMATION

The following additional sources of project management information may be used to complement this chapter's topic material. This material complements and expands on various concepts, practices, and the theory of project management as it relates to areas covered here.

- Dan Ono, "Project Evaluation at Lucent Technologies," chap. 27 in David I. Cleland (ed.), *Field Guide to Project Management* (New York: Van Nostrand Reinhold, 1997).
- Robert H. Kohrs and Gordon C. Weingarten, "Measuring Successful Technical Performance: A Cost/Schedule/Technical Control System," in David I. Cleland, Karen M. Bursic, Richard J. Puerzer, and Alberto Y. Vlasak, *Project Management Casebook,* Project Management Institute (PMI). (Originally published in *Proceedings,* PMI Seminar/Symposium, Montreal, Canada, 1986, pp. 158–164.)
- ANSI Standard ANSI/EIA 748-98, *Earned Value Management Systems,* American National Standards Institute, 1998. This standard establishes the criteria for implementing an earned value management system and administering it. The 32 criteria for an accepted earned value management system are documented here.
- Quentin W. Fleming and Joel M. Koppelman, *Earned Value Project Management,* 2d ed., Project Management Institute Publications, September 2000. This book is a distillation of information from various government agencies to put the concept of EVM into perspective. The authors have included the essentials for project management and incorporated extensive experience in how to implement and use EVM in an effective manner. It is easy to understand and apply the principles of EVM because of the authors' presentation of the information.
- Wayne F., Abba, "How Earned Value Got to Primetime: A Short Look Back and Glance Ahead," *Proceedings,* Project Management Institute Annual Seminar/Symposium, Houston, September 7–16, 2000, pp. 763–764. This article is an overview of EVMS and its history, and includes those countries interested in EVM as a system for their projects. It touches on the evolution of EVMS from its early days in the 1960s to the current interest in EVMS as a valuable tool for organizations to use to manage by data. The author cites major Department of Defense programs that have benefited from the use of EVMS.

- Gary Humphreys, "Industry Ownership of Earned Value Management Systems (EVMS)," *Proceedings,* Project Management Institute Annual Seminar/Symposium, Houston, September 7–16, 2000, pp. 739–748. This article describes the EVMS and some planning considerations essential to establishing an EVMS. It contains the formulas and metrics of EVMS coupled with descriptions of the metrics. The article contains some excellent explanations of the operation and meaning of critical data reported in the EVMS as well as its meaning for the project leadership.

14.11 DISCUSSION QUESTIONS

1. What are the merits of using trend analysis to confirm that a project is on schedule within the budget or behind schedule and over budget?
2. What are some side benefits of invoking EVMS in a contract for a project other than the ability to track cost and schedule data?
3. What assurance does EVMS give that the technical performance is progressing along with the cost and schedule?
4. In implementing EVMS within an organization, what changes might be expected in the project planning phase of the project's life cycle?
5. What additional data collection would be required to implement an EVMS on a project?
6. When would it be disadvantageous to implement EVM on a project? What criteria would apply to the use/nonuse of EVM?
7. What effect would a change of scope have on the EVMS functions during the project execution phase?
8. Who is responsible for the operation and effective use of the EVMS during project execution? Why?
9. What differences are there between a project with an EVMS and one without an EVMS?
10. Are there elements of EVM that could be used to manage the cost, schedule, and technical performance of a project without using all the capabilities?
11. When there is an adverse trend in both cost and schedule on a project that is 40 percent complete, what are some of the actions that might be taken to correct the situation?
12. What is your opinion of implementing EVM in just one project? Is it possible to do that without the parent organization becoming involved?

14.12 USER CHECKLIST

1. Why use an EVMS when there are controls currently within the projects being pursued in an organization? What added value is there to using an EVMS?

2. When the project's scope is ill defined, what impact would this have on project planning for EVMS?
3. What are the benefits of an EVMS that justifies its cost for planning and implementation on major projects?
4. What is the shortest-duration project that would benefit from implementing EVMS when the data collection and reporting cycle is 12 days?
5. What probable action would be appropriate if an EVMS reported that the cost and schedule variances are at 20 percent, 23 percent, and 29 percent for three consecutive weekly report periods?
6. What changes would need to be made to the planning activities in your organization if EVMS was implemented and how would this affect the organization?
7. How would you implement EVMS in your organization for a large government project that is pending award on the basis of assessment of your EVM capability?
8. You are awarding a major contract to an organization that will be paid incremental monthly payments on the basis of earned value calculations? What assurances from earned value do you need to make the payments?
9. During an audit, you find that your organization does not use a WBS to divide the work into manageable pieces (work packages). What difficulty will your organization have in implementing an EVMS without a WBS approach to project definition?
10. Schedules have historically been summary-level graphics without detail or good estimates of task durations. How does this affect implementation of EVMS in your organization?
11. As the project manager for a new, major project, you have been asked to brief senior management on implementation of EVM. What top-level requirements would you place in your briefing to show how you will use EVM?
12. Your EVMS has been qualified and meets all criteria, but about halfway through a project the data show radical changes to the trend through "discovered" new work. What actions would you take to validate the situation to confirm that the project is out of control?

14.13 PRINCIPLES OF PROJECT MANAGEMENT

1. Detailed project planning is essential to scope the project and to construct the foundation for effective implementation of EVMS on any project.
2. EVMS is only as accurate as the data collection system supporting it.
3. EVMS does not control a project; it only provides information to decision makers for corrective action.

4. EVMS is not a police system; it is a system for measuring the cost-schedule performance of a project.
5. Manager understanding and interpretation of the information is a critical part of the EVMS.

14.14 PROJECT MANAGEMENT SITUATION—IMPLEMENTING EARNED VALUE

Several senior managers at Goflex Corporation attended a seminar where earned value management was presented as a topic and promoted as the best system going to keep track of progress on major projects. As both a prime and major subcontractor for contracts requiring the use of earned value management, it was necessary for Goflex to have an earned value management system in place and report cost and schedule data.

Goflex president, Michael Lyons, gave his explanation of the earned value management system. "It is easy to understand, the baseline always starts in the lower left corner and extends to the upper right corner. Never saw it done any differently. The baseline has a name like, bekweekus, and there are a couple of other lines filled in as the information is put in the system. As long as these lines don't vary too much from the baseline, we are okay. The project manager's job is to be sure to generate the data so that it follows the lines. Nobody understands what it means anyway."

The senior managers did not feel comfortable trying to explain to Mr. Lyons the reason for the earned value management system or the impact that it has on managing the projects. Mr. Lyons didn't realize that progress payments are made on the basis of the progress shown by the figures in the system. Any "adjustments" to the figures, as may have been implied by Mr. Lyons, would be in direct violation of the contracts with customers.

Goflex's project managers were generally familiar with an earned value management system, but often felt that they did not have the support of senior management when it came to devoting time to planning the project to meet the needs for an effective earned value management process. Collectively, the project managers noted some weaknesses in the planning and execution that would preclude full implementation of an earned value management system.

- The WBS was never completed in full for any project. A summary-level WBS was considered sufficient.
- The schedule used many milestones for completion of tasks, but the tasks were not planned with resources and associated costs.
- Cost estimates were made at summary levels because the work was primarily "level of effort."
- Configuration control for the project was nonexistent. Changes to the project scope were made informally without documentation.

- The project management information system was not capable of providing information on a timely basis.

In the past, project managers just provided information on earned value parameters that was believable. No one questioned the data and no one checked to see where the data came from. At the end of projects, there was a reconciliation that was covered by "discovered" work or rework to justify the late deliveries and cost overruns.

14.15 STUDENT/READER ASSIGNMENT

1. The corporation has less than a full understanding of the earned value system, and the first step is to raise the level of understanding in senior management. Prepare a briefing that explains the fundamental graph with BCWS, BCWP, and ACWP lines to be used to brief Mr. Lyons and associates.
2. Identify the components of project planning that support the EVMS. Describe the planning activities and give criteria for the adequacy of each activity to ensure support to the EVMS.
3. What benefits for project effectiveness and efficiency do you see as a result of an effective EVMS being implemented in Goflex? What is the additional cost (dollars, labor, other) to Goflex?
4. What type of training do you think is needed for the project managers if an effective EVMS is implemented?
5. What type of training is needed for Goflex's senior management on an effective EVMS? Identify the key points that must be made.

CHAPTER 15
PROJECT TERMINATION

"If you can look into the seeds of time, and say which grain will grow and which will not."
WILLIAM SHAKESPEARE, 1564–1616
Macbeth, Act I, line 58

15.1 INTRODUCTION

All projects end, either as a casualty because of excessive cost and/or schedule overruns, or failure to accomplish their performance expectations. Also, a project may be terminated if its expected results no longer have a "strategic fit" in the purposes of the enterprise. In addition, a project that completes its life cycle and transfers its results to the operations of the sponsoring enterprise is in a form of termination that project team members hope to have. Whatever the form of "termination," such termination should come out of a project monitoring, evaluation, and control process.

This chapter starts off with the question, Why terminate? The types of project termination as briefly described in the previous paragraph are suggested. The strategic implications of a project termination are dealt with, and a description of projects in trouble is given. How some projects seem to have an endless life, even though they may be "losers," is considered. Some termination strategies are suggested, along with an examination of termination possibilities. A set of procedures is suggested for terminating a project. The chapter ends with a description of typical posttermination activities.

15.2 WHY TERMINATE?

Projects usually are terminated for two basic reasons: project success or project failure. Senior managers, who "own" the project and who see the project as a building block in the design and execution of organizational strategy, must create a cultural ambience that encourages projects to be successful, but also allows a project to fail if it has lost its strategic fit in the organization's plans for the future.

Project success means that the project has met its cost, schedule, and technical performance objectives and has been integrated into the customer's organization to contribute to the customer's mission. A successful project means that the organization has been successful in positioning itself for the future; a specific strategy has been designed and implemented.

Project failure means that the project has failed to meet its cost, schedule, and technical performance objectives, or it does not fit in the organization's future. Failure and success are relative factors.

Project termination comes about for several reasons:

- The project results have been delivered to the customer. If appropriate, service and maintenance contracts can be negotiated and consummated.
- The project has overrun its cost and schedule objectives and/or is failing to make satisfactory progress toward attaining its technical performance objectives.
- The project owner's strategy has changed such that the project no longer has a strategic fit in the owner organization's future.
- The project's champion has been lost, thereby putting the continued application of resources on the project in doubt.
- Environmental changes have emerged which adversely influence the project's future.
- Advances in the state of the art hoped for in the project (such as in research and development) have not been realized, and therefore further funding is not forthcoming.
- The project's priority is not high enough to survive in competition with higher-priority projects.

Of course, the lines of demarcation between the projects falling into these situations are not always clear. These situations merely provide a framework for approaching inevitable project termination questions.

Many projects never survive the entire life cycle to become a new product, service, or organizational process. For example, Chrysler scrapped a development project for a new luxury car. Code named LX, the car would have been the flagship of the Chrysler fleet. Clearly, Chrysler had the intellectual and financial wherewithal to complete this project. Requiring an investment of around $300 million, the car would have been built on an existing platform. Why was the project taken around the corner and shot between the eyes? The margins were going to go away because there were so many competitors coming into the same point of the market.[1]

Most projects do not have a sharp beginning. As the project enters its life cycle and as the true cost, time, and performance parameters of the project become better known, senior management at some point must consider the inevitable decision of whether to discontinue the project. Project termination should not be viewed as a failure but rather as a strategic decision implemented when a project does not or may not support the organizational strategies. Adams and Dirlan note that large

[1]Jerry Flint, "The Car Chrysler Didn't Build," *Forbes,* August 12, 1996, pp. 89–91.

firms that are known for their leadership role in successful innovations are also the ones that have undertaken a large proportion of unsuccessful projects.[2] The organization that does not make the often difficult, yet necessary, decision to terminate a project at the appropriate time can incur significant losses.

Sometimes projects fall on hard times, such as in the case of the National Ignition Facility (NIF). This project involved the building of a large laser, which would allow bomb makers to study the physics of nuclear weapons without exploding them, thus helping assure the reliability of the U.S. nuclear stockpile. In addition, the project, originally estimated to be about $2.2 billion when completed, would have enabled engineers to explore the possibility of commercial power plants based on a type of "laser fusion" and would offer scientists the chance to probe matter under conditions never before created in a laboratory. Revelations of sky-high costs, technical issues, and overly optimistic sales messages in late summer of 2000, damaged Livermore's credibility. The General Accounting Office, the U.S. congressional auditing agency, reported that the true costs of the project would be at least $4 billion, and that it would likely not be finished before 2008. An internal review at Livermore turned up the problems after numerous external reviews missed them.[3]

The Boeing Company recently canceled a project to build a pair of "superjumbo" jet models in favor of shifting a strategy to battle arch rival Airbus Industry in offering smaller and higher-performance versions of its existing aircraft. The planned superjumbo jets were envisioned to be a version of the 747 that could carry 500 or more passengers for distances of as many as 10,000 miles. Boeing had reached a threshold on the project, which would have required it to commit to spending $5 to $7 billion or even more on the new superjumbos. The life cycle for a complex aircraft such as the 747 superjumbo requires 4 or 5 years of resource commitment before first deliveries because of the complexity of such systems as landing gear, cockpit systems, and engines requiring long periods of development and certification. Part of the reason for cutting out the project was insufficient customer interest to justify the development expenditure.[4]

15.3 TYPES OF PROJECT TERMINATION

Spirer addresses project termination as consisting of two broad types: first, a natural termination when the project goals have been met, and second, an unnatural termination when some project constraints have been violated, performance is inadequate, or the project goals are no longer relevant to some overall needs. Emotional issues with a project termination include:

- Fear of no future work
- Loss of interest in task remaining

[2] W. Adams and J. B. Dirlan, "Big Steel, Invention and Innovation," *Quarterly Journal of Economics,* vol. 80, May 1966, pp. 167–189.

[3] James Glanz, "A Great Hope of Physics Falls on Hard Times," *The New York Times* (*Science Times*), September 26, 2000, pp. D1–D2.

[4] "Investors Cheer Boeing's Cancellation of New 747s," *The Wall Street Journal,* January 22, 1996, p. A-33.

- Loss of project-driven motivation
- Loss of team identity
- Selection of personnel to be reassigned
- Reassignment methodology
- Division of interest

The natural termination of a project is concerned with:

- Identification of remaining deliverable end products
- Certification needs
- Identification of outstanding commitments
- Control of charges to the project
- Screening of uncompleted tasks not needed
- Closure of work orders and work packages
- Identification of physical facilities assigned to the project
- Identification of project personnel
- Accumulation and structuring of project historical data
- Disposing of project material
- Agreement with client on remaining deliverable end products
- Obtaining needed certifications
- Agreement with suppliers on outstanding commitments
- Communicating closures
- Closing down physical facilities
- Determining external requirements for audit trail data[5]

Spirer suggests the use of diagrams and checklists as analytical tools for the management of project termination.[6] In addition, he suggests a work breakdown structure for the problems of project termination, as shown in Fig. 15.1.

15.4 STRATEGIC IMPLICATIONS

When the project is overrunning its costs and schedule, then termination should be considered. When termination is being considered, the project should be evaluated from its strategic context: Does the project continue to have a strategic fit in the design and execution of organizational strategies? By asking and seeking complete

[5]Herbert F. Spirer, "Phasing Out the Project," in D. I. Cleland and W. R. King (eds.), *Project Management Handbook* (New York: Van Nostrand Reinhold, 1983), pp. 254–255.
[6]Ibid., pp. 256–260.

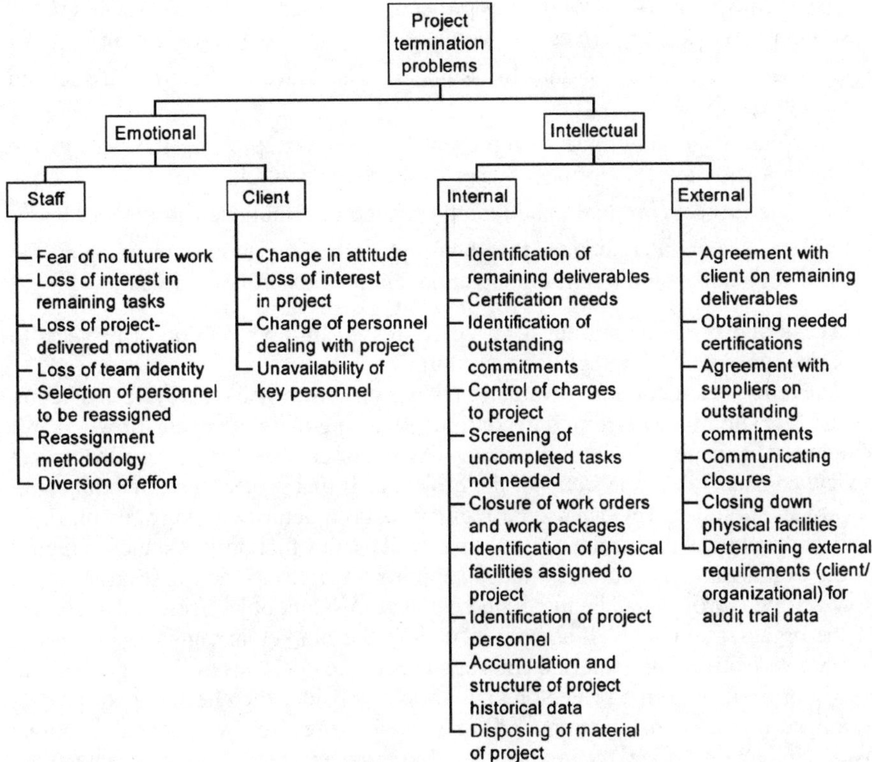

FIGURE 15.1 Work breakdown structure for project termination. [*Source: Herbert F. Spirer, "Phasing Out the Project," in D. I. Cleland and W. R. King (eds.),* Project Management Handbook *(New York: Van Nostrand Reinhold, 1983), p. 248.*]

and candid answers to the following questions, insight into the project's strategic context can be gained:

- Does the project continue to fit in the strategic plans of the organization?
- Does the project continue to complement a strength of the organization?
- Correspondingly, does the project avoid a dependence on a weakness of the organization?
- Are the project results likely to be consistent with the strategy of the sponsoring organization?
- Will the project continue to help that organization achieve its objectives?
- Will the completion of the project help that organization accomplish its goals?
- If the project results are put into an operating mode, will these results provide a competitive advantage to the sponsoring organization?

- Is the project consistent with other projects and programs that are related to the strategic mission, objectives, and goals of the sponsoring organization?
- Can the project owner continue to assume the financial and other risks associated with the project?
- Does the project continue to represent a specific step along the way to the accomplishment of the project owner's objectives and mission?
- Does the project continue to be directly related to established strategies?
- Does the project team believe that the project continues to have a strategic fit in the design and execution of organizational strategies? If not, why not?

These questions are similar to those addressed in the selection of projects to support a company's strategy, discussed in Chap. 4.

The crucial element of the evaluation of a project during its life cycle is whether the project should be permitted to continue maturing in its life cycle, or whether it should be terminated in an orderly fashion. All too often a regularly scheduled project review considers only the status of the project as if that project were standing still. A project should be reviewed on a regular basis to determine its status on time, cost, and technical performance factors, as well as how that project stands in regard to the organizational strategies that it supports. A bottom-line question is to ask how well that project can be integrated with the mission, objectives, and strategies of the organization, as well as how it fits into the market strengths, weaknesses, comparative advantage, internal consistency, and key policies of the organization.

Major reviews of the project always should consider the alternative of project termination. Such considerations and analysis will be received with mixed emotions by the project owner and the project team. Experienced project management people can think of times when a project should have been canceled. The Lockheed L1011 project and the Washington Public Power Supply System debacle (often referred to as WHOOPS) are striking examples of projects that might better have been terminated when early difficulties were encountered. In the nuclear power plant construction industry, there are plants that have had awesome difficulties and, after many years of effort with billions of dollars invested, have not become operational. Even some that have been completed have not had their operating license approved.

15.5 CONTINUING THE "LOSERS"

Why are some projects continued even though the project may be supporting a dying cause? The L1011 Tri-star Jet Program of Lockheed was known to be unlikely to earn a profit. For more than a decade it accumulated enormous losses. Because the program was Lockheed's reentry into commercial aviation, it became a symbol, which broadened Lockheed's image beyond simply being a defense contractor.[7]

---

[7]Barry M. Staw and Jerry Ross, "Knowing When to Pull the Plug," *Harvard Business Review,* March–April 1987, p. 71.

Staw and Ross suggest that there are several factors that encourage decision makers such as project managers and project owners to become locked into continuing strategies that cause the organization to lose: the project itself, managers' motivations, social pressures, and organizational pushes and pulls.[8]

Within the project itself, short-term problems are not likely to discourage the continuation of the project because such problems are usually looked on as necessary costs or investments for achieving the project's purposes. Project managers easily can view each setback as a temporary situation that can be corrected with more time and money.

A project's termination costs also may impede withdrawal. Managers sometimes don't fully fathom closing costs and salvage value when making the initial decisions to start the project. New projects are supposed to succeed rather than fail. Psychological factors influence the way the manager gathers information about the project and interprets and acts on that information. Managers often are rewarded for coping with short-run problems, "staying the course," and persevering to success.

People often fail to recognize when a project is beyond hope. We have an uncanny ability to see only what is in accord with our beliefs and commitments. If a manager is convinced that a project will produce results, there is a predisposition to slant original estimates of costs and other data. Sometimes when managers recognize that a project is in real difficulties, they may choose to invest more resources in the project rather than accept failure.

Social pressures include the managers' unwillingness to expose their shortcomings and mistakes to others. If the project's success is tied to a perceived loss of power or loss of a job, then hanging on even in the face of grave difficulties makes sense to the decision maker. When an individual becomes identified with a project, he or she tends to defend the venture despite mounting losses and doubts about its feasibility. Then, too, managers are expected to "weather the storm" like Winston Churchill and Lee Iacocca and hang tough until they are successful. We see managers who are persistent as a sign of leadership; withdrawal is a sign of weakness. Given these perceptions, why would we expect managers to back off from a losing project?

Organizational pushes and pulls include the inertia that impedes withdrawal from losing projects. Sometimes it's just easier not to rock the boat. Organizational politics sometimes prevents a project termination. A project that supports a long-standing organizational strategy or company identity is not easily terminated even in the face of declining sales or profitability.

Why do people get hooked into supporting a project far beyond the point where the project should have been terminated? Staw and Ross comment on the difference between an objective evaluation of the project's possible and probable future success. They are concerned with how this objectivity can be clouded by those people who manage the project and become emotionally attached to seeing that it is successful—no matter what the odds or the evidence suggests.[9]

---

[8]Ibid.

[9]Ibid., p. 68.

Projects are supposed to succeed—not fail. Yet a project that is a recognized failure can be a benefit to the organization. A company that cuts its losses by terminating a failing project has made an important and timely strategic decision, a decision that should be approached with as much discipline and analysis as the decision to launch the project. Project termination should not be considered a failure, but rather a necessary key decision in the design and execution of organizational strategies. Project termination is an integral part of the divestment phase of the project to stop work on the project and transfer the results to the owner organization for use in the owner's ongoing business.

Orderly termination, for whatever reason, is a strategic management responsibility of senior managers. Termination is an option to be considered when one or more of the following conditions exist:

- There are serious cost and schedule overruns.
- Technical performance is compromised, or technical risks are too great.
- The project does not have a strategic fit in the sponsoring organization's future.
- The customer's strategy has changed.
- Competition has made or threatens to make the project results obsolete.
- The purposes for which the project was originally established have changed.

15.6 PROJECTS IN TROUBLE

When a project is experiencing serious cost and schedule overruns, termination may be a real issue, particularly from the strategic management perspective of the corporation. At such times, an outside audit should be conducted by either outsiders within the owner organization or an outside agency such as a consulting firm. Such audits can give insight into the reasons for the cost and schedule overrun and help determine whether the project has become a termination candidate. At such times, senior managers' most important task is to create a cultural ambience that incorporates honest and frank disclosure.

If the project continues to play a vital role in the design and execution of organizational strategies and if it has sustained cost and schedule overruns, an important step must be taken to reevaluate the project as comprehensively as possible. Such a review will enable senior managers to determine where the basic responsibility for cost and schedule overruns lies. If the review determines that the responsibility lies in an inadequate initial design, then engineering and redesign leading to a recosting and rescheduling of the project make good sense. During such a redesign, the doubtful or not fully understood parts of the project should be analyzed. The project team may not be able to terminate a project simply because that project team may have an understandably subjective view of the project. An audit should be conducted by an independent team, one that is not involved in the project. Such a team's effort could go beyond the audit and on into a full design review, recosting, and development of remedial strategy to get the project back on track. Senior managers who own the project also must be aware that a cultural ambience that encourages projects to succeed must also allow projects to fail if those projects fill no strategic role.[10]

Davis suggests several questions that should be asked when a project overrun threatens:

- Does the project involve pioneer technology?
- Is it a new project, and what experience has the project manager had in implementing this type of project before?
- Is it a bigger project than the company has handled before?
- Was the costing done on the project before the design was completed?

If the answers to any of the first three questions on the project's newness are yes, Davis opines that it is probable that you have a serious overrun. If the fourth question is answered in the affirmative, then it is likely that you have a serious overrun arising from incorrect estimates.[11]

15.7 TERMINATION STRATEGIES

If a project is to be terminated, some senior managers will replace the project manager with an individual who is skilled at closing out projects. Such a termination project manager would be wise to conduct an immediate review of the status of the work packages, along with the funding, schedule, and technical performance parameters. Several other things must be done:

- Ensure that all project deliverable end products have been provided to the project owner and that all project functional work is finished, along with any closeout of records.
- Review the status of all contracts to ensure that requirements have been met or provisions made if such requirements have not been duly satisfied.
- Work with the project team in developing and distributing a closeout plan that provides guidance for an orderly termination of all elements of the project.
- Maintain an ongoing surveillance of the closeout activities, including the closeout of all records and the disposition of materials.
- Notify relevant stakeholders of the termination.
- Ensure that all financial matters on the project have been satisfactorily terminated.
- Assist members of the project team to find other work in the organization.
- Prepare the project history, particularly a "lessons learned" report, so that future teams in the organization can benefit from the experiences of the project.

[10] Paraphrased from David Davis, "New Projects: Beware of False Economies," *Harvard Business Review*, March–April 1985, pp. 95–101.

[11] Ibid., pp. 96–97.

- Conduct a postcompletion audit of the project to identify strengths and weaknesses in the management of the project, what impact mistakes have had, how such mistakes can be avoided in future projects, and how that organization was affected, positively or negatively, by the project.

15.8 EVALUATION OF TERMINATION POSSIBILITIES

Termination is always an option on every project. A strategy for management to use in dealing adequately with the ever-present termination issue should include these steps:

- Review the project and its strategic context on a regular, disciplined basis.
- Recognize the psychological and social forces that motivate one to "stay the course."
- Recognize that there are a prevailing belief and cultural force that encourage the commitment of more resources to solve current difficulties and assure that success is "just around the corner."
- Define, with senior management participation, what constitutes both success and failure on the project. This definition is needed at the start of the project and should be reinforced at major review points during the project life cycle.
- Listen carefully to the concerns of others about the project. What are the project stakeholders saying? Are they saying that project termination is a good strategy?
- Evaluate the real ability of the project team to listen to and hear bad news. Does such news carry important information about the project's health, continuation, or termination?
- Ask whether the managers "bet too much of the farm" on the project where a termination would "break the bank," resulting in a perception of both organizational and personal failures.
- Determine if the project manager feels that a lot of people will have their futures adversely affected if the project is terminated.
- Step back and evaluate the project from an outsider's perspective. Use of an internal task force or audit team can help in getting such an outsider's viewpoint.
- Encourage project team members always to provide accurate information, even if that information contains messages that are not palatable and might suggest that project termination is an alternative worthy of full consideration.
- Consider replacing key members of the project team with new people who can bring a perspective less influenced by the project and past events. Consider replacement of the project manager.
- Build an organizational culture that supports the philosophy that projects are experimental, temporary uses of resources to support organizational strategies and require constant surveillance to guard against a project becoming a permanent fixture in the organization.

15.9 TERMINATION PROCEDURES

It is beyond the scope of this book to provide detailed termination procedures. Two books provide excellent guidance in this respect. Archibald suggests some comprehensive checklists as aids in planning and controlling the work necessary to terminate a project. He concludes that there are benefits of using such checklists to do the following:

- Clearly indicate the closeout functions and responsibilities, reducing ambiguity and uncertainty.
- Reduce overlooking of important factors.
- Permit closeout progress to be monitored.
- Aid project team members with little or no experience in closing out a project.
- Inform project team members about the activities of others during the closeout phase.[12]

Kerzner and Thamhain suggest a sample listing of typical activities in six areas to consider in managing the affairs of project closeout and transfer:

- Documentation
- Contract administration
- Financial management
- Program management
- Marketing
- Final management review[13]

15.10 POSTTERMINATION ACTIVITIES

When projects are terminated, frequently certain posttermination activities are necessary to the project. Continuing service, maintenance, and logistics support may be an opportunity for future work. In some industries the follow-up service and maintenance contract work can be more profitable than the work of completing the project itself.

Audits can be a postcompletion activity. It is essential that the project managers ensure that adequate records are retained to support any postaudit activity that is initiated. British Petroleum (BP) conducts a postproject appraisal (PPA) of its major projects. British Petroleum's PPA strategies were also discussed in the control context in Chap. 13. Since the inception of PPAs in 1977, the company has

[12] Russell D. Archibald, *Managing High-Technology Programs and Projects* (New York: Wiley, 1976), p. 264.

[13] Paraphrased from Harold Kerzner and Hans J. Thamhain, *Project Management Operating Guidelines* (New York: Van Nostrand Reinhold, 1986), pp. 454–455.

appraised more than 80 of its project investments worldwide, including onshore and offshore construction projects, acquisitions, divestments, project cancellations, research projects, diversification plans, and shipping activities. The appraisals are done to improve company performance. Some characteristics of British Petroleum's postappraisal are as follows:

- Members of the appraisal team must have at least 15 years of broad-based experience with BP and have no affiliation with the projects they appraise.
- Project periods are examined from their conception usually until after they have become operational.
- Project records are reviewed, and all people involved in the project are interviewed. During the interviews, evaluators attempt to understand the psychology of the project members and the managers.
- Final appraisal reports are submitted to senior managers. Managers starting similar projects can review the reports.
- Four main lessons have been taught to BP management from their postproject appraisals: Determine costs accurately, anticipate and minimize risk, use a formal method to evaluate contractor selection and performance, and improve project management by setting up a "projects department" that helps project managers develop and use project management techniques.[14]

As projects continue to be recognized as building blocks in the design and execution of organizational strategies, termination of those projects that do not further the purposes of the organization becomes important. Earlier in this book the point was made that senior managers can maintain surveillance over the stream of product, service, and process projects as a way to determine if future strategies are evolving adequately. Of course, the principal purpose of such surveillance is to determine the schedule, cost, and technical performance promises of the department. Equally important is to view the stream of projects to determine which ones have a strategic fit with the enterprise's future purposes; also equally important is to identify those projects that should be eliminated if they are not consistent with the enterprise's evolving strategies.

In any project termination, it has to be clear what the project accomplished or did not accomplish. The project termination may be viewed differently by the customer, the contracting organization, the project manager, and the project team members. None of the stakeholders have the same stake in project completion and termination. The project termination manager must understand that. This requires that the project manager offer inducements of some sort to those stakeholders for whom termination is not seen as desirable. Clearly communicated project objectives and goals at the outset of the project, reinforced during the project's life cycle, can help facilitate a harmony of viewpoints among the project stakeholders, particularly during termination time.

---

[14]Frank R. Gulliver, "Post-Project Appraisals Pay," *Harvard Business Review,* March–April 1987, pp. 128–132.

15.11 TO SUMMARIZE

The major points that were communicated in this chapter include:

- Some projects succeed, and some projects are destined for failure. It is important to know the difference.
- Project success and failure are relative factors. Some criteria were given to use in determining when a project is likely to be in a failure mode.
- An enterprise should develop specific criteria to use in considering the question of whether or not an ongoing project should be reviewed to determine if it is likely to "fail."
- Examples from the Chrysler Corporation and the Boeing Company were cited to show the rationale used in these organizations for terminating their projects.
- Project termination consists of two broad types: (1) a natural termination when the project's objectives have been met and (2) an unnatural termination when some project constraints have been violated.
- When a project is terminated before it has reached the end of its life cycle, emotional issues will be involved. A sample of these emotional issues was cited in the chapter.
- A work breakdown structure for the challenges of project termination was presented in the chapter.
- A series of questions were presented in the chapter to help in gaining insight into the project's likely strategic fit in the enterprise's purposes.
- In any decision to terminate a project, consideration should be given to the alternatives that should be considered to replace the planned technical performance of the project.
- Subtle factors are at play, which can cause a project that should be terminated to continue.
- The project manager and the project team are unlikely and unsuitable candidates to make the decision of whether or not to terminate a project. The decision to terminate rests with the manager(s) who sponsored the project.
- Projects are supposed to succeed, not fail, yet termination of a project would make sense if such a project does not promise to contribute to the operational competence or strategic effectiveness of the enterprise.
- In the chapter a series of questions were presented that can help gather information to use in a decision of whether or not to terminate.
- Sometimes it makes sense to replace the existing project manager with a "termination" project manager. A series of actions and steps that such a project manager should take were suggested in the chapter.
- A posttermination project audit should be conducted to develop a profile of "lessons learned" concerning the effectiveness of the management of the project

and what might have been done differently in the management of the project that could have ensured the survival in the strategies of the enterprise.

- If a project is likely to be a candidate for termination, it is best to find this out as soon as possible during the life cycle of the project.
- It is no fun to terminate a project, but if the operational and strategic purposes of an enterprise are not being served by the project, then take it "around the corner and shoot it between the eyes."

15.12 ADDITIONAL SOURCES OF INFORMATION

The following additional sources of project management information may be used to complement this chapter's topic material. This material complements and expands on various concepts, practices, and the theory of project management as it relates to areas covered here.

- C. Lewis Penland, "Giving Mother Nature a Helping Hand," and Francis M. Webster, Jr. (ed.), "The Space Shuttle Challenger Incident," in David I. Cleland, Karen M. Bursic, Richard J. Puerzer, and Alberto Y. Vlasak, *Project Management Casebook,* Project Management Institute (PMI). (Originally published in *PM Network,* July 1994, pp. 14–22; and *Project Management Journal,* June 1987, pp. 41–68.)
- J. Rodney Turner, *The Handbook of Project-Based Management* (London: McGraw-Hill Book Company, 1999). This book provides a touch of realism. It is about new approaches to the practice of general management based on projects. The book is about managing change through the use of projects. The book maintains a single point contact—the emphasis is on the project manager. The book is intended for anyone engaged in the management of change and the management of supporting projects to facilitate the change. The reader should pay attention to chap. 15, "Finalization and Closeout," for guidance on the management of the termination of a project.
- Joan Knutson, *Project Management for Business Professionals* (New York: Wiley 2001). This book focuses on both the technical and human sides of project management. The experts who contributed to this book address the procedures and processes for managing projects. Included in this examination is an exploration of relations and the political and organizational considerations that can impact a project. The people who have contributed to this book include the foremost practitioners and researchers from academia, consulting, and industry.
- Terry Cooke-Davis, "Project Closeout Management: More than Simply Saying Good-bye and Moving On," in Joan Knutson (ed.), *Project*

Management for Business Professionals (New York: Wiley, 2001). In this chapter the author presents the activities that project managers generally advocate and that should be a part of the project closeout management. He builds a strong case as to why closeout management is vitally important to project stakeholders. In addition, the author reminds the reader of the factors that prevent organizations from managing effective project closeout strategies. The paper ends with a citation of the steps that organizations can take to improve closeout management.

- Carl L. Pritchard, "Project Termination: The Good, the Bad, the Ugly," in David I. Cleland (ed.), *Field Guide to Project Management* (New York: Van Nostrand Reinhold, 1998). The author suggests that project-termination models fall into three very general categories: positive, negative, and premature. He further notes that although termination has some standard practices, each type of termination requires specific attention. A valuable figure is provided at the end of the material, which portrays a "closeout methodology."
- Harry M. Staw and Jerry Ross, "Knowing When to Pull the Plug," *Harvard Business Review,* March–April 1987. Portions of this article have been paraphrased in the chapter. However, the article is of sufficient importance in considering the forces involved in continuing a project, even to the point of irrationality, that a reading of the entire article is worthwhile.

15.13 DISCUSSION QUESTIONS

1. Discuss various situations in which projects may fail that indicate the need for project termination. How can managers recognize these situations?
2. List and discuss quantitative and qualitative factors that should be assessed when you consider project termination.
3. What are some of the emotional issues involved in project termination?
4. Describe some of the project manager's tasks with respect to project termination.
5. The strategic fit is an important aspect of major projects. What questions can managers ask in order to evaluate strategic fit?
6. Why are some projects continued even when failure is obvious? What role do psychological factors play?
7. What steps can managers take in order to comprehensively reevaluate an ongoing project so as to address the termination question?
8. Discuss some of the termination strategies described in the chapter.
9. Discuss the importance of an outsider's perspective on ongoing projects.
10. What are some of the steps involved in the termination procedure?
11. What posttermination activities are important? Why is generation of "lessons learned" important?

12. Customer acceptance is an important part of project termination. What can management do to ensure that the customer is satisfied with project results?

15.14 USER CHECKLIST

1. Does the cultural ambience of your organization encourage project success and also allow for failure? Describe some examples to explain your answer.
2. Are any of the ongoing projects within your organization in a situation that indicates project termination? Are these projects being terminated? Why or why not?
3. Do the managers of your organization usually recognize the need for project termination? Who has responsibility for eliminating those projects that have no further value for the organization?
4. What qualitative and quantitative factors are assessed by the project managers of your organization in order to make a termination decision?
5. How are the emotional issues of project termination addressed in your organization? Are project team members usually satisfied with termination decisions? Why or why not?
6. What tasks are performed by project managers in your organization during project termination?
7. Do the project leaders periodically assess the strategic fit of projects? What questions are posed?
8. Do projects in your organization, which seem destined to fail, often continue beyond where they should have been terminated? Explain.
9. Do the necessary information and control systems exist to enable project managers to make a termination decision? Explain.
10. What termination strategies are used by project managers in your organization? How?
11. Do termination procedures exist for projects in your organization? What steps are involved?
12. Do project managers compile a list of lessons learned after the completion of a project? What other posttermination activities are done?

15.15 PRINCIPLES OF PROJECT MANAGEMENT

1. The expected project results should support the "choice elements" of the enterprise.
2. The criteria of project "success" and project "failure" should be established by the enterprise.

3. Each project in the "project portfolio" of the enterprise should be periodically evaluated to determine whether to continue or terminate the project.
4. There are organizational, political, and social issues likely to support the continuation of a project in an enterprise.
5. The termination cycle of a project can be treated as a project unto itself.
6. A postproject appraisal is an effective means of developing a "lessons learned" profile on a project.

15.16 PROJECT MANAGEMENT SITUATION—SHUTTING DOWN A PROJECT

Termination of a project involves bringing a project to a planned and orderly conclusion. Project termination should be planned with care and attention as are other project life-cycle phases. One of the purposes of a project termination plan is to discourage the loss of key people when termination is imminent, leaving others to "clean up" the project affairs. An important part of project termination is to provide assistance and guidance to the members of the project team who are being demobilized. The project manager should consider the following actions:

- Prepare and issue a project termination plan.
- Hold project termination meetings with the project team members to finalize remaining tasks.
- Be available for counseling with project team members about reassignment opportunities, emotional issues, and career opportunities.
- Determine how the final documentation will be distributed, and to whom.
- Work with team members to assure clear phaseout protocols in terms of individual and team responsibilities.
- Meet with human resource individuals, functional managers, and line managers as needed to identify project termination personnel needs, and assist team members in scheduling interviews, orientation of remaining members, and if appropriate, incoming personnel who are brought in for the termination of the project.
- Have a final meeting to thank everyone—and recognize the distinguished contributions of project team members.

Another important part of project termination involves: (1) procedural issues and phaseout administrative procedures; (2) transfer of responsibilities; (3) cost closure activities, and (4) preparation of relevant documentation.

The following items should be addressed in the project closeout report:

- Final cost, schedule, and technical performance accomplishments
- Final cost report
- Closeout approvals

- Finalization of permits, licenses, and/or environmental issues
- Closeout of contracts
- Results of a "project termination audit" with a citation of the major lessons learned
- Final supporting documentation
- Turnover and acceptance of official project files
- Assistance in the development of operational phase planning and resource allocation
- Development and finalization of an official closeout agreement(s)

15.17 STUDENT/READER ASSIGNMENT

The student/reader is asked to evaluate the project closeout strategies as suggested in the foregoing material. What additional major closeout issues might be involved?

P · A · R · T · 5

INTERPERSONAL DYNAMICS IN THE MANAGEMENT OF PROJECTS

CHAPTER 16
PROJECT LEADERSHIP

"To be a leader of men one must turn one's back on men."
HAVELOCK ELLIS, 1859–1939

16.1 INTRODUCTION

Leadership has been studied and written about extensively in the twentieth century. There have been thousands of articles and papers authored about the subject. Several hundred books have also appeared that examine the function of leadership in many different forms of organizations—governmental, industrial, and political to mention a few. One of the more recent examinations of the leadership function in modern organizations is a comparison of leaders and managers. One notable researcher, Warren Bennis, described a leader as someone "who does the right thing," and the manager as someone "who does things right." This is an interesting viewpoint that is explored more fully in this chapter.

This chapter tries to put leadership into a project perspective. Leadership is defined; some of the principal studies of leadership are presented. The nature of a leadership style that is appropriate for leading a project team is described. Leadership is offered from the perspective of the management of attention, management of meaning, management of trust, and management of self, and should arouse the personal interest of the student and reader. Project leadership is differentiated from other forms of leadership—and the competencies that project leaders should have are described. A clear differentiation between project leaders and project managers cannot be made because that differentiation involves an understanding of a complex set of variables. As the study and publication about leadership—vis-à-vis managership—continues, perhaps the authors will be able to offer further insight into these ideas in the future.

16.2 CONCEPT OF LEADERSHIP

The concept of leadership is surely as old as the concept of organized activity. Organizations have been led with varying degrees of effectiveness by people called "leaders" since the development of organized societies. Leadership in the political, social, military, legal, economic, and technological environments has been celebrated and studied throughout history. Today these studies continue to try to understand what both separates and harmonizes leaders and followers.

Our view of a project leader is that individual who leads a project team during the project life cycle and accomplishes the project objective on time and within budget. Project leadership is defined as a presence and a process carried out within an organizational role that assumes responsibility for the needs and rights of those people who choose to follow the leader in accomplishing project results. In this chapter the concept of project leadership will be presented. But before discussing project leadership, we will present some ideas about the general nature of leadership.

16.3 WHAT IS LEADERSHIP?

There are many definitions of leadership. Fiedler cites nearly a dozen different definitions with varying connotations and degrees of emphasis on elements. He defines a leader as "the individual in the group given the task of directing and coordinating tasks in relevant group activities or who, in the absence of a designated leader, carries the primary responsibility for performing these functions in the group."[1] Peter Drucker opines that effective leadership is based primarily on being consistent.[2] Arthur Jago defines leadership as both a process and a property. He states that leadership is the use of noncoercive influence to direct the activities of the members of an organized group toward the accomplishment of group objectives. He considers leadership in the context of a set of qualities or characteristics attributed to those who are perceived to successfully employ such influence.[3]

Jago notes that leadership is not only an attribute but also what the person does. Under this definition, then, anyone on a project team may take on a leadership role. In his research, Professor Hans J. Thamhain has identified that project management requires skills in three primary areas of abilities leadership: interpersonal, technical, and administrative. He goes further and offers some suggestions for developing project management skills needed for effective project management performance.[4]

The subject of leadership has received much attention, yet we do not have a universally accepted definition of the term. James McGregor Burns's book, *Leadership,* cites one study with 130 definitions of the term.[5] Another book notes over 5000 research studies and monographs on the subject. The editor of a handbook concludes that there are no common factors, traits, or processes that identify the qualities of effective leadership.[6] Most books tend to equate leadership with

[1]Fred E. Fiedler, *A Theory of Leadership Effectiveness* (New York: McGraw-Hill, 1967), p. 8.

[2]Peter Drucker, "Leadership: More Doing than Dash," *The Wall Street Journal,* January 6, 1989.

[3]Arthur Jago, "Leadership: Perspectives in Theory and Research," *Management Science,* vol. 28, no. 3, March 1982. Copyright © 1982 by the Institute of Management Sciences.

[4]See Hans J. Thamhain, "Developing Project Management Skills," *Project Management Journal,* September 1991, pp. 39–44.

[5]J. M. Burns, *Leadership* (New York: Harper & Row, 1978), p. 2.

[6]B. M. Bass, *Stogdill's Handbook of Leadership: A Survey of the Theory and Research,* rev. ed. (New York: Free Press, 1981).

a "hero person." Others view leadership as characterized by personal characteristics such as charisma, intelligence, energy, style, commitment, and so on. Other theorists view leadership as depending on anything from task conditions to subordinate expectations.[7]

Literally thousands of studies have explored leadership traits. Some of the traits relate to physical factors, some to abilities, many to personality, and others to social characteristics. Of all the traits that have been described, there appears to be most support for the roles of activity, intelligence, knowledge, dominance, and self-confidence.

16.4 STUDIES OF LEADERSHIP

Other views of leadership have taken a new direction, looking not only to the individual's traits but also to the behavior of the leader and those people who are being led. McGregor summarizes some of the generalizations from recent research on leadership, which portray "leadership as a relationship." According to McGregor:

> There are at least four major variables now known to be involved in leadership: (1) the characteristics of the leader; (2) the attitudes, needs, and other personal characteristics of the followers; (3) characteristics of the organization, such as its purpose, its structure, the nature of the tasks to be performed; and (4) the social, economic, and political milieu. The personal characteristics required for effective performance as a leader vary, depending on the other factors.

McGregor's viewpoint is important because "it means that leadership is not a property of the individual, but a complex relationship among these variables."[8]

Other approaches have explained leadership as a type of behavior, for example, dictatorial, autocratic, democratic, and laissez-faire. The operating styles of these types of leaders consist of getting work done through fear (the dictator), centralizing decision making in the leader (the autocrat), decentralizing decision making (democratic leadership), and allowing the group to establish its own goals and make its own decisions (laissez-faire leadership).

Still other approaches deal with leadership as specific to the particular situation in which it occurs. Therefore, leaders have to be cognizant of the groups of people (superiors, subordinates, peers, etc.) to which they are related as well as the organizational structure, resources, goals, time variables, and so forth. This suggests the real complexities of trying to understand what leadership is all about. A basic conclusion can be drawn: Successful leaders must be adaptive and flexible, always aware of the needs and motivations of those whom they try to lead. Leaders must be aware of how they perform as leaders and how their behavior affects the performance of those on whom they depend. As leaders try to change the behavior of their people, a change in their behavior also may be needed.

---

[7]Louis B. Barnes and Mark P. Kriger, "The Hidden Side of Organizational Leadership," *Sloan Management Review*, Fall 1986.

[8]Douglas McGregor, *The Human Side of Enterprise* (New York: McGraw-Hill, 1960), pp. 179–189.

Gadeken found in his research in the DOD environment that when project managers take on large, complex, or one-of-a-kind projects, technical and management skills are not sufficient to ensure leadership skills become the predominant focus.[9]

Hauschildt, Keim, and Medcof, in a study of 257 successful and 191 unsuccessful projects, found that project success is much more dependent on the human factor (project leadership, top-management support) than upon project management.[10]

Research into the matter of leadership continues, because we really know little of what it is and why it works sometimes and fails at other times. Project managers who strive to improve their leadership abilities should read in this area of study.

16.5 LEADERSHIP STYLE

An important part of leadership is the " style" with which the leader carries out the role. Much has been written about leadership style. Also, the characteristics of successful leaders have been examined in detail. What follows is just a sample of the abundant views on the subject.

John E. Welch, Jr., CEO of General Electric Company and a superb leader, will no longer tolerate autocratic, tyrannical managers in leadership positions. In a letter to shareholders, CEO Welch discussed management techniques and goals. According to him, GE "cannot afford management styles that suppress and intimidate" subordinates. Welch has categorized managers into several types:

- A leader who delivers on commitments financial or otherwise and shares the values of the company has onward-and-upward prospects.
- A leader who does not meet commitments but shares the company values will get a second chance, preferably in a different environment.
- A leader who doesn't meet commitments and doesn't share values is soon gone from the company.
- The fourth type is the most difficult to deal with. This individual delivers on commitments, makes all the numbers, but doesn't share the values. This individual is typically the one who forces performance out of people rather than inspires it, the autocrat, the big shot, the tyrant.

In today's environment, where it is necessary to have good ideas from every person in the organization, those people whose management styles suppress and intimidate are not needed. GE proclaims high priorities for focusing on customers, resisting bureaucracy, cutting across boundaries, thinking globally, demonstrating enormous energy, and being able to energize and invigorate others.

[9]See Owen C. Gadeken, "What the Defense Systems Management College Has Learned from Ten Years of Project Leadership Research," Proceedings, PMI Research Conference, June 2000, pp. 247–256.

[10]See Jurgen Hauschildt, Gasche Keim, and John W. Medcof, "Realistic Criteria for Project Manager Selection and Development," *Project Management Journal,* September 2000, pp. 23–32.

GE has sent many managers to observe highly successful Wal-Mart Stores, Inc., where the leadership factors of speed, the bias for action, and utter customer fixation have helped drive this high-discount store to success.[11]

Leadership style is in general of two types: people-centered, described as democratic, permissive, consensus-seeking, participative, follower-oriented, and considerate, and task-centered, described as structured, task-dominated, restrictive, directive, autocratic, and socially distant. Task-oriented leadership style usually is associated with productivity but may depress follower satisfaction, whereas people-centered leadership tends to enhance group cohesiveness but not consistently increase productivity.

Leaders, except for the likes of Napoleon, Alexander, or Ghengis Khan, whatever their style, all have their "superiors" to whom they must subordinate their wishes, or both they and their organization will probably fail. Managers (and leaders) who are successful in being promoted up through the organizational hierarchy have demonstrated an ability to lead their followers and to follow their superior leader. Emotionally and intellectually, a leader is wed to the conviction that an organizational unit cannot accomplish its mission without some degree of obedience at every level in the hierarchy. A leader needs obedience and discipline from followers and thus accords it to superiors. To demand obedience from one's followers but to withhold it from higher authority would constitute an inconsistency that would jeopardize the fundamental discipline of authority-responsibility-accountability, which holds an organization together.

We have all seen or heard of mavericks, who "march to a different drummer" and cannot function in a hierarchical organization and strike out on their own. Upon leaving and starting a new "business," these mavericks usually end up creating some form of a leader-follower organizational structure.

Effective leadership, then, is usually preceded by effective followership. A successful project leader doubtlessly has performed successfully as a follower. This success provided the basis for that individual's opportunity to become a successful leader. Followers provide the opportunity and legitimacy to the leadership role. In the changing, complex world of project management, we believe that tomorrow's project managers cannot successfully emerge without having learned the skills and developed the attitudes of followership.

The following examples can help emphasize the differences of leadership style depending on both the leader and the circumstances.

Peter Ueberoth, "project manager" for the 1984 Olympics, is described as having a management style that ensures that he is in control at all times. His ego and inner toughness helped him promote and successfully conclude a project with a global objective. His abilities in cultivating the stakeholders were instrumental in raising the money required for the Olympics. Ueberoth believes that a leader's role is to inspire people to greater efforts. He believes that authority is 20 percent given and 80 percent taken. If someone faltered on his project, he made a change and put someone in who could handle the job.[12]

---

[11]James C. Hyatt and Amal Kumar Naj, "GE Is No Place for Autocrats, Welch Decrees," *The Wall Street Journal,* March 3, 1992.

[12]Paraphrased from "Master of the Games," *Time,* January 7, 1985.

Leadership style should not necessarily be consistent in all activities. On the contrary, project leaders should be as flexible as possible, gearing their leadership style to the specific situation and the individuals involved, that is, within the key elements of any leadership situation the leader, the led, and the situation. Figure 16.1 shows a continuum of leadership behavior with the basic ingredient being the degree of authority used by a manager versus the amount of freedom left for subordinates. Autocratic, democratic, and laissez-faire leadership styles can be identified across this continuum from boss-centered leadership to subordinate-centered leadership.

It is difficult to generalize about the characteristics of leaders. They come in all shapes, colors, and sexes. Some are brilliant, some are dull, some are articulate, some can write, some are proactive, and others are laid back. However, a few characteristics are found in leaders who have proven track records:

- They have their act together; their personal ambition drives them to succeed and the organization they are leading to succeed.
- They are visible to their people; they are not "absentee landlords." There is no doubt in anyone's mind that they are in charge and on top of everything.
- They are available to their people to listen, debate, and gather the necessary facts; at the same time they are ready to say, "Let's do it."
- They are decisive and in the long run make the decisions that turn out to be right. They know when to stop gathering information and recommendations into a decision and say, "OK, go do it."
- They see the best in the people with whom they work, not the worst. The leaders see winners and praise and develop these winners for higher levels of performance.

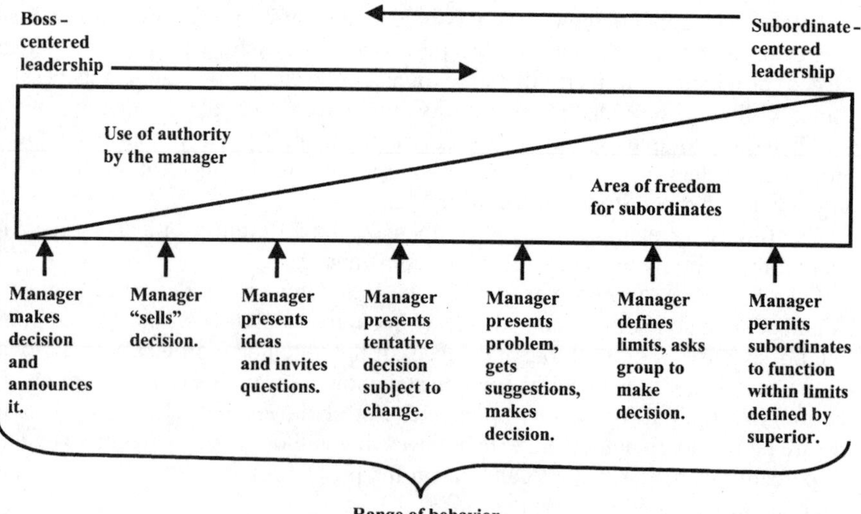

FIGURE 16.1 Continuum of leadership behavior. (*Source: Robert Tannenbaum and Warren H. Schmidt, "How to Choose a Leadership Pattern,"* Harvard Business Review, *March–April 1958, p. 96. Copyright © 1958 by the President and Fellows of Harvard College. Reprinted by permission, all rights reserved.*)

- They are simplistic and avoid making things complex. Good leaders make things simple, coming at people from different approaches until people are convinced, or people convince them that this or that is the way to do things.
- They are fair and patient, and usually they have a sense of humor that tides them and their people over the rough spots that come to any enterprise. Humility is a mark found in good leaders who recognize that they are leaders only because the followers have allowed them to remain in a leadership role.
- They work hard at leadership, at providing the people with the resources needed to do the job, and follow up to see if people are doing those jobs.

16.6 MANAGEMENT VIS-À-VIS LEADERSHIP

What's the difference between leadership and management? Management is usually considered to be a more broadly based activity including functions other than leading. According to Davis:

> Leadership is a part of management, but not all of it. A manager is required to plan and organize, for example, but all we ask of the leader is that he gets others to follow. Leadership is the ability to persuade others to seek defined objectives enthusiastically. It is the human factor, which binds a group together and motivates it toward goals. Management activities such as planning, organizing, and decision making are dormant cocoons until the leader triggers the power of motivation in people and guides them toward goals.[13]

The factor that empowers the project team and ultimately determines which projects fail or succeed is the leadership brought to bear on the project at all levels in the enterprise. When a project is undertaken in the implementation of an enterprise strategy, the key to making that project happen is the quality of leadership. Lynn Crawford, a notable PMI member, in her research has noted that project leadership "appears consistently in the highest ranking category amongst Project Manager Competence factors."[14]

Warren Bennis, a prolific writer in the field of leadership,[15] offers an intriguing differentiation between these two roles: "a leader does the right thing (effectiveness); a manager does things right (efficiency)." In this context, the project leader develops the vision for the project, assembles the resources, and provides the inspiration and motivation for working with project stakeholders in doing the right thing to accomplish the project's objectives; completing the project so that its technical performance, cost, and schedule objectives are attained and so that the project results have a place in the future of the enterprise.

[13]Keith Davis, *Human Relations at Work,* 3d ed. (New York: McGraw-Hill, 1967), pp. 96–97.

[14]Lynn Crawford, "Profiling the Competent Project Manager," *Proceedings,* PMI Research Conference, June 2000, pp. 3–15.

[15]Warren Bennis, "Good Managers and Good Leaders," *Across the Board,* October 1984, pp. 7–11.

Bennis's description of leaders doing the right thing and managers doing things right is a useful way of looking at the differences between leadership and managership. Taking this difference and fleshing it out a bit, one could come up with something like the following to show the difference:

Leadership

- Finds and develops a vision for the project, and sells that vision to the project team and other stakeholders.
- Copes with operational and strategic change involving the project.
- Builds networks of interest with key stakeholders and develops strategies to ensure their support for the project.
- Sets the general direction of the project and its work.
- Provides the conditions surrounding the project that motivate and commit stakeholders to support the project.
- Watches for broad patterns and relationships that have the potential for impacting the project and leads the way to ensure that these patterns and relationships provide positive support to the project.
- Does the right thing in providing leadership of the stakeholders so that they support the project's purposes.
- Becomes a symbol of the project and its purposes.
- Builds political support for the project among all stakeholders and other vested interests.
- Is concerned with effectiveness in the use of project resources.

Managership

- Copes with the complexity of designing, developing, and operating the management systems to support the project.
- Maintains oversight over the efficiency and effectiveness of the resources being used on the project and the management systems used to support those resources.
- Designs and executes the operation of the planning, organizing, motivating, and control functions used to support the project.
- Is concerned with order and efficiency in the use of resources on the project.
- Keeps key stakeholders informed on the progress or lack of progress being made on the project.
- Reassigns resources as needed to provide support to the project team.
- Monitors and evaluates the ability of the individuals on the project team and the team itself to contribute meaningfully to the project purposes.
- Ensures that the communication system used to support the project is working satisfactorily.

- Provides monitoring, facilitating, teaching, and other means to develop the individual and collective competencies of the project team.
- Does things right.

Marino opines, "Today's mega corporations are so encumbered with rules, regulations, and traditions that they are designed to be managed rather than led." He suggests another way of looking at the difference between managers and leaders (see Table 16.1).[16]

TABLE 16.1 Managers versus Leaders

Managers	Leaders
• Managers cuss	• Leaders discuss
• Managers stew	• Leaders do
• Managers resolve	• Leaders involve
• Managers spare	• Leaders share
• Managers preach	• Leaders teach
• Managers depress	• Leaders impress
• Managers detect	• Leaders respect
• Managers haze	• Leaders praise
• Managers control	• Leaders extol
• Managers remand	• Leaders expand
• Managers react	• Leaders enact
• Managers yank	• Leaders thank
• Managers bray	• Leaders pay
• Managers follow rules	• Leaders make them
• Managers dread failure	• Leaders learn from it
• Managers are afraid to make mistakes	• Leaders turn mistakes into new business
• Managers do things right	• Leaders do the right things
• Managers do things that translate into action	• Leaders do things that translate into vision
• Managers do things that demand results	• Leaders do things that expand opportunities
• Managers do things that protect the status quo	• Leaders do things that make their companies grow
• Managers think and work inside the box	• Leaders are happier thinking and working outside the box
• Managers ask, What's wrong with the company?	• Leaders ask, What's right with the company?

[16]Sal Marino, "STRAIGHT TALK: The Difference between Managing and Leading," *Industry Week,* June 1, 1999, p. 30.

Project managers must not only manage but also lead. Project leadership should be appropriate to the project situation because leadership is a continuous, flexible process. There are no consistent characteristics of leadership to point to and flatly state: "That's what makes a leader." Decisiveness is often cited as a desirable leadership characteristic, yet if a leader makes a wrong decision, the organization can suffer. Each management situation, the people involved, the times, the characteristics of the followers, the urgency of the decision, and so on, all influence both the leaders and the followers.

Project leadership is an interpersonal and strategic process, which seeks to influence the project stakeholders to work toward closure of the project purposes. Project leadership takes place through interaction, not in isolation.

Project managers, like most managers in modern organizations, are both leaders and followers, operating in a culture where both formal and informal networking relationships proliferate. In such relationships networking goes beyond the project manager's formal authority, often leading to the use of influence over peers and superiors to affect the outcome of the project.

A project manager's leadership position encompasses three fundamental roles: an interpersonal role, which includes figurehead and leader in liaison functions; an informational role, which entails disseminating information and acting as a spokesperson; and a decision-maker role, in which the project manager acts as entrepreneur, resource allocator, and negotiator.[17]

Project leaders are the people who do the right thing; project managers are people who do things right. Bennis recognizes that both roles are important in management, but they differ profoundly. There are, according to Bennis, people in senior positions in organizations doing the wrong thing well. Part of the fault for having people do the wrong thing may well lie with our schools of management, where we teach people to be good management technicians but we fail to train people for leadership.[18] Bennis goes further to identify the competencies found in people who exhibit effective leadership in their proven track records:

- Management of attention
- Management of meaning
- Management of trust
- Management of self

We will take these competencies and adapt them to the art of project leadership.

Management of Attention

An apparent trait in most successful leaders is their ability to draw others to follow them toward a purpose. Often this purpose takes an early form of a dream or a

[17]These fundamental roles are described in Henry Mintzberg, "The Manager's Job: Folklore and Fact," *Harvard Business Review,* July–August 1985, pp. 54–56.

[18]Paraphrased from Bennis, op. cit., p. 8.

vision. Because of the leaders' extraordinary commitment to their dreams, they communicate a commitment, which attracts people to them. People enroll in the leaders' visions. A project manager who is effective usually knows what he or she wants to be done on the project and reflects this in a project plan.

The first leadership competency is a set of intentions, a vision, or a direction in the sense of the project objective, goals, and strategies. A project manager who has developed a project plan has taken an important step in becoming a project leader. The project leader's role is not merely to state the project objective, goals, and strategies, but to create a meaning for the project that team members can rally around. No matter how grand the project objective, the effective project leader must use a word, a model, or a slogan to make the vision clear to others. The project leader's goal is to go beyond a mere clarification or explanation, to the creation of what the project objective means in satisfying a customer or owner. The project manager integrates customer facts, concepts, and needs, and project team capabilities into a meaning of the project for the project team and the stakeholders. At times the leader's most important task may be to communicate the project's objective to key stakeholders whose viewpoint could be supportive of, contrary to, or indifferent to the project objective.

Management of Meaning

An important responsibility of senior managers as leaders is to communicate the meaning of the project results within the strategic fit of the corporate mission. When the project team members can sense that the project is a building block in the design and execution of enterprise strategies, an important step has been made toward giving the project meaning.

To make the meaning of the project apparent to the members of the enterprise, key enterprise managers must work at communicating the vision of the project through meetings and conversations.

Management of Trust

Trust is an assured reliance on the character, ability, strength, or truth of someone or something. A project leader is someone in whom confidence is placed. An ambience of trust is essential to all organizations; the main determinants of trust are reliability and commitment. Reliability deals with the maintenance of a stable project management approach and a project culture that sets and expects high performance standards. Commitment means that the project team members, and the senior managers responsible for supporting the team, are pledged to making the project results happen. Team members want to follow a team leader they can count on, even if they disagree with the leader's viewpoint. A project leader who shifts positions frequently stands to lose the confidence and commitment of the people. One cannot emphasize enough the significance of constancy and focus in the management of a project.

Management of Self

The first challenge most of us face is to manage ourselves. If we have knowledge and skills and are sufficiently motivated, then we stand a good chance of deploying the knowledge and skills effectively. Management of self is critical. Without it we can do more harm in a management position than good. Bennis notes:

> Like incompetent doctors, incompetent managers can make life worse, make people sicker and less vital. Some managers give themselves heart attacks and nervous breakdowns; still worse many are "carriers," causing their employees to be ill.[19]

Management of self means that a project leader and the team members can make mistakes, but they get those mistakes out of the way as soon as possible, learn as a result of them, and move ahead.

Project leadership can be recognized by outsiders and felt by the project team. Project leadership gives power and significance to the project effort. As noted by Bennis, in organizations with effective leaders, four themes are evident:

> *People feel significant.* Everyone feels that he or she makes a difference to the success of the organization. The difference may be small such as the prompt delivery of potato chips to a mom-and-pop grocery store or development of a tiny but essential part for an airplane. But where they are empowered, people feel that what they do has meaning and significance.
>
> *Learning and competence matter.* Leaders value learning and mastery, and so do people who work for leaders. Leaders make it clear that there is no failure, only mistakes that give us feedback and tell us what to do next.
>
> *People are part of a community.* Where there is leadership, there is a team, there is a family, there is unity. Even people who do not especially like one another feel the sense of community. When Neil Armstrong talks about the *Apollo* explorations, he describes how a team carried out an almost unimaginably complex set of interdependent tasks. Until there were female astronauts, the men referred to this feeling as "brotherhood." I suggest they rename it "family."
>
> *Work is exciting.* Where there are leaders, work is stimulating, challenging, fascinating, and fun. An essential ingredient in organizational leadership is pulling rather than pushing people toward a goal. A "pull" style of influence attracts and energizes people to enroll in an exciting vision of the future. It motivates through identification, rather than through rewards and punishments. Leaders articulate and embody the ideals toward which the organization strives.[20]

---

[19]Ibid., p. 10.
[20]Ibid., p. 11.

16.7 PROJECT LEADERSHIP

In any discussion of project leadership, it is essential to recognize the types of challenges found in running a project, figuring out what to do and how to manage the creation of something that currently does not exist, despite uncertainty, diversity, and an enormous number of problems and challenges, and getting a management job done through a diverse set of people despite having little direct control over most of them.

During a project manager's first months on the job, a lot of time is spent establishing where the project is and what plans should be established to get the project moving. In addition to setting plans, effective project leaders usually allocate significant time and effort to begin developing a network of cooperative relationships with the people they believe are needed to bring the project to a favorable outcome. This networking takes up considerable time as cooperative relationships with and among the stakeholders are developed. Project leaders can use a wide variety of face-to-face methods with the stakeholders, such as:

- Doing favors for them in expectation of reciprocity
- Stressing the project manager's role
- Encouraging and convincing the stakeholders' identification with them
- Nurturing their professional reputation with others
- Maneuvering to make stakeholders feel they are dependent on the project manager for resources to use in their organizational elements

Project leaders can call on peers, corporate staff people, or professionals anywhere in the organization, when necessary, to support the project. Excellent project leaders ask, encourage, cajole, praise, reward, demand, manipulate, use politics, and generally work hard at motivating others through effective interpersonal skills in gaining and holding support for the project.

What should senior managers do to develop leadership capabilities in their project managers? Putting someone in the job because that person is a "good manager" can be risky. Unless the project is easy to learn, it can be difficult for an individual to learn fast enough to develop a sound leadership approach; it would also be difficult to build a project stakeholder network quickly enough to effectively manage the project. The best approach would probably be to "grow" your own project managers. These are some of the things that senior executives can do:

- Identify the emerging young professionals who show a potential for leadership and management and put them in charge of one of the many small projects usually found in an organization. This would test the potential capability of the professionals, expose them to the unstructured, ambiguous ambience of a project, and provide some experience in networking.
- Send the neophyte to a formal training course to open some vistas; however, leadership cannot be fully learned in a classroom environment. Too often

courses deal with overly simplistic people relationships or suggest some tools such as time management, quantitative methodology, behavioral science, or "how to run meetings" courses, which give little insight into the realities of leading a dynamic project. Nevertheless, some formal course work would probably be useful. An alternative would be to encourage the leader-to-be to begin and continue a professional reading program on leadership. By reading several good books on leadership, at least some questions can emerge in the reader's mind as to what knowledge, skills, and attitudes are usually found in effective leaders.

- One of the best things that senior management can do is to maintain ongoing surveillance of the project, evaluate performance, coach the project leader when required, and ensure that contemporaneous project management techniques are being used. Indeed, senior managers can set an example of outstanding management, including the integration of the project in the design and execution of enterprise strategies.

16.8 LEADERSHIP COMPETENCIES

Leadership competencies are critical to the success of any organization and at all levels within the enterprise. Many aspects of leadership are stated or implied in this book because they apply to all managers. Project management is a subset of the enterprise's need for leadership at an operative level, whereas the project is defined by senior managers exhibiting leadership traits and characteristics. This section focuses on the project manager and the competencies desired for the best chance of project success.

The project manager's success as a leader is dependent on a set of personal capabilities. *The project manager should understand the technology involved in the project.* The word technology is used in the sense of a method followed to achieve a practical purpose. Under this broad definition, any of the means used to provide useful products and services have a "technology" content. Engineering, science, computers, manufacturing, and after-sales service have obvious elements of technology. Government, educational, military, and ecclesiastical organizations employ elements of technology in meeting their mission and objectives. The technology involved in an engineering school is different from the technology of a school of business. Projects developed within these organizations would require a project manager who had an understanding and mastery of different technologies. A project to develop a new computer would require a different understanding of that technology than the development of a new airplane.

How much depth of understanding should the project manager have in the technology related to the project? There is a risk factor to be considered. If the project manager spends too much time keeping up with the technological details, the managerial aspects of the project could be neglected. A rule of thumb is suggested: The project manager should have enough understanding of the technology of the project to ask the right questions and know if correct answers are being given. If there are detailed or in-depth questions to be asked and answered, then the project manager should seek the guidance of the appropriate member(s) of the project team. In some cases technological issues may be so important to the project that a task force of experts should be appointed to study the issues and make appropriate recommendations to the project manager.

Another key capability of the successful project manager is to have that blend of interpersonal skills to build the project team and to work with the team and other project stakeholders so that there is a culture of loyalty, commitment, respect, dedication, and trust. Simply stated, be a leader of the project stakeholders. The ability to work with and through people—and in so doing win their support—is crucial to the success of the project. As discussed earlier, the project manager must depend on many people from different parts of the project to support the project and to work diligently to give quality results on the work packages for which they are responsible. The ability to function as a *leader* within the context of an acceptable and admired interpersonal style is important. All of us tend to do better when we work with a manager or leader whom we respect, who treats us with respect and dignity, and who is a pleasant person to work with. Also remember, the single greatest cause for failure of managers is their lack of interpersonal skills, a fact that has been demonstrated many times.

An understanding of the management process is another critical skill required of the project manager. This means that the project manager must know the fundamentals of the management functions of planning, organizing, motivation, direction, and control. The fundamentals that the project manager must know, as well as the project team members, include the fundamentals outlined in this book. People who become project managers and who come from one of the technologies of the project have a particular challenge here. For example, a successful engineer who is appointed to a project manager position will usually have a strong tendency to try to keep fully abreast of engineering considerations in the project. She or he will probably tend to pay more attention to the engineering aspects of the project because that is the portion most easily understood. All that attention to engineering considerations means that less time is available for oversight of the other project work packages and less time is available to truly perform in a managerial and leadership position on the project.

The ability to see the systems context and strategic context of the project is another characteristic of the skillful project manager. This means that the project is viewed as a set of subsystems or interrelated elements. These interrelated elements operate so that actions taken on one of the elements may produce a change in any of the other elements within the total project system. For example, the model of a project management system given in Chap. 4 carries this important message: The systems framework of managing a project implies that consideration be given to myriad relationships, which have both formal and informal implications. A partial or parochial approach to the management of the project will set the stage for difficulties or at best the chance that efficiency and effectiveness will be reduced. In the systems view, the project manager should consider the probable impact of forces outside the project on the project's outcome. In Chap. 6 the idea of project stakeholders was introduced, and the advice was given that the potential impact of stakeholder actions on the project should be one of the considerations in the planning for and execution of decisions on the project. Also, each project is part of a larger system. These larger systems include the political system, the economic system, the technological system, the legal system, the social system, and the competitive system. Within these larger systems there are always forces that can help, or hinder, or even defeat project purposes. A project manager who takes the systems approach has improved chances for identifying forces outside the project that could influence the project and its outcome.

The project manager needs to know how to make and implement decisions within the systems context of the project. Making a decision requires the management of such fundamentals as:

- Defining the decision problem
- Developing the databases required to evaluate the decision
- Considering the alternative ways of using the resources to accomplish the project purposes
- Undertaking an explicit assessment of the risk and cost factors to be considered
- Selecting the appropriate alternative
- Developing an implementation strategy for the selected alternative
- Implementing the decision

The making of decisions is much more complex than is implied in the above fundamentals. However, if these fundamentals are used as a philosophical guide to the making and execution of decisions, the probability of having more timely and successful decisions is enhanced.

Of course, the key characteristic of the successful project manager is the ability to produce results—the delivery of the project's technical performance results on time, within budget, and so that a contribution is made to the strategic purpose of the enterprise.

The characteristics of successful project managers/leaders are not much different from those of any manager—except that project managers/leaders must be able to put their *knowledge, skills,* and *attitudes* to work in the distinctive culture of the management of an ad hoc project. Also, it must be remembered that a project leader has to work with many different stakeholders on the project over whom he or she has no delegated authority.

Keep in mind that the project manager manages *the processes* involved in the project work packages arising out of the appropriate functional entities and disciplines supporting the project. As an integrator-generalist, the project manager provides a focus and synergy for the totality of the project work packages—the embodiment of the work packages of the project into an entity that supports organizational strategic purposes. Indeed, the project manager is the *general manager* of the project as far as the enterprise is concerned.

Leaders come in all sizes, shapes, and colors. Although leadership has been exhaustively studied, we still are able to make only some generalizations about its process and attributes. Every person who has responsibility for other people has a leadership role to play. The style of leadership depends on the leader, those being led, and the environment. In general, some of the key characteristics of successful leaders are known.

By drawing on Warren Bennis's ideas, we define project leaders as those people who do the right thing (selecting the objectives, goals, and strategies) and project managers as those who do things right (building the project team and making it work). A person who manages a project should be both a leader and a manager, developing competencies in the management of *attention, meaning, trust,* and *self.*

What are some of the modus operandi of a successful project manager or project leader? And for the aspiring project-manager-to-be, how can these characteristics and manner be developed in the individual? Indicated below are a few of the more important desirable characteristics and the manner of the successful project manager.

First, a capability to conceptualize the likely deliverables of the project—particularly in providing something of value to the customer and translating those deliverables into the technical performance objective of the project—usually expressed in a product, service, or organizational process needed by a customer. This capability may first express itself in the context of the development of a vision for the project, later to be transferred into a technical performance objective for the project and supported by meaningful and accurate cost and schedule estimates.

Second, an optimistic attitude that prevails and causes the project manager to deal effectively with the bad news and fear of failure that buffets any project. Many successful people have traveled a path to success that has had many opportunities for failure. For those who endure, a failure is looked at as a stepping-stone to improvement followed by success. It has been said that the person who knows failure has a better chance of success and that failures enable us to learn and move to successful ventures in our lives.

Third, having a tough skin and accepting the inevitable faultfinding or blame that can come in the management of a project. This is not the normal constructive criticism that is to be expected in the management of a project because stakeholders view the project from their parochial perspective and hope to influence the project work to their advantage. A project manager who does not find some griping about the project should be suspicious, because any strategy in the management of a project can be criticized.

Fourth, to empower the project team members and, as appropriate, the project stakeholders as well. For the project team policies, procedures, and linear responsibility charts are ways to delegate the authority to the team members to empower them to perform their best on the project work.

Fifth, the ability to assume risks in the management of the project. A development project for a new product, service, or organizational process has risk and the more the state of the art is being pushed, the higher the risk. Life is full of risks as individuals drive our highways, fly on airlines to attend project meetings, or negotiate contracts with vendors and subcontractors. What is important is the ability to study and discern the principal risks endemic to the project and be able to identify those experts who can properly assess and evaluate the risks and work with other experts in designing a strategy to reduce or eliminate the risk. Risks are always with us, and creativity and innovation are the handmaidens of risk.

Sixth, the ability and the courage to make decisions involving the project. The "buck" stops with the project manager, who has the authority and responsibility to make the decisions involving the project, which includes the inevitable trade-offs involved in the use of resources. Of course, some decisions may not be the prerogative of the project manager—such as the decision to terminate a project before it is completed because of a lack of "strategic fit" in the enterprise. However, the successful project manager is one who should recommend decisions by a higher authority if needed in the management of the project and help motivate those in higher levels of the organization who need to make a decision involving the project. To fail to make a decision is a decision in itself—that is, to do nothing—with the risk of continuing to use resources without any rationale for such use because someone is procrastinating on her or his decision responsibility.

Seventh, tenacity—or persistence in holding to a position regarding the project. Of course, tenacity regarding the project can be carried too far. Most innovative products are the result of many failures by the developers, but those who have unusual perseverance and toughness in sticking to their work and objectives eventually come forth with a winner. Tenacity can bridge the gap between dreams and reality.

Eighth, the ability to mentor, teach, coach, and guide the people using resources on the project. This means that the project manager is less of a manager and more of a facilitator who works as a focal point with the team members—and other stakeholders—to pull everything on the project. This means that the project manager is a role model in these matters and sets the stage for the members of the project team to carry out their responsibilities through mentoring, teaching, and coaching the people who work on the project work packages.

16.9 TO SUMMARIZE

Some of the major points that have been expressed in this chapter include:

- There have been many studies about leadership. In the last few years several writers have tried to differentiate between what a leader does and what a manager does.
- There have been many definitions of leadership. Literally thousands of studies have explored leadership traits.
- McGregor's viewpoint of leadership is important because he states that leadership is not a property of the individual but a complex relationship among many variables.
- Effective leadership is interdependent with effective followership.
- The leadership style practiced by an individual may not be considered in all activities.
- A few manager and leader characteristics are presented in this chapter that are found in leaders who have proven track records.

- Bennis notes that a leader does the right thing and a manager does things right. In this chapter a few differences between leadership and managership were described.
- Project leadership is an interpersonal and strategic process that seeks to influence the project stakeholders to work toward closure of the project purposes.
- The project manager's success as a leader is dependent on a set of personal capabilities.
- In this chapter a few of the more important desirable characteristics and the manner of the successful project manager were presented.
- Leaders have existed in human affairs ever since people banded together for some common purpose.
- In modern organizations where traditional and nontraditional project teams are used, there are many opportunities for individuals to try out their leadership capabilities.
- A critical leadership role that the project manager must carry out is that played with the project stakeholders. Keeping these individuals and organizations informed of what is going on with respect to the project will make it easier to call on them when their support is needed for the project purposes.
- It is important that members of the project team understand the difference between what a leader does and what a manager does so that they can support the project manager as he or she carries out these roles.
- The study of leaders and leadership vis-à-vis managers and management will likely continue as the theory and practice of the management discipline continue to evolve.
- In recent times, there is a growing interest in examining the differences and similarities between project managers and project leaders.

16.12 ADDITIONAL SOURCES OF INFORMATION

The following additional sources of project management information may be used to complement this chapter's topic material. This material complements and expands on various concepts, practices, and the theory of project management as it relates to areas covered here.

- Jeffrey K. Pinto, "The Elements of Project Success," chap. 2 in David I. Cleland (ed.), *Field Guide to Project Management* (New York: Van Nostrand Reinhold, 1997).
- D. E. Knutson and J. K. McClusky, "The Environment and Molecular Sciences Laboratory Project Continuous Evolution in Leadership," in David I. Cleland, Karen M. Bursic, Richard J. Puerzer, and Alberto Y. Vlasak, *Project Management Casebook,* Project Management Institute (PMI). (Originally published in *Proceedings,* Project Management Institute, 25th Annual Seminar/Symposium, Vancouver, Canada, October 17–19, 1994, pp. 923–929.)

- Wendy Briner, Michael Gelles, and Colin Hastings, *Project Leadership* (England: Gower, 1990). This book begins by explaining why the concept of project leadership is being adopted so widely. The first part of the book examines the project leader's task. It outlines the six key elements of the role, and looks at the processes used by effective project leaders. The second part of the book provides advice on how to handle the issues that emerge at each stage of a project's life. The book is based on a combination of research and experience.
- Steven W. Flannes, Ph.D., and Ginger Levin, DPA, *People Skills for Project Managers* (Vienna, Va.: Management Concepts, 2001). This book is filled with methods and tools for handling people challenges that involve communication, motivation, performance, behavior, crises, and so forth. The book is practical and relevant to the individual who works as a project leader in dealing with many stakeholder issues. The book is filled with real-life scenarios, augmented with concrete approaches and solutions supported by the latest research. The authors conclude that their book is "a must-have resource for strong, effective project leadership."
- Peg Thomas and Jeffrey K. Pinto, "Major Leadership: A Question of Timing," *Project Management Journal,* vol. 30, no. 1, March 1999, pp. 19–26. This article explains the relationship between successful project leadership and the importance of project leaders developing appropriate and complementary temporal skills for their work. The authors make the important point that project leaders have to be able to think and talk about the future of the project to inspire and influence the behavior of various project stakeholders. The authors also recognize the growing body of literature that recognizes that project managers need to function as visionaries, technical experts, motivators, team leaders, negotiators, and so forth.
- Gary Gemmill and David Wilemon, "The Hidden Side of Leadership in Technical Team Management," *Research Technology Management,* November–December 1994. The authors recognize that project managers must deal with the functions of management on the projects to which they are assigned, but also cope with a multitude of interpersonal issues always present in team-oriented work environments. The article is based on a field study of 100 project leaders, which examines several interpersonal issues found in most technical teams and suggests some options for managing them. If the project leaders deal only with the technical dimensions of the project and are not competent in dealing with the hidden interpersonal issues, the management of the project can be undermined.
- Thomas W. Zimmerer and Mahmoud M. Yasin, "A Leadership Profile of American Project Managers," *Project Management Journal,* March 1998, pp. 31–38. This paper reports the results of a survey of senior project managers. The results identify positive success and negative leadership as the cause of project failure. The characteristics of leadership are identified, as well as the project management tools that are most useful and most often used. The article concludes that project managers need to combine technical competency with the ability to develop and display leadership.

16.11 DISCUSSION QUESTIONS

1. Discuss the various definitions of leadership. What traits embody these definitions?
2. What are some of the approaches used in researching leadership? How do these differ?
3. Why has leadership been so difficult to define?
4. Leadership style should not necessarily be consistent in all activities. Explain.
5. Discuss some of the personal characteristics of successful project leaders.
6. What is the difference between leadership and management? What characterizes each?
7. Discuss the fundamental roles of a project manager's leadership position. How do these roles interact?
8. What is meant by management of attention, management of meaning, management of trust, and management of self?
9. Discuss the four themes that are often evident in organizations with effective leaders.
10. How can project managers establish a network among project stakeholders? Why is this network so important?
11. Discuss the participatory approach to project management. What assumptions define participative management?
12. Discuss some of the characteristics of successful leadership that are documented in the literature.

16.12 USER CHECKLIST

1. How does your organization define leadership? What traits characterize successful leaders in your organization? Explain.
2. How does your organization build leaders? Are young professionals given the opportunity to learn project management and leadership?
3. What style do leaders of your organization use? Is this style effective? Why or why not?
4. Do the project managers of your organization change their leadership styles to fit various activities? How?
5. Do the managers in your organization understand the difference between leadership and management? Do they develop skills in both areas? Explain.
6. Do project managers in your organization demonstrate competencies in management of attention, meaning, trust, and self? Why or why not?
7. Are the four themes, which are evident in organizations with effective leaders, evident in your organization? Explain.

8. Do the project managers in your organization form networks with project stakeholders? How?
9. What participatory approaches to management are used in your organization? Are these effective leadership tools? Explain.
10. Which of the core dimensions for effective project management are apparent in the leaders of your organization?
11. Where do the project managers and other managers of your organization fall in Table 16.2?
12. How would you judge the overall effectiveness of the leadership in your organization? What could be done to improve the leadership process?

16.13 PRINCIPLES OF PROJECT MANAGEMENT

1. Project leadership is a presence and process carried out in the context of a project.
2. There is not a universal definition of leadership.
3. Project leadership is not a property of the project leader, but a complex relationship among the many variables found in the project stakeholder community.
4. The project leader's style strongly influences the manner in which the project stakeholders perceive such a leader.
5. Effective leadership is usually preceded by effective followership.
6. There are distinct characteristics found in leaders who have proven track records.
7. A leader does the right thing; the manager does things right (attributed to Warren Bennis).
8. The project leader's success is dependent to a large degree on a set of personal capabilities.
9. In modern organizations there are abundant opportunities for individuals to try out their leadership capabilities.

16.14 PROJECT MANAGEMENT SITUATION—BEING A PROJECT LEADER

At a meeting of experienced senior project managers in a large "systems" company, chaired by the author, nine of the participants were asked to write down a phrase, word, or sentence describing the characteristics of good project leaders whom they had encountered in their careers. Another eight of these project managers were asked to describe the characteristics of poor project leaders they had known in their careers. The responses of these project managers are shown in Table 16.2. The contrast between good and poor project leaders is evident. You, the reader,

TABLE 16.2 Good and Poor Project Leaders

Good project leader	Poor project leader
• Positive attitude; recognition; knowledgeable.	• Uses authority-position title to direct knowledgeable.—does not try to understand or solve. Does not listen effectively—ignores or rejects input not politically acceptable. Changes scope or direction at will while blaming others for doing the wrong things.
• Interest in personal aspects of employees (family situations, etc.); anticipates concerns (problems) before they become evident. Excellent role model; decisive.	• Does not ask for help; does not set an example for the followers; does not know the technical aspects of the process.
• Clearly communicates a vision of what is to be accomplished, who challenges and motivates. Key is that manager gives measurable parameters by which to chart progress. A "results-oriented" manager.	• Lets the managers run the business in an undisciplined manner. Does not stay on top of the problems when they arise. Cares only about the bottom line. Does not commend, only criticizes.
• Can see a vision of the future for the business, communicates it to the people involved in the business, and allows the personnel involved to make contributions toward the goals. Exhibits trust, support, and willingness to take blame and suffer disappointment, yet still trust and support.	• Does not translate the vision (if there is one)—does not explain why. Pays little attention to implementation—"That's just this month's buzzword."
• Helps subordinates set a direction for work and allows them to grow toward that goal. Is a mentor, not a master. Knows where everyone is going and why everyone wants to get there—and is able to convince others to follow.	• Does not listen to others' ideas. Does not know how to constructively criticize. Expects perfection. Does not recognize or compliment a job well done. Discourages creative thought and new ideas.
• Sensitive to the effects of decisions on everyone involved. Emphasizes teamwork. Recognizes individuals and groups for contributions. Tries extra hard to relate to subordinates.	• Totally focuses on self-promotion. Unenthusiastic. Cannot communicate vision or ideas.
• Listens to thoughts and ideas of subordinates. Does not antagonize, but offers criticism. Brings harmony to historically battling departments. Does not just sit in office, but goes out into the field.	• Is not people-oriented; shows lack of interest; is not forceful enough; has no vision and/or way to implement a vision.
• Recognizes a good job and how to do a better job. Notes problems or flaws created.	• Communicates through the combination of yelling, waving and pointing hands, and a dissatisfied look on his or her face. When something does go right, says, "That's not bad, but just make sure you don't do this." In essence, speaks in negative terms only. A true believer in theory X, but does not even know it.
• Treats coworkers as human beings, as people rather than just another cog. Asks for input and thought on problems; allows subordinates to spend time on projects of their choice. Does not point a finger to place blame, but says, "The problem cannot be undone—what can be done to prevent it in the future?"	

should ask yourself how you would describe your own leadership style—and whether you fall under the good or poor leadership column.

16.15 STUDENT/READER ASSIGNMENT

The students-readers should ask how they would describe their own leadership style. Would they fall under the good or poor leadership column?

CHAPTER 17
PROJECT COMMUNICATIONS

"What we've got here is failure to communicate."

FRANK R. PIERSON
Cool Hand Luke, 1967

17.1 INTRODUCTION

How many problems and frustrations exist today because of the failure of people to communicate—not only in their personal lives, but in their professional lives as well? The best of project management strategies will not work unless there is free communication among the project team—as well as with other stakeholders. There are many forces that impact the way that people on the project team communicate, not only with themselves, but also with higher-level managers. There have been many articles and books written about how to communicate. In this chapter, we offer a brief snapshot of some of the essentials for effective communication within the project context.

This chapter begins with a citation of some common communication problems, and then progresses to the process of communication and informal communication. Listening, often missed when the communication process is described, is offered as an essential for effective communication in an environment. Nonverbal communication, written communications, and advice on how to better communicate in meetings are described. The role of modern technology and its impact on communication are presented, along with a description of common communication links. Some basic ideas will be presented to bring attention to the role that the art of communication plays in the management of projects. We will also discuss some things that can be done to improve communication within the project teams.

17.2 IMPORTANCE OF COMMUNICATION

The dictionary defines communication as the process by which information is exchanged between individuals through a common system of symbols, signs, or behavior.[1] The importance of communications to the individual engaged in management is singled out by Peter Drucker, who stated that the ability to communicate heads the list of criteria for success. He notes that one's effectiveness depends on the ability to reach others through the spoken or written word when working in large organizations, and this ability to communicate is perhaps the most important of all the skills an individual can possess.[2]

A survey of U.S. corporations by the *Harvard Business Review* gives some insight into the ability to communicate as a factor in promotability.[3] A portion of this study consisted of a listing of personal attributes and their importance in promotion. Technical skill based on experience placed fourth from the bottom in the list of 22 attributes. Other attributes, such as ambition, maturity, capacity for hard work, ability to make sound decisions, ability to get things done with and through people, flexibility, and confidence, all ranked high. Ability to communicate was at the top of the list.

In project management, the importance of communication is keynoted by Sievert, who says:

> A high percentage of the frictions, frustrations and inefficiencies in our working relationships is traceable to poor communication. In almost every case, the misinterpretation of a design drawing, a misunderstood change order, a missed delivery date or a failure to execute instructions is the result of a breakdown in communication.[4]

17.3 COMMUNICATION PROBLEMS

Poor communication involving an error in or misinterpretation of a design drawing can reverberate through a project. One of the most notorious of such errors and subsequent reverberations was on the Diablo Canyon nuclear power plant project in San Luis Obispo, California.[5] One critical part of this project is presented below. The result of the breakdown in communications on this project should be apparent to the reader.

> On September 22, 1981, the Pacific Gas & Electric Company (PG&E) received its low-power testing license from the Nuclear Regulatory Commission (NRC) for the Diablo Canyon nuclear plant. After 13 years of planning, design, and construction, PG&E expected to begin commercial operation of the plant. This was not to be. On September 27, 1981, PG&E discovered that a serious error had been made which subsequently has been referred to as the "mirror-image error." To better understand the events that led to the mirror-image error and other errors, it is necessary to go back at least four years.[6]

[1]*Webster's New Collegiate Dictionary* (Springfield, Mass.: Merriam, 1977), p. 228.

[2]Peter F. Drucker, "How to Be an Employee," *Fortune,* May 1952, p. 126.

[3]"What Helps or Harms Productivity," *Harvard Business Review,* January 1964.

[4]Richard W. Sievert, Jr., "Communication: An Important Construction Tool," *Project Management Journal,* December 1986, p. 77.

[5]California Public Utilities Commission, Public Staff Division, Diablo Canyon Rate Case, vol. 1, Background, Diablo Canyon Project History, prepared testimony of Bruce Deberry (Public Staff Division), March 1978, pp. 25–27.

[6]U.S. Nuclear Regulatory Commission, Order Suspending License, CLI-81-30, November 19, 1981, PSD Exhibit no. 10,033.

On March 8, 1977, PG&E transmitted unverified, unlabeled, handwritten sketches of the Unit 2 containment geometry to Blume Associates in place of Unit 1 data. The diagrams were not labeled, but Blume correctly interpreted that the sketches were for Unit 2. However, Blume personnel believed Units 1 and 2 were aligned in the same way. That is, Blume assumed that both units had all components facing the same direction. Blume then performed its seismic analysis for Unit 1 on this basis. Blume then returned the information from this analysis to PG&E labeled "Unit 1," when in fact the analysis was really applicable to Unit 2, not Unit 1. PG&E accepted the analysis representing Unit 1, and knowing that the units were "mirror-image" units, flipped the diagrams to be applicable to the mirror-image unit, Unit 2. In truth, the data were now applicable to Unit 1, not Unit 2. As a result, the seismic analysis for both units was incorrect.[7] This error went undetected for over 4½ years until it was discovered by a PG&E engineer while reviewing various drawings related to the plant.

The events that followed the initial discovery and reporting of this error became increasingly broad in scope and effect. After the initial report of the design error on September 27, 1981, PG&E informed the NRC on October 9, 1981, that both Units 1 and 2 had been designed incorrectly.[8] PG&E committed to the NRC staff to postpone fuel loading and to reanalyze a limited sample of the seismic design of the plant.

The initial review was performed by Robert Cloud Associates, who indicated that the design problems were more pervasive than first thought.[9] The results of this review were orally presented to the NRC at a meeting on November 3, 1981, and in a written report on November 18, 1981. In the Cloud analysis, additional design errors had been discovered.[10] In response, the NRC staff conducted special inspections at the offices of both PG&E and Blume and found that PG&E's quality assurance program did not effectively control the review and approval of design information passed between PG&E and Blume. The investigation also discovered that the design work by Blume had not been covered by a quality assurance program prior to July 12, 1978.[11]

In response to these findings, the NRC commissioners on November 18, 1981, issued an order suspending the low-power license for Diablo Canyon Unit 2 pending satisfactory completion of certain actions, including an independent design verification program (IDVP).[12] This order found that, contrary to PG&E's statements in its operating license application, Diablo might not have been properly designed and violations of the NRC's regulations might have occurred. The NRC indicated that had this information been available previously, the license would not have been issued.[13]

---

[7]B. H. Faulkenberry et al., Related Report of Seismic Related Errors at DCNPP, Units 1 and 2, NUREG-0862, U.S. Nuclear Regulatory Commission, Region V, November 1981, PSD Exhibit no. 10,034.

[8]Letters from Crane to Engelken, October 12, 1981, and November 5, 1981, PSD Exhibit no. 10,035.

[9]R. L. Cloud Associates, Inc., Preliminary Report, Seismic Reverification Program, November 21, 1981, PSD Exhibit no. 10,036.

[10]Ibid.

[11]Inspection Reports 50-275/81-29 and 50-323/81-18, Special Inspection of Seismic Related Errors at DC, Units 1 and 2, U.S. Nuclear Regulatory Commission, PSD Exhibit no. 10,037.

[12]U.S. Nuclear Regulatory Commission, Order Suspending License, CLI-81-30, November 19, 1981, PSD Exhibit no. 10,033.

[13]Letter from Denton to Furbush, November 19, 1981, PSD Exhibit no. 10,038.

The IDVP was a massive effort. The Bechtel Corporation was hired by PG&E to manage the DVP and complete the plant. The Public Staff Division of the California Public Utilities Commission recommended in its review of the prudence of design and construction of the Diablo Canyon project that $2.484 billion be disallowed as unreasonable costs on the project because of mirror-image and other design errors and the design verification program that followed during 1981 to 1985.[14]

Project managers and professionals often fail to recognize that communication on a project takes many forms: verbal in-group and individual exchanges of information, and documentation such as design drawings, reports, contracts, work orders, and the like. Lack of quality assurance and control in engineering documentation creates the opportunity for errors by those who use the documentation in their work, as in the situations demonstrated by the Diablo Canyon project.

A former CEO of an automobile manufacturing enterprise, recalling his failed efforts to upgrade the company's manufacturing capabilities by using project management techniques, felt that his major failure was in not doing a better job of communicating with the people in the organization. His failure to communicate made it impossible for the people to share his vision for the company. The people did not understand—or appreciate—why the organization was being torn apart and realigned, why certain plants had been targeted for closing, or why other changes were under way. He lamented that his inability to communicate caused him to be way out ahead of the people in the company in trying to institute change. But, because his people were still at the bottom, trying to decide whether or not to go along with the change, the change was not effectively carried out on a timely basis.

Thomas, Tucker, and Kelly note that the lack of effective communications continues to be a major obstacle to project success.[15] During the communication process information is shared—information which is needed to make and implement decisions. In the retailing business, Wal-Mart's outstanding management is keyed to information. Samuel Moore Walton, one of the great showmen of retailing, was applying such concepts as a flat organization, empowerment, and gain sharing long before anyone used those terms. He shared information, right down to single-store results, with the "associates," as Wal-Mart called its employees. Profit sharing, equal to 5 to 6 percent of an associate's earnings, extends to the lowest levels.

Sam Walton had an insatiable hunger for information which in turn facilitated the quick decision making typical of the Wal-Mart culture. Managers gather information from Monday to Thursday, exchange ideas on Friday and Saturday, and implement decisions in the store on Monday.

Walton often spoke with genuine admiration for his competitors—he knew them intimately and copied their best ideas.[16]

---

[14]Executive Summary, Review of the Costs of PG&E Diablo Canyon Nuclear Power Plant Project and Recommendations on the Amount of Costs Reasonable for PG&E to Recover from Its Customers, by the Diablo Canyon Tear, Public Staff Division, California Public Utilities Commission, San Francisco, May 14, 1987, p. 14.

[15]See Stephen R. Thomas, Richard I. Tucker, and William R. Kelly, "Compass: An Assessment Tool for Improving Project Team Communications," *Project Management Journal,* December 1999, vol. 10, no. 4, pp. 15–23.

[16]Bill Saporito, "What Sam Walton Taught America," *Fortune,* May 4, 1992, pp. 104–106.

17.4 THE PROCESS OF COMMUNICATION

A project manager uses communication more than any other force in the project environment to ensure that team members work together on project problems and opportunities. The means and channels of information include:

- Plans
- Policies
- Procedures
- Objectives
- Goals
- Strategies
- Organizational structure
- Linear responsibility charts
- Leader and follower style
- Meetings
- Letters
- Telephone calls
- Small group interaction
- Example set by the project manager

The above partial list strongly suggests that an important function of the project manager is to manage the process of communications with the project stakeholders. However, to manage the communications process, one must understand the nature of that process.

Communication is the process by which information is exchanged between individuals through a common system of symbols, signs, or behavior. People communicate with each other by three principal means: by an actual physical touch, such as a tap on the shoulder, a pat on the back, or the ritualistic expression of the handshake; by visible movements of some portions of their bodies, such as the pointing of a finger, the wink of an eye, a smile, a nod, or a grimace; and by the use of symbols, spoken or portrayed, which have some meaning based on experience.

A few elementary considerations about communication have a broad application across many different organizations. First is that the sender of a message recognize some of the basic concepts of communication theory and practice, which are:

- Be as specific and forthright as possible about the information to be transmitted.
- Know who the receiver is and what the expectations of that receiver are in being the object of some communication media: verbal, written, or nonverbal.
- Design and develop the message with the receiver in mind, considering the potential limited view of that receiver, his or her likely perceptions, and the role he or she plays in the project.

- Select the means or medium for the message, giving careful attention to how the receiver will likely react to a particular medium.
- Plan for the timing of the communication in order to include considerations of the criticality of the message to the receiver(s).

Most present-day failures in communication can be traced to misunderstandings of the symbols that play an important part in the process of human communication. Such misunderstandings come about largely because of our inadequacies in creating, transmitting, and receiving these symbols, both written and spoken. People on a project team will readily recognize the value that symbols have in communicating phenomena in the engineering disciplines. Unfortunately, the symbols that are used in communication in the management discipline are not so precise in their meaning.

Words and combinations of words give us more difficulty in communicating than other kinds of visual symbols. Take the word manage as an example. One dictionary[17] defines *manage* as follows:

> 1. To direct or control the use of; handle, wield, or use (a tool, machine, or weapon). 2. To exert control over; make submissive to one's authority, discipline, or persuasion. 3. To direct or administer (the affairs of an organization, estate, household, or business). 4. To contrive or arrange; succeed in doing or accomplishing, especially with difficulty: *I'll manage to come on Friday—intr. 1. To direct, supervise, or carry on business affairs; perform the duties of a manager. 2. To carry on; get along: I don't know how they manage without him.*

The term *management* is defined as:

> 1. The act, manner, or practice of managing, handling, or controlling something. 2. The person or persons who manage a business establishment, organization or institution. 3. Skill in managing; executive ability.

The real meaning of words—or symbols—depends primarily on how the reader or listener perceives them. A word such as management would have one meaning for a corporate executive and another meaning for a union executive or a worker. The meaning would depend on the image that the word holds for the individual. Such meaning is something we have inside ourselves. Words and phrases that an individual uses may not evoke the same image in someone else's mind; knowing this, one should be as specific as possible in using words. Assuming that everyone knows what you are talking about is usually a poor assumption. Lewis Carroll's way of hinting at the dilemma of spoken communication points out how many of us feel about our use of the spoken word in our communication attempts: "When I use a word," Humpty Dumpty said in rather a scornful tone, "it means just what I choose it to mean—neither more nor less."[18]

---

[17] *American Heritage Dictionary of the English Language* (Boston: Houghton Mifflin, 1976), p. 792.

[18] Lewis Carroll, *Alice's Adventures in Wonderland and Through the Looking Glass*, 4th printing (New York: Macmillan, 1966).

Communication is a two-way process: between the sender and the receiver(s). Taken in this light, the receiver should not be considered a passive recipient but rather a destination that is likely to have a purpose in getting the message and to be influenced by the perceptions and beliefs of those people who send the message. A few guidelines include the following:

- Have the interest and motivation to listen actively and carefully to the message.
- Be sensitive to the sender, to include an awareness of who is sending the message, why, and the likely reason for the message to be sent.
- Influence the likely means by which the message is being transmitted, whether verbally, in writing, or by nonverbal means.
- Plan for and initiate appropriate and timely feedback to include immediate receipt of the message, and a response for providing the requested information and action or presenting a schedule as to when a response can be given to the message.
- Be sure to ask for clarification if there is any problem in understanding the message and its likely intent.

Some writers have used models in explaining the communication process. For example, Gibson, Ivancevich, and Donnelly use the model depicted in Fig. 17.1. The elements of their model include:

Source. The originator of the communication

Encoder. The oral or written symbols used to transmit the message

Message. What the source hopes to communicate

Channel. The medium used to transmit the message

Decoder. Interpretation of the message by receiver

Receiver. Recipient for whom the message is intended

Feedback. Information used to determine the fidelity of the message

Noise. Anything that distorts, distracts, misunderstands, or interferes with the communication process

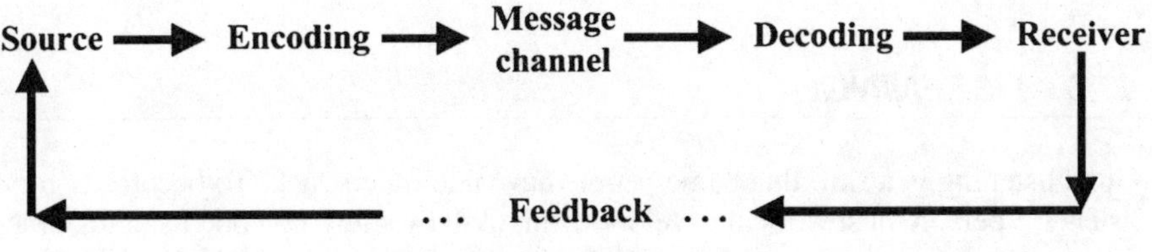

FIGURE 17.1 A communication model. (*Source: James L. Gibson, John M. Ivancevich, and James H. Donnelly, Jr.,* Structure, Process, Behavior, *Dallas Business Publications, 1973, p. 166.*)

The model shows the elements involved in the communication process. A project manager should realize that communication is the capstone of the management functions of planning, organizing, motivating, directing, and controlling; without effective communications these management functions cannot be planned or carried out adequately.

17.5 INFORMAL COMMUNICATION

The above model portrays the formal communication that is carried out usually through the organization portrayed by the traditional organizational chart and by the linear responsibility chart. But there also is an informal organization. Membership in this informal organization is dependent on common ties, such as friendship, kinship, social status, and so on. The need for an informal as well as a formal organization lies in both the psychological and the social needs of human beings and their desire to accomplish personal and organizational objectives. People join informal groups at their place of work for social contact, companionship, emotional support, and such things of value coming out of a particular community to which an individual belongs.[19]

What can the project manager do about the communications within the informal organization? His or her interpersonal style of management will probably have some effect on these informal communications. A few suggestions are offered in this matter:

- Accept the notion of informal communication in the project and what such communication can do and can not do for enhancing project effectiveness.
- Find ways of getting feedback from the informal organization. Identify the "informal leaders," and spend time listening to what they have to say.
- Use these informal leaders as a source for testing technical approaches, ideas, strategies, administrative actions, reorganizations, and other things whose acceptance by the members of the project team is required for success.
- Recognize that much of the cultural ambience of the project is reflected in the attitudes and behavior of people in the informal organization. Insofar as possible, work with the informal organization in support of organizational purposes.[20]

17.6 LISTENING

Good listening is a skill that some people have and others lack. To become a good listener, a person must work at developing this skill by studying good listening practices and applying them to his or her own conscious program of self-development.

[19]See Ross A. Webber, *Management* (Homewood, Ill.: Irwin, 1975), chap. 21, for a thorough discussion of groups and informal organizations.

[20]Some of the material in this section and the preceding section has been paraphrased from D. I. Cleland and D. F. Kocaoglu, *Engineering Management* (New York: McGraw-Hill, 1981), pp. 124–125.

The importance of listening is underscored graphically by a Sperry advertisement that appeared in *The Wall Street Journal.* The ad read, in part, "It's about time we learned how to listen. [There's] the problem of people not knowing how to listen. Most of us spend about half our waking hours listening. Yet research studies show that we retain only 25 percent of what we listen to. Because listening is the one communication skill we're never really taught."[21]

A project is tied together by its system of communications. Perceptive project managers are discovering that communication hinges on both the spoken word and the ability to listen to what the other person is saying. Often members of the project team overlook listening. It is the most important link in project communications, and it is usually the weakest one. Even with good listening, retention is a problem. Right after we listen to someone talk, we remember only about half of what has been heard. Even after we have learned something, we tend to forget from one-half to one-third of it within 8 hours. Part of the problem of poor listening is a lack of training in listening skills. Yet experience demonstrates that if one is a careful listener, one starts getting some answers. Also, there are some emotionally based reasons for not ignoring the need to develop better listening skills:

- Listening may uncover some unexpected problems; it is more comfortable not to listen, to ignore unsavory news that might make the project manager fearful of the project's status.
- Team members may withhold bad news in the futile hope that the problem will work itself out.
- People, managers included, really do not want to tell a superior bad news. This probably accounts for the fact that bad news simply does not flow uphill in organizations. If the manager communicates in any way that he or she does not want to listen to bad news, no bad news will be reported to him or her.
- People really don't want to listen to anything that is contrary to their preconceived ideas or prejudices.
- People can think faster than they can listen. In conversations, particularly heated ones, we tend to race ahead, thinking through a response, and in so doing we lose touch with what the speaker is saying.
- For some reason people listen to get the "facts" and miss the main idea or ideas of what the speaker is saying.
- We mentally turn off what we do not want to hear—our emotions act as aural filters.
- When someone says what we really want to hear, we are "all ears."
- When we hear something that opposes our deeply rooted preconceived notions, our brains become involved in planning a rebuttal or a response intended to put the speaker down or on the defensive.

It is much easier and more comfortable to talk down to subordinates than it is to talk up to superiors. Conversely, it is much easier to listen to superiors (because

[21]*The Wall Street Journal,* September 11, 1979, p. 11.

we have to) than to subordinates (because we may feel that we do not really have to listen to them). Often we do so only to be polite.

Project communication—including listening—is carried out with project professionals, work package managers, functional managers, general managers, service managers, customers, subcontractors, and other stakeholders. The building and maintenance of alliances with these stakeholders require that avenues be opened so that messages can flow. Perhaps the most obvious and effective method is the human chain of people talking to people. The human chain has potential, but often it does not work for these reasons:

- Without good listeners, people do not talk freely and there is not an effective flow of communications.
- Only one bad listener is required to impair the flow of communication.
- There may be a flow of messages, but they can be distorted because of noise along the communication network.

Whereas Fig. 17.1 depicts a generic communication model for one-on-one communication, Fig. 17.2 is a model that shows the many different communication channels that a project manager must maintain. This model demonstrates the need for the project manager to be an excellent communicator.

One of the most influential issues in communication is the interpersonal barriers—such as ethics, morals, beliefs, prejudices, politics, biases, and those things that condition our behavior in both our personal and professional lives. Knowing of the potential and often real existence of these influences, people who are trying to communicate should keep the following factors in mind when sending or receiving messages:

- Hearing only what you want to hear, rather than what the sender hoped that you would hear.
- Emotional involvement and outlook regarding the subject matter being transmitted.

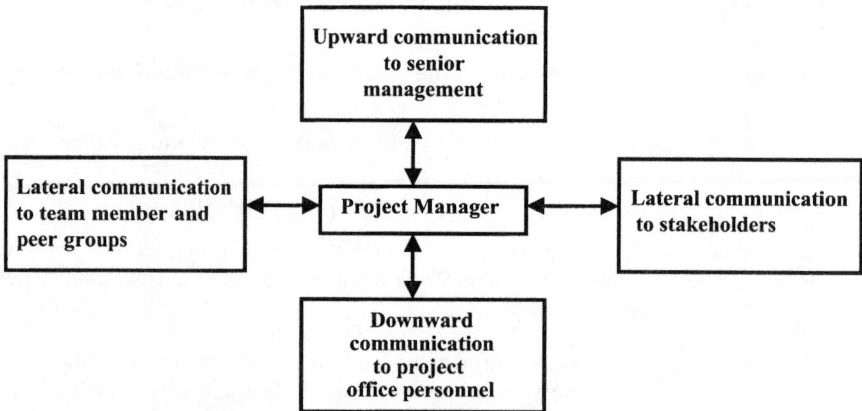

FIGURE 17.2 Project communications arising out of role dependencies.

- Ignoring the contents of a message that runs counter to what we want to believe—often found when bad news is transmitted.
- A preconceived image of the receiver, perhaps tainted by past instances of both "good news" and "bad news."
- Different meaning of words—a likely situation when doing business in the global marketplace.
- A preconceived evaluation of the source to include positive, negative, and indifferent feelings as to the credibility of the source.
- An ego involvement that restricts the extent to which the receiver and the sender are willing to send or receive a message that might impact negatively on their ego involvement with the project.

Listening can be disturbing, because it may uncover problems and cast doubt about an existing strategy. Consider the ill-fated space shuttle *Challenger* that was destroyed on January 28, 1986.

In the aftermath of the accident, the presidential commission chaired by William Rogers conducted a wide-ranging investigation of the accident and of NASA decision making leading up to the accident.[22] Although the immediate attention was directed at the cause of the accident, the commission soon expanded the focus of the investigation from equipment to NASA management decision making.

In the hearings, a significant amount of information was made public which showed that the engineering staffs of NASA and the Morton Thiokol Corporation (prime contractor for the solid-fuel rocket booster) expressed doubts about both the design of and the effects of cold weather on the O-ring seals. Similar concerns had been expressed by Rockwell International Corporation's engineers and management (the builders of the shuttle orbiter) about the effects of cold and ice on other elements of the shuttle system. NASA and Morton Thiokol reports have since surfaced which document a suspected problem with the seals dating back over 8 years prior to the ill-fated launch.

The Rogers commission concluded that sufficient evidence had existed to postpone that *Challenger* launch, and that a critical problem existed with the design of the seals. Testimony was presented which indicated that NASA middle managers either disregarded or did not give sufficient credence to the warnings of contractors and engineers. Furthermore, higher levels of NASA managers, who had responsibility for the final decision to launch, were not even informed of the significant technical concerns that had been expressed. Thus the commission concluded that there were serious flaws in the managerial decision-making and communication process and that safety concerns were deemphasized in favor of preserving the flight schedule. If more effective communication had existed on that program, perhaps the accident could have been prevented. One thing is clear—the key managers on that program were not good listeners. Perhaps they did not want to be.

---

[22]Report of the Presidential Commission on the Space Shuttle *Challenger* Accident, Washington, D.C., June 6, 1986.

Sometimes it is more comfortable not to listen, to ignore those subtle hints that might indicate a problem. For example, in the interest of completing a project quickly, a project owner may choose not to hear the conceptual estimator's claim that the cost of implementing the project design may not meet the budget constraints. Eager to proceed with the project, the owner may choose not to evaluate why the project costs may exceed the budget. A burning desire to complete the project may blind the owner to the possibility that the overall project may not be economically feasible.[23]

The principal reasons why communication on a project can be a problem for the project manager are as follows:

- People withhold information on a problem in the hope that the problem will go away.
- Team members are reluctant to share information, which might be critical to the success of the project. They want to protect their territory and perhaps their jobs.
- The project manager maintains one-way contact with the team members, speaking but neglecting to listen. He or she issues orders, gives briefings, and submits direction for contract charges, but fails to seek feedback to see if anyone understands and is committed to what is going on.
- The project review meetings, which should maximize the two-way flow of information, turn out to be one-person shows, with the project manager talking, not listening. If the functional and work package managers practice the same style, one can imagine the lack of communication on the project.
- If the people, at whatever level on the project, do not understand the communication process, then faulty communication is bound to exist.
- There is a lack of trust among the project team members. Hartman, in his research, found that (1) effective communication is easier and more likely to be complete between people who trust each other, (2) teams work better together if the people in them can trust each other, and (3) we are more likely to be accepted as manager of a project—if others can trust us to do our jobs well.[24]

17.7 NONVERBAL COMMUNICATION

A part of communication that is all too often ignored is the subtle hidden messages that people send out through nonverbal means. As we consider nonverbal communication, a whole series of physical gestures come to mind: facial expressions, nodding, hand and body movements, eye movements, and so forth. Those of us who have visited the modern version of a burlesque theater would surely recognize the value of nonverbal communications in certain situations.

It would be difficult to portray nonverbal communication as formal or informal, because it depends on the context in which it is carried out. All of what we say, or

[23]Sievert, op. cit., p. 77.

[24]See Francis T. Hartman, "The Role of TRUST in Project Management," *Proceedings,* PMI Research Conference, June 2000, pp. 23–28.

do, and our style of listening play a role in our communication, which is subject to interpretation by others. Even a failure to act is a way of communication.

Porter divides nonverbal communication into four categories:

- *Physical.* The personal type of communication, including facial expressions, tone of voice, sense of touch, sense of smell, and body motions
- *Aesthetic.* Creative expressions, such as playing instrumental music, dancing, painting, and sculpturing
- *Signs.* Mechanical communication, such as the use of signal flags, the 21-gun salute, horns, and sirens
- *Symbolic.* Religious, status, or ego-building symbols[25]

Team leaders need to be aware of nonverbal communication. By observing nonverbal cues, leaders can have insight into the success that they are having in interacting with the team members. Team members display attitudes and feelings through nonverbal communication. If the project leader and the team members are aware of the physical gestures in communication, the chances improve for having an open, honest team that can deal with conflict in a forthright manner.

Regardless of your position in the project team, it is important to develop an appreciation of the uses and limitations of nonverbal communication. Being aware of the possibility of what nonverbal cues can transmit can keep the project team working toward more effective understanding of what the project team is really saying.

For example, a project director in a huge aerospace company called a meeting of higher-management people who supported his research project. He wanted them to fund development of a new project internally. Early in the meeting, as he began to outline the sizable costs involved, he sensed their disapproval from facial expressions and body postures. His intuition told him that if they were asked to make an explicit decision on the project, it would be negative. So he changed his line of argument and began stressing the possibilities for external rather than internal funding of the project. And he assiduously avoided asking for a funding decision at that time.[26]

Our physical actions, dress, manner, and language all communicate things to others about us. The project manager's interactions with the project stakeholders are affected by what they behold in her or his imagery, physical setting, and body language—and by what the project manager learns about stakeholders through their actions.

17.8 WRITTEN COMMUNICATIONS

In project management, written communications include proposals, reports, plans, policies, procedures, letters, memoranda, and other forms of transmitting

[25]George W. Porter, "Non-Verbal Communications," *Training and Development Journal,* June 1969, pp. 3–8.

[26]Michael B. McCaskey, "The Hidden Messages Managers Send," *Harvard Business Review,* November–December 1979, p. 145.

information. Writing effectively is an art and a skill, to be practiced at all times. Writing is a highly developed and very complicated part of communication. All people on a project team work at writing something to convey meaning to the reader. To serve this purpose, the message must be easily understood when it is quickly read. A well-written document reflects the writer's knowledge of the subject. The message is simple, clear, and direct.

The field of writing is so huge that we have neither the space nor the time to put much together on this subject. All professionals and managers should be able to find ample readings and courses on how to overcome their writing deficiencies. In this chapter we have provided only a few key suggestions on how to improve the writing deficiencies all too common in project management documentation.

The most important question that the project people can ask themselves is, Have we written our message clearly? There is nothing more important to a project team than being understood. A message should be easy to understand and informative.

Project proposals are one of the most important communication tools. Reports run a close second in importance. A proposal or report which uses simple, understandable language and uses tables, bar charts, pie charts, and graphs effectively will be more understandable to readers than one filled with technical jargon, vague concepts, and ambiguous language.

Effective writing depends on adequate preparation, based on selecting, analyzing, and organizing ideas required to communicate the intended message. Many processes have been suggested for preparing a written document; we have found the following steps useful.

First, think through the ideas you want to express in the written document. Build a "model" of the message in your mind. Do not start serious writing until you have a reasonably final model of what you want to say. On major documents such as reports, plans, and proposals, this step requires the project manager to work closely with the project team in "getting their arms around" the problem or opportunity.

Second, establish the basic purpose of the message. What are the general and specific purposes? Is the general purpose to direct, inform, question, or persuade? The specific purpose may be obvious or require thought and analysis. It is important that you take adequate time to define the specific purpose clearly; otherwise, it may be difficult to transmit a clear message to the intended receiver. The basic purpose of the message can be clarified in the next step.

Third, collect and analyze the qualitative and quantitative database of facts and assumptions bearing on the purposes of the message. Depending on the purpose of the message, detailed analysis may be required.

Fourth, organize the material into topics and subtopics in some logical sequence. Examine the way the material hangs together: Does the grouping make sense? Is the sequencing proper? If not, rework.

Fifth, prepare the first draft of the message. For short messages, this should be easy. For longer messages, including letters, papers, reports, and so on, this will require considerable effort. Concentrate on one section of the draft at a time and revise as required; do not try to write the entire message and then revise it in its entirety. After each section is finished, sit back and reflect on how it all fits together. Then set the first draft aside and let it "incubate" before beginning the first revision. Ask yourself these questions:

Is it objective?
Is it logical?
Are there any fallacies in the reasoning behind the message?
Did you say what you intended?
Is there too much (not enough) detail?
Does the main text of the message flow smoothly in a clear and logical manner?

Sixth, check out the message for acceptable grammar, spelling, punctuation, format, numbering, abbreviations, and the use of the right words and phrases.

Seventh, if the message is a report, does it follow this conventional structure?

1. Summary
2. Introduction
3. Discussion
4. Observations, conclusions, and recommendations

Finally, send the message. If you have carried out these steps diligently, the chances are good that the receiver will get the message.

17.9 PROJECT MEETINGS

The effectiveness of the meetings that the project team joins will tell a good deal about the sensitivity of the project manager to communications. Most meetings are poorly conceived and run; managers and professionals dislike meetings, particularly if the meetings are called by someone else. All would agree that well-run meetings are an effective focus around which the project can be managed. Many important matters on a project can be resolved by the project manager's working individually with the team members by telephone, conversation with individuals, or a brief ad hoc meeting with several team members. There is, however, value in getting all the team members together at scheduled times to talk and listen about the project.

Purpose

Project meetings serve several important functions[27]:

- The meeting helps define the major team players and the project. Those attending belong to the project team and other stakeholders; those who are absent do not.

[27]These functions are paraphrased from Antony Jay, "How to Run a Meeting," *Harvard Business Review,* March–April 1976, pp. 120–134.

- The meeting provides an opportunity to revise, update, and add to the store of knowledge that the team possesses. This knowledge includes facts, perceptions, experience, judgment, and folklore. The information that the team has acquired separately or in smaller meetings is important to the cohesiveness and strength of the team's role of knowing the cost, schedule, and technical performance status of the project.
- Meetings can help the team members know where the individual parts fit into the general collective aim of the group, and where individual success can contribute to team success.
- A meeting also helps team members become and remain committed to the project. By participating in decisions affecting the project team members, they will feel an obligation to accept the decision even though they may have argued against it during the team meeting. Many times opposition to a project decision arises out of not being consulted rather than the decision itself. Then, too, a project team decision is much harder to challenge than a decision by an individual.
- A project team meeting may be the only time where the team actually exists and works as a group and the only time that the project leader has visibility as the project manager leading the team.
- A meeting is a status arena, one place for team members to play out their roles. A meeting may be the only time when members get the chance to find out their relative standing on the team.
- Finally, a meeting provides the collective opportunity to pull together all the information on the project to see what action should be taken and to provide information on an individual work package.

By its nature, the project management process requires many meetings. Successful meetings do not just happen; many meetings suffer from:

- Domination of the discussions by certain vocal members.
- The lack of an agenda around which to focus the discussions; consequently, the discussions ramble.
- Domination of the discussion by the chairperson of the meeting, who fails to draw out the more reticent participants.
- The lack of firm starting and stopping times.
- People reading or talking among themselves, distracting others, and becoming nonparticipants in what is going on in the meeting.
- Avoidance of the meeting because it is emotionally upsetting to some team members.

How to Manage Project Meetings

It is believed that project managers spend at least 90 percent of their time communicating with team members, superiors, peers, colleagues, stakeholders, and other who have—or believe they have—a vested interest in the project. The increasing use of project management, particularly in the context of global work, has posed new challenges for communicating about the project.

Communication issues involved in the management of a global project can be particularly challenging because of the different languages and cultures and the behavior of people. Other issues include:

- The assumptions about how people behave as they do.
- Different mores, traditions, practices, and meaning of nonverbal communication.
- The geographic and time differences involved in communicating.
- The inherent stress and frustration of communicating in an acquired language.
- Likely misunderstanding arising out of working with people of a different culture and language.
- Differing management and business practices.
- Lack of experience in working in a foreign country having a different language and a different culture.

Videoconferencing is a recent innovation that promises to expand the ability of the project team to better communicate. The ability to communicate during project meetings is an important attribute for the project manager and other team members.

A project meeting can be productive. It is an efficient way to share information, obtain immediate feedback on issues or questions, or clarify an unclear point. Meetings can save time that would be spent otherwise sending and answering memoranda, making telephone calls, or waiting for inquiries. By far the greatest value of a meeting is to bring collective judgment to project problems and opportunities.

The chairperson of the meeting has the responsibility to plan, host, and lead the meeting to establish the proper climate. That climate and the feeling conveyed to the participants will have a great impact on the outcome of the meeting. The chairperson must guide, stimulate, clarify, control, summarize, and evaluate the outcome in terms of the meeting objective, keeping in mind her or his responsibility for managing the project.

A meeting, even a conference with another person, is a managerial activity and can be looked at from the standpoint of the management functions of planning, organizing, and controlling. Using these management functions as a guide, we can suggest a strategy to prepare for and conduct a meeting. Planning and organizing a meeting consist of several actions, which include:

- Determining the objective, or expected output, of the meeting
- Preparing the agenda
- Selecting and inviting the participants
- Determining the timing and physical arrangements
- Considering matters of protocol such as seating, introduction of newcomers, notification of attendees, and so on
- Preparing and distributing materials required for participant study

A key question to consider when you are planning a meeting is simply this: Is this meeting really necessary? In this context Jay noted: "The most important question you should ask is: 'What is this meeting intended to achieve?' You can ask it in different ways—'What would be the likely consequences of not holding it?' 'When it is over, how shall I judge whether it was a success or a failure?'"[28]

Often the need for a meeting can be eliminated by judicious analysis of a problem before people are called together to seek some solution. A practical process to reduce the number of meetings is to prepare a brief informal memorandum to yourself that addresses these questions:

- What is the issue (specific problem or opportunity) for which the meeting should be held? Many times meetings are held without any definition of what the output of the meeting will be.
- What are the facts? Problems or opportunities do not exist in an information vacuum. There are some facts that bear on the situation, that cause the problem or suggest the opportunity.
- What are the potential alternatives or solutions—and the associated costs and benefits that relate to these alternatives? Even a cursory thought about alternatives can prove useful in deciding whether a meeting should be held.
- What specific recommendations can be proposed to deal with the problem or opportunity at this time, and could be suggested to the meeting participants?
- What will happen if the meeting is not held?

Answering all these questions should help you find a solution and possibly eliminate the need for the meeting. Even a project team member who wishes to meet with the project manager will find that trying to answer these questions will reduce the frequency with which he or she needs to consult with the team leader.

Part of the planning for a meeting is a plan for the organization of the meeting. Motivating, leading, and controlling the meeting are the chairperson's responsibility. Given the agenda as a standard, the activities to be carried out by the meeting leader include the following:

- Seeking points of agreement or disagreement
- Limiting discussion
- Encouraging all to participate
- Periodically summarizing points of agreement or disagreement
- Identifying action items to be investigated by individual members
- Adhering to time limits for starting, stopping, and dealing with agenda items
- Reinforcing the objective and expected outcome of the meeting
- Encouraging and controlling disagreements
- Taking time during the meeting to assess how well things are going and what might be done to improve the effectiveness of the discussion

---

[28]Ibid., p. 47.

- Making sure each agenda item has a time allocation
- Providing for all participants to express their ideas and recommendations without interrupting or degrading their comments
- Listening to everything
- Playing down irrelevant issues, perceptions, and personal speculations as soon as possible, before they can become disruptive
- Stopping the discussion and redirecting the meeting as needed
- Always being patient
- Making the decisions that the meeting is to bring into focus

One of the clearest signals that the chairperson is not doing the job is that he or she does most of the talking. The chairperson's greatest influence stems from the participants' perception of his or her commitment to the objectives of the meeting and skill and efficiency in helping the participants in meeting that objective.

If minutes are to be kept of the meeting (certainly recommended during the project planning, organizing, and evaluating meetings), then brevity is desirable. The summary minutes should include these facts:

- Times started and ended, date, place of meeting, and list of participants with their project role
- Agenda items discussed and decisions reached or held for further study
- Enumeration of action items and person(s) responsible for follow-up and reporting back to project team
- Time, date, place, and instructions for the next meeting

Meetings are an essential process of management. Depending on how they are run, they can improve or impair communications, promote or discourage cooperation, and encourage or discourage people. The value of an effective meeting is that it can serve as a cornerstone for successful team building and for planning and evaluating progress on the project, as well as a communications link between the project stakeholders. Technology is changing the way we communicate.

17.10 THE ROLE OF TECHNOLOGY

Teleconferencing is a growing use of technology to facilitate the management of a project, particularly a project where the key team members are geographically separated. Some project managers have found that teleconferencing is a useful substitute for business travel. When teleconferencing is used in lieu of face-to-face meetings, there tends to be far less socializing and chitchatting before the meeting starts, and the participants tend to stick with the agenda better.

Electronic mail has become a medium to unite people rather than through meetings and paperwork. Electronic mail has gained increasing acceptance as a substitute for regular mail, memoranda, and other means of written and verbal communication.

Because e-mail saves time and storage space and is much faster than regular written communication, it is, if properly used, a means of increasing productivity.

Today, thousands of businesses build their sites on the Internet. Finding out how the Internet can transform your business or your project is challenging. Hundreds and perhaps thousands of seminars are being given to explain how the Internet can be used to gain access to the information superhighway.

Another technological application is the electronic bulletin board. An electronic bulletin board enables a message originator to post a message on the "board" so that anyone connected with the computer network can see it. Readers can read the message and then express their opinions by posting their own messages.

Groupware—computer software explicitly designed to support the collective work of teams—can facilitate the discussions carried out in project teams and accomplish the project work in much less time than needed to have team members attend meetings. In one case at Boeing Company, a team composed of engineers, designers, machinists, and manufacturing managers used Team Focus software from IBM to design a standardized control system for complex machine tools in several plants. Normally, such a job would have required more than a year—with electronic meetings it was done in 35 days.

At IBM the development of the 9370 mainframe computer was carried out through organizational networking. The objective was to develop, build, and get the complex machine to market quickly with minimal use of the corporate hierarchy. Many IBM people collaborated from the research laboratories, manufacturing plants, suppliers, marketing groups, and distribution centers around the world. The project team members were linked by electronic mail, teleconferencing, and other communications channels into ad hoc project teams, multidisciplinary conferences, and the like.[29]

Consider that the typical manager spends somewhere between 30 and 70 percent of the day in meetings. Even in the best-run meetings, you seldom get all the latent, best ideas out of people. Some people remain shy, feel they are juniors in the meeting, feel intimidated, or are just too polite to say anything that might be adversarial or might challenge a superior. Then, too, research has shown that 20 percent of the people in a meeting do 80 percent of the talking. In electronic meetings comments can be kept anonymous. This can be a powerful incentive to speak out.

Groupware is helping teams function more efficiently and can help move the company toward a more team-focused organization. Groupware, when properly used, can improve the quality and productivity of meetings. A team of people can reach a genuine consensus; members become more committed and feel more a part of the substantive process of what is going on in the team electronic meeting.

The strategy seems to be to use groupware to integrate people into team-based organizations. It can be used for any type of project that requires groups of people to work together, as in product-process design teams engaged in simultaneous engineering. When a dramatic change in the market requires immediate and coordinated strategies to develop a countervailing strategy, group computing can

---

[29]Fred V. Guterl, "Goodbye Old Matrix," *Business Month,* February 1989, pp. 32–38.

achieve a consensus, even when people are geographically separated, in much less time. A strategic planning meeting that used to take 2 days can be done in 3 or 4 hours. When group computing is used, it is a "nondiscriminating" meeting—you don't know if the messages are coming from juniors, seniors, men, women, minority persons, or those people who have vested interests in what is being discussed. The technology of group computing makes people equal not in terms of power but in terms of being heard.

There are drawbacks: The verbal cues that can indicate how a person feels about an issue are lacking. Without the "eyeball-to-eyeball" contact, people's attention can wander. By putting so much on the electronic networks, management can learn a lot more about what is going on in the organization and how people are feeling. For some people this could be viewed as an invasion of personal privacy.

Whatever the shortcomings, group computing can open new avenues for productivity and consensus building in the organization. As computers become smaller, electronic meetings can be held anywhere—wherever you have your computer. All this could lead to drastic population dispersal—keep in mind that cities and the transportation to those cities have been built so that people can physically work together. Now, with electronic meetings people can get together and communicate intellectually through the computers. A lesson to consider: In the future when you go on vacation, do not take your computer with you![30]

17.11 COMMUNICATION LINKS

In the most general sense, the project manager needs to maintain communication links with all the project stakeholders. However, certain stakeholders require direct and ongoing communication:

- Customers (owners, users)
- Project team members
- General managers
- Functional managers
- Regulatory agencies
- Subcontractors

Communication with all these stakeholders is important; the customer stands out as the most important of all. There are a few key objectives in communicating with the customer that the project manager and the project team should keep in mind:

- Never surprise the customer. Keep the customer informed of anything that has affected or could affect the schedule, cost, or technical performance objectives of the project.

[30]David Kirkpatrick, "Here Comes the Payoff from PCs," *Fortune,* March 23, 1992, pp. 93–102.

- Don't depend on formal reports to keep the customer fully informed. An excellent practice is for the project manager to keep in touch on a regular basis by telephone and visits with the customer. During these periods any items of progress or lack thereof can be brought to the customer's attention. If a problem has developed on the project, let the customer know even if a solution is not apparent. The customer will appreciate this news and might be able to help in designing a solution.
- Always follow up any customer question or concern with an action item, with someone designated as the action person to follow up and report back to the project manager. When an answer or solution is available, inform the customer. During the interim keep the customer informed of the progress.
- Remember that the customer has to be happy with the project's progress and its results. Build your customer communication philosophy around this objective.

Management of the project team depends so much on information and communication—and knowledge flows through people who are dealing with the different technologies needed to bring the project objective into focus. Given the knowledge, skills, and attitudes of the project team, the willingness of team members to engage in ongoing conversation about the project is necessary to keep everyone informed of the project's status. Also, an important benefit is that this bonds the team together. Conversation is important within the team itself but also with the project stakeholders who are external to the team's organization. Customers, suppliers, regulators, local community leaders, and other important stakeholders have to be engaged in continuous conversation to bond them to the project as well as keep them advised of what is going on and how their interests are being affected and are likely to be affected by the project itself. Keeping the project stakeholders informed at all times can do much to reduce the fear that they might have of the compromise of their stakes in the project. Conversation also helps build trust and project team loyalty.

Information and conversation do much to penetrate the organizational and disciplinary boundaries that exist in any project. The larger the project, the greater the likelihood that there will be more complex boundaries.

17.12 TO SUMMARIZE

Some of the key messages that were presented in this chapter include:

- Communication is the process by which information is exchanged between senders and receivers through a common system of symbols.
- The ability to communicate is an important criterion for promotability.
- A high percentage of frictions, frustrations, and inefficiencies in our working relationships with other people is traceable to poor communication.

- Most problems in organizations can be traced to people—and the inability to communicate with people is often the cause of such problems.
- A couple of examples were given of failures in communication in major projects that caused extraordinary delays and increases in costs.
- There are many formal and informal means or channels for communicating with project stakeholders. In this chapter a few of these means and channels are listed.
- Most present-day failures in communication can be traced to misunderstandings of the symbols that play an important part in the process of human communication.
- Communication is a two-way process between a sender and the receiver(s).
- Informal communication is carried out through the informal organization. Membership in this organization is dependent on common ties and interests.
- Listening is an important part of communication. It is a skill that can be learned and is important to the project manager and all project stakeholders in finding out what is going on in a project.
- There can be many impersonal barriers to communication, such as ethics, morals, beliefs, prejudices, politics, biases, and other things that condition behavior.
- Nonverbal communications are carried out through physical gestures such as facial expressions, nodding, hand and body movements, eye movements, and so forth. The prudent project manager learns to look for nonverbal communications when working with the project stakeholders.
- There is an abundance of written communications involved in the management of a project. All project stakeholders should work diligently at improving their ability to communicate through writing.
- Project meetings, both formal and informal, are an extremely important means for communicating about the project's status.
- Most meetings are not very well planned or executed.
- The application of management theory and practice to a meeting can be done through the use of planning, organizing, motivation, directing, and control processes.
- A prescription of how to plan and execute excellent project review meetings was described in the chapter.
- The advancement of technology has provided extraordinary benefits to improving the ability to communicate in the business world and to manage projects. Such means as telecommunications, groupware, and the Internet can provide the project manager with enhanced communication capabilities, if properly planned and executed.
- A policy on how to communicate with the customer is an important matter and should be planned and diligently executed.

17.13 ADDITIONAL SOURCES OF INFORMATION

The following additional sources of project management information may be used to complement this chapter's topic material. This material complements and expands on various concepts, practices, and the theory of project management as it relates to areas covered here.

- Francis M. Webster, Jr., and Stephen D. Owens, "How to Get the Right Message Across," chap. 20 in David I. Cleland (ed.), *Field Guide to Project Management* (New York: Van Nostrand Reinhold, 1997).
- Steven A. Goldstein, Gwen M. Pullen, and Daniel R. Brewer, "Can We Talk? Communication Management for the Waste Isolation Pilot Plant, a Complex Nuclear Waste Management Project"; Michael Newell, "Communication Risk Management in Municipal Government Projects: City of New Orleans Computer-Aided Dispatch System Project"; and Jhan Schmitz, "Communicating Constraints: Schedule Baseline and Recovery Measures on the Hong Kong Airport Projects," in David I. Cleland, Karen M. Bursic, Richard J. Puerzer, and Alberto Y. Vlasak, *Project Management Casebook,* Project Management Institute (PMI). (Originally published in *Proceedings,* PMI Seminar/Symposium, Denver, 1995, pp. 572–581; *Proceedings,* PMI Seminar/Symposium, Denver, 1995, pp. 224–233; and *Proceedings,* PMI Seminar/Symposium, Denver, 1995, pp. 121–128.)
- R. Max Wideman, *Comparative Glossary of Common Project Management Terms,* Home Page and Index, The Project Management Forum, Wideman Comparative Glossary of Common Project Management Terms, vol. 2.1 (e-mail: pmg%maxwideman.com). One of the challenges in communicating effectively in any profession is to understand the specialized language of that discipline. In this glossary, Wideman has assembled and defined the principal project management terms. In so doing, he has provided the project management professional with an important document that he or she can use in improving the ability to communicate in the project management community.
- *Harvard Business Review on Effective Communication,* a *Harvard Business Review* paperback, Boston, Mass., 1999. This concise paperback contains eight chapters about communication, including how to run a successful meeting, change employees' behavior, build effective management teams, and other communication-related topics. The authors of the chapters are all notable professionals in their field. The topics of the chapters should have special appeal to a project manager and those people associated with project management.
- Martin D. Hynes, III, "Information Management: For the Project Manager in an Information Age," in Joan Knutson (ed.), *Project Management for Business Professionals* (New York: Wiley, 2001). The author reviews project and portfolio information from a general perspective, and then offers examples of information management tools. The author starts off by describing the great abundance of

information and its availability in modern societies. As today's project manager manages projects, this context of abundance of information should not be forgotten. He further notes that people associated with a project want to receive specific information needed to do their job. In the author's belief, the challenge for the project manager is to communicate project-related information effectively within an environment where information overload is a growing force.

- Peg Thomas, "Project Team Motivation," in Jeffrey K. Pinto (ed.), The Project Management Institute, *Project Management Handbook* (San Francisco: Jossey-Bass, 1998). An important, desirable outcome of communication within the project stakeholder community is the motivation of these stakeholders to support the project ends. In this chapter the author explores some of the leading theories of motivation with applications to project teams. Thomas explains strategies that have been proven to boost motivation.
- Thomas R. Drury, "Team Communications in Complex Projects," *Journal of Engineering and Technology Management,* vol. 4, issue 4, 1988. The author makes a strong case that engineering managers who are faced with a system development project need to develop expertise in communications ability. The work activity of a system development team may be characterized in terms of a group of coupled conversations. He outlines a model that offers a graphic presentation of team resources and a metaphor for interactive communication activity. According to the author, the use of such a model encourages complete consideration of the needs for a given conversation, and helps document successful interaction patterns.

17.14 DISCUSSION QUESTIONS

1. Explain why the ability to communicate is among the most often cited attributes of successful project managers.
2. Describe a project management situation from your work or school experience. What communication problems affected the project? How?
3. What are some of the means by which information is communicated on a project? In what kinds of situations is each of these most effective?
4. Define communication in terms of the elements in the communication process. Define each of the elements.
5. Listening is often the most important aspect of good communication. Why is this factor often ignored? What can managers do to increase their listening skills?
6. Discuss nonverbal communication. How can a manager use an understanding of nonverbal communication in determining what project team members are really saying?
7. Describe some of the important steps in the preparation of a written document. What role do written documents play in the communication process?

8. Why are team meetings important for project communications? What steps can a project manager take in order to hold effective meetings?
9. What planning must be done in order to ensure an effective meeting process?
10. How might technology improve the effectiveness of communication in the management of a project?
11. What are some of the communication links that a project manager must control? What can be done to manage these links?
12. What are other things that management can do to ensure good communication throughout the organization?

17.15 USER CHECKLIST

1. How would you judge the ability of the project managers in your organization with respect to communications? Explain.
2. What kinds of communication problems are often experienced on projects within your organization? How are these problems usually managed?
3. What methods do the project managers in your organization use to communicate information? Are the methods used usually effective for the particular situation? Why or why not?
4. Do the project managers of your organization listen to the problems and suggestions of project team members? How can they improve their listening skills?
5. What nonverbal communications are often used by project team members? Do project managers recognize and interpret nonverbal communications? Explain.
6. Are the written documents (policies, reports, etc.) from project managers effective at presenting information? What is lacking in these written reports? What can be done to improve their effectiveness?
7. How effective is the communication process during project team meetings? Do project leaders usually control project meetings? Explain.
8. How often are team meetings held? Do project team members understand the purpose of team meetings and their roles in the meetings? Why or why not?
9. Are project team meetings planned and organized ahead of time? Why or why not?
10. Has the organization fully exploited the use of technology to improve communications in the management of the project?
11. Do project managers manage the various communication links within and external to the organization? What are some of these links?
12. How would you judge the overall effectiveness of the communication process in your organization? How can the process be improved?

17.16 PRINCIPLES OF PROJECT MANAGEMENT

1. There is a direct relationship between project success or failure and the effectiveness of the communication pattern carried out on the project.
2. Communication is the principal force used by a project manager to ensure that the project stakeholders work together.
3. Communication is the process by which information is exchanged between individuals through a common system of symbols, signs, or behavior.
4. The real meaning of words, or symbols, depends primarily on how the reader or listener perceives them.
5. Communications in the informal organizations can be a major influence in the management of the project.
6. The ability to listen effectively to project stakeholders is an important part of the communications in the management of the project.
7. The degree of trust that exists among the project stakeholders can influence the effectiveness of the management of the project.

17.17 PROJECT MANAGEMENT SITUATION—HOW TO COMMUNICATE

Communication in the management of a project flows downward, upward, horizontally, and to outside receivers, such as the stakeholders. The purpose of project communications is to provide information to the project stakeholders. Project information is distributed through a variety of means such as project meetings, hard copy document distribution, shared access to networked electronic databases, fax, electronic mail, voice mail, videoconferencing, and individual and group discussions. Project management communications are also carried out through software, such as correspondence, memos, reports, and documents describing a project's planning, organization, evaluation and control strategy, and progress.

A project communications plan may be formal or informal and highly detailed or broadly framed. The plan should be based on the needs of the project, its stakeholders, and the sponsoring organization.

The project progress report provides an agreed weekly, monthly, or quarterly project status report that includes all costs, activities, and progress, through the last working day of a previous reporting period. Each report should be so designed that it is sufficiently comprehensive and self-explanatory so that the reader can grasp and understand the status of project activities, especially those concerning established baselines: scope, schedule, and cost.

Communications planning determines the information and communications needs of stakeholders: who needs what information, when do they need it, and how is it provided to them. The communication plan should provide the following:

- A collection, filing, and distribution schema for the information.
- A distribution structure that details to whom information will go and information types, such as status reports, data, schedule, technical documentation, and so forth.
- A description of the information to be distributed, including format, content, level of detail, and if needed, a security classification for the information.
- A schedule for when the different types of information will be produced and distributed.
- A schema on how information regarding the project can be accessed between scheduled distributions.
- A strategy on how the communications management plan is updated and refined as the project progresses.
- A protocol for reporting emergency or vitally important information that arises on the project.
- A schema for the information that is required for the project's status report.

17.18 STUDENT/READER ASSIGNMENT

The students/readers are requested to select a project known to them and determine what information should go into the project's status report, and to whom such information should be distributed.

To start, the project's status report should include a narrative of accomplishments by major activities, the status of each major component of the project, and the project manager's overall assessment of the project.

What other information should go into the project's progress report?

CHAPTER 18
EFFECTIVE PROJECT TEAMWORK[1]

"Our technology, reach and resources aren't enough to make us the global best. It's all about people, nurturing, energizing and inspiring them to search for ideas and to cooperate. It's about creating a culture that brings everyone into the game across the organization."

JEFF R. IMMELT, CEO, GENERAL ELECTRIC COMPANY

18.1 INTRODUCTION

Dieter Zetsche replaced James P. Holden as CEO of Chrysler in November 2000. Holden was fired with less than 1 year on the job because of severe financial problems that led to a $2 billion loss in 2001, and a nonprofit projection until 2003. Since the merger with Daimler-Benz in 1998, Chrysler's fortune was shifting rapidly. The company owed much of its success to the ability to avoid head-on competition by creating such innovative, high-margin vehicles as the minivan. But, as more efficient rivals, such as Honda and Toyota, moved swiftly into the U.S. market with similar vehicles, prices began to fall and Chrysler started to suffer. Now Zetsche has to turn the company around. Part of his strategy is to overhaul the product development process, a highly interdisciplinary 3-year team effort involving all functional areas across Chrysler, plus hundreds of suppliers, contractors, government agencies, and the Daimler-Benz parent company. Zetsche is determined to wring synergism out of the $36 billion merger by combining German engineering with American marketing. He is devoting a great deal of effort toward an agreement with Daimler to make diesel engines and other M-class SUV parts available for Chrysler's Cherokees and PT-Cruisers. Going even further, Zetsche plans to install a wide array of Mercedes parts in a large number of Chrysler cars, ranging from the Grand Cherokee to the Dodge Intrepid. Starting in 2004, the new Crossfire, a two-seat roadster, will be built with Mercedes transmissions,

[1]This chapter was written by Hans J. Thamhain, Professor of Management at Bentley College in Waltham, Massachusetts. He has held engineering and project management positions with Verizon/GTE, General Electric, Westinghouse, and ITT. Dr. Thamhain is well known for his research and writings in project management. He has written over 70 research papers and 5 professional reference books in project management. Dr. Thamhain is a certified project management professional and the recipient of the IEEE Engineering Manager of the Year 2000 Award.

axles, and engines. In addition Chrysler will work with Mitsubishi to develop small and mid-size cars.

Zetsche admits that a "not-invented-here" syndrome kept Chrysler and Mercedes from sharing technology in the past. In fact, it took a group of senior executives from both companies several months to work out a cooperative technology agreement. Although the policy draft of this executive agreement is still heavily debated by several newly formed committees in both companies, "a momentum of cooperation is developing."

With increasing competition, Zetsche needs to keep customers coming through the door without offering big price discounts. This will require combining new styles and features with thrift and agility in the market. In overhauling the new vehicle developing process, Zetsche counts on W. Bernhard, his chief operating officer, to put more focus on the early stages of the car development. By pulling together teams from all areas of the company, including design, styling, engineering, manufacturing, marketing, purchasing, and field services, he hopes to increase resource efficiency and quality, without diminishing Chrysler's creativity and market responsiveness. A key ingredient in this new process is a system of "quality gates," borrowed from Mercedes. This will allow Zetsche and his team to monitor and review a vehicle via 11 checkpoints throughout its 3-year development cycle. Zetsche is fully aware of the organizational complexities and team challenges involved in bringing Daimler-Chrysler into black numbers: "To my knowledge, no one has ever turned such a grim situation around without a near-death experience." However, Zetsche has no choice. If successful, he can pretty well name his price. Otherwise, his name will likely join the list of casualties of the $36 billion merger experiment.[2]

18.2 THE NEED FOR EFFECTIVE TEAMWORK

The complexities and challenges faced by Chrysler are quite common in today's business environments that involve a wide array of project-related issues (Armstrong, 2000; Dillon, 2001; Gray and Larson, 2000; Thamhain and Wilemon, 1999). As indicated in the Chrysler scenario, the team effort often spans organizational lines, including an intricate functional spectrum of assigned personnel, support groups, subcontractors, vendors, partners, government agencies, and customer organizations. Uncertainties and risks introduced by technological, economic, political, social, and regulatory factors are always present and can be an enormous challenge to organizing and managing the project teams. Yet another challenge is the establishment of effective linkages among the various team factions and support groups for proper communications, decision making, and control. This involves broad-scale team building toward interorganizational alliances and cooperation. It also requires team leaders to cross functional lines and deal with resource personnel over whom they have little or no formal authority, and to deal effectively with resource sharing, multiple reporting relationships, and accountabilities.

[2]Website www.DaimlerChrysler.com and "Can This Man Save Chrysler," *Business Week,* September 17, 2001.

Teamwork: A Managerial Frontier

Most managers consider effective cross-functional teamwork a key determinant of project performance and success in today's competitive world of business.[3] Virtually all managers recognize the critical importance of effective teamwork and strive for continuous improvement of team performance in their organizations. Yet only 1 in 10 of these managers has a specific metric for actually measuring team performance. Obviously, this creates some tough challenges, especially in project-based environments where teamwork is crucial to business success. In these organizational environments, work teams must successfully integrate multidisciplinary activities, unify different business processes, and deal with cross-functional issues, such as innovation, quality, speed, producibility, sourcing, and service. Managerial principles and practices have changed dramatically. Not too long ago, project management was considered to a large degree "management science." Project leaders *could* ensure successful integration for most of their projects by focusing on properly defining the work, timing, and resources, and by following established procedures for project tracking and control. Today, these factors are still crucial. However, they have become threshold competencies, critically important, but unlikely to guarantee project success by themselves. Today's complex business world requires *project teams* that are fast and flexible, and can dynamically and creatively work toward established objectives in a changing environment (Bhatnager, 1999; Jasswalla and Sashittal, 1999; Thamhain, 2001; Thamhain and Wilemon, 1999). This requires effective networking and cooperation among people from different organizations, support groups, subcontractors, vendors, government agencies, and customer communities. It also requires the ability to deal with uncertainties and risks caused by technological, economic, political, social, and regulatory factors. In addition, project leaders have to organize and manage their teams across organizational lines. Dealing with resource sharing, multiple reporting relationships, and broadly based alliances is as common in today's business environment as e-mail, flextime, and home offices.

Because of these complexities and uncertainties, traditional forms of hierarchical team structure and leadership are seldom effective and are being replaced by self-directed, self-managed team concepts (Barner, 1997; Thamhain and Wilemon, 1999). Often the project manager becomes a social architect who understands the *interaction of organizational and behavioral variables,* facilitates the work process, and provides overall project leadership for developing multidisciplinary task groups into unified teams and fostering a climate conducive to involvement, commitment, and conflict resolution. Typical managerial responsibilities and challenges of today's project team leaders are summarized in Table 18.1. This table can also be used as a tool for self-assessment, including establishing performance measures, training, and organizational development needs.

---

[3] "Field survey on team leadership, conducted between 1999 and 2000, involving a sample of 560 managers in 38 technology-oriented companies," working paper, H. Thamhain, 2000. One survey question was, "What factors and conditions do you consider most critical for effective business performance?" Seventy-two percent of the respondents ranked teamwork among the top three factors crucial to success.

TABLE 18.1 Responsibilities and Challenges of Project Team Leaders

- Bringing together the right mix of competent people which will develop into a team
- Building lines of communication among task teams, support organizations, upper management, and customer communities
- Building the specific skills and organizational support systems needed for the project team
- Coordinating and integrating multifunctional work teams and their activities into a complete system
- Coping with changing technology requirements and priorities while maintaining project focus and team unity
- Dealing with anxieties, power struggles, and conflict
- Dealing with support departments; negotiating, coordinating, and integrating
- Dealing with technical complexities
- Defining and negotiating the appropriate human resources for the project team
- Encouraging innovative risk taking without jeopardizing fundamental project goals
- Facilitating team decision making
- Fostering a professionally stimulating work environment where people are motivated to work effectively toward established project objectives
- Integrating individuals with diverse skills and attitudes into a unified work group with unified focus
- Keeping upper management involved, interested, and supportive
- Leading multifunctional task groups toward integrated results in spite of often intricate organizational structures and control systems
- Maintaining project direction and control without stifling innovation and creativity
- Providing an organizational framework for unifying the team
- Providing or influencing equitable and fair rewards to individual team members
- Sustaining high individual efforts and commitment to established objectives

18.3 WHAT WE KNOW ABOUT PROJECT-ORIENTED TEAMWORK

Concept and Process

Teamwork is not a new idea. The basic concepts of organizing and managing teams go back in history to biblical times. In fact, work teams have long been considered an effective device to enhance organizational effectiveness. Since the discovery of important social phenomena in the classic Hawthorne studies (Roethlingsberger and Dickinson, 1939), management theorists and practitioners have tried to enhance group identity and cohesion in the workplace (Dyer, 1977). Indeed, much of the *human relations movement* that followed Hawthorne is based on the group concept. McGregor's (1960) theory Y, for example, spells out the criteria for an effective work group, and Likert (1961) called his highest form of management the participating group or system 4. However, the process of team building becomes more complex and requires more specialized management skills as bureaucratic hierarchies decline and horizontally oriented teams and work units evolve.

Redefining the Process

In today's more complex multinational and technologically sophisticated environment, the group has reemerged in importance as the *project team* (Fisher, 1993; Nurick and Thamhain, 1993; Thamhain and Wilemon, 1999).

> *Teambuilding can be defined* as the process of taking a collection of individuals with different needs, backgrounds, and expertise and transforming them into an integrated, effective work unit.

In this transformation process, the goals and energies of individual contributors merge and focus on specific objectives. When describing an effective project team, managers stress consistently that high performance, although ultimately reflected by producing desired results on time and within budget, is a derivative of many factors which are graphically shown in Fig. 18.1.

Team building is an ongoing process that requires leadership skills and an understanding of the organization and its interfaces, authority, power structures, and motivational factors. This process is particularly crucial in environments

FIGURE 18.1 Characteristics of a high-performing project team.

where complex multidisciplinary or transnational activities require the skillful integration of many functional specialties and support groups with diverse organizational cultures, values, and intricacies (Oderwald, 1996). Typical examples of such multidisciplinary activities requiring unified teamwork for successful integration include:

- Establishing a new program
- Transferring technology
- Improving project-client relationships
- Organizing for a bid proposal
- Integrating new project personnel
- Resolving interfunctional problems
- Working toward major milestones
- Reorganizing mergers and acquisitions
- Transitioning the project into a new activity phase
- Revitalizing an organization

Because of their potential for producing economic advantages, work teams and their development have been researched by many. Starting with the evolution of formal project organizations in the 1960s, managers in various organizational settings have expressed increasing concern with and interest in the concepts and practices of multidisciplinary team building. As a result, many field studies have been conducted, investigating work group dynamics and criteria for building effective, high-performing project teams. These studies have contributed to the theoretical and practical understanding of team building and form the fundamental concepts discussed in this chapter. Prior to 1980, most of these studies focused just on the behavior of the team members, with limited attention given to the organizational environment and team leadership. Although the qualities of the individuals and their interaction within the team are crucial elements in the teamwork process, they represent only part of the overall organization and management system which influences team performance, a fact that was recognized by Bennis and Shepard as early as 1956. Since 1980 an increasing number of studies have broadened the understanding of the teamwork process (Dumaine, 1991; Tichy and Urlich, 1984; Walton, 1985). These more recent studies show the enormous breadth and depth of subsystems and variables involved in the organization, development, and management of a high-performing work team (Gupta and Wilemon, 1996). These variables include planning, organizing, training, organizational structure, nature and complexity of task, senior management support, leadership, and socioeconomic variables, just to name the most popular ones. Even further, researchers such as Dumaine (1991), Drucker (1996), Peters and Waterman (1987, 1997), and Moss Kanter (1989) have emphasized the nonlinear, intricate, often chaotic, and random nature of teamwork, which involves all facets of the organization, its members, and environment. These teams became the conduit for transferring information, technology, and work concepts across functional lines quickly, predictably, and within given resource restraints.

Team Life Cycle

The life cycle of a project team spans across the complete project, not just a particular phase. For example, the Chrysler team responsible for creating a new model integrates activities ranging from assessing an opportunity to product research, feasibility analysis, development and engineering, technology transfer to manufacturing, marketing, and field service. This work may also involve bid proposals, licensing, subcontracting, acquisitions, and offshore manufacturing. The need for close integration of activities across the entire project life cycle requires that these multidisciplinary teams stay together as a unified, effective work group for most of the product life cycle, rather than just for a particular phase of core activities. For example, the primary mission of the product development team may focus on the engineering phase, but the team also supports activities ranging from recognition of an opportunity to feasibility analysis, bid proposals, licensing, subcontracting, transferring technology to manufacturing, distribution, and field service. This creates managerial challenges in dealing effectively with resource leveling, priority conflicts, and long-range multifunctional commitment

Toward Self-Direction and Virtual Teams

Especially with the evolution of contemporary organizations, such as the matrix, traditional bureaucratic hierarchies have declined and horizontally oriented teams and work units became increasingly important to effective project management (Fisher, 1993; Marshall, 1995; Shonk, 1996). Increasingly, the team leader's role as supervisor has been diminished in favor of more *empowerment and self-direction* of the team, as defined in Table 18.2.

TABLE 18.2 Self-Directed Teams

Definition. A group of people chartered with specific responsibilities for managing themselves and their work, with minimal reliance on group-external supervision, bureaucracy, and control. Team structure, task responsibilities, work plans, and team leadership often evolve on the basis of needs and situational dynamics.

Benefits. Ability to handle complex assignments, requiring evolving and innovative solutions that cannot be easily directed via top-down supervision. Widely shared goals, values, information, and risks. Flexibility toward needed changes. Capacity for conflict resolution, team building, and self-development. Effective cross-functional communications and work integration. High degree of self-control, accountability, ownership, and commitment toward established objectives.

Challenges. A unified, mature team does not just happen, but must be carefully organized and developed by management. A high degree of self-motivation, and sufficient job, administrative, and people skills must exist among the team members. Empowerment and self-control might lead to unintended results and consequences. *Self-directed* teams are *not* necessarily *self-managed;* they often require *more* sophisticated external guidance and leadership than conventionally structured teams.

In addition, advances of information technology made it feasible and effective to link team members over the Internet or other media, creating a *virtual team* environment, as described in Table 18.3.

Virtual teams and *virtual project organizations* are powerful managerial tools, especially for companies with geographically dispersed project operations, and for linking contractors, customers, and regulators with the core of the project team.

These contemporary team concepts are being applied to different forms of project activities in areas of products, services, acquisition efforts, political election campaigns, and foreign assistance programs. They are also found in specialty task groups such as venture teams, *skunk works,* process action teams, and focus groups. For these kinds of highly multifunctional and nonlinear processes, researchers stress the need for strong integration and orchestration of cross-functional activities, linking the various work groups into a unified project team that focuses energy and integrates all subtasks toward desired results. Although these realities hold for most team efforts in today's work environment, they are especially pronounced for efforts that are associated with risk, uncertainty, creativity, and team diversity such as high-technology and/or multinational projects. These are also the work environments that first departed

TABLE 18.3 Virtual Teams

Definition. A group of project team members, linked via the Internet or media channels to each other and various project partners, such as contractors, customers, and regulators. Although physically separated, technology links these individuals so they can share information and operate as a unified project team. The number of elements in a virtual team and their permanency can vary, depending on need and feasibility. An example of a virtual team is a project review conducted among the team members, contractors, and customer over an Internet website.

Benefits. Ability to share information and communicate among team members and organizational entities of geographically dispersed projects. Ability to share and communicate information in a synchronous and asynchronous mode (application: communication across time zones, holidays, and shared work spaces). Creating unified visibility of project status and performance. Virtual teams, to some degree, bridge and neutralize the culture and value differences that exist among different task teams of a project organizations.

Challenges. The effectiveness of the virtual team depends on the team members' ability to work with the given technology. Information flow and access are not necessarily equal for all team members. Information may not be processed uniformly throughout the team. The virtual team concept does not fit the culture and value system of all members and organizations. Project tracking, performance assessment, and managerial control of project activities are often very difficult. Risks, contingencies, and problems are difficult to detect and assess. Virtual organizations often do not provide effective methods for dealing with conflict, power, candor, feedback, and resource issues. Because of the many limitations, more traditional team processes and communications are often needed to augment virtual teams.

from traditional hierarchical team structures and tried more self-directed and network-based virtual concepts (Fisher, 1993).

18.4 MEASURING PROJECT TEAM PERFORMANCE

"A castle is only as strong as the people who defend it." This Japanese proverb also applies to organizations. They are only as effective as their unified team efforts. However, team performance is difficult to measure because of it involves highly complex, interrelated sets of variables, including attitudes, personal preferences, and perceptions that are difficult to quantify. Yet, in spite of the existing cultural and philosophical differences among organizations, research shows[4] that a general agreement exists among managers on certain performance measures for project teams and their results.

Project Performance Measures

Starting with the "bottom line" of *team results* and *project success,* a considerable agreement exists among managers on the following metrics of nine measures:

1. Project success according to agreed-on results
2. On-time performance
3. On-budget performance

An estimated 90 percent of project managers include these factors among the three most important measures of project success. The majority of managers rank these factors in the shown order. In addition, other factors are often mentioned as important to project success. They include:

4. Overall customer or sponsor satisfaction
5. Responsiveness and flexibility to customer requirements and changes
6. Dealing effectively with risk and uncertainty
7. Positioning the project for future business
8. Stretching beyond planned goals
9. Organizational learning to benefit future projects

Team Effectiveness Measures

Team characteristics drive project performance. However, this relationship is not "linear." Moreover, project performance is influenced by many "external" factors, such as technology, socioeconomic factors, and market behavior, making it difficult

[4]In fact over 90 percent of the project managers interviewed during a survey by Thamhain and Wilemon (1997) mentioned three factors, (1) project success according to agreed-on results, (2) on-time performance, and (3) on-budget performance, among the most important criteria of team performance.

to determine exactly how much the team characteristics influence project performance. Yet, lessons from field research[5] strongly suggest specific factors, such as those graphically shown in Fig. 18.1, characterizing high-performing teams. More specifically, Table 18.4. breaks the characteristics of a high-performing team into four categories: (1) work and team structure, (2) communications and control,

TABLE 18.4 Benchmarking Your Team Performance

Work and Team Structure

- Team participates in project definition, and work plans evolve dynamically
- Team structure and responsibilities evolve and change as needed
- Broad information sharing
- Team leadership evolves on the basis of expertise, trust, and respect
- Minimal dependence on bureaucracy, procedures, and politics

Communication and Control

- Effective cross-functional channels and linkages
- Ability to seek out and process information
- Effective group decision making and consensus
- Clear sense of purpose and direction
- Self-control, accountability, and ownership
- Control is stimulated by visibility, recognition, accomplishments, autonomy

Team Leadership

- Minimal hierarchy in member status and position
- Internal team leadership based on situational expertise, trust, and need
- Clear management goals, direction, and support
- Inspires and encourages

Attitudes and Values

- Members are committed to established objectives and plans
- Shared goals, values, and project ownership
- High involvement, energy, work interest, need for achievement, and pride self-motivated
- Capacity for conflict resolution and resource sharing
- Team building and self-development
- Risk sharing, mutual trust, and support
- Innovative behavior
- Flexibility and willingness to change
- High morale and team spirit
- High commitment to established project goals
- Continuous improvement of work process, efficiency, and quality
- Ability to stretch beyond agreed-on objectives

[5]For more detailed discussions of the field research see H. J. Thamhain and D. L. Wilemon, "A High-Performing Engineering Project Team," in R. Katz (ed.), *The Human Side of Managing Innovation* (New York: Oxford Press, 1997).

(3) team leadership, and (4) attitude and values. These broad measures can provide a framework for benchmarking.

Teams that score high on these characteristics are also seen by upper management as most favorable in dealing with cost, quality, creativity, schedules, and customer satisfaction. They also receive favorable ratings on the more subtle measures of team performance, such as flexibility, change orientation, innovative performance, high morale, and team spirit.

The significance of determining the association between team characteristics and project performance lies in two areas. First, it offers some insight as to what an effective team environment looks like, providing the basic framework for team assessment, benchmarking, and development. Second, a better understanding of how team characteristics affect project performance provides building blocks for further research on organization development, such as defining drivers and barriers to team performance. It also provides a framework for leadership style development.

18.5 A MODEL FOR TEAM BUILDING

Figure 18.2 provides a simple model for organizing and analyzing the variables, which influence the team's characteristics and its ultimate performance, as baselined in Fig. 18.1 and Table 18.1. The influences shown in Fig. 18.2 are divided into four sets: (1) drivers and barriers to high team performance; (2) managerial leadership style, including components of authority, motivation, autonomy, trust, respect, credibility, and friendship; (3) organizational environment, such as working conditions, job content, resources, and organizational support factors; and (4) social, political, and economic factors of the firm's external business environment. All four sets of variables are intricately interrelated. However, using the systems approach allows researchers and management practitioners to break down the complexity of the process, thus helping in analyzing team performance and in developing leadership effectiveness.

Drivers of and Barriers to High Team Performance

Management tools such as benchmarking and root-cause analysis can be helpful in identifying the drivers and barriers toward effective teamwork. *Drivers* are factors that influence the project environment favorably, such as interesting work and good project leadership. These factors are perceived as enhancing team effectiveness, and therefore correlate positively with team performance. *Barriers* are factors that have an unfavorable influence, such as unclear objectives and insufficient resources, therefore impeding team performance. Based on field research,[6] the 10 strongest drivers and 15 strongest barriers are listed in Table 18.5.

---

[6]Studies by Thamhain and Wilemon (1999) into work group dynamics clearly show significant correlation and interdependencies among work environment factors and team performance. These studies indicate that high team performance involves four primary factors: managerial leadership, job content, personal goals and objectives, and work environment and organizational support. Kendall-tau rank-order correlation was used to measure the actual correlation of 60 influence factors to (1) project team characteristics and (2) team performance. Statistical significance was defined at a confidence level of 95 percent or better.

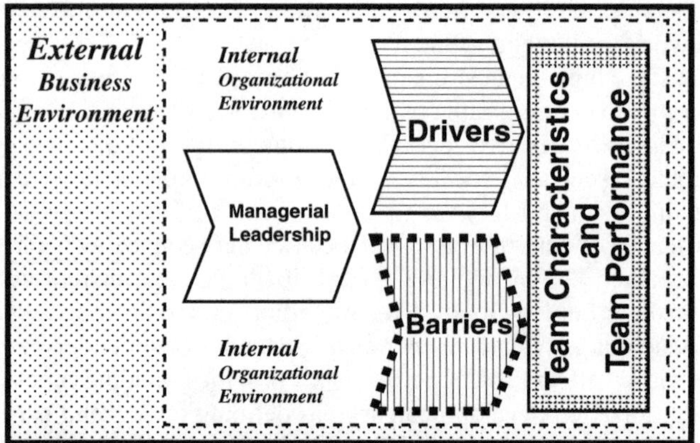

FIGURE 18.2 Model for analyzing team performance.

TABLE 18.5 Strongest Drivers of and Barrier to Project Team Performance

Drivers	Barriers
1. Clear project plans and objectives	1. Communication problems
2. Good interpersonal relations and shared values	2. Conflict among team members or between team and support organizations
3. Good project leadership and credibility	3. Different outlooks, objectives, and priorities perceived by team members
4. Professional growth potential	4. Poor qualification of team/project leader
5. Professionally interesting and stimulating work	5. Poor trust, respect, and credibility of team leader
6. Project visibility and high priority	6. Insufficient resources
7. Proper technical direction and team leadership	7. Insufficient rewards
	8. Lack of project challenge and interest
8. Qualified, competent team personnel	9. Lack of senior management support, interest, and involvement
9. Recognition of sense of accomplishment	10. Lack of team definition, role conflict, and confusion
10. Management involvement and support	11. Lack of team member commitment
	12. Poor project team/personnel selection
	13. Shifting goals and priorities
	14. Unclear team leadership and power struggle
	15. Unstable project environment, poor job security, and anxieties

All of these factors have been listed alphabetically to avoid conclusions drawn too narrowly. Although the actual statistics yielded different performance correlation levels for different drivers and barriers, they represent the strongest association observed in the field studies. All correlations are statistically significant at $p = .05$ or better. Collectively, they explain over 85 percent of the variance in project team performance.

It is further interesting to note that many of the factors in Table 18.5 are, to a large degree, based on the perception of team members. That is, team members *perceive* "good personal relations" or "communication problems." Because this perception is the reality that influences the team behavior, management must deal with the conditions as seen by the people and foster a project environment conducive to the needs of the team. Such a favorable work environment not only enhances the drivers and minimizes the barriers to project performance, but is also associated with the *15 measures that characterize a high-performing project team,* as discussed earlier and shown in Fig. 18.1.

These field research–based observations provide some focus and support to the broad range of field observations and criteria for effective team management discussed throughout this chapter. An important lesson follows from the analysis of these observations. Managers must foster a work environment supportive to their team members. Creating such a climate and culture conducive to quality teamwork involves multifaceted management challenges, which increase with the complexities of the project and its organizational environment. No longer will technical expertise or good leadership alone be sufficient, but excellence across a broad range of skills and sophisticated organizational support is required to manage these project teams effectively. Thus, it is critically important for project leaders to understand, identify, and minimize the various barriers to team development.

18.6 BUILDING HIGH-PERFORMING TEAMS

What does all this mean to managers in today's work environment with high demands on efficiency, speed, and quality? Project teams are becoming increasingly more important. However, exploiting the team potential can rarely be done "top-down." Given the realities of today's business environment, its technical complexities and cross-functional dependencies, and the need for innovative performance, more and more project leaders have to rely on information and judgments by their team members for developing solutions to complex problems. Especially with decision processes distributed throughout the team and solutions often evolving incrementally and iteratively, power and responsibility are shifting from managers to the project team members who take higher levels of responsibility, authority, and control for project results. That is, these teams become *self-directed,* gradually replacing the more traditional, hierarchically structured project team. These emerging team processes are seen as a significant development for orchestrating the multifunctional activities that come into play during the execution of today's complex projects. These processes rely strongly on group interaction, resource and power sharing, group decision making, accountability, self-direction, and control. Leading such self-directed teams also requires a great deal of team management skill and overall guidance by senior management. In addition, managers must realize the organizational dynamics involved during the various phases of the team development process. A four-stage model, originally developed by Hersey and Blanchard, and graphically shown in Fig. 18.3, is often being used by management researchers and practitioners as a framework to analyze the team development process. The four stages are labeled: (1) *team formation,* (2) *team start-up,* (3) *partial integration,* and (4) *full integration.* These stages are also known as *forming, storming, norming,* and *performing,* giving an indication of team behavior at each of the stages.

No work group comes fully integrated and unified in its values and skill sets, but needs to be skillfully nurtured and developed. Leaders must recognize the professional interests, anxieties, communication needs, and challenges of their team members and anticipate them as the team goes through the various stages of integration. Moreover, team leaders must adapt their managerial style to the specific situation of each stage.[7] That is, team leaders must recognize what works best at each stage, and what is most conducive to the team development process. Many of the problems that occur during the formation of the new project team or during its life cycle are normal and often predictable. However, they present barriers to effective team performance. The problems must be quickly identified and dealt with. Early stages, such as the *team formation* and *start-up,* usually require a predominately directive style of team leadership. Providing clear guidelines on the project mission, its objectives, and its requirements and creating the necessary infrastructure and logistics support for the project team are critically important in helping the team pass through the first two stages of development quickly. During the third stage, *partial integration,* or *norming,* the team still needs a considerable amount of guidance and administrative support, as well as support in dealing with the inevitable human issues of conflict, power and politics, credibility, trust, respect, and the whole spectrum of professional career and development. This is

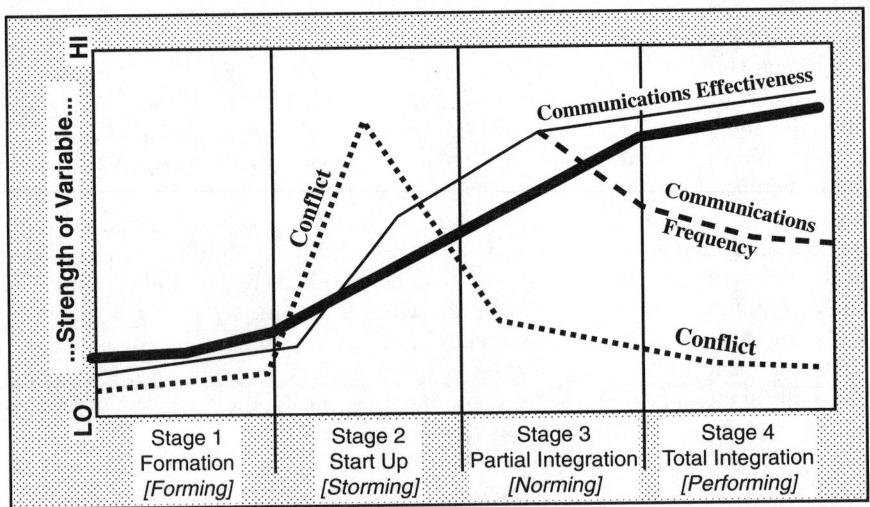

FIGURE 18.3 Four-stage team development model.

[7]The discussions on leadership style effectiveness is based on field research by H. J. Thamhain and D. L. Wilemon, published in "A High-Performing Engineering Project Team," in R. Katz (ed.), *The Human Side of Managing Innovation* (New York: Oxford Press, 1997).

the stage where a combination of *directive and participative leadership* will produce the most favorable results. Finally, a team that reaches the fully integrated stage, by definition, becomes "self-directed." That is, such a fully integrated, unified team can work effectively with a minimum degree of external supervision and administrative support, as described earlier in Table 18.2. While at this stage, the team often appears to have very little need for external managerial intervention, it requires highly sophisticated external leadership to maintain this delicate state of team effectiveness and focus.

In summarizing the criteria of effective team management, *three propositions* can be stated to highlight the organizational environment and managerial style conducive to high project team performance:

- *Drivers and barriers.* The degree of project success seems to be primarily determined by the strength of specific driving forces and barriers that are related to *leadership, job content, personal needs, and the general work environment* as shown in Table 18.4.
- *Team environment.* The strongest single driver of team performance and project success is a *professionally stimulating team environment,* characterized by interesting and challenging work, visibility and recognition of achievements, growth potential, and good project leadership.
- *Work challenge dividend.* A professionally stimulating team environment is also favorably associated with *low conflict, high commitment, highly involved personnel, good communications, willingness to change, innovation, and on-time and on-budget performance.*

To be effective in organizing and directing a project team, the leader must not only recognize the potential drivers and barriers but also know when in the life cycle of the project they are most likely to occur. The effective project leader takes preventive actions early in the project life cycle and fosters a work environment that is conducive to team building as an ongoing process.

The effective team leader is a social architect who understands the interaction of organizational and behavioral variables and can foster a climate of active participation and minimal dysfunctional conflict. This requires carefully developed skills in leadership, administration, organization, and technical expertise. It further requires the project leader's ability to involve top management to ensure organizational visibility, resource availability, and overall support for the new project throughout its life cycle.

18.7 RECOMMENDATIONS FOR EFFECTIVE TEAM MANAGEMENT

A number of specific recommendations may help managers in cultivating productive working conditions for multidisciplinary task integration and in building high-performing project teams. The sequence of recommendations follows to some degree the chronology of the team development process shown in Fig. 18.3.

Negotiate the Work Assignment. At the outset of any project, team leaders should discuss with their team members the overall task and its scope and objectives. Involvement of the people during the early phases of the assignment, such as bid proposals and project and product planning, can produce great benefits toward plan acceptance, realism, buy-in, personnel matching, and unification of the task team. A thorough understanding of the task requirements comes usually with intense personal involvement which can be stimulated through participation in project planning, requirements analysis, interface definition, or a producibility study. In addition, any committee-type activity, presentation, or data gathering will help involve especially new team members and facilitate integration. It also will enable people to better understand their specific tasks and roles in the overall team effort. Senior management can help develop a "priority image" and communicate the basic project parameter and management guidelines.

Communicate Organizational Goals and Objectives. Management must communicate and update the organizational goals and project objectives. The relationship and contribution of individual work to overall business plans and their goals, as well as of individual project objectives and their importance to the organizational mission, must be clear to all team personnel.

Plan the Project Effectively. An effective project definition and involvement of potential team members early in the life cycle of a project or specific mission will have a favorable impact on the work environment, enthusiasm of the team toward the assignment, commitment toward the project objectives, team morale, and, ultimately, team effectiveness. Because project leaders have to integrate various tasks across many functional lines, proper planning requires the participation of all stakeholders, including support departments, subcontractors, and management. Modern project management techniques, such as phased project planning and stage-gate concepts, plus established standards, such as the Project Management Body of Knowledge (PMBOK®),[8] provide the conceptional framework and tools for effective cross-functional planning and organizing the work toward effective execution.

Staff and Organize the Project Team. Project staffing is a major activity, usually conducted during the project formation phase. Because of the pressures on the project manager to produce, staffing is often done hastily and prior to properly defining the basic work to be performed. The results are often suboptimal matches of personnel with the job requirements, resulting in conflict, low morale, suboptimum decision making, and ultimately poor project performance. For best results, project leaders should define the project organization, principle tasks, and work processes before finalizing the formal task assignments to candidates. Ideally, candidates are involved in the front-end feasibility analyses and project planning activities. Interviews for specific task and leadership assignments should always be conducted one-to-one.

[8]The Project Management Body of Knowledge, PMBOK®, was developed by the Project Management Institute to provide a standardized framework for project planning and execution. PMBOK has been adopted by ANSI and other institutions worldwide to provide a standard framework for project management practice and research.

Define the Project Organization, Interfaces, and Reporting Relations. The keys to successfully building a new project organization are clearly defined and communicated responsibilities and organizational relationships. The tools for systematically describing the project organization come, in fact, from conventional management practices, including (1) *charter* of the program or project organization; (2) *project organization chart,* defining the major reporting and authority relationships; (3) *responsibility matrix* or *task roster;* and (4) *job description.*

Build a High-Performance Image. Building a favorable image for an ongoing project, in terms of high priority, interesting work, importance to the organization, high visibility, and potential for professional rewards, is crucial for attracting and holding high-quality people. Senior management can help develop a "priority image" and communicate the key parameters and management guidelines for specific projects. Moreover, establishing and communicating clear and stable top-down objectives helps in building an image of high visibility, importance, priority, and interesting work. Such a pervasive process fosters a climate of active participation at all levels, helps attract and hold quality people, unifies the team, and minimizes dysfunctional conflict.

Define Work Process and Team Structure. Successful formation and development of a project team requires an infrastructure conducive to teamwork. The proper setup and communication of the operational transfer processes, such as concurrent engineering, stage-gate process, CAD/CAE/CAM, and design-build, is important for establishing the cross-functional linkages necessary for successful project execution. Management must also define the basic team structure for each project early in its life cycle. The project plan, task matrix, project charter, and operating procedure are the principal management tools for defining organizational structure and business process.

Build Enthusiasm and Excitement. Whenever possible, managers should try to accommodate the professional interests and desires of their personnel. Interesting and challenging work is a perception, which can be enhanced by the visibility of the work, management attention and support, priority image, and the overlap of personnel values and perceived benefits with organizational objectives. Making work more interesting leads to increased involvement, better communication, lower conflict, higher commitment, stronger work effort, and higher levels of creativity.

Ensure Senior Management Support. It is critically important that senior management provides the proper environment for a project team to function effectively. At the onset of a new project, the responsible manager needs to negotiate the needed resources with the sponsor organization, and obtain commitment from management that these resources will be available. An effective working relationship among resource managers, project leaders, and senior management critically affects the credibility, visibility, and priority of the engineering team and their work.

Define Effective Communication Channels and Methods. Poor communication is a major barrier to teamwork and effective project performance. Management can facilitate the free flow of information, both horizontally and vertically, by work space design, regular meetings, reviews, and information sessions. In addition, modern technology, such as voice mail, e-mail, electronic bulletin boards, and videoconferencing, can greatly enhance communications, especially in complex organizational settings.

Build Commitment. Managers should ensure team member commitment to their project plans, specific objectives, and results. If such commitments appear weak, managers should determine the reason for such lack of commitment of a team member and attempt to modify possible negative views. Anxieties and fear of the unknown are often a major reason for low commitment. Managers should investigate the potential for insecurities, determine the cause, and then work with the team members to reduce these negative perceptions. Conflict with other team members and lack of interest in the project may be other reasons for such lack of commitment.

Conduct Team-Building Sessions. A mixture of focus team sessions, brainstorming, experience exchanges, and social gatherings can be powerful tools for developing the work group into an effective, fully integrated, and unified project team. Such organized team-building efforts should be conducted throughout the project life cycle. Intensive team-building efforts may be especially needed during the team formation stage. Although formally organized and managed, these team-building sessions are often conducted in a very informal and relaxed atmosphere to discuss critical questions such as (1) How are we working as a team? (2) What is our strength? (3) How can we improve? (4) What support do you need? (5) What challenges and problems are we likely to face? (6) What actions should we take? and (7) What process or procedural changes would be beneficial?

Ensure Project Leadership. The project management and team leadership positions should be carefully defined and staffed at all project levels. To build and lead a project team, especially in a dynamic or self-directed work environment, requires credibility, trust, and respect of the project leader, a quality that usually comes from the image of a sound decision maker with a good track record.

Create Proper Reward Systems. Personnel evaluation and reward systems should be designed to reflect the desired power equilibrium and authority/responsibility sharing of an organization. A quality function deployment philosophy, where everyone recognizes the immediate "customer" for whom a task is performed, helps focus efforts toward desired results and customer satisfaction, both for company internal and external customers. It also helps foster a work environment that is strong on self-direction and self-control.

Manage Conflict and Problems. Project managers should focus their efforts on problem avoidance. That is, managers and team leaders, through experience, should recognize potential problems and conflicts at their onset, and deal with them before they become big and their resolutions consume a large amount of time and effort.

Ensure Personal Drive and Involvement. Project managers and team leaders can influence the team environment by their own actions. Concern for their team members, the ability to integrate personal needs of their staff with the goals of the organization, and the ability to create personal enthusiasm for a particular project, all can foster a climate of high motivation, work involvement, open communication, and ultimately high engineering performance.

18.8 TO SUMMARIZE

The major points that have been covered in this chapter include:

- The increasing complexities of today's project environment, both internally and externally, prompt enormous managerial challenges for directing, coordinating, and controlling teamwork. Especially with the expansion of self-directed team concepts, additional managerial tools and skills are required to handle the burgeoning dynamics and infrastructure.
- Effective teamwork is a critical determinant of project success, along with the organization's ability to learn from its experiences and position itself for future growth.
- To be effective in organizing and directing a project team, the leader must not only recognize the potential drivers and barriers to high-performance teamwork, but also know when in the life cycle of the project they are most likely to occur.
- Four major conditions must be present for building effective project teams: (1) professionally stimulating work environment, (2) good project leadership, (3) qualified personnel, and (4) stable work environment.
- Effective project leaders take preventive actions early in the project life cycle and foster a work environment that is conducive to team building as an ongoing process.
- The new business realities force managers to focus increasingly on cross-boundary relations, delegation, and commitment, in addition to establishing the more traditional formal command and control systems
- Project leaders must involve top management, to ensure organizational visibility, resource availability, and overall support for the new project throughout its life cycle.
- Building effective project teams involves the whole spectrum of management skills and company resources, and is the shared responsibility between functional managers and the project leader.
- The effective team leader is a social architect who understands the interaction of organizational and behavioral variables and can foster a climate of active participation, minimal dysfunctional conflict, and effective communication, and is open to change, commitment, and self-direction.

- By understanding the criteria and organizational dynamics that drive people toward effective team performance, managers can examine and fine-tune their leadership style, actions, and resource allocations toward continuous organizational improvement.

18.9 ADDITIONAL SOURCES OF INFORMATION

The following additional sources of project management information may be used to complement this chapter's topic material. This material complements and expands on various concepts, practices, and the theory of project management as it relates to areas covered here.

- Jeffrey K. Pinto, "The Elements of Project Success"; Charles J. Teplitz, "Making Optimal Use of the Matrix Organization"; Jimmie L. West, "Building a High-Performing Project Team"; Robert J. Yourzak, "Motivation in the Project Environment"; David I. Cleland, "New Ways to Use Project Teams"; Dale E. Knutson, "Benchmarking Tools for Operation Teams"; Preston G. Smith, "Concurrent Engineering Teams"; and Karen M. Bursic, "Self-Managed Production Teams"; chaps. 2, 14, 18, 19, 29, 31, 32, and 33 in David I. Cleland (ed.), *Field Guide to Project Management* (New York: Van Nostrand Reinhold, 1997).
- David I. Cleland, "The Benfield Column Repair Project," in David I. Cleland, Karen M. Bursic, Richard J. Puerzer, and Alberto Y. Vlasak (eds.), *Project Management Casebook,* Project Management Institute. (Also in *PM Network,* February 1996, pp. 25–30.)
- Jon R. Katzenbach and Douglas K. Smith, *The Wisdom of Teams: Creating the High Performance Organization* (New York: Wiley, 1994). This book contains a wealth of information on the art of building teams for high-performance results. The authors provide a balance of real and artificial scenarios that offer useful ideas for handling work responsibilities, executive egos, communications, and people skills. This book also captures the benefits of teams at all levels in an organization.
- Cynder Niemaela and Rachael Lewis, *Leading High Impact Teams: The Coach Approach to Peak Performance* (Guerneville: High Impact Publications, 2001). This book offers a range of guidelines from the conceptual to the practice of team building and team leading. The book describes "Top 10 High-Impact Team Practices" for the reader as well as covering team development by phase and many practical examples of "how to." The book is recommended for executive managers as well as team members.
- Jim Partlow and Don Wynes, "Teamwork Puts Troubled Project Back on Track: A Case Study in Relationship Building," *Information Strategy,* Winter 2002, pp. 12–21. This article tells a story of a high-energy project team that started a

project well but got diverted through changes to an organization. The results of a new project manager's style raised concerns of trust, fear, suspicion, and doubt. What happened to the team and the project is interesting.

- Mike Voight and John Callaghan, "A Teambuilding Intervention Program: Application and Evaluation with Two University Soccer Teams," *Journal of Sport Behavior,* December 2001, pp. 420–425. This article describes team building in a sports environment and the seven stages in the framework to develop a team. The authors aptly describe a team-building process that can be used in a project environment. Important are the measures of results to evaluate individual and team performance with the supposition that improved individual performance can enhance team performance.
- Anonymous, "Don't Just Do It," *Software Development,* December 2001, p. 17. This article discusses the "care and feeding" of a software development team so the product can get out the door. Senior managers worry about product delivery, but the project manager must ensure that the team is cared for to permit them to do their work. There are tips that differ from the normal advice one receives and which are designed to give the team as much time as possible to do their work.

18.10 DISCUSSION QUESTIONS

1. Using the lead-in case scenario for this chapter, identify Chrysler's organizational challenges and issues that require effective project teamwork for resolution.
2. Identify and profile the type of leadership style that is needed for effectively managing a new product team at Chrysler.
3. Write a charter for the manager in charge of new car development at Chrysler. Give a definition of a charter. Why is *team building* such an important and challenging toll for today's managers?
4. What are some of the characteristics of a fully integrated team? Develop a list of performance measures or criteria for a specific team you know and evaluate this team against your benchmark list.
5. Identify specific project situations in which team building is critical. Why?
6. What are some of the organizational variables that influence project team characteristics? How can you as project leader influence these variables?
7. List and discuss some of the task- and people-related qualities of successful project teams. How can you measure these qualities? How can you influence these qualities?
8. Discuss the importance of a project manager's understanding of the drivers of and barriers to team performance.
9. Do you accept the three propositions related to project success given in this chapter? Explain.

10. What are the characteristics of a self-directed, self-managed team? As a project manager, how can you promote such team behavior?
11. How do you develop team leadership skills?
12. How do you identify the "best" team members for a newly formed project?

18.11 USER CHECKLIST

1. How do you unify a newly formed team and move it toward the "performing" stage fast?
2. How do you develop a well-performing project team further?
3. Identify a work group that you know. Define the metrics for measuring team performance.
4. What means and methods, other than money, would you consider to motivate your team toward high project performance?
5. Is it more challenging to lead project teams today than 25 years ago? Why, or why not?
6. How does the team leadership style change over a typical project life cycle?
7. Why is team commitment and ownership important to team performance?
8. As a team leader, how can you build team commitment to the project objectives?
9. As a project team leader, how can you integrate senior management into your team?
10. Two team leaders have identical projects in the same organizational environment. One team perceives the project as highly interesting and important with a lot of opportunities for career advancement, whereas the other team perceives the project as boring, routine work with little opportunity for career advancement. What do these two project leaders do differently?
11. How can the proper application of conventional project management tools and techniques help in building high-performing teams?
12. How can senior management help in building a high-performing project team?

18.12 PRINCIPLES OF PROJECT MANAGEMENT

1. A *project team* is a work group organized for the purpose of executing a specific set of project activities. Team members are often brought together from different functional areas, contractors, and agencies. Frequently, they also include sponsors, customers, and management personnel.
2. For a work group to transform into a team, its members must be unified by a common set of values, norms, and objectives; these group members must also have mutual dependencies, trust, and respect.

3. *Team building* is the process of taking a collection of individuals with different needs, backgrounds, and expertise and transforming them into an integrated, effective work unit.
4. *Project team performance* is defined by an intricate set of interrelated variables. Team performance is often measured against an agreed-on set of project performance parameters that, in part, are driven by the personality traits of the project team and the project work environment.
5. *Team leadership* is a core competency vital to effective project management; it comes in various forms and levels of sophistication. In its very basic form, team leadership is the ability to transform a project objective into tangible results via team efforts.
6. Team leadership is difficult to define in a functional statement, but vision and inspiration are at its central focus.
7. Often differentiated from *management,* team leadership is associated with the ability to understand the complex interaction of organizational and behavioral variables, and to foster a team environment of active participation, mutual trust, respect, commitment, and innovative team behavior. Team leadership is needed at every task level of every project.

18.13 PROJECT MANAGEMENT SITUATION—USING TEAMS TO MEET CHALLENGES

Acme Manufacturing Company's products have been diminishing in sales for the past year and the future does not look any better. Products have sold well in the past without a lot of competition from either domestic manufacturers or foreign manufacturers. The Marketing Department advises that the products are not meeting current customer needs, customers are buying the competition's products, and Acme is losing market share. This loss of sales is attributed to better products being developed by competitors and the competitors' products being viewed as new, innovative, and having a futuristic style.

The R&D Department has not designed a new product in 3 years, and only minor changes have been made to Acme's existing product line. The thinking in the R&D Department is that the customers will return to Acme because the product line is unequaled by others and the "quality" is high. The competitors' products do not compare in reliability, durability, and usability.

Acme's senior management decides to build a team of people at the director's level to review the products and the competition to determine how Acme can improve its position in the marketplace. The team is identified, and many of the directors claim they have important work to do. Using this "reason," many of the directors fail to make the first meeting. This kickoff meeting is viewed as a failure without support from the directors, and senior management is frustrated.

One of the long-time directors, who is within a month of retiring from Acme, explains his view of what happened to the kickoff meeting. Most of the directors view the charter of the team to be someone else's responsibility and feel the team is doomed to failure. Any director participating on the team will be considered part of the team's failure. Finding an excuse to not attend, and possibly be relieved from the team, is seen as the best solution. Further, Acme's directors have always had autonomy in what they did and this team effort appears to erode that autonomy of action.

Senior management is puzzled. The directors are key players in the organization and have the best chance to find the right solution to the products' declining sales. They have performed well in the past when given tasks. What is the difference between what they have done in the past and working within a team? There has been some competition among the directors previously, but no director has refused to cooperate or work with another director. Senior managers were just recently informed by a consultant that "teams" provide the best chance for success.

18.14 STUDENT/READER ASSIGNMENT

The project management situation describes a situation where senior management is making a first attempt to solve an organizational problem through a team concept using the talents of several directors. This use of teams seems to be a total failure and does not provide a solution to Acme's declining sales.

The information provided in this chapter describes what should be done to form teams and the criteria for team success. There are the barriers to and drivers for teams as well as discussion on team performance. Team leadership is also considered important to successful team performance.

Using what you have learned from this chapter as supplemented by readings, analyze Acme's situation and develop the important points about what happened and what could be done to get the directors interested and involved. Motivating factors should focus on positive means of changing attitudes and values.

REFERENCES

Armstrong, David, "Building Teams across Boarders," *Executive Excellence,* vol.17, no.3, March 2000, p. 10.

Barner, R., "The New Millennium Workplace," *Engineering Management Review* (IEEE), vol. 25, no. 3, Fall 1997, pp. 114–119.

Bhatnager, Anil, "Great Teams," *The Academy of Management Executive,* vol. 13, no. 3, August 1999, pp. 50–63.

Dillon, Patrick, "A Global Challenge," *Forbes Magazine,* vol. 168, September 10, 2001, pp. 73+.

Drucker, Peter F., *The Executive in Action: Managing for Results, Innovation and Entrepreneurship* (New York: Harper, 1996).

Dumaine, Brian, "The Bureaucracy Buster," *Fortune,* June 17, 1991.

Dyer, W. G., *Team Building: Issues and Alternatives* (Reading, Mass.: Addison-Wesley, 1977).

Fisher, Kimball, *Leading Self-Directed Work Teams* (New York: McGraw-Hill, 1993).

Gray, Clifford F., and Erik W. Larson, *Project Management* (New York: Irwin/McGraw-Hill, 2000).

Gupta, A. K., and D. L. Wilemon, "Changing Patterns in Industrial R&D Management," *Journal of Product Innovation Management,* vol. 13, no. 6, November 1996, pp. 497–511.

Jassawalla, Avan R., and Hemant C. Sashittal, "Building Collaborate Cross-Functional New Product Teams," *The Academy of Management Executive,* vol. 13, no. 3, August 1999, pp. 50–63.

Katz, Nancy, "Sports Teams as Model for Workplace Teams," *The Academy of Management Executive,* vol. 15, no. 3, August 2001, pp. 70–72.

Kostner, Jaclyn, *Bionic eTeamwork: How to Build Collaborative Virtual Teams at Hyperspeed* (Dearborn, Mich.: Kaplan, 2001).

Likert, R., *New Patterns of Management* (New York: McGraw-Hill, 1961).

Maccoby, Michael, "Building Cross-Functional Capability," *Research Technology Management,* vol. 42, no. 3, May/June 1999, pp. 56–58.

Marshall, Edward, *Transforming the Way We Work* (New York: AMACOM, 1995).

McGregor, D., *The Human Side of Enterprise* (New York: McGraw-Hill, 1960).

Moss Kanter, Rosabeth, "The New Managerial Work," *Harvard Business Review,* November–December 1989.

Nurick, A. J., and H. J. Thamhain, "Project Team Development in Multinational Environments," in D. Cleland (ed.), *Global Project Management Handbook* (New York: McGraw-Hill, 1993).

Oderwald, Sylvania, "Global Work Teams," *Training and Development,* vol. 5, no. 2, February 1996.

Parker, Glenn, Jerry McAdams, and David Zielinski, *Rewarding Teams: Lessons from the Trenches* (San Francisco: Jossey-Bass, 2000).

Peters, Thomas J., and Robert H. Waterman, *In Search of Excellence* (New York: Harper & Row, 1987, 1997).

Rees, Fran, *How to Lead Work Teams* (San Francisco: Jossey-Bass, 2001).

Robbins, Harvey, and Michael Finley, *Why Teams Don't Work* (San Francisco: Berrett-Koehler, 2000).

Roethlingsberger F., and W. Dickerson, *Management and the Worker* (Cambridge, Mass.: Harvard University Press, 1939).

Senge, Peter, *The Fifth Discipline: The Art and Practice of the Learning Organization* (New York: Doubleday/Currency, 1990, and audiocassette, 1994).

Shonk, J. H., *Team-Based Organizations* (Homewood, Ill.: Irwin, 1996).

Thamhain, H. J., "Team Management," chap. 19 in J. Knutson (ed.), *Project Management Handbook* (New York: Wiley, 2001).

Thamhain, H. J., "Managing People," chap. 68 in M. Kutz (ed.), *Mechanical Engineer's Handbook* (New York: Wiley, 1998).

Thamhain, H. J., "Managing Technology: The People Factor," *Technical and Skill Training,* August/September 1990.

Thamhain, H. J., and D. L. Wilemon, "Building Effective Teams in Complex Project Environments," *Technology Management,* vol. 5, no. 2, May 1999.

Thamhain, H. J., and D. L. Wilemon. "Building High-Performing Engineering Project Teams," in R. Katz (ed.), *The Human Side of Managing Technological Innovation* (New York: Oxford Press, 1997).

Tichy, Noel, and David Ulrich, "The Leadership Challenge—Call for the Transformational Leader," *Sloan Management Review,* Fall 1984, pp. 59–69.

Verma, Vijay K., *Managing the Project Team* (Newton Square, Pa.: Project Management Institute, 1997).

Walton, Richard, "From Control to Commitment in the Workplace, *Harvard Business Review,* March–April 1985.

Zenger, John H., Ed Musselwhite, Kathleen Hurson, and Craig Perrin, *Leading Teams* (Homewood, Ill.: Business One Irwin, 1994).

P • A • R • T • 6

THE CULTURAL ELEMENTS

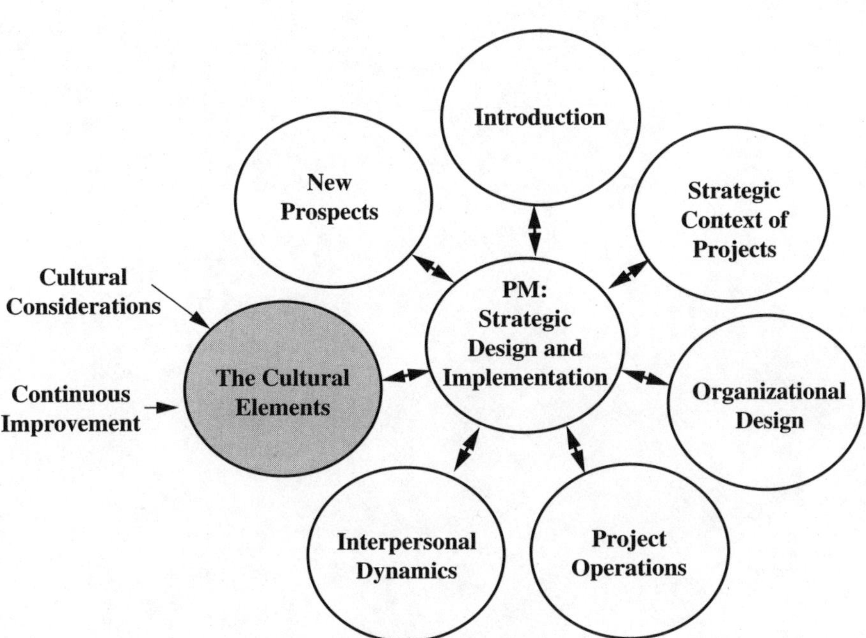

CHAPTER 19

CONTINUOUS IMPROVEMENT THROUGH PROJECTS

"...everything is in constant motion and every change seems an improvement."
ALEXIS DE TOCQUEVILLE, 1805–1859

19.1 INTRODUCTION

In order to compete in today's marketplace—much of which is overseas—a company must have a means of continuously improving its products, services, and organizational processes. In order to bring about such improvements, an interdisciplinary and interorganizational approach is required. The use of project teams has helped in the conceptualization and realization of enterprise improvements. As the enterprise develops its future through the use of the "choice elements," described in Chap. 1, and ongoing changes happen in its social, competitive, technological, political, and economic "systems," a philosophy of how to deal with change—sometimes minor change—is needed. Project management can help in this regard.

In this chapter, some examples of change brought about by project teams will be indicated. Survival through small changes will be described through management innovation, continuous improvements in productivity, and product quality through projects. Trendsetters that set a pace for change will be described, as well as the role of product integrity. Continuous improvement in manufacturing through the use of productivity gains, manufacturing philosophies, computer-integrated manufacturing, and just-in-time manufacturing will be provided. The basic message of the chapter is that projects, although small in nature (as described in Chap. 3), can be used as an organizational strategy for continuous improvements in organizations.

19.2 WHY CONTINUOUS IMPROVEMENT?

Continuous improvement of products, services, and organizational processes is becoming the hallmark for success in the global marketplace. Yet there still remains the opportunity for major technological breakthroughs—even of the transistor, jet engine, or computer magnitude. On balance, however, the predominant competitive advantage will be gained by a company's ability to provide a cultural ambience where creativity leads to innovation followed by continuing small improvements in products, services, and organizational processes to the customer.

The reader might properly ask, Why is the topic of continuous improvement dealt with in a book about project management? The answer lies in the strategic management responsibilities of the senior executives of the enterprise. Continuous improvement covers the design, development, and implementation of many innovations in the products, services, and processes making up the organization. Senior managers who are concerned with maintaining the competitiveness of their enterprise have little choice but to pursue a strategy of continuous improvement—a strategy that should encompass all elements of the organization. The realities of global competition offer no alternatives: Either compete by being able to offer customers improved products and services, or go out of business. Senior managers have the residual responsibility to prepare the enterprise for its future. This book has repeatedly made the point that projects are building blocks in the design and execution of organizational strategies. Projects, which have as their objective the continuous improvement of products, services, and processes, are important blocks in preparing the enterprise for its competitive future. Indeed senior managers have little choice: Either provide for a stream of projects directed to bringing about continuous improvement in the enterprise, or prepare to cease to exist. Competitors will not allow you to maintain a competitive edge for long in products and services. They will develop their own continuous improvement projects and move ahead of the complacent status quo–directed enterprise.

As modern enterprises cope with the need to continuously change their provision of value to customers, they must accept the watershed changes underway on how modern companies operate. As one CEO noted: "No longer do we operate in rigid hierarchical structures where directives and orders are dictated by supervisors to subordinates. The new style is collaborative, horizontal, and connected. Work gets done by networks of individuals who together design, create, build, and solve problems. Networks of thousands and networks of two. Networks as close as the next cubicle and as far as a distant country."[1]

The Wall Street Journal recognized the changes impacting contemporary organizations. This publication noted that project managers are becoming the vanguard of more and more work in America.[2]

19.3 SOME EXAMPLES

Some examples of continuous improvement through projects follow:

1. In the early 1990s, at Lynchburg, Virginia, the local community college operated a program to help businesses with challenges to their productivity and

[1]Timothy M. Donahue, President & CEO, Letter to Shareholders, "Let's Network," *NEXTEL 2000 Annual Report.*
[2]Bernard Wysocki, Jr., "Flying Solo," *The Wall Street Journal,* August 19, 1990, p. A-1.

effectiveness. The program was based on the principle that internal workers can identify and solve issues associated with lost productivity as well as outside consultants. What was needed for the businesses to improve their situation would be a short training course for a team of approximately five people in each business.

Several companies took advantage of the training and joined in the effort to improve their productivity through "self-analysis" of business practices. Each contributed one team for their project and for the training. The results were dramatic in terms of increased productivity and effectiveness. All teams made some progress and others may indeed have saved their respective business from serious financial problems.

All teams were trained in how to look at situations and how to look for improvements. The only specific requirement for the composition of the teams was for each team to have a financial analyst to provide the expertise in the cost savings that may be realized by the efforts of the team. All other members of the team were selected from operations, production, maintenance, and other functional areas. The majority of the members were high school graduates and a few had undergraduate degrees.

One company made parquet floor tiles for a single customer and was the only vendor of parquet floor tiles to the customer. The contract provided a reasonable return for sales of the floor tiles. The customer notified the company that their tiles were expensive and an alternate supplier was being sought for negotiations on the price. This placed the company in serious jeopardy with its one customer. The productivity improvement team was given the issue, How can we become more competitive and retain our customer? A five-person productivity improvement project team initiated the search for solutions and found the following problems:

- There was only a 43 percent yield on the parquet floor tile. This statement indicated there was a 57 percent loss of material during the manufacturing process, which was pure waste in the entire process from purchase of materials to preparation for delivery.
- The type of wood being used for the tiles was less expensive than another type, but the lower-cost materials showed a higher rate of failure. The difference in cost was about 12 percent, but the loss of materials, not including labor and other processing costs, resulted in losses of nearly 30 percent.
- The process was found to be randomly applied for the final bonding procedure, which may have caused as much as 25 percent loss of yield. Changes to the processes and materials resulted in a significant improvement in the yield. New materials and changes to the bonding process resulted in an actual yield of slightly more than 85 percent—nearly doubling the output of parquet floor tiles at only a slight increase in material cost. The company's ability to produce tiles at a highly competitive price allowed it to retain its current customer base.

2. The US Airways' 1999 Annual Report gave a brief description of the work done by project team members in the choosing of a cabin configuration. The

design and acceptance of this configuration required the team to work with individual departments, functions, customers, and crews.[3]

3. The Boeing company has initiated a "moonshine shop" out of which "moonshine teams" work to develop new technology and manufacturing processes to reduce manufacturing cost. These interdisciplinary teams, to include workers, are so named because they work outside traditional channels—to help drive out costly, traditional manufacturing techniques. These teams assisted in the development of strategies and processes to facilitate the use of innovative "lean manufacturing" processes. The time that it takes to assemble the Boeing 777 was dropped to 37 days from 71. The company has initiated assembly lines for the Boeing 737 from the traditional bays in which planes were parked among fixed catwalks and other machinery for days at a time. Once the wings and landing gear are attached, each plane is moved through to completion at 2 inches a minute for two shifts a day. Other innovations include the design and operation of a just-in-time inventory system for the aircraft components, to include engines. It is clear that these "nontraditional" project teams are making significant contributions in the strategy to find new ways to build airplanes.[4]

Given the inevitability of continuous improvement in competitor offerings, a chapter that ties together continuous improvement and project management is a needed addition to this work. As you peruse this chapter, consider this question: How does the use of project management facilitate a continuous improvement strategy in the enterprise?

19.4 SURVIVAL THROUGH CHANGE

Competitive survival requires that the company develop the ability to pursue a comprehensive and prolonged program of continuous improvement. Unceasing action must be under way to raise the quality and productivity of the organizational processes, which will lead to improved products and services. Undesirable situations and strategies in the enterprise that negatively affect a strategy of continuous improvement need to be reevaluated. Everything has to be done better. Change has to be managed for the continuous betterment of the enterprise through a strategy of adding something worthy to that which already exists. One senior industry executive has noted that the result of many small improvements is the surest way, in most industries, to increase competitive advantage. Through such continuous improvement, and the attitude required to assume the risks of such improvement, enhanced competition is realized, and the stakeholders—customers, suppliers, employees, creditors, and so forth—find increasing value in continuing an association with the enterprise.

---

[3]Stephen M. Wolf, chairman, and Rakesh Gangwal, CEO, "Letter to Shareholders," *US Airways 1999 Annual Report,* pp. 2–11.

[4]J. Lynn Lunsford, "Lean Times—with Airbus on Its Tail, Boeing Is Rethinking How It Builds Planes," *The Wall Street Journal,* September 5, 2001, p. A1.

Continuous improvement can take many forms. Increasing the efficiency of current operations through better utilization of assets is one way. Developing strategies that result in greater effectiveness—doing the right thing—is another way of realizing continuous improvement. A common base of continuous improvement through the use of state-of-the-art technology in organizational processes ultimately improves the products and services offered by the enterprise. Caution is required to make sure that the introduction of technology is consistent with all the systems that will be impacted by that technology.

By using teams as the organizational design alternative for working at continuous improvement in the organization, an interdisciplinary and interorganizational perspective of the need and remedial strategy for continuous improvement can be provided. Such interdisciplinary teams provide an ambience where there is a high probability that all facets of the problem or opportunity will be exposed and considered by the team. In the operation of such teams, a viewpoint by one team member is likely to cause a reaction by another team member, who would not have reacted if there had not been the stimulus provided by work in the team situation. In the material that follows, additional examples are given of how continuous improvement through teams has been achieved:

- General Electric Company initiated the concept of a "workout" where interdisciplinary teams work at improving productivity through eliminating wasteful paperwork, duplication, unnecessary approvals, and other bureaucratic impediments to efficiency and productivity. At the company's Schenectady turbine plant, a team effort under workout strategy has improved productivity beyond anything that was ever envisioned. In the steam turbine bucket machinery center, teams of hourly employees now operate, without supervision, a $420 million new milling machine that the team members have selected, tested, and approved for purchase. The cycle time for the operation has dropped 80 percent. CEO Jack Welch has stated: "It is embarrassing to reflect that for probably 80 to 90 years we've been dictating equipment needs and managing people who knew how to do things much better and faster than we did."[5]
- A manufacturer, as part of a new manufacturing strategy, established a project team with the objective of making one of its key plants a model for high-capacity manufacturing efficiency and round-the-clock operation. Several key subobjectives had to be met to attain the overall project objectives:

  Hire, train, and deploy a third shift of workers to meet the round-the-clock objective.

  Develop a master schedule to provide maintenance support in between the rotating shift of production operation.

  Assess, design, install, and check out increased automation of the production process.

  Develop and implement just-in-time (JIT) inventory management policy and procedures to include transport of finished products on an accelerated schedule.

[5]John F. Welch, Jr., chairman of the board and chief executive officer, *1991 Annual Report,* General Electric Company, pp. 1–5.

Conduct a feasibility study for the development of flexible manufacturing systems to deal with an anticipated future mix of new product models, fewer parts, and more standardized assembly procedures.

By using teams to accomplish these objectives, the performance of the company's manufacturing plant was improved. Subsequent teams were used to transfer the improved strategy to other plants.

- In the early 1990s, Boeing Company needed to reduce the cost of building a plane by no less than 25 to 30 percent. Turning the company into an efficient manufacturer was an awesome challenge the CEO faced. Overcoming the hierarchical orientation of the company and reinventing the company in the way it designs and builds its customized, high-quality products were challenges that had to be met. Cost savings realized through the continuous improvement process teams were used to substantially lower prices and fund the innovations needed to reduce customers' maintenance and fuel costs. Improvements included:

  Reduction of inventory costs through just-in-time strategies

  Reduction of the time required to manufacture a plane from more than a year to just 6 months

  Cutting engineering hours

  Eliminating raw material waste

  Minimizing expensive tooling

  Using concurrent engineering during the design phase of new products

  Empowering product design teams to approve or alter a design change

  Using computer-aided design[6]

Continuous improvement springs from the creative act that leads to improvements—the introduction of something new in the organization's products, services, and processes. Creative thought is required, which, properly formulated and implemented, leads to acts of change in the enterprise. "New blood" is introduced to the organization, providing for a change to be developed and introduced into products, services, and processes.

19.5 MANAGEMENT INNOVATION

Innovation has been studied primarily in the context of product-process innovation. Ray Stata, writing in the *Sloan Management Review,* offers a refreshing view of innovation in the context of management innovation. In the next few paragraphs, his viewpoints are paraphrased and augmented.[7]

> Innovation is usually thought of in the context of product-process innovation. The opportunities for innovation go beyond just products and processes—the key to

[6]Shawn Tully, "Can Boeing Reinvent Itself?" *Fortune,* March 18, 1993, pp. 66–73.

[7]Ray Stata, "Organizational Learning–the Key to Management Innovation," *Sloan Management Review,* Spring 1989, pp. 63–74.

progress in companies also rests with management innovation. Japan's success and rise to an industrial power was based on management innovation, not technological innovation in the traditional sense. Management innovation comes about through the use of management technology.

Systems thinking and dynamics can be used to solve complex problems in organizations. The combination of individual and group learning means that organizational learning also takes place. The rate at which individuals and organizations learn is one key to sustainable competitive advantage. This is so because:

- Organizational learning occurs through shared insights, knowledge, and mental models.
- Organizations can learn only as rapidly as the weakest link in the organizational membership.
- Learning builds on past knowledge and experience—on memory. Organizational mechanisms such as policies, strategies, plans, and procedures are used as means to retain organizational learning.

Systems thinking is required—the recognition that a change in one part of the organization will have impact throughout the organizational system. A basic characteristic of a system is the delay time between cause and effect, such as when an order is received and when the finished products are shipped. Another example is the time between the start of the design of a product and when that design is finished and committed to manufacturing. Anything that can be done in the decision process to motivate a systems viewpoint and to reduce the time delay in the linkages of the system will raise the chances of improvements in products and processes.

In the design of any system, such as a planning system, the process that people go through is just as valuable as the output. When a task force is appointed to do strategic planning for the organization, the output of the objectives, goals, and strategies of the organization is valuable. Another valuable "product" of the planning process is the learning, both individual and collective, enjoyed by the people working together on a project team to understand and carry out the development of objectives, goals, and strategies. For example, at Analog Devices, Inc., strategic planning was initiated through the appointment of 15 corporatewide product, market, and technology task teams that pulled together 150 professionals. These teams worked for 12 months to come up with nine imperatives for improvement through change to include specific recommendations for how to bring about these changes. The professionals serving on these teams acquired an understanding of corporate beliefs and assumptions that had served the company well in the past. A significant outcome that became clear to both senior managers and project team members was the need to coordinate technological development across divisions and to centralize certain aspects of manufacturing. The need to better coordinate product planning to capitalize on the company's strengths became clear to the people engaged in this planning effort.

Other project teams were formed in the company to consider strategies to reduce the percentage of orders shipped late. The use of a team approach in dealing with this problem and other problems was helpful in improving interdepartmental communication and helped get people to think about problems and issues in an interdisciplinary and objective fashion as well as subjectively and politically. The teams facilitated the abandonment of parochial departmental thinking—to separate the vital few problems from the trivial many and focus organizational resources on solving them. Indeed, teamwork was elevated to a virtue in the culture of Analog Devices, Inc.

> In order to carry forth the advantages gained from teamwork, openness, and objective thinking, these attributes were included in the company's performance appraisal process and in the criteria for hiring and promotion. The company found that the best way to introduce knowledge and modify behavior is by working with small teams that have the power and wherewithal to bring about change.
>
> Management innovation is an important part of industrial competitiveness, and it will surely become even more of a factor in the future. Management innovation requires new knowledge and skills, new technology, and then the means for the technological transfer of that innovation to the management community. The bottom-line question is this: Are the United States and U.S. companies investing enough in management innovation?

Ray Stata's viewpoint of the need for management innovation, which is interpreted as innovation in how managerial functions of planning, organizing, directing, and control are carried out, has great appeal for continuing improvement through the use of projects.[8]

19.6 CONTINUOUS IMPROVEMENT IN PRODUCTIVITY

Recent increases in manufacturing productivity in the United States came about through the innovation of small changes leading to continuous improvement.

How did these increases come about? We "worked smarter," closed obsolete plants, downsized and restructured organizations, trained employees, got more worker involvement through the use of self-autonomous teams, improved management of manufacturing, developed total quality management processes, eliminated waste, and gained the leverage of updated equipment. A policy of continuous improvement was followed.

To maintain the growth in productivity, U.S. manufacturers will have to keep the same pace of broad continuous improvement strategy across the board for all hardware, software, and the utilization of people. Continuous improvements in *process* technology will be a must.

Productivity improvements through continuous improvement projects are coming in the office. These improvements are the reason behind many of the layoffs, with office staff getting leaner and more productive. The majority of the job cuts in offices will likely be permanent. Improved technology in offices has replaced people, and the offices will be more productive with leaner payrolls.

Demand will grow for information and computer specialists not doing routine administration but doing creative design work that leads to better, more timely information for managers and professionals. More office people will truly be involved in doing "knowledge work," leading to continued improvements in quality and productivity.

---

[8]Ibid., pp. 63–64.

Some suggestions are paraphrased and augmented from *Fortune* magazine on how to facilitate continuous improvement:

- Avoid protecting short-term profit to the detriment of long-term innovation.
- Use a company strategy that encourages both incremental improvement and "breakthrough" innovative strategies.
- Get interest rates down and provide for a reduction in capital gains taxes.
- Recognize that innovation is a process that must be managed through the use of all employees and all the resources of the enterprise, facilitated through the use of project teams.
- Dismiss the popular view that creativity and innovation are mysterious, divinely inspired miracles. Most come from long, hard work immersed in the product-process technology that is improved.
- Accept the notion that only about 1 of every 20 to 30 new product ideas becomes a successful product; 1 of 10 to 16 becomes a "hit." This means that failure in innovation should not be punished—the effort should be rewarded.
- Foster a corporate culture that sustains and rewards innovation—even that which fails—for out of such failures will ultimately come successes.
- Use concurrent engineering to foster innovation, leading to earlier commercialization of products and better performance of organizational processes.
- Work closely with the customer—lead that customer. In satisfying the customer's needs, watch for how all organizational processes can be improved. A good starting point is to do benchmarking against the competitors and other industry best performers.
- Encourage all employees to adopt a professional reading program. Not only will they pick up ideas that can lead to innovation, but also they will gain insight into how the parts of the organization and the disciplines will fit together to form a useful synergy in products and processes.
- Avoid the inevitable delayers, debaters, coordinators, and others who fear change and are comfortable with the status quo. These plodders can kill creativity and innovation in the best-managed organizations.
- Take risks—get the product reasonably well designed, and get to the marketplace ahead of the competitors. Then develop a rigorous strategy of continuous improvement of products and strategies to keep ahead of the customer. Once a market base is established and the customers have accepted the product, then fine-tuning of the product can be done during the follow-on strategy of continuous improvement.[9]

Once an innovative strategy is undertaken for product-process change through the use of project teams, there is an unending struggle to keep up a continuous improvement posture. There is little rest, even for the innovative person and organization. Competition will always be relentless.

---

[9]Paraphrased in part from Brian Dumaine, "Closing the Innovation Gap," *Fortune,* December 2, 1991, pp. 56–62.

19.7 PRODUCT QUALITY THROUGH PROJECTS

The drive for improvement in product quality has resulted in U.S. manufacturers catching up with international competitors. But new dimensions and applications of total quality improvement are happening. U.S. carmakers are getting their quality up to the level of the Japanese cars, but the Japanese are expanding their concept of quality to a new concept called *miryokukteki hinshitsu*—translated to English, it means "things gone right." By entering the second phase of quality, the "personality" of the car is dictating additional quality improvements. The Japanese believe that quality in automobiles is now taken for granted and that in a defect-free product it is the fine touches that will impress consumers. Many of these fine touches have a technological base, such as:

- Computer-driven hydraulics to cushion jolts
- Equal pressure for stereo buttons, door locks, and turn-signal levers
- Electronically activated, liquid-filled engine mounts that dissipate engine vibration when the car is idling
- Aluminum body and suspension, which improve performance by slashing weight
- Air bags
- Seat belts that use pressurized gas to cinch tightly and automatically in a crash
- Windshield wipers tuned to speed up as the car accelerates
- Electronically adjustable suspension
- Other sundry performance goals[10]

19.8 TRENDSETTERS

Innovative acts by entrepreneurs and companies have changed the way organizations function as well as the products they offer. There have been some notable change-makers that set a new trend:

- Ray Kroc and his fast-food restaurants leading to McDonald's
- Ted Turner, in CNN and UHF television, who discovered satellites and changed the way the world gets its news
- Federal Express
- Alexander Graham Bell
- The Mustang automobile
- The Chrysler minivan

[10]David Woodruff et al., "A New Era for Auto Quality," *Business Week,* October 22, 1990, pp. 82–96.

- Hewlett-Packard's LaserJet printer
- At GE Jack Welch and his vision of a less hierarchical company
- E-mail
- World Wide Web
- Free Trade
- International teams

Once a trend has been set, the opportunity for continuous improvement is also set. Without continuous improvement, obsolescence is very real. Project teams can help keep this from happening.

Change is necessary to survival—a trite statement, but one that is ignored by many people and organizations. Resistance to change is very real, and to bring about change, you need to have change champions—a role that can be admirably performed by project teams—and provide adequate resources and the commitment and support of senior management.

Sometimes the act of continuous improvement can be as simple as changing the way people work. In research laboratories and in the professional work carried out by product design engineers there will usually be found adequate opportunity for improvement through assessing how, and where, people spend their time. Studies show that researchers rarely spend more than one-quarter of their day in the laboratory, and design engineers work on design only 20 percent of the time. Alleviating the impediments can be as simple as drawing attention to them, and to someone who can do something about them.

Historically, the way to be competitive has been to concentrate on improving the *product* (the result) through continuous improvement, scrap and repair costs, 100 percent inspection, warranty costs, return of defective products, and assessment of complaint notices from customers, that is, essentially concentrate on the results that the organization produces after the product or service has been delivered to the customer. The *processes* that have been utilized in the producing of the production or services have not been given their proper due.

When the managers and professionals who want to improve the overall competitiveness through continuous improvement restrict their creative and innovative work to the *results* of the organization, they are neglecting a major source of improvement—the processes that are undertaken in the enterprise to produce the results. Such processes include product and service design, marketing, manufacturing, after-sales service, quality, supplier relationships, and finances. A willingness to explicitly study *product, service,* and *process design and implementation* can provide enormous benefits to the enterprise if carried out effectively. When teams are utilized to *concurrently* consider all the organizational functions required in conceptualizing, making, and delivering of a product or service to the customer, better competitiveness is realized. A project team provides for the integration of the many disciplinary considerations involved in meeting and exceeding the competition. Quality changes will happen in the marketplace—and the competitors will benchmark your products and your processes in their efforts to match and exceed your competitive products and services.

Focusing on the organizational processes needed to create and deliver quality products and services to the customer makes good sense. Improvement of organizational processes is needed. But in managing such improvement, the integrity of the products and services should not be forgotten.

New products brought about through the support of efficient organizational processes are at the center of global competition. Key strategies in the global marketplace include the development of high-quality products at lower cost and faster commercialization.

19.9 PRODUCT INTEGRITY

Product integrity means that the product has more than just a basic functionality or performance character; it has additional characteristics which complement the customers' values and lifestyles. Industrial products match the existing work flows and production systems of the customer. Product integrity has two components: internal integrity and external integrity. *Internal integrity* refers to the unity between a product's function and its construction—the parts fit smoothly, the components match well, and the product works well, and the layout maximizes the available space. From an organizational viewpoint, product integrity means that a focus has been achieved through interdisciplinary teams working with customers and suppliers. *External integrity* refers to the unity between the product's performance and the customers' expectations. Product integrity can be enhanced by having an organizational approach which provides focus and management to the product management activities and includes how people do their work, the way decisions are made, the effectiveness of information flow and use, and the way that supplier and customer considerations are integrated into both the management of the project and the technical aspect of the product.[11]

Most people in the enterprise work in the manufacturing or operations side of the business. Improvement in manufacturing systems technology has become a key global competitive consideration.

19.10 CONTINUOUS IMPROVEMENT IN MANUFACTURING

Continuous improvement through the introduction of technology to an organization has to be done with great care. Despite the billions of dollars that General Motors invested in factory robots in the United States, GM failed to take into consideration the human issues that would arise as the result of introduction of automated factory systems in its plants. The company had to slow down the start-up of its automated factories until the workers were ready and trained to assume their roles in the production systems. Even though GM invested in advanced manufacturing

[11]Kim B. Clark and Takahiro Fujimoto, "The Power of Product Integrity," *Harvard Business Review,* November–December 1990, pp. 107–118.

systems to improve productivity, reduce costs, and improve quality, the company's eroding domestic market share was not saved.

Automobile manufacturers have moved to "lean manufacturing" strategies and techniques that were developed in Japan. This was made clear in a 5-year study of the automobile industry by the International Motor Vehicle Program at the Massachusetts Institute of Technology (MIT). This study showed that the techniques of lean manufacturing developed in Japan have made traditional mass production methods as obsolete as the Model T. In lean manufacturing, teams of skilled workers use flexible manufacturing systems to produce customized products of endless variety to exacting quality standards, quickly, and at low cost. In Europe, carmakers have concentrated too much on improving the efficiency of mass production techniques—the enhancement of an existing technology whose time has passed.

Continuous improvement in the automobile industry led by the Japanese manufacturers has caught the attention of automakers in Europe. Competitive survival will require these manufacturers to change, not only through a continuous improvement strategy, but also through more radical change.

Change is under way in European automobile plants. These automakers are struggling to meld local workers and Japanese manufacturing and management techniques. This change is not without pain. Automobile manufacturing executives have taken longer to understand that the new innovations will not work without radical shifts in labor relations. Japanese management techniques such as just-in-time inventory management and total quality management are complex and require careful consideration of the human issues involved as well as close employee cooperation. To meet competition, the work culture in France and other European countries will have to change. The major impact will be on labor relations. In the use of modern manufacturing and automation, workers have to abandon their narrow job classifications and learn more jobs, solve problems, and work in less predictable and programmed ways. Customers demand a greater variety of products and options. This means complex, harder jobs for employees. After decades of working in traditional ways, workers in Europe are having difficulty accepting the new philosophy.

To compete, European companies have had to downsize and restructure. Older, big factories were closed and new, smaller ones set up to use modern management concepts. Workers will have to be closely screened to ensure that they have the aptitude and the right values to work in production teams, quality improvement teams, product-process design project teams, and such innovative organizational design alternatives as are being used in the modern factory. The opportunities for project management have never been better.[12]

Change is difficult in all organizations. The challenge for small and midsize companies is particularly critical because these companies often lack the resources to adequately cope with the needed change. Financial resources are scarce. Continued legislation that puts additional administrative burdens on these companies ties up managerial and professional time. Many of these small enterprises are poorly capitalized.

---

[12]Paraphrased from E. S. Browning, "Europe's Auto Makers Struggle to Meld Local Workers and Japanese Techniques," *The Wall Street Journal,* November 22, 1991.

19.11 MANUFACTURING PHILOSOPHIES

A manufacturing management philosophy for continuous improvement means many things. It means that truly a systems viewpoint has to be taken in the seeking of improvement in the products and processes of the enterprise. Total quality management, just-in-time inventory, flexible manufacturing systems, and other innovative techniques have to be integrated as a way of life in continuous improvement in manufacturing. All the organization's people have to be involved in the process, and the message has to get across that the improvement is not just another program but is a set of related programs and projects working together. The formation of project teams to study and make recommendations for improvement not only brings in the multifunctional nature of the change, but gets people involved at different levels and from different functions, which sends an important message to the organization of the systems nature of the changes.

Some basic principles for continuously improving manufacturing processes include the following:

- Reduce the number of parts by combining or eliminating them.
- Minimize assembly surfaces, and ensure that all processes are on one surface and completed before moving on to the next one.
- Design for top-down assembly, thus gaining the advantage of gravity and reducing the number of clamps and fixtures.
- Improve assembly access by increasing unobstructed vision and ensuring adequate clearance for standard tooling.
- Maximize parts compliance by providing adequate grooves, guide surfaces, and specifications for marrying parts. This can reduce misalignments and poor quality.
- Maximize part symmetry, which makes parts handling easier. If symmetry is not possible, then design in the asymmetry or alignment features that are possible.
- Make effective use of rigid parts rather than flexible ones. Rigid parts are easier to handle. Provide adequate surfaces for mechanical gripping, and design in barriers to tangling, nesting, or interlocking, which take time to correct.
- Minimize, or avoid, separate fasteners. By using standard fasteners in components, such as snap-fits, the assembly process can be simplified.
- Provide self-locking parts such as tabs, indentations, or projections on mating parts to identify them and their orientation through final assembly.
- Move toward modular design for common functional requirements and standard interfaces for easy interchangeability of modules. This will improve testing and service on the product, provide more options, and offer faster, continuous improvement of the product.[13]

---

[13]Paraphrased from Theresa R. Walter, "Design for Manufacture and Assembly," *Industry Week*, September 4, 1989, p. 82.

19.12 COMPUTER-INTEGRATED MANUFACTURING (CIM)

CIM has become a key strategy in improving U.S. competitiveness. The decision to introduce CIM into manufacturing operations can improve efficiency and reduce product costs. But CIM by itself will not ensure competitive success. Customers have to buy the products; competitors will be trying to beat you out in the marketplace. Order entry procedures have to be effective; sales and sales promotion take on new significance. Customers may require financing, and the work force may see CIM as a threat. The use of CIM has to be blended into other functions and activities of the enterprise—thus a decision to implement CIM has potential reverberations throughout the company.

Responsibility for the development of a CIM strategy is not limited to manufacturing; it encompasses all functional elements of the business. The enterprise's functional and specialized areas have to interact through project teams in answering the questions about the CIM strategy: How will that strategy affect our business? What are its short- and long-term results? What will be done with extra production time? How will the improvement in manufacturing productivity affect the rest of the business? What forces are developing in the marketplace that have a potential influence on our manufacturing strategies? If new products are developed, how will the manufacture of those products be changed by our current strategy in CIM? These are a few of the questions that have to be answered in considering the use of CIM.

Market needs and customer-use considerations have to be evaluated. What will be the cost of shared resources and overhead allocations? How does the mission of the information system change? How will manufacturing deal with shorter runs of more specialized products?

The introduction of CIM is a strategic decision that affects all business activities, a decision whose impact will be felt across the traditional functional lines of the enterprise and will extend to suppliers, customers, regulatory bodies, local communities, and unions, to name a few. CIM implies that decisions concerning what you make, how you make it, where it will be made, and how the products will be supported have to be made in unison and in harmony. CIM should come out of the development of systems and strategic project planning in the enterprise—basically as part of a broader business strategy. Thus CIM does not just lower manufacturing costs, it is part of a larger business strategy that has to be compatible with globally changing technologies, competitive realities, and unforgiving markets. It is difficult to imagine CIM without the catalyst of project management.

JIT

19.13 JUST-IN-TIME (JIT) MANUFACTURING

JIT manufacturing is revolutionary in the way that inventories are managed in today's organizations. The improvement in manufacturing technology through the use of JIT strategies has created systems changes. JIT concepts and processes have been adequately described in the literature. There is the opportunity in organizations to take

JIT manufacturing beyond the manufacturing plant to other situations in the enterprise. Billesback and Schniederjans have suggested the application of JIT techniques to the administration activities of the enterprise. The following section, paraphrased from these authors, provides a refreshing opportunity for the transfer of JIT technology:

> Consider the application of the JIT concept to administrative activities. Administrative activities are any activities not related to the production of goods: scheduling, billing, order entry, accounting, plant maintenance, and financial tasks, to name a few. Manufacturing activities include such things as assembling a component, stamping, welding, milling, sanding, grinding, and cutting metal; transporting work-in-progress is also considered a manufacturing activity.
>
> Think of JIT as a philosophy for the elimination of all waste in administrative activities. Waste does not add value to the product. Some activities simply add cost to the product or process.
>
> Storing, moving, expediting, scheduling, and inspection soon reach diminishing returns in adding value, as do stacking, filing, mailing, transmitting, "rush ordering," routing, and proofing.
>
> Some project-directed strategies that can be employed to improve administrative just-in-time (AJIT) techniques include the following:
>
> - Provide employees with time for identifying problems and solutions to improve productivity.
> - Improve layout to facilitate flow of work.
> - Locate workers whose work is related close together.
> - Allow workers to see the total of work to be performed insofar as possible.
> - Look for barriers to communication.
> - Specialized and quality training and performance help to reduce waste.
> - Consider standardization of activities and pooling of responsibility for similar activities.
> - Use worker-centered total quality management.
> - Trace performance to a specific individual.
> - Decentralize authority to make the appropriate decisions.
> - Monitor the process from start to finish to see what factors slow down or halt the processes.
> - Reduce the number of workers until the processes slow down or come to a halt.
> - Look for extra workers, procedures, policies, processes, and backup equipment that mask organizational weaknesses.
> - Consider a better grouping of related organizational functions and processes.
> - Look for bureaucratic structure or processes that stifle information flows, decision making, and orderly procedures, or create lethargy in the flow of work.
> - Provide for an organizational design (steering committee) to facilitate the AJIT technique review and facilitate the transfer of processing technology.
> - And the bottom line—use cross-functional project teams to seek improvement.[14]

[14]Thomas J. Billesback and Marc J. Schniederjans, "Applicability of Just-in-Time Techniques in Administration," *Production and Inventory Management Journal,* 3d quarter 1989, pp. 40–44.

JIT philosophies and techniques can extend from the purchases of components and materials throughout the entire manufacturing process and out to the customer to include after-sales services to customers. Within organizations there are many different forms of "inventory" that include information, housekeeping supplies, time, equipment, machine and equipment time, material—indeed any resource that can provide value to organizational products and processes. Strategies can be employed to reduce the amount of inventory required to support the organization. USAir dealt with this kind of opportunity through a project team in a very explicit way.

Reduction of inventory was the reason that the CEO of USAir gave for standardizing the interiors of aircraft in the fleet. The postmerger fleet of the airline had 5 colors of seat belts, 5 colors of curtains, 16 colors of seat covers, and 3 different kinds of carpet. The company was stocking 180 different seat covers, which proved too costly from an inventory standpoint. Furthermore, the requirements of nine types of aircraft, each with varying shapes of seat backs and bottoms, with some having in-arm food trays, differently shaped window seats, and so on, added complexity to the inventory management challenge for the airline. Planners at USAir worked with designers and suppliers on selecting patterns and colors for the redone aircraft fleet, keeping in mind the need for a corporate identity and ease of care and management of the inventory.[15]

19.14 TO SUMMARIZE

In this chapter the following key messages were presented:

- In today's competitive marketplace a strategy of continuous improvement of organizational products, services, and processes is required to survive and grow.
- Project teams and other alternative teams are the principal means for organizing the resources of the enterprises to bring about continuous improvements in those things of value created for customers and other stakeholders.
- Many examples were given of contemporary organizations that have successfully pursued a strategy of continuous improvements.
- Continuous improvements include the ongoing small incremental advancements in the enterprise's products, services, and organizational processes. "Breakthroughs" in technology leading to marked improvements are also included under the umbrella of "continuous improvements."
- In this text the point has been made repeatedly that products are building blocks in the design and execution of organizational strategies. Continuous improvement projects and other team initiatives are no different.
- Continuous improvement initiatives take many forms, from the use of advanced technology through improved performance of the management functions.

[15] *USAir Magazine,* October 1991, p. 7.

- Innovation in the theory and application of management processes and practices is an important source for continuous improvement.
- The emergence and use of project management processes and techniques are in themselves significant management improvements.
- There are manifold opportunities for continuous improvement in the manufacturing or production function of the enterprise through the use of teams. By having workers serve on such teams, an important base of knowledge and skill is brought to the teams' endeavors.
- In the chapter a few of the "trendsetters" in products and services were noted. Once a trendsetter is successful, the competition will move in to capture some of the trendsetter's market.
- Some of the philosophies that can be used as a fulcrum from which to leverage continuous improvement were noted by drawing on some examples from the automobile industry to include computer-integrated manufacturing and just-in-time manufacturing.
- It is difficult in modern organizations to separate continuous product, service, and process improvement and project/team management.

19.15 ADDITIONAL SOURCES OF INFORMATION

The following additional sources of project management information may be used to complement this chapter's topic material. This material complements and expands on various concepts, practices, and the theory of project management as it relates to areas covered here.

- David I. Cleland, "New Ways to Use Project Teams," chap. 29 in David I. Cleland (ed.), *Field Guide to Project Management* (New York: Van Nostrand Reinhold, 1997).
- Jerry B. Baxter, "Responding to the Northridge Earthquake," and John R. McMichael, "Boeing Spares Distribution Center: A World-Class Facility Achieved through Partnering," in David I. Cleland, Karen M. Bursic, Richard J. Puerzer, and Alberto Y. Vlasak, *Project Management Casebook,* Project Management Institute (PMI). (Originally published in *PM Network,* November 1994, pp. 13–22; and *PM Network,* September 1994, pp. 9–19.)
- H. James Harrington, *Business Process Improvement: The Breakthrough Strategy for Total Quality* (New York: McGraw-Hill, 1991). This book, although published over 10 years ago, provides valuable information on how to improve organizational productivity while reducing cycle time and cost. It provides step-by-step guidance on how to determine and meet customer needs, create business improvement teams, train team leaders, and ensure that appropriate process changes are implemented. The book also offers a prescription on how to eliminate layers of bureaucracy, simplify processes, and carry out benchmarking of the best-of-the-breed performers.

- William E. Conway, *Winning the War on Waste: Changing the Way We Work* (Nashua, N.H.: Conway Quality, 1994). This book is a "how-to" approach to improving an organization's need for change in the way work is accomplished for better productivity and reduced waste. It is an excellent guide on how to identify and assess the different operations that may be contributing to wasted materials, time, and capital. Conway also guides the reader through work analysis, as well as the human aspects of continuous improvement.
- Brian K. Schimmoller, "The Changing Face of Project Management," *Power Engineering,* May 2001. The author makes the claim that the time and money elements of modern projects have elevated project management as an enterprise perspective. Although written from the perspective of an electrical power plant construction, the article has general application, and describes a philosophy of project management that is much more comprehensive and involved than just scheduling strategies.
- Mark C. Maletz and Nitin Nohria, "Managing the Whitespace," *Harvard Business Review,* February 2001. The authors enter the plea that modern people in business are being told to operate in the organizational white space where speed and flexibility reign. Indeed—in a project context—most projects are initiated within an organization's formal structure. Managers are encouraged to shift to white space if there is great uncertainty over an opportunity. There are organizational politics that block normal organizational processes. White space using project management requires a changed way of thinking about how work gets completed, measured, and recognized in modern organizations.
- Karen M. Bursic, "Self-Managed Production Teams," in David I. Cleland (ed.), *Field Guide to Project Management* (New York: Van Nostrand Reinhold, 1998). In the last couple of decades, the field of project management has moved from the fringes of the management discipline to the forefront. The author notes this growth of project management and looks at the project implications involved in the use of self-managed productions. Production teams are comprised of people from a particular area of work, often called a work cell, that have broad authority and responsibility to manage the day-to-day activities of a manufacturing activity.

19.16 DISCUSSION QUESTIONS

1. What does a philosophy of continuous improvement mean?
2. Why should senior managers be concerned about the process of continuous improvement?
3. It has been said that global competitiveness and continuous improvement are relentlessly linked. Why does such a linkage exist?
4. What are some of the forms that continuous improvement can take on in an enterprise?
5. Why does the use of project teams as an organizational design for continuous improvement make sense?

6. If project teams are not used as an organizational design for continuous improvement, what alternative organizational designs might be used?
7. What are some of the common themes of continuous improvement that can be deduced from a perusal of the examples given in the chapter?
8. What are some of the opportunities for continuous improvement to be found in the area of management innovation?
9. What are some of the benefits to be realized from continuous improvement in productivity and quality considerations?
10. Which companies have set high standards in the area of continuous improvement?
11. What are some of the key gains for continuous improvement in manufacturing? Are there any opportunities for the "technology transfer" of these gains to service-related enterprises?
12. If we could, by some deliberate and diabolic stroke, eliminate all projects and all project management literature, how could we (would we) deal with change in the world?

19.17 USER CHECKLIST

1. Does your company have any philosophy and strategy of continuous improvement? If not, why not?
2. Is the climate in your company compatible with the development of a deliberate strategy for the design and implementation of continuous improvement initiatives?
3. After a perusal of the examples given in this chapter of organizations that have pursued a strategy of continuous improvement, are there any areas where such a strategy might be developed for your company?
4. What challenges does your company currently face that suggest the need for continuous improvement leading to increased productivity and quality?
5. Are there opportunities in your company for the use of alternative teams for continuing improvement initiatives?
6. What might be learned if the senior management of your company were willing to appoint a project team to examine the opportunities for continuing improvement in manufacturing?
7. Are there any attitudinal issues that might mitigate against the development and implementation of successful continuous improvement initiatives in your company?
8. What are the major strategic and operational issues facing your company at present? Given these issues, would a strategy of continuous improvement initiatives hold any promise for producing positive results for the enterprise?

9. Are there any opportunities for administrative just-in-time improvements in your company? Why or why not?
10. What might be the outcome if each of the major functions within the enterprise were examined to see what specific opportunities might exist for designing and launching a continuous improvement strategy?
11. Are any of your company's competitors using continuous improvement initiatives in their organizational strategy? If so, could any of these initiatives be benchmarked to gain insight into the competitor's advantages?
12. The only thing that is permanent today is change. Furthermore, any company's strategy in products and organizational processes is aging. Given these conditions, do the senior managers of the enterprise possess the leadership qualities to cope with change through the development and implementation of continuous improvement strategies?

19.18 PRINCIPLES OF PROJECT MANAGEMENT

1. An enterprise's existing products, services, and processes will change, largely because of environmental and competitive forces.
2. Continuous improvement of enterprise products, services, and organizational processes is necessary for survival in today's competitive environment.
3. Changes in enterprise products and services must be accompanied by an assessment of the requirement for organizational process change.
4. Project teams are the key organizational designs to bring about the evaluation and continuous improvement of organizational processes.
5. All organizational processes are likely candidates for a continuous improvement strategy.

19.19 PROJECT MANAGEMENT SITUATION—CHALLENGES FOR THE NEW MANAGERS

The growing use of teams in today's organizations has caused marked changes in the manner in which managers "manage." The managers of today cannot function effectively in the new team-driven culture without some major changes in philosophy, attitude, and method of managing. These new managers are leaders, mentors, facilitators, coaches, sponsors, advocates, "chaplains," comforters, trainers, teachers, team players, entrepreneurs, problem finders, and problem solvers through the teams supporting the organization. The new manager is a Socratic manager who asks questions that will get people thinking about their work and moving in the right direction within the larger organization context in which the people exist.

Some of the competencies required of these new managers include:

- Lifetime learning will be required of the people in the organization.
- Openness will be demanded of the new manager. Crucial performance information about the organization and the team's work will have to be made available to the people doing the work.
- The new managers will not be the traditional watchdog, controller, bureaucrat, police officer, or slave driver.
- The new managers will require interpersonal skills to work with the many stakeholders who believe that they have some claim about the product, service, or processes under way in the enterprise.
- The new patterns of authority, responsibility, and accountability found in the team-driven enterprises will require major changes in the philosophy of the new managers.

The notion that organizations must innovate and create new ways of improving organizational products, services, and organizational processes means that all people associated with the organization will have a major challenge to improve their knowledge, skills, and attitudes.

19.20 STUDENT/READER ASSIGNMENT

Given the description of the competencies of the new managers in today's organizations, what resources and initiatives should these new managers provide for the support of the team-based strategies in the enterprise?

Suggest a strategic way to help people work across organizational boundaries and functional territories to get the job done.

CHAPTER 20

CULTURAL CONSIDERATIONS IN PROJECT MANAGEMENT[1]

"A culture is in its finest hour before it begins to analyze itself."
ALFRED NORTH WHITEHEAD, 1861–1947

20.1 INTRODUCTION

The concept of a culture arose from studies performed by anthropologists who studied ancient societies. A culture of a particular social group—such as a nationality, race, or, in modern terms, an organization, is essentially a "way of life" in that society. In this chapter the term *culture* is used to described the collective values, beliefs, customs, knowledge, practices, attitudes, mores, and history of a business organization—more particularly in a project-driven organization. As an organization becomes involved in the use of project management to manage change in that organization, the existing culture is modified.

This chapter provides further insight into the nature of an organizational culture. Some of the influences expressed in cultural terms of a strategic-project linkage in the enterprise are noted. In addition, the cultural features of the project-driven organizational design are viewed. Other cultural factors than can impact a project team and other stakeholders are noted. The trust factor among the project team members and the extensions of the culture of the project team within an organization are recognized. Another contribution of the chapter is insight into how conflict impacts the culture of the project team. The chapter ends with a description of how a project's success—or failure—can be influenced by the prevailing culture.

In other chapters of this book, we have presented the concepts of project management. This chapter develops the concept of an organizational culture as the ambience within which project management exists.

[1]This chapter contains substantial material from the author's article, "The Cultural Ambience of Project Management—Another Look," *Project Management Journal,* June 1988, pp. 49–56.

20.2 DEFINING CULTURE

Culture is a set of refined behaviors that people have and strive toward in their society. Culture, according to anthropologist E. B. Taylor, includes the totality of knowledge, belief, art, morals, law, customs, and the other capabilities and habits acquired by individuals as members of a society.[2]

Anthropologists have long used the concept of culture in describing primitive societies. Modern sociologists have borrowed this anthropological concept of culture and used it to describe a way of life of a people. Here the term is used to describe the synergistic set of shared ideas and beliefs associated with a way of life in an organization.

The interest in a company's culture is illustrated by efforts that were under way at Du Pont Company. Du Pont was trying to create a new culture driven by profits, not just research prowess. The new chairman, Edgar S. Woolard, Jr., recognized that the company took too long to convert research into products that can benefit customers. The company tried to restructure a "bloated bureaucracy." About 30 percent of its research budget, or more than $400 million a year, was being shifted toward speeding new products to customers. A "skunk works" antidote was tried by several Du Pont departments to speed the new-product process. These departments created small, interdisciplinary project teams to study all new-product ideas. These teams included research, manufacturing, and sales representatives. Du Pont was also working more closely with customers.

Another problem being addressed at Du Pont was how to reduce the debugging of a new product after the product is launched. Interdisciplinary teams worked to think through how to prevent such debugging after product delivery.

The emphasis was shifted from only product development to improving the processes used to manufacture new products. The potential savings and resultant products from improvement in manufacturing at Du Pont at more than 100 plants worldwide were huge. At the polymers division, 60 percent of the research budget was being spent on improving processing and only 40 percent on new products. A few years ago, 70 percent of the budget went to products and only 30 percent to processing.

One of Du Pont's plants, located in Wilmington, Delaware, had gained productivity improvements, some of which had come from involving workers in the process. To encourage the workers to think about improvements, the plant manager gave cost figures on any of the operations down to the newest worker on a maintenance team. More discipline was coming in the manner in which projects were approved. Research managers said that they think a lot harder before approving projects without an obvious payoff. But still there remains a strong commitment to fundamental research at the senior management level of the company.[3]

---

[2]Paraphrased from E. B. Taylor, *The Origins of Culture* (New York: Harper & Row, 1958) (1st ed. published in 1871).

[3]Scott McMurray, "Changing a Culture: Du Pont Tries to Make Its Research Wizardry Serve the Bottom Line," *The Wall Street Journal,* March 27, 1992.

20.3 THE NATURE OF AN ORGANIZATIONAL CULTURE

An organizational culture is the environment of beliefs, customs, knowledge, practices, and conventionalized behavior of a particular social group. Every organization, every corporation has its distinct character. People make organizations work, and the culture of the corporation ties the people together, giving them meaning and a set of principles and standards to live and work by. Arnold and Capella remind us that achieving the right kind of corporate culture is critical and that businesses are human institutions.[4] One author has defined organizational culture as:

> The pattern of basic assumptions that a given group has invented, discovered, or developed in learning to cope with its problems of external adaptation and internal integration, and that have worked well enough to be considered valid, and, therefore, to be taught to new members as the correct way to perceive, think, and feel in relation to those problems.[5]

A given group is described as a set of people who have been together long enough to have shared significant problems, solved those problems, observed the effects of the solutions, and taken in new members.[6] Project teams in the life cycle of a project meet the definition of a "given group" and, therefore, can be seen as developing a distinct culture, one that is influenced by the culture of the organization to which the team belongs.

GE, one of the premier corporations of our times, has developed and implemented a proactive strategy in its organizational culture. A few of the building blocks of such strategy follow[7]:

> ...the most significant change in GE has been its transformation into a Learning company.

> ...by finding, challenging, and rewarding these people, by freeing them from bureaucracy, by giving them all the resources they need—and by simply getting out of their way—we have seen them make us better and better each year.

> ...where people are free to dress as they wish and encouraged to act and to take risks.

> Integrity—it's the first and most important of our values.

---

[4]D. R. Arnold and L. M. Capella, "Corporate Culture and the Marketing Concept: A Diagnostic Instrument for Utilities," *Pubic Utilities Fortnightly,* October 17, 1985, pp. 32–38.

[5]Reprinted from Edgar H. Schein, "Coming to a New Awareness of Organizational Culture," *Sloan Management Review,* Winter 1984, p. 3, by permission of the publisher. Copyright© 1984 by the Sloan Management Review Association. All rights reserved.

[6]Ibid., p. 5.

[7]John F. Welch, Jr., Jeffrey R. Immelt, Dennis D. Dammerman, and Robert C. Wright, Letter, "To Our Customers, Share Owners, and Employees," General Electric Company 2000 Annual Report, February 9, 2001.

Relishing change—we've long believed that when the rate of change inside an institution becomes slower than the rate of change outside, the end is in sight.

Focus on the customer.

Using size—we fight every day to create the quickness and spirit of a small company.

Annihilating bureaucracy—we cultivate the hatred of bureaucracy in our company.

Self-confidence, simplicity and speed.

Training—we've always had great advanced training programs at GE.

Informality—informality is not generally seen as a particularly important cultural characteristic in most large institutions, but it is in ours.

The GE of the future will be based on the cherished values that drive us today; mutual trust and the unending, insatiable, boundaryless thirst for the world's best ideas and best people.

Corporate culture usually is explained in terms of organizational values and beliefs and the behavior of members of the corporation. In the corporate setting, the value orientation and leadership examples set by senior managers greatly influence employee behavior.

Two Stanford professors argue that a strong, almost cultlike culture, which, among other things, supports a willingness to experiment and discard whatever didn't work, is innovative. The people in such organizations share such a strong vision that they know in their hearts what's right for their company.[8]

But a cultural unit has many subcultures: the company, the work group, and the project teams. Together these also help determine individual behavior. Managerial behavior is affected by the culture in which the manager operates. This culture in turn is reflected in the subordinate elements of the organization. Culture influences managerial philosophy that in turn affects the organizational philosophy. The organizational culture can be affected by the lack of a management philosophy on which plans, policies, procedures, guidelines, rules, and basic values important to the growth and survival of the organization are based.

The culture of an enterprise and the culture of a project within that enterprise are mutually interdependent, influencing each other as the two organizations work together. Indeed, the establishment of a project to design and develop a new product, service, or organizational process for the enterprise influences the existing culture. The management of a project to create something for the enterprise that did not previously exist modifies in some way the existing culture for the organization.

[8]James C. Collins and Jerry I. Porras, "Why Great Companies Last," *Fortune,* January 16, 1995, p. 129.

Thus it could be said that cultural change is a reverberation in the enterprise set in motion by the forces of change needed to successfully develop, produce, and turn the project results over to the customer. A benchmarking project, which seeks to learn something about competitors and the "best in the industry" performers, will probably raise issues relating to the local culture. A reengineering project which results in downsizing, restructuring, and realignment of the way in which organizational processes are managed will impact the local culture, perhaps in adverse ways as people are reassigned, lose their jobs, or have to learn new duties in managing the organizational processes.

When a project is set up to develop and use self-managing production teams in manufacturing, there will be major changes in reporting relationships, management styles, management processes, and the authority and responsibility relationships within the enterprise. Managers will find that the very nature of the use of projects to manage change in the enterprise can have a marked impact on the local culture. The real challenge facing the project manager and other organizational managers is to assess the impact that the project will likely have on the organizational culture and make provisions for senior managers to evaluate how that impact will influence how people think and react throughout the enterprise. Project managers then need to understand how and why their project will impact the project team and the culture of the local organization, and work with specialists and managers in the enterprise to ensure that proper planning is under way to understand how cultures will be impacted and if people will need orientation and training to determine what the overall impact will be on the local culture.

20.4 THE STRATEGIC MANAGEMENT LINKAGE

The importance with which strategic management, which embodies the use of projects as building blocks in the design and execution of organizational strategies, is held in the enterprise depends to a large degree on how the culture of the organization supports the use of strategic management as a philosophy of management in the stewardship of the enterprise. The culture of the enterprise in turn is dependent on the culture of the society in which the strategic management is being carried out. A project in another country, such as in a country of the Far East, will probably take substantially more time and resources than for a project done in the United States. Culture forces in any society cannot be ignored in the management of an international project. Indeed, the influence of a national culture has a greater impact on an individual's attitude than does an organizational culture. Some of the considerations to be evaluated when taking on a project in another country or society are the following:

- Are the project team members able to disagree with the project manager?
- Are the project team members comfortable in being involved in the project decisions?

- Do the project team members expect the project manager to be a participative-style manager—or conversely, do the team members expect the project manager to be more of an autocratic manager?
- Do the project team members expect to have specific instructions as to their authority and responsibility in the management of the project affairs?
- Are the project team members able to subordinate their individual interests to the overall interest of the project?
- Do the team members expect the project manager to manage them as individuals—or more in the collective sense as an entire team?
- Are the individual team members satisfied with recognition of the team's efforts, or do they expect to have individual recognition as well?
- What are the key ingredients of the "philosophy of management" held by the key decision makers in the enterprise?
- Can there be resistance to change among the members of the project team?

The growing importance of understanding the national culture in which a project is being carried out is reflected in such catchwords as "multicultural projects," "multicultural teams," and "multicultural project management." Although the concept and process of project management are generic and universal, project management practice depends on the culture in which it is carried out. For example, a construction industry project application would be somewhat different than one carried out in the U.S. Department of Defense. A project dealing with basic research would have processes that would be different than one carried out for the updating of a production facility. Project team members with different cultural circumstances would likely interpret the same project management concepts and processes differently.

20.5 ADDITIONAL EXAMPLES OF CORPORATE CULTURE

Values, the basic concepts and beliefs of an organization that often are reflected in the documentation of a corporation, contribute to the heart of the corporate culture. For example, at Harnischfeger Engineers, Inc., the attitude toward project management is communicated in a project management mission statement:

> Professional management for projects through eminently qualified personnel, using state-of-the-art project management techniques resulting in satisfied customers and achieving Harnischfeger Engineers' profit objectives.

This mission is printed on a business-size card with a statement of the company's project management strategy on the reverse side:

- *To ensure* that realistic achievable schedules and cost estimates are developed for projects.

- *To ensure* effective communications and rapport with HEI management and project-related groups within and outside the company.
- *To coordinate* technical support groups to ensure optimum installation sequencing and acceptance testing of project systems and components.
- *To complete* projects on schedule, within budget, and in compliance with technical and other specifications.

Such a card, and the clear message it conveys, has obvious value in influencing the corporate culture and communicating the corporate philosophy to project stakeholders.

Motorola, Inc., a giant organization, was more nimble than other large corporations. In part, Motorola's nimbleness comes from its ability to change, to foster a participative culture, and to use teams as a way to organize workers and professionals to do productive, quality-driven work. It has an elaborate corporate culture that kindles rather than stifles conflict and dissent, finds promising but neglected projects, and generates a constant flow of information and innovation from thousands of small teams which are held to quantifiable goals.

Intelligence gathering in Motorola is done through a department that has as its mission the reporting of the latest technological developments, gleaned from conferences, journals, rumors, and such. Intelligence gained from many sources helps build "technology road maps" that assess where breakthroughs are likely to occur and how these breakthroughs can be integrated into new products and processes. The culture of conflict helps identify and fix mistakes quickly, unmasks and eliminates weak or illogical efforts, and keeps senior managers abreast of problems and opportunities in the marketplace.[9]

Traditional manager bureaucrats may fight more for their turf than for what is right for the enterprise. This fighting slows decision making and prevents people from trying anything new. These turf battles contributed to IBM missing the early markets for laptop computers, notebook computers, and workstations. IBM also sat back and procrastinated while competitors exploited mail-order distribution of personal computers. Even in more mature product lines, IBM came months late to the market, a failure to capitalize on what concurrent engineering can do to commercialize products sooner. Part of IBM's previous problems came out of their early success in the 1980s, when the company had so many successes that the managers began to lose touch with competitive realities. With the success of the 1980s, IBM's culture became complacent, and the "measure of success" became how high someone could rise in the company. The highest compliment someone could pay a rising star was: "He's good with foils, the transparencies used on overhead projectors in all IBM meetings. Foils became such a part of the culture that senior executives started having projectors built into their beautiful rosewood desks."[10]

Duke Power Company has managed its nuclear power plant projects effectively and has avoided the construction cost overruns that have been experienced by many other utilities. This success is attributable to an excellent strategic and project

[9]G. Christian Hill and Ken Yamada, "Staying Power," *The Wall Street Journal,* December 12, 1992.
[10]Paul B. Carroll, "The Failures of Central Planning at IBM," *The Wall Street Journal,* January 28, 1993.

management process inseparable from and executed within a supportive corporate culture having these distinguishing features:

- Tight control of construction with little dependence on outsiders
- An in-house engineering and construction staff
- Procurement mostly through a subsidiary
- Clear-cut responsibility for problems—no outside engineering firms, consultants, or contractors with which to share the blame
- Operation of the company by engineers with hands-on experience
- Hiring and promoting of local talent with local community ties in areas where labor unions are weak
- Utilization of computer tracking for flexible job assignments to obviate idleness
- Internal competition among plants and departments to counter any trends toward mediocrity[11]

Morty Lefkoe, president of a consulting firm that specializes in helping corporations reshape their cultures, believes that the most common cause of failure of mergers is a "clash of corporate culture."[12] He further believes that:

- Behavior in an organization is determined more by its culture than by directives from its managers or any other factor.
- It is almost impossible to implement any strategy that is inconsistent with an organization's culture.
- Culture has a greater impact on a company's success than anything management can do.

An organization's culture can impact its effectiveness. Two researchers have found a relationship between culture and long-term economic performance. These researchers have also documented the cultural traits that successful companies share.[13]

20.6 CULTURAL FEATURES

An organization's culture consists of shared explicit and implicit agreements among organizational members as to what is important in behavior, as well as attitudes expressed in values, beliefs, standards, and social and management practices. The culture that is developed and becomes characteristic of an organization affects strategic planning and implementation, project management, and all else.

[11]Ed Bean, "Going It Alone," *The Wall Street Journal,* October 17, 1984.
[12]*Fortune,* July 20, 1987, p. 113.
[13]John P. Kotter and James L. Heskett, *Corporate Culture and Performance* (New York: Free Press, 1992).

It is possible to identify common cultural features that positively and negatively influence the practice of management and the conduct of technical affairs in an organization. Such cultural features develop out of and are influenced by:

- The management leadership-and-follower style practiced by key managers and professionals
- The example set by leaders of the organization
- The attitudes displayed and communicated by key managers in their management of the organization
- The managerial and professional competencies
- The assumptions held by key managers and professionals
- The organizational plans, policies, procedures, rules, and strategies
- The political, legal, social, technological, and economic systems with which the members of an organization interface
- The perceived and/or actual characteristics of the organization
- Quality and quantity of the resources (human and nonhuman) consumed in the pursuit of the organization's mission, objectives, goals, and strategies
- The knowledge, skills, and experiences of members of the organization
- Communication patterns
- Formal and informal roles

The policies of an organization reflect its overall cultural climate. Two examples of how this climate can positively affect a project are seen in the approaches taken by the Florida Power and Light Company and the Arizona Power Service Company on their respective nuclear plants. Florida Power and Light established a special office in Bethesda, Maryland, near NRC headquarters and staffed it with engineers to facilitate exchange of information with the NRC during the St. Lucie Unit 2 nuclear plant licensing process. Senior management of Arizona Power Service (APS) established the following policies concerning the NRC:

> Don't treat NRC as an adversary; NRC is not here to bother us—they see many more plants than the licensee sees; inform NRC of what we (APS) are doing and keep everything up front; and nuclear safety is more important than schedule.[14]

This type of corporate attitude prompted the following conclusion from the NRC: A characteristic of the projects that had not experienced quality problems was a constructive working relationship with and understanding of the NRC.[15]

In the projects studied by NRC there appeared to be a direct correlation between the project's success and the utility's view of NRC requirements. More

---

[14]Improving Quality and the Assurance of Quality in the Design and Construction of Nuclear Power Plants, NUREG-1055, U.S. Nuclear Regulatory Commission, Washington, D.C., May 1984, pp. 3–21.

[15]Ibid.

successful utilities tended to view NRC requirements as a minimum level of performance, not maximum, and they strove to achieve increasingly higher, self-imposed goals. This attitude covered all aspects of the project, including quality and quality assurance.

Some of the actions taken for the improvement of the ambience for project management in IBM follow[16]:

- IBM's corporate executives launched a worldwide corporate initiative in 1996 to increase the core competence in project management and to ensure that IBM took a consistent approach to project management across all of its business processes around the world.
- A corporate Project Management Center of Excellence (PMCOE) was established.
- A new worldwide project management curriculum was developed for project managers at basic, intermediate, and advanced levels.
- IBM developed and deployed project management training for all project team members, senior managers, and executives.
- IBM benchmarked the best project management practices with other corporations and government agencies to ensure new approaches were tried and adopted.
- An internal project manager certification process was established.
- A project management mentoring program was initiated.

20.7 THE PROJECT CULTURE

Each project has a distinct culture reflecting in part a universal culture found in all projects. Some insight into this universal culture can be found in what follows.

The project team is an organizational entity devoted to the integration of specialized knowledge for a common purpose: the delivery of the project results on time and within budget to support organizational strategies. The project team must be organized for creativity and innovation to emerge and grow; the team must be organized as a force for continuous improvement and constant change in positioning the enterprise for dealing with its changing products/services and processes in a changing global marketplace. Appointment and empowerment of a project team are an explicit recognition that creativity and innovation to bring about change in products/services and organizational processes are both possible and essential to organizational survival.

In a review of the culture of several highly successful teams, it was found that these elite teams talked about working together all the time, and they performed almost without individual egos. Shared interests, a lack of individual egos, and the power of team trust and absolute loyalty characterized high-performing teams such as the teams that capped the raging Kuwaiti oil wells after the Gulf War. This

[16]Sue Guthrie, "IBM's Commitment to Project Management," *Project Management Journal,* vol. 29, no. 1, March 1998, pp. 5–6.

team acted as one, no one bossed everybody around, and the teamwork was at a rarefied level a group of people acting as one.[17]

Every organization has to provide for the means to maintain surveillance over the real and potential changes in its environment—and then it has to design the means for the organization to manage the change needed to remain competitive. What this means is that the organization has to have the discipline to consider abandoning those products, services, and processes that are currently successful, and to provide the means for an orderly, disciplined, and systems strategy to develop new organizational initiatives in those products and services provided to customers, and in the organizational processes by which those initiatives come forth. Several strategies are needed to bring a project focus to the management of change in the enterprise: First, enhance the organizational culture so that people at all levels and in all specialties are encouraged to bring forth ideas for improvement in their areas of responsibility. Second, develop an organizational culture that seeks to abandon that which has been successful through the continuous improvement of existing products/services and processes. Third, become a "learning" organization through the explicit recognition that all organizational members will have to retrain and relearn new technologies to escape obsolescence. Fourth, organize the enterprise's resources so that explicit opportunity is available to bring an organizational focus (a project focus) to the development and implementation of new organizational initiatives that will bring forth new products/services and processes. Fifth, provide a strategic management capability by which organizational leadership is proactive in providing the resources, the vision, and the discipline to strategically manage the future through the use of product and process projects.

The project team, a "body of companions" dedicated to the creation of something that does not currently exist in the enterprise, provides for a way of decentralizing the organization of resources to deal with change. The team members represent those different specialties needed to create value to satisfy the needed change in products, services, and processes. The nature of the task needed to bring about the change determines the organizational membership on the "body of companions." The organizational membership in turn influences the culture of the project team and, to a certain extent, that of the participating stakeholders as well. Because each member of the project team comes from specialized areas of the enterprise and represents parochial areas of expertise, the objectives, goals, and strategies of the project team must be unequivocal and crystal-clear to all the project stakeholders. Only focused project objectives, goals, and strategies will hold the team together and enable it to create something that does not currently exist in the enterprise, in an efficient and effective manner.

The culture of the project team is a powerful force to hold the team together through its life cycle and through the inevitable pressures that buffet the team as new people join the team, old ones leave, and relationships with the project stakeholders vacillate. A project team culture is a community, a pattern of social interaction arising out of shared interest, mutual obligations, cooperation, friendships, and work challenges.

---

[17]Kenneth Labach, "Elite Teams Get the Job Done," *Fortune,* February 19, 1996, pp. 90–99.

What can the project manager do to improve the culture of the project team? Of course, there are no magical solutions, but a few suggestions are in order to help strengthen the team's culture:

- Keep the team members regularly informed on the status of the project, including both the good and the bad news. This should be done at the regular review of the project teams' work.
- Promote the sharing of ideas, problems, opportunities, and interests among the team members particularly with those team members who are new to the project. Give these new team members a sense of belonging at the earliest time.
- Have social activities for the team, such as informal lunches, coffee breaks, dinners, and trips to contractor's plants or to competitive projects. Do not overdo this and do not interfere with the personal "off-duty" time of the team members.
- Cultivate the use of first names or nicknames on the team.
- Limit the use of language and demeanor that puts a hierarchical stamp on the team and its work.
- As a team leader, advise, coach, mentor, prompt, and facilitate as much as possible a team environment which comes across as one in which people can be supported, encouraged, rewarded, and challenged with work and social interactions.
- Keep the people informed on what competitors are doing and what their competitive threat could mean to the project team.
- Work at creating a sense of importance and urgency to the project and its work. Make the most of having senior executives visit the project team and be briefed by the team members on the work they are doing.
- Reduce the formality in dealing with the team members.

The project manager should be constantly aware that he or she has to maintain a balance within the team's culture that produces winning results, keeps people motivated and reasonably happy, and allows the people to accomplish their individual goals and aspirations as well as the objectives and goals of the project. Projects and cultures change. But sometimes people do not want to change.

20.8 WHY CHANGE?

Individual and group behavior in an organization is controlled as much by the basic relationship that people have in the organizational culture as by anything else. It includes organizational policies and procedures, and how perceptions, rules, and expected behavior are carried out.

Businesses are organized to exploit the profitability of their products and services. The process of innovation can help improve that profitability. But there is a larger dimension of innovation—the abandonment of old ways and the creation of new products, services, and organizational processes needed to bring something new to the customers. To innovate means to challenge the existing order, the prevailing viewpoint, and to assume the risk and uncertainties and the enmity of those who wish to preserve the status quo, who encourage the traditional viewpoint, who postpone evaluation of existing strategies, who tolerate mediocrity and even failure because of a fear of what change might bring. We have a litany of old saws that serve to protect us and rationalize the status quo:

- Don't rock the boat.
- The way to get along is to go along.
- Why change?
- I'm only a couple of years from retirement.
- What we are doing now is good enough.
- I like things the way they are.
- The good old days were the best days.

Innovation cannot be fostered in the same way that improvement in organizational efficiencies can be fostered. This is a mistake that many managers make. Managers have to run an efficient business, make a profit, and at the same time provide for the concurrent development of new products, services, and organizational processes. Responses of managers in their attempts at remaining competitive include the following diverse strategies:

- Do nothing, with the hope that the traditional products and services offered will be adequate.
- Respond with defensive strategies of cutting costs, reducing the number of products and even product lines, and emphasizing the most profitable products and services.
- Make innovation a way of life in the organization, and energize and empower every member of the enterprise to look for new and better ways of doing things, both in doing things for the internal customers and in providing products and services to outside customers.
- Work at building a culture that encourages and rewards creativity and innovation from the highest to the lowest levels of the organization.
- Have senior management provide a leadership model in encouraging creativity and innovation, and tune the organizational management and cultural systems to make innovation a way of life to be accepted by everyone at every level in the enterprise. Everyone in the organization is protected from any criticism of any sort for brainstorming and being on a constant quest for new ways of working and serving the organization.

20.9 THE CONSTANCY OF CHANGE

Corporations and other types of organizations are seeing the beginning of the end for doing business under the traditional "command-and-control" management style. Organizational hierarchies with explicit chains of command are being redesigned and realigned. Managers in organizations today are seeing their traditional roles being challenged. An article in *Fortune* magazine brought into focus some of the current changes affecting managers and their roles. According to *Fortune,* call these new nonmanagers "sponsors, facilitators anything but the 'M' word. They're helping their companies and advancing their careers by turning old management practices upside down."[18]

Rapid change and the demands of organizational stakeholders to include customers, suppliers, workers, unions, local communities, and such vested groups have helped change the role of managers to one of being able to provide an organizational context in which decisions are made and executed through a "consensus and consent" management style rather than the traditional and antiquated command-and-control management style.

Table 20.1 summarizes the changes in management and leadership of the old world of "command and control" and the new world of "consensus and consent." Organizational structures are becoming flatter, and many "middle managers" have found their supervisory roles and their work in gathering, processing, and transmitting of information becoming superfluous partly through the emergence of computers and more sophisticated information systems. Challenges from stakeholder groups, such as institutional investors, have started a trend of greater involvement by outside directors who want corporate managers to be more responsive to stakeholders and less responsive to some internal, traditional order of managing the company. Managers in the future will likely see continued changes in their roles, motivated by pressures from key stakeholders through the board of directors, and from the workers who can and will contribute to the management of the enterprise.

The new nonmanagers who are emerging will be called something different than managers. New titles are coming forth, such as leader, facilitator, coach, sponsor, mentor, and adviser, titles that suggest a role far from the traditional command-and-control model now becoming obsolete.

The evolving theory and practice of project management have played a major role in demonstrating that a management philosophy of consensus and consent is workable and is more in tune with what people want in today's organizations. Organizations that have used project management over many years have seen a subtle change in attitudes about what managers should be responsible for and what form their exercise of authority has taken. Project managers who have had to operate in the context of the matrix organizational design have had to develop strength in their exercise of de facto authority, because many times their de jure or legal delegation of authority has been insufficient to get the job done.

20.10 PROJECT MANAGEMENT ACTIONS

In a very real sense of the word, project managers have to be change managers, and at the same time participate with other organizational managers in designing and facilitating a culture that brings out the best in people.

[18]"The Non-Manager Managers," *Fortune,* February 22, 1993, pp. 80–83.

TABLE 20.1 Changes in Management and Leadership Philosophy

The old world: command and control	The new world: consensus and consent
• Believes "I'm in charge."	• Believes "I facilitate."
• Believes "I make decisions."	• Believes in maximum decentralization of decisions.
• Delegates authority.	• Empowers people.
• Executes management functions.	• Believes that teams execute management functions.
• Believes leadership should be hierarchical.	• Believes that leadership should be widely dispersed.
• Believes in theory X.	• Believes in theory Y.
• Exercises de jure (legal) authority.	• Exercises de facto (influential) authority.
• Believes in hierarchical structure.	• Believes in teams and matrix organizations.
• Believes that organizations should be organized around functions.	• Believes that enterprises should be organized around processes.
• Follows an autocratic management style.	• Follows a participative management style.
• Emphasizes individual manager's roles.	• Emphasizes collective roles.
• Believes that a manager motivates people.	• Believes in self-motivation.
• Stability.	• Change.
• Believes in single-skill tasks.	• Believes in multiple-skill tasks.
• Believes "I direct."	• Believes that a manager leads, as opposed to directs.
• Distrusts people.	• Trusts people.

Source: David I. Cleland, *Strategic Management of Teams* (New York: Wiley, 1996), p. 249.

The project manager who is able to function as a project leader along with the other managers (leaders) of the parent organization is responsible for arranging conditions conducive to a creative and disciplined culture supportive of project teamwork. Certain actions can help develop and maintain such a culture.

First, design and implement an ongoing disciplined approach in planning, organization, and control of the project management system. This is a fundamental first so that team members have a model to use in managing the project. This is one of the first task-related actions to let people know where they stand and what is expected of them on the project team.

Provide as much leeway as possible for the project team members to try new ways of getting their jobs done. This includes the encouragement of experimentation without fear of reprisals if mistakes are made.

Give team members a reasonable amount of attention through project reviews, strategy meetings, and checking in on a regular basis to see how things are going. Too much attention can be counterproductive and might be interpreted as meddling.

Too little attention might be construed as disinterest on the part of the project manager. Make sure the members of the team understand in specific terms their authority, responsibility, and accountability so that they know what's expected of them on their work package. If team members know their assigned work packages, there is less likelihood that they will become overburdened with the minute details of their jobs. Creativity requires the opportunity to reflect on the totality of the job being done. If members are too busy with details, it's easy for them to miss the big picture.

Give project team members part ownership in the decisions affecting the project. When the team members know that their opinions are valued on project matters, their self-confidence is bolstered and the chances for creative thinking are enhanced. By encouraging participation in decision making on the project, the general culture of the project team will be improved. When team members see their work on the project work packages as challenging and the goals as realistic, they are more likely to exhibit creative behavior and be happier in their work.

Maintain proper oversight of the project. The project manager maintains oversight of the project by watching and directing the major activities and course of action of the team. Most team members who have maximized their creative potential would prefer a low level of oversight by the project manager. The oversight effort should be focused on those activities most directed to achieving project results.

Encourage the use of creative brainstorming approaches to solve the many unstructured problems that arise during the project's life cycle. Many different types of unstructured solutions will be needed to solve these problems. The use of a single routine problem-solving approach would be inappropriate because such an approach too often assumes there is one "best and correct" solution. Projects create something new. Innovation and creativity are musts to deal with something so new.

Provide timely feedback to the project team. In this way, the project manager will encourage open communication in the project's culture. If the team members sense that key information is being withheld or that the project leader is less than candid with them, dissatisfaction and disenchantment could result, thus adversely affecting the project culture. If feedback is provided too late for the team members to make adjustments, the project could suffer and the individuals could be discouraged.

Provide the resources and support to get the job done. This is another fundamental and positive contribution that the project manager can and must make to the project culture. Adequate resources are required to do the job and to facilitate a creative and innovative culture. A shortage of resources may allow people to use their innovative and creative skills, but in the long run adequate resources are needed to ensure a supportive culture.

Finally, recognize the key "people-related" cultural factors and utilize them. These people-related factors include:

- Rewarding useful ideas
- Encouraging candid expression of ideas
- Promptly following up on team and member concerns
- Assisting in idea development

- Accepting different ideas; listening to that team member who is "marching to a different drummer"
- Encouraging risk taking
- Providing opportunities for professional growth and broadening experiences on the project
- Encouraging interaction with the project stakeholders so that there is an appreciation by the team members of the project's breadth and depth

The growing institutionalization of project management means that the concepts, processes, and techniques of this discipline are accepted and practiced as key strategic behavior in the enterprise. Project management is simply a way of life in dealing with needed changes in organizational products, services, and processes. A stream of projects in varying phases of their life cycle flowing through the enterprise utilizes resources to prepare the enterprise for survival and growth in its marketplace. There is an explicit acceptance of the use of a project management system as described in Chap. 4. The culture of the organization accepts and supports the use of project management and displays the following characteristics:

- There is an excitement about project management as a way to deal with changes in the enterprise as contemporaneous project management concepts, processes, and techniques are being used throughout the organization.
- Products and services have been commercialized sooner, at lower cost, and are of higher quality, leading to customers who have a high degree of satisfaction with the products and services being used in their organizations.
- Appropriate organizational strategies, policies, procedures, and organizational design initiatives have been developed, have been communicated, and are understood by the members of the organization.
- There have been extraordinary efforts to clarify the relative authority, responsibility, and accountability endemic to the matrix organization. Workshops to clarify the individual and collective roles in the matrix organization are conducted as new project teams are appointed and start to carry out their work in the enterprise.
- Senior managers and other managers of the enterprise recognize the value of project teams in dealing with the cross-functional and cross-organizational opportunities that face the organization. Such managers are fully committed to support such teams through the assignment of appropriate resources. In addition, such managers review, on a regular basis, the results being attained by the teams.
- Managers, team members, and professionals work hard at communicating about the strategies and the results that are being carried out on the stream of projects in the enterprise. People are free to express both their ideas and their concerns involving the projects. There are few if any "hidden agendas" during the project review meetings and the atmosphere in such meetings is informal, comfortable, and relaxed.

- A proactive educational and training program is under way on a continual basis to upgrade the knowledge, skills, and attitudes of people in the theory and practice of project management, as well as the use of alternative teams in the management of change.
- Appropriate merit evaluation systems are in place that recognize and reward individual and team performance in the design and development of projects in the organization.
- Project management is so pervasive in the culture of the enterprise that it is simply recognized as "the way we do things around here!"
- Excellent performance as a project manager and experience on a project team are recognized as vital steps in progressing from a professional to the management career field.

20.11 THE TRUST FACTOR

A key challenge to the project leader is to manage the team members and the other project stakeholders so that one of the key characteristics in the team's culture is trust—a security that one feels concerning the integrity, ability, and character of people associated with the project. To trust is to have confidence in the abilities and personalities of the team. To trust the team is to feel that team members will be responsive and responsible in the making and implementing of decisions affecting the team, the project, and the other stakeholders. Trust must exist between the team and higher management, and these managers must have a vision of how the project fits into the larger goals and objectives of the enterprise.

Trust is a condition in a relationship that takes years to develop, and then it can be damaged or destroyed by a single act of imprudence. Trust is easy to violate; it requires that members of the project team "open up" to each other and let each other know "where they stand." Trust is particularly challenging to develop and maintain on a project team, where people from different disciplines have to pull together for the common project goals and projects. Jack Welch, CEO of General Electric, said in GE's 1991 annual report that the corporation will be built on mutual trust and respect and that "trust and respect take years to build and no time at all to destroy." The bottom line of trust is that a person's word is his or her bond.

High-performance teams consciously develop a strong foundation of professional trust. They:

- Trust one another.
- Count on each other.
- Rely on constant top-quality commitment.
- Promise only what they can deliver.[19]

[19]Jaclyn Kostner and Christy Strbiak, "How to Get Breakthrough Performance with Teamwork," *PM Network*, May 1993, pp. 24–26.

By becoming sensitive to the importance of trust, the project manager can enhance the positive aspects of the culture.

20.12 CULTURE AND PROJECT EXTENSIONS

Sometimes projects are extended even when they might better be abandoned. These unwise extensions often are caused by the cultural factors present in the parent organization and on the project team. Managers who have had track records of success will hang onto a near losing or fatal project simply because they are used to winning and don't want the project to fail. With such an attitude, resources will be poured into the project to make it work. Also, people tend to see what supports their beliefs, even to the point of biasing cost and schedule estimates to support their views. Investing more resources in the project is perceived as a preferable alternative to admitting failure. To fail would be to admit to others that the project could not be handled; hanging on in the face of mounting project losses seems to make more sense. When a person has become a project champion, it's easy to rationalize the defense of that project despite growing concerns about its feasibility and eventual outcome. Social and cultural pressures also tend to encourage managerial persistence to "stay the course" and to "stick to your guns" to exhibit strong leadership.

Administrative inertia also can occur when the cancellation of the project and the divestment of resources that would follow are perceived as politically threatening. Institutionalization of the project in the organization's strategy can be another powerful persuader for continuance. Indeed, an organization can become so enamored of a project that the costs of terminating it are perceived as much greater than a persistent continuation even though the project's linkage with the organization's strategic purpose has become spurious. Managers concerned with projects should be aware of the psychological, social, and cultural forces that can influence their viewpoint and can hamper their rational judgment on when it's best to "pull the plug" on the project.[20]

20.13 INFLUENCING THE TEAM'S CULTURE

A project manager can help develop a supportive culture for the team through the team-building process. Team building aims at developing a team's work competencies, such as meeting objectives, goals, and schedules. Other important parts of team building are developing the team's ability to resolve conflicts, building trust with stakeholders, and encouraging effective communication. A supportive team culture helps members execute these competencies and thus enhance team performance.

[20]Material in this section has been drawn from Barry M. Staw and Jerry Ross, "Knowing When to Pull the Plug," *Harvard Business Review,* March–April 1987, pp. 68–74.

What can the culture do for the project team? Several writers have noted the following:

- Culture creates social ideals which help guide behavior.[21]
- Culture sends messages to insiders and outsiders about what the organization stands for.[22]
- Culture helps align individual and organizational goals and values.[23]
- Culture serves to control, monitor, and process beliefs and behavior in the organization.[24]

Somehow, the project manager has to sell a dream of the project's objectives to the project team and, in so doing, help facilitate a supportive, successful project management culture.

20.14 CONFLICT

When people are working together, circumstances are ripe for controversy, disagreement, opposition, and intellectual struggles as team members pursue their individual and collective roles on the team. How this inevitable conflict is dealt with will impact the project and organizational culture.

Conflict is an inevitable force to be contended with in any organizational effort. On a project team composed of people with different specialist skills, the opportunity for conflict is ready-made. Disagreements over the use of functional input to the project can occur; people who are fluent in a functional specialty can have problems in being able to communicate with other functional specialists who have their functional parochialism and beliefs. Interpersonal conflict where people don't want to get along because of personal prejudices, ethics, morals, value systems, and the like can be a basis for ongoing conflict.

Managers and professionals work in an environment in which conflict is to be expected, as internal and external stakeholders seek support from the resources that are dedicated to the project. One study estimated that the average manager spends over 20 percent of his or her time in dealing with conflict.[25]

Inherent conflict in the management of the project team can be worked to an advantage through a subtle forcing of discussion and debate in the resolution of the disagreements causing the conflict. Out of the discussion can come the opportunity

---

[21]M. R. Louis, "Organization as Cultural Bearing Milieux," in L. R. Pondy et al. (eds.), *Organizational Symbolism,* vol. 1: *Monographs in Organizational Behavior and Industrial Relations* (Greenwich, Conn.: Jai Press, 1983), pp. 39–54.

[22]T. C. Dandridge, "Symbols' Function and Use," in L. R. Pondy et al. (eds.), *Organizational Symbolism,* vol. 1: *Monographs in Organizational Behavior and Industrial Relations* (Greenwich, Conn.: Jai Press, 1983), pp. 69–79.

[23]R. Harrison, "Strategies for a New Age," *Human Resource Management,* vol. 22, no. 3, Fall 1983, pp. 209–235.

[24]A. L. Wilkins, "Organizational Stories as Symbols," in L. R. Pondy et al. (eds.), *Organizational Symbolism,* vol. 1: *Monographs in Organizational Behavior and Industrial Relations* (Greenwich, Conn.: Jai Press, 1983), pp. 81–92.

[25]K. W. Thomas and W. H. Schmidt, "A Survey of Managerial Interests with Respect to Conflict," *Academy of Management Journal,* vol. 10, 1976, pp. 315–318.

to evaluate the alternative options to be considered in the resolution of the conflict and to learn better how to work together as an effective team. Other general benefits of conflict include the development of a team culture in which there is a motivation to work together in seeking consensus in resolving conflict and in the management of the project resources. Other benefits include a better understanding of individual and collective roles in the organization of the project team and other stakeholders. Also, the resolution of conflict can help get people acclimated to the dynamic nature of the project and the demands that alternative stakeholders can place on the project, which will at times be contrary to each other.

During the formative phases of the project team, an important issue should be raised: How will we deal with and resolve conflicts on this team? By getting members of the team to talk about how they would like to deal with the inevitable conflicts, there is a greater chance that the conflicts will be properly managed and resolved. Conflicts that are successfully resolved at lower levels of the team are, in general, the way to go. Senior management involvement should be infrequent in part because the senior managers would not be familiar with the details of the conflict. Only when the team is unable to resolve the conflict or the conflict has higher implications in the organization should senior management become involved.

20.15 CODE OF ETHICS FOR PROJECT PROFESSIONALS

It is an accepted fact that all professional organizations and their members have a code of ethics by which individuals may be guided to the correct behavior in professional dealings with others. Professional organizations that have certification programs will also have a code of ethics—rules of expected behavior—for those individuals involved in the professional designation.

These codes of ethics provide guidance to the professional on the basis of obligations to others when conducting business in one's field of endeavor. These guiding tenets are typically broad in nature without detailed descriptions of different situations. Professionals must maintain an awareness of their respective obligations and meet those obligations through mature judgment and application of professional conduct.

A code of ethics may overlap the law of the country or community, but it can never conflict with the law. A code of ethics may often conflict with the accepted norm of professional conduct when applied to individuals in different countries. For example, it is often expected that a person will make some form of payment to obtain work or to conduct work within a country. This payment by U.S. and western European standards would be considered a "bribe" that contributes nothing to a project or its product. In many countries this payment is culturally accepted and expected.

A code of ethics provides a sharp focus on those owed an obligation and avoids establishing relationships between the professional and other individuals or groups. It would diffuse the intent of a code of ethics to insert such requirements as "one

must donate 10 percent of the salary to charity to maintain one's certification as a pilot." Charity, although commendable, is totally irrelevant to a person's obligations as a pilot and the qualifications that a pilot must have.

Table 20.2 is the code of ethics for the American Society for the Advancement of Project Management. This sample outlines what the society believes is important in order to be a member and the standards of professional behavior that one should strive to achieve.

TABLE 20.2 asapm Code of Ethics

We, the members of asapm, in recognition of the importance of our profession, in affecting organizational value to customers, principals, public, or shareholders throughout the world, and in accepting a personal obligation to our profession, its members, and the communities we serve, do hereby commit ourselves to the highest ethical and professional conduct and agree:

1. to maintain high standards of integrity and professional conduct, and to accept responsibility for our actions;
2. to avoid real or perceived conflicts of interest whenever possible, and to disclose them to affected parties when they do exist;
3. to be honest in representing our project management capability, and realistic in the application of project management;
4. to reject bribery in the conduct of our professional responsibilities;
5. to improve and promote the understanding of project management, its appropriate application, and potential consequences;
6. to maintain and improve our project management competence;
7. to seek, accept, and offer honest criticism for the purpose of enhancing project management body of work, and to credit properly the contributions of others;
8. to treat all persons fairly regardless of such factors as race, religion, gender, disability, age, national origin, or those not mentioned;
9. to avoid injuring others, their property, reputation, or employment by false or misleading words or action;
10. to assist colleagues and co-workers in their professional development and to support them in following this code of ethics.

Source: American Society for the Advancement of Project Management (asapm), Colorado Springs, Colorado. Permission granted to reproduce in its entirety.

A code of ethics should be used to establish the cultural norm as it documents obligations to others. The code may be used in training individuals to show what is considered acceptable behavior. It may also be used to discuss situations that challenge one's ethical practices and to identify means of avoiding unethical actions.

When dealing with others, customers for example, the code of ethics can serve as a statement of "this is what you can expect from us." Any violations or apparent violations of the code of ethics will erode the customer's confidence in the person as well as the trust in the professional's adherence to an adopted standard of conduct. Variances from a code of ethics will surely result in an expectation not met to the person or organization owed that obligation.

20.16 TO SUMMARIZE

Key points made in this chapter include:

- A culture is the set of refined behaviors that people have in the society to which they belong, whether it is a nationality, a family, an enterprise, or a project team.
- More specifically, an organizational culture is the environment of beliefs, customs, knowledge, practices, and conventionalized behavior of a particular social group.
- Organizational cultures are expressed in slogans, plans, policies, procedures, ethics, morals, and beliefs, to name a few, as well as in the leader and follower style practiced in the organization.
- The culture of an enterprise and the culture of the project team in that enterprise are usually mutually supportive.
- A few examples of organizational cultures were given to show the reader how some organizational leaders make a deliberate effort to design and reinforce the culture of their enterprise.
- A representative list of the cultural features that are developed out of the strategies of an enterprise was presented.
- Elite project teams reflect a distinct culture, reinforced by certain behavior that builds team cohesiveness, integrity, trust, and loyalty.
- A few suggestions were offered to help the project manager improve the culture of the project team. It was pointed out that these suggestions were not magical solutions but if followed could help reinforce a positive culture in the project team.
- Attitudes toward product, service, and organizational process change in the enterprise can be influenced by the leadership style of the managers.
- A comparison of the changes in management and leadership philosophy in the "command-and-control" and the "consensus-and-consent" modes was presented.
- The growing institutionalization of project management in organizations does impact the culture of the organization.
- Trust, the ability to rely on the integrity, ability, and character of a person or thing, is a much-desired characteristic of a project team.
- High-performance teams consciously develop a strong foundation of professional trust.
- Conflict is to be expected in the operation of any team and should be dealt with in a forthright and empathetic manner, leading to a solution that is a "win-win" situation for all concerned.

- The social and intellectual characteristics of a project team are the basis for how well that team embodies trust, conviction, commitment, and loyalty into the team's way of working.
- Benjamin Disraeli stated that "all power is a trust; that we are accountable for its exercise." It is no different on a project team.

20.17 ADDITIONAL SOURCES OF INFORMATION

The following additional sources of project management information may be used to complement this chapter's topic material. This material complements and expands on various concepts, practices, and the theory of project management as it relates to areas covered here.

- Jimmie L. West, "Building a High Performing Project Team," chap. 18 in David I. Cleland (ed.), *Field Guide to Project Management* (New York: Van Nostrand Reinhold, 1997).
- B. Baker and R. Menon, "The Power of Politics: The Fourth Dimension of Managing the Large Public Project,"; W. B. Derrickson, "St. Lucie Unit 2: A Nuclear Plant Built on Schedule,"; and Michael S. Lines, "Learning the Lessons of Apollo 13," in David I. Cleland, Karen M. Bursic, Richard J. Puerzer, and Alberto Y. Vlasak, *Project Management Casebook,* Project Management Institute. (Originally published in *Proceedings,* PMI Seminar/Symposium, Vancouver, Canada, October 1994, pp. 830–833; *Proceedings,* PMI Seminar/Symposium, Houston, 1983, pp. V-E-1 to V-E-14; and *PM Network,* May 1996, pp. 25–27.)
- Edgar H. Schein, *Organizational Culture and Leadership,* 2d ed. (New York: Jossey-Bass, 1997). This second edition of a basic book on organizational culture contains new research and new case examples. The book defines organizational culture and expands on the concept and its application to the challenges of corporate management. The concept put forth by Schein sheds light on the workplace. He has been able to transform the abstract concept of culture into a practical tool that managers and students can use to gain an appreciation and understanding of the dynamics of organizational change and cultural considerations.
- John P. Kotter and James L. Heskett, *Corporate Culture and Performance* (New York: Free Press, 1992). This book describes a study by the authors on culture and the role it plays in the capacity of major corporations to succeed or fail in the marketplace. The study is based on empirical rather than anecdotal evidence, gathered from more than 200 blue-chip enterprises in 22 industries. The authors argue that an adaptive culture that aligns an organization's interest with those of key constituencies has a better chance of succeeding in a competitive environment. As an example the authors point out that Kmart's lack of a customer-service ethos cost it dearly in competition with Wal-Mart. Kotter and

Heskett provide the first comprehensive assessment of how the culture of an enterprise can influence its economic performance.

- Drew B. Fetters and John Tuman, "Project Management—Agent for Change: Building a New Culture for a Nuclear Engineering Organization," Project Management Institute Seminar Symposium, October 1989, pp. 589–599. This paper examines the impact of dramatic change on a company's nuclear engineering department, and examines the role of project management in helping build a world-class nuclear organization. The company brought about major changes in the technical, economic, political, and cultural elements of its nuclear organization. Four items were identified in the culture of the enterprise that needed changing and reinforcement: (1) values; (2) communications; (3) change in project management, and (4) training initiatives.
- Dragon Z. Milosevic, "Echoes of the Silent Language of Project Management," *Project Management Journal,* vol. 30, no. 1, March 1999, pp. 27–39. This paper examines the influence of cultural values on project management, interprets the silent language, and shows how to use the language for successful multicultural project management. Milosevic suggests that a strategy to deal with multicultural project management is to understand your own culture and silent language, understand the culture and silent language of your team members, identify cultural and language gaps, and avoid problems or resolve the gaps.
- Richard Bauhaus, Peggy Bauhaus, and Shawna Bauhaus, "Cultural Communication on Global Project Teams," Project Management Institute Seminar/Symposium, Atlanta, Ga., October 1989, pp. 432–440. The authors' purpose is to recognize the challenges and roadblocks caused by the culture variable on global teams. They further identify the best practices used by managers and team members to communicate across cultures represented by the team members. They further suggest some strategies used to consider cultural differences in order to accomplish successful project completion. The authors conclude that managers and participants on global teams must include the cultural dimension in developing and implementing projects.

20.18 DISCUSSION QUESTIONS

1. Define culture in terms of its use in describing an organization.
2. How do organizational beliefs and values affect corporate culture? How do senior managers' values and beliefs influence employee behavior?
3. What kinds of corporate documentation can assist in understanding the culture of an organization? How?
4. Describe an organization from your work or school experience. What was its culture like? Explain.
5. How is the project management culture exhibited in organizations? Explain.

6. Discuss some of the factors that influence cultural features. What role does each factor play in defining culture?
7. Discuss some of the cultural factors that affect teamwork. How can management develop an effective team? What people-related factors must be considered?
8. In addition to the project team culture, what other cultures must the project manager be concerned about? Why?
9. What can culture do for the project team in terms of meeting individual as well as organizational goals? Explain.
10. What are some of the cultural characteristics of successful teams?
11. How can the project team's management of the first major problem help the project manager assess the potential of the team? Explain.
12. Discuss the importance of the role of project culture in the overall effectiveness of the project.

20.19 USER CHECKLIST

1. Define the culture of your organization.
2. What senior management values and beliefs have affected the definition of culture in your organization? How?
3. What corporate documentation exists that has influenced the culture of your organization? Explain.
4. Define the culture of the various projects within your organization. Do project cultures differ from the overall organizational culture? Why or why not?
5. What features (such as leader-and-follower style) have influenced the culture of your organization? How?
6. Do project managers pay enough attention to individual members of the project team? How has this affected the operating culture?
7. What factors play a role in the creation of effective teamwork in your organization? What inhibits teamwork?
8. What other cultures must project managers understand? Do the project managers of your organization interact effectively with the various cultures they encounter? Why or why not?
9. Does the project culture in your organization combine individual and organizational goals? Why or why not?
10. Compare some of the successful and not so successful projects within your organization. What were the cultural characteristics of these projects? Explain.
11. How do project team members handle conflict on the project? What does this indicate about the effectiveness of the project team? Why?

12. Do the project and senior managers of your organization understand the importance of the role of culture in the overall success of a project? Why or why not?

20.20 PRINCIPLES OF PROJECT MANAGEMENT

1. A project culture is the environment of beliefs, customs, knowledge, practices, and conventionalized behavior of a particular social group.
2. Project cultures are usually explained in terms of values and beliefs, and the behavior of members of the sponsoring organization and the project team itself.
3. The culture of an enterprise and the culture of a project within that enterprise are mutually interdependent.
4. It is possible to identify common cultural features that positively and negatively influence the management of a project.
5. Specific measures can be taken to influence the culture of the enterprise and the projects within that enterprise.
6. A major element of a project culture is the trust team members have, as well as the trust of the sponsoring organization.
7. There are key characteristics to be found in successful and in failing projects.
8. Conflict is an inevitable force to be contended with on any project.
9. Elite project teams reflect a distinct culture, reinforced by certain behavior that builds team cohesiveness, integrity, trust, and loyalty.

20.21 PROJECT MANAGEMENT SITUATION—CONDUCTING A CULTURAL ASSESSMENT

In this chapter, E. B. Taylor's definition of culture was given to describe the knowledge, beliefs, art, morals, laws, customs, and other capabilities and habits acquired as a member of a society. Culture can be equated with all those human behaviors that are transmitted from generation to generation through a learning process. Another way of looking at a culture is that it is all part of a system of action in a society.

Every organization, to some extent, has a culture that is inherent in that organization. We suggest that an enterprise that uses projects as building blocks in the design and execution of strategies has a distinct project culture, which is a reflection of the culture of the sponsoring enterprise.

What are the components of a culture? We suggest that there are three such components: (1) concepts, (2) activities, and (3) results. Culture is the learning of concepts and the transmission of such concepts through activities that produce desired results. Adopting these components to the management of a project, we have the following:

1. Concepts which are embedded in the strategy of the enterprise.
2. Activities carried throughout the management of the project.
3. Results, through the delivery of the project's objectives on time and within budget, which make a contribution to the strategic purposes of the enterprise.

What would be the rationale for trying to determine the culture of a project? We believe that the manner in which the project is managed is directly related to the culture of the sponsoring enterprise, as well as the past and current manner in which the project is being managed.

An effective, enduring change in the management of a project requires that the relevant cultures also be changed. How is a project's culture determined? An understanding of the prevailing culture of a project can be determined through the assessment of the following:

1. Concepts and ideas expressed in the policies, procedures, protocols, and strategies of both the sponsoring organization and the project itself. For example, the degree to which the sponsoring organization has defined and delegated authority and responsibility to a project team, and in the accuracy and completeness of the project plan. In addition, the degree to which contemporaneous state-of-the-art project management literature is reflected in the management of the project.
2. How the activities involved in the utilization of the project resources are carried out. How well a regular and complete review of the project's progress is accomplished. How effectively the organizational design for the project is being utilized.
3. Are the planned results for the project being attained? Are the project work packages being accomplished on time and within budget? Is the project likely to have a strategic fit when the results become part of the inventory of products, services, or organizational processes being offered by the sponsoring organization?

20.22 STUDENT/READER ASSIGNMENT

The student/reader is asked to select a project of their choice and perform a cultural assessment of that project through the collection of information about that project. Put the information in the appropriate components of a culture as indicated above. A couple of suggestions to get started are offered below:

1. *Concepts.* Definition of how the organizational design for the project is reflected in organizational policies and procedures. Is the project being managed from a project management system perspective?

2. *Activities.* How well the project team is working as an integrated body of professionals supporting the project.
3. *Results.* How well the work packages are being accomplished on time and within budget.

After the information about the project has been collected in the relevant components of the project's culture as indicated, above, prepare a description of the culture of the selected project. Good luck!

P · A · R · T · 7

NEW PROSPECTS

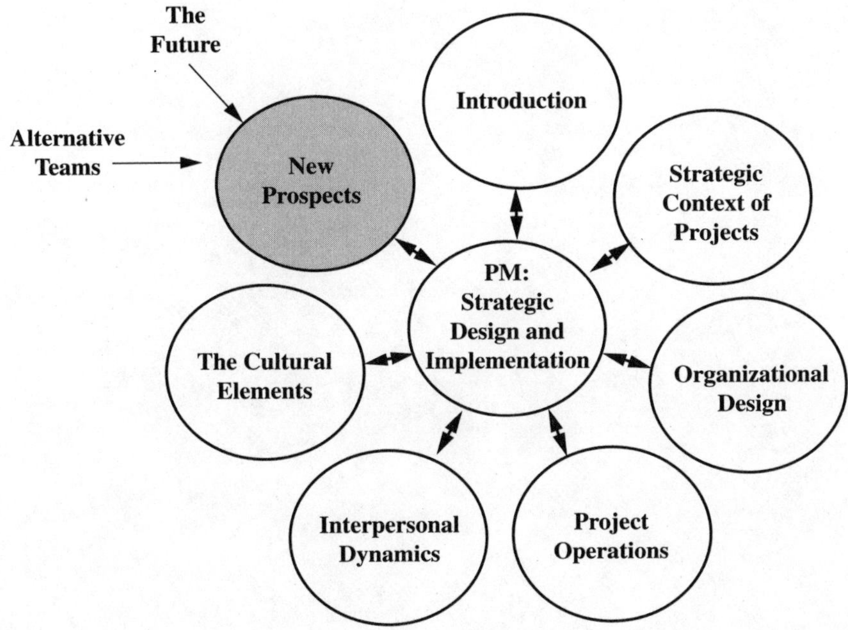

Teams

CHAPTER 21
ALTERNATIVE PROJECT TEAMS[1]

"The formation and use of teams is an art form for corporate America."
Business Week, *November 1, 1993, p. 160*

21.1 INTRODUCTION

Some uses of alternative project teams are described in Chap. 19, Continuous Improvement through Projects. Today, an approach to the management of many initiatives in the enterprise can be facilitated by the use of a series of different projects and project teams that serve different purposes for dealing with change in an organization. Some of these teams are similar to project teams; others go beyond the traditional use of project teams.

In this chapter, both traditional and nontraditional project teams are reviewed. The use of such teams centers around such matters as market assessment, competitive analysis, ascertaining organizational strengths and weaknesses, stakeholder evaluation, and senior-level decision making, to name a few. The chapter concludes with a look at how some of the traditional roles carried out by people in the organization have been changed by the use of alternative project teams.

21.2 A PLACE IN BUSINESS FOR ALTERNATIVE TEAMS

Teams are becoming commonplace. The use of teams in dealing with operational and strategic change is an idea whose time has come. Alternative teams are becoming key elements in the design of contemporary organizational forms. Cross-functional and cross-organizational work has become the norm for dealing with the development and execution of product, service, and process changes in today's enterprise. The use of alternative teams has come from the success that

[1]In preparing this chapter the authors drew from David I. Cleland, chap. 1, *Strategic Management of Teams* (New York: Wiley, 1996).

project management has had in the past and continues to have today in a wide variety of organizations. Industrial, military, governmental, educational, social, and ecclesiastical entities have a common need to know how to deal with the inevitable changes impacting their well-being and how to enhance their ability to survive in the unforgiving "marketplace" that they face.

The use of teams has modified the theory and practice of management. For individuals, the ability to serve as a contributing member of alternative teams and to provide leadership while serving on these teams has become a core competency directly related to their careers in today's enterprises. Professionals at all levels in today's organization are afforded extraordinary opportunities to learn about management and leadership, and the opportunities have never been better for these individuals to try out their mettle as a manager—and leader—in some capacity in today's organizations. Change, survival, and growth are the demands that condition everyone's behavior in today's organization, from the worker on the production line to the executive who must appreciate and maintain oversight over the work of different alternative teams in the enterprise.

The basic purpose of this chapter is to present a paradigm of the alternative teams that exist today in most well-run organizational entities. According to *Business Week,* the formation and use of teams has become a critical management skill. An editorial in the magazine noted that "those companies that learn the secrets of creating cross-functional teams are winning the battle for global market share and profits. Those that don't are losing out."[2]

In the preceding chapters of this book alternative forms of teams have been briefly described, usually within the project management context. Indeed, many of these teams have characteristics quite similar to those found in more traditional project teams. But there are subtle differences. In this chapter these alternative teams will be described more fully, particularly in their application. The reader who senses that such teams are primarily of the project management ilk is correct in that perception. On the other hand, some of these teams are different and are used for different purposes—yet all fall into the general category of a distinct organizational design, which helps the organization deal with the inevitable change impacting the enterprise.

21.3 TRADITIONAL AND NONTRADITIONAL PROJECT TEAMS

A traditional project team is one in which the mode of thought and behavior of the team is established by custom and usage derived from past practice, primarily from the construction and defense industries. These traditional project teams can be described in the following context:

- A substantial body of knowledge exists which describes how and why such teams exist and how they can be used for the development of products in varied industries.

[2]*Business Week,* November 1, 1993, p. 160.

- These projects typically involve the design, development, and production (construction) of physical entities that have a specific role in supporting customer products and infrastructures. A new weapon system, a new highway, and a new facility are examples of such physical entities.
- A distinct life cycle is found in these projects, starting with an idea and progressing through design, development, production (construction), and turnover to customers. These projects usually involve the development of after-sales support and service initiatives.
- Substantial financial, human, and other resources have to be marshaled for the conduct of these projects.
- Construction projects found in the world today are examples of these traditional projects.
- There tends to be a long history in the informal practice of using project management processes and techniques in conceptualizing and bringing the results of these projects into the customer's organization.
- Finally, when people think of teams in contemporary organizations, they tend to think of traditional project teams, because these are most visible in today's enterprises.

A nontraditional team has many of the characteristics of traditional teams. Yet these nontraditional teams have a life of their own and can be described in the following context:

- The organizational element with which these teams deal is already in existence and the purpose of the team is to improve the efficiency and effectiveness of that element, such as through process reengineering.
- The teams improve the efficiency and effectiveness of the organizational element, usually in terms of changing the processes that are involved.
- Although a conceptualization process is involved in setting up these teams, the work of such teams begins immediately in dealing with existing problems and opportunities that already exist in the enterprise.
- Although hardware considerations may be involved, the team typically deals with the improvement of the use of organizational resources through processes in meeting enterprise purposes.
- The "deliverables" of these alternative teams may be a report which recommends the design and implementation of resources committed to the improvement of the use of resources in accomplishing enterprise mission, objectives, goals, and strategies reflected in a new or improved process, policy, procedure, or plan of action.
- These alternative teams use much of the theory and practice of project management, but change its use, particularly in its application to diverse enterprise purposes.
- These alternative teams have vital linkages with the design and execution of both operational and strategic initiatives in the enterprise.

- Some of the alternative teams have brought about significant changes in the individual and collective roles carried out by people in modern organizations, which include major career opportunities for young professionals.
- These teams are having a marked influence in the cultures of organizations to which they belong.

John Tuman of Management Technologies Group, Inc., defines nontraditional projects using the following descriptors: First, the objectives of the project tend to be forced on top management. Second, the cost, schedule, and technical objectives of the projects tend to develop over the life cycle as work on the projects continues. Third, nontraditional projects typically deal with organizational change initiatives directed at improving processes, procedures, systems, structures, and culture. Nontraditional projects have a unique life cycle. Fourth, the resources on such projects are allocated on an as-needed basis. Fifth, the management of nontraditional projects tends to operate outside the corporate infrastructure, as vested stakeholders are dealt with on the project. Finally, the use of nontraditional projects tends to contribute to the development of a unique culture arising out of the workings and relationships of the project team.[3]

The use of these nontraditional teams has become a means of improving the global competitiveness of the enterprises to which they belong.

21.4 THE TYPES OF ALTERNATIVE TEAMS

Table 21.1 provides a classification of teams that differentiates between whether they are ad hoc or ongoing elements of the organization.

The work carried out by teams is described below:

- *Market assessment.* The discernment and development of an understanding of the possible and probable changes in the markets in which the enterprise competes.
- *Competitive assessment.* An examination and evaluation of the strengths, weaknesses, and probable strategies of competitors in their product, service, and organizational processes.
- *Organizational strengths and weaknesses.* An assessment of the strengths and weaknesses of the enterprise, particularly regarding its ability to create and deliver competitive products and services through effective and efficient organizational processes.
- *Benchmarking.* Review of the performance of the "best-in-the-industry" enterprises to determine what operational and strategic abilities enable them to perform so well.
- *Established strategic performance standards.* Identification, development, and dissemination of the performance criteria by which the enterprise's ability to produce results is assessed.

[3]John Tuman, "Project Management for Nontraditional Projects," Workshop Lecture, 12th INTERNET World Congress on Project Management, Oslo, Norway, June 9–11, 1994.

TABLE 21.1 Classification of Teams

Type team	Output contribution	Time frame
Reengineering team	Handles business process changes	Ad hoc
Crisis management team	Manages organizational crisis	Ad hoc
Product and product development team	Handles concurrent product and process development	Ad hoc
Self-directed production team	Manages and executes production work	Ongoing
Task force and problem-solving team	Evaluates/resolves organizational problems/opportunities	Ad hoc
Benchmarking team	Evaluates competitors/best-in-industry performance	Ongoing
Facilities construction project team	Designs/develops/constructs facilities/equipment	Ad hoc
Quality team	Develops/implements total quality initiatives	Ongoing
General purpose project team	Develops/implements new initiatives in the enterprise	Ad hoc
Audit team	Evaluates organizational efficiency and effectiveness	Ad hoc
Plural executive team	Integrates senior-level management decisions	Ongoing
New business development team	Develops new business ventures	Ad hoc

Source: David I. Cleland, *Strategic Management of Teams* (New York, Wiley, 1996), p. 10.

- *Vision quest.* The discovery of what the future should be for the enterprise, that is, the general direction that the enterprise should follow to become what its leaders want it to be.
- *Stakeholder evaluation.* The discovery, development, and maintenance of common boundaries with those people and institutions that have an ongoing vested interest in those things being created by the project.
- *Market research.* Assessment of the possibilities and probabilities of the marketplace, to include the potential for specific products and/or services for markets which may not currently exist.
- *Product-service-process development.* The use of concurrent engineering techniques and processes to develop new products and services along with needed organizational processes to support competitive performance in the marketplace.

- *Business process reengineering.* Used to bring a fundamental rethinking and radical redesign of business processes to achieve extraordinary improvements in enterprise performance.
- *Crises management.* Teams that act as organizational focal points in dealing with crises that may arise in the organization's activities.
- *Self-managed production initiatives.* Improved quality and productivity in manufacturing and production operations through the use of self-managed production teams.
- *Resolution of short-term initiatives.* Ad hoc groups used to solve short-term organizational problems or exploit opportunities for the enterprise.
- *Quality improvement.* Total quality management accomplished through the use of quality teams (sometimes called quality circles) that utilize cross-functional and cross-organizational designs to develop and integrate quality improvement efforts.
- *Audit processes.* Teams that evaluate the competency of organizations, programs, projects, and functions to deliver quality products, services, and processes.
- *Senior-level decision making.* The use of teams of senior managers in the enterprise, which provides for a top-level synergy in its strategic management.
- *New business development initiatives.* Teams that are used to explore the design and development of new business ventures for the enterprise.

From the foregoing list of teams it should be clear that teams are an organizational design that can be used to manage a wide variety of operational and strategic initiatives. Such teams have had an impact on the success of contemporary enterprises. "The ability to organize employees in innovative and flexible ways and the enthusiasm with which so many American companies have deployed self-managing teams is why U.S. industry is looking so competitive."[4]

In the material that follows, more information on the role of these teams, including examples of their use, is presented.

21.5 ALTERNATIVE TEAMS: MORE ABOUT THEIR ROLES

Alternative teams can play important roles in the organization when properly focused on the enterprise's mission and the need for selected activities not otherwise assigned to either a project or a functional manager. The authors have identified 17 different tasks for alternative teams, which are described in the following paragraphs.

Market Assessment

One company used "headlight teams" to evaluate a preliminary set of industry discontinuities or drivers that had been developed by senior management that were

[4]Rahul Jacob, "Corporate Reputations," *Fortune,* March 6, 1995, pp. 54–64.

likely to affect the company. The teams evaluated each discontinuity in depth, seeking to discover how the trend might impact current customers and current economics in the company. Also, the teams evaluated the dynamics of the trends and the probable factors that might accelerate or decelerate these trends. Finally, a summary was created of which companies were likely to gain or lose from these trends. As the assessment by the teams began to emerge, other teams in the company composed of business unit managers and corporate managers reviewed the strategic importance the trends might have for the company. After the teams had completed their work, a penetrating insight into industry changes likely to impact the company had been done.[5]

Competitive Assessment

No enterprise can exist without being aware of its competitors. In the global marketplace companies watch each other closely in order to determine what new or improved products and services they are developing to add to their inventory of products and services in the marketplace. A major company in the aerospace industry uses competitive assessment teams to do an explicit assessment of its competitors whenever the company elects to form a proposal team and submit a proposal to the Department of Defense for a new aircraft. These competitive assessment teams have the objective of finding out as much as possible about the strategy likely to be used by other aircraft manufacturers that are likely to bid on the proposal for a new aircraft. The teams establish what needs to be known about the competitor's strategies, their strengths and weaknesses, and the probable bid strategies they are likely to pursue, ranging from their technical proposal, cost considerations, pricing and bid strategy, and any likely distinct edge that the competitor might have. A large number of people in the company are engaged in collecting information about the proposal competitors—and all of this information is sent to the assessment team that has been formed in the marketing department for integration, evaluation, and a prediction of the probable strategies to be employed by the competitors. The principal question that the team is expected to answer is: Who are the major competitors on this proposal, and what are their strengths, weaknesses, and probable strategies? Given this knowledge about the competitors, what strategy should the company pursue to enhance its chances of winning this proposal?

Organizational Strengths and Weaknesses

Concurrent with the development of an assessment of an enterprise's competitors, an evaluation needs to be done through the medium of an interdisciplinary team of the company's strengths and weaknesses vis-à-vis its most probable five or six competitors. A toy manufacturer has a sophisticated process for determining what

[5]Reported in Gary Hamel and C. D. Prahalad, "Seeing the Future First," *Fortune,* September 5, 1994, pp. 64–70.

its competitors are likely to do in designing and bringing out innovations in the toy business. Once a clear strategy of a competitor's product development effort has been determined, a team drawn from the different disciplines of the company is appointed to evaluate what the competitor's product might do in the marketplace—and how well the company is able to meet the competitor in that marketplace. An explicit analysis of the company's strengths and weaknesses is carried out and then passed on to the key decision makers in the company who are charged with the responsibility for developing a remedial product strategy to counter what the competitors are doing.

Benchmarking

Benchmarking is defined as an ongoing strategy of measuring organizational products, services, and processes against the most formidable competitors and industry leaders. The results of benchmarking, once determined, can help in the making of decisions about what should be changed in the enterprise. In addition, benchmarking results provide a standard against which organizational performance can be judged. Benchmarking is usually used in three different contexts: (1) competitive benchmarking of the five or six most formidable competitors; (2) best-in-the-industry benchmarking where the practices of the best performers in selected industries are studied and evaluated; and (3) generic benchmarking in which business strategies and processes are studied that are not necessarily appropriate for just one industry. A couple of examples of benchmarking follow:

- At General Motors benchmarking is becoming a major strategy in the company's drive to improve its products, services, and organizational processes. Every new operation must be benchmarked against the best in the class, including looking beyond the car manufacturing industry. General Motors has a core group of about 10 people whose responsibilities are to coordinate its worldwide benchmarking activities.[6]
- At Ma Bell's Global Information Solutions (GIS), a reorganization of the company has reduced staff by 20 percent, has assembled several hundred cross-functional teams, and has emphasized that all objectives must clearly link to key results: customer or shareholder satisfaction and profitable growth. The CEO refers to the people reporting to him as the "16 people I support." GIS constantly benchmarks itself against rivals and surveys customers constantly, turning the results into a measure of "customer delight" on a scale of one to seven. Key characteristics of the culture at GIS include vision, trust, rewards, and compassion—all the tools of enlightened leadership.[7]
- Union Carbide's Robert Kennedy used benchmarking to find successful businesses, determine what made them successful, and then translate their successful strategy to his own company. The benchmarking team at Union Carbide looked to L. L. Bean to learn how it runs a global customer service operation out of one

[6]Joyce E. Davis, "GM's $11,000,000,000 Turnaround," *Fortune*, October 17, 1994, pp. 54–74.
[7]Thomas A. Stewart, "How to Lead a Revolution," *Fortune*, November 28, 1994, pp. 60–61.

center in Maine. By copying L. L. Bean, Union Carbide teams were able to consolidate seven regional customer service offices, which handled shipping orders for solvents and coatings, into one center in Houston, Texas. By giving employees more responsibility and permitting them to redesign their work, 30 percent fewer employees were able to do the same work—including the analysis of processes to reduce paperwork to less than half. For lessons on global distribution, Union Carbide looked to Federal Express, and for tracking inventory via computer, Union Carbide borrowed from retailers such as Wal-Mart.[8]

The process of benchmarking can be carried out either in the operational or strategic context. The most popular view of benchmarking is that done to evaluate the enterprise's ability vis-à-vis its competitors and best-in-the-industry performers in the performance of its short-term context. Evaluation of a firm's ability to manufacture, market, and finance its current products and services is considered to be operational benchmarking. Strategic benchmarking, on the other hand, is carried out to evaluate the management's ability to *strategically* manage the firm—its ability to develop next and future generations of products, services, and organizational processes. Carried out in this context, strategic benchmarking examines the enterprise's *choice elements* as performance standards that must be continually sustained to provide a pathway for the future. To survive and grow in the future, an enterprise must outdo its competitors and the best-in-the-industry performers.

To outperform its competitors, a firm needs to develop and sustain a presence of central competencies that support the current and, when modified, future states of performance. It must build central competencies that will sustain competitive advantage, target a shift in choice elements to support future purposes such as entering new markets with new products and services, and develop organizational processes that will sustain the creation of improved capabilities. This means that the firm has to:

- Have organizational processes better than its competitors.
- Know its customers better than do the competitors.
- Know the likely shift in technologies that will improve organizational processes.
- Respond quickly to changes in competitor behavior, market shifts, and the development and application of technology.
- Use organizational processes better than do the customers, and at least equal to what the best in the industry are able to do.

A company that wishes to initiate a strategic benchmarking initiative should talk to others in the industry to gain their cooperation in doing reciprocal benchmarking. The benchmarking could be carried out in areas that are not "company proprietary" such as facilities management, internal auditing practices, human resource practices, employee safety and health, compensation benefits, training, executive development, and perhaps even quality programs.

---

[8]Sina Moukheiber, "Learning from Winners," *Forbes,* March 14, 1994, pp. 41–42.

Benchmarking makes sense as a means of gaining insight into how well the enterprise compares with its competitors and the best in its industry. Once the comparison has been carried out, performance standards for the enterprise can be set up.

Established Strategic Performance Standards

The strategic performance standards for an enterprise are reflected in Chap. 1 and Fig. 1.2. The elements in this model include mission, objectives, goals, strategy, structure, roles, style, systems, and resources that are the standards against which enterprise performance can be judged.

Contemporary enterprise managers, faced with growing, unforgiving competition in the marketplace, are using strategic planning teams of managers and professionals to identify and study the strategic alternatives available to the enterprise and make recommendations concerning which are the most promising alternatives for the enterprise to pursue. By reviewing the results of the work of other teams—such as benchmarking and competitive analysis teams—the strategic planning teams have a better chance of finding and selecting those strategic alternatives that best fit the enterprise's strengths and weaknesses. Dynamic markets, rigorous competition, and changing environmental conditions have made the strategic planning process much too complicated for a central planning staff to carry out. Although such a staff can provide valuable facilitative services for such processes, senior managers, who have the final responsibility for selecting the strategic direction of the enterprise, need all the help they can get. Strategic planning teams that can study and recommend strategies for the long-term use of the enterprise fill a critical need in today's fast-moving companies that are meeting and beating the competition in the marketplace.

Vision Quest

In Chap. 1, the concept of a vision for an enterprise and a project was discussed. Because of the elusive nature of finding a vision for an organization, teams of people with proven track records in creativity leading to innovative new products, services, and organizational processes can be used to do the analysis and "brainstorming" usually required to see and bring a meaningful vision into play. For example, an aircraft manufacturer appointed an interdisciplinary team to examine the potential for the expansion of the company's after-sales service business. The company's superior after-sales support was a major reason that customers purchased aircraft from the manufacturer. After deliberating for several months, the team developed a market plan that included a vision for the expansion of the company's after-sales service capabilities, supported by a vision that called for the company to offer superior after-sales service that would consistently outperform what its competitors are able to do. A food-processing company had a team to work on the development of a vision for growth based on critical mass in large product categories, geographic diversity, brand leadership, and marketing innovation.

Stakeholder Evaluation

Stakeholders are becoming more important in the management of the enterprise and in the management of a project. An important responsibility of a project team is to identify the project's stakeholders and develop a strategy for the management of such stakeholders. The reader should revisit Chap. 6, where examples of successful project stakeholder management are given.

Market Research

A major food processor appointed a project team to evaluate the global potential for its line of prepared foods. Over a 1-year period the team traveled extensively to assess local country markets, talk with subsidiary managers in the countries visited, and in general collect information concerning the eating habits of people in both developed countries and those countries that were undergoing both social and economic development. Several major general findings came out of the work of this project team:

- The demand for processed convenience foods would continue strong in the developed countries and spread to those developing countries where discernible increases in the living standard of the citizenry are evident.
- As the income level rose in the developing countries, new markets would open to include sales for pets as well as humans.
- Major markets that are expected to continue and in some cases to accelerate include prepared-food supplies to food service organizations, infant foods, and dietary and weight-control foods.
- The social and economic changes that have occurred throughout the world will likely not be without social and political upheavals in certain areas of the world.
- Technological innovation in the growing of crops and the manufacture and processing of food products will continue, giving a strategic advantage to those enterprises that are able to keep up with or lead technological improvements in food-processing initiatives.

Product-Service-Process Development

The use of concurrent engineering teams has been described elsewhere in this book. Revisiting this subject again, we can note that by using concurrent engineering teams, significant benefits can be realized such as:

- Reduction of engineering change orders of up to 50 percent
- Reduction of product development time between 40 and 50 percent
- Significant scrap and rework reduction by as much as 75 percent
- Manufacturing cost reduction between 30 and 40 percent

- Higher quality and lower design costs
- Fewer design errors
- Reduction and even elimination of the need for formal design reviews because the product-process development team provides for an ongoing design review
- Enhanced communication between designers, managers, and professionals in the supporting processes
- Simplification of design, which reduces the number of parts to be manufactured, creates simplicity in fixturing requirements, and allows for ease of assembly
- Reduction in the number of surprises during the design process
- Greater employee involvement on the concurrent engineering teams, leading to enhanced development of their knowledge, skills, and attitudes

The powerful idea behind concurrent engineering is that it has started more cooperative work among enterprises and this cooperation is even extending to competitors. Through working together with key stakeholders—like customers and suppliers—and even with direct competitors in changing technical standards and shared research, there is a greater good for a greater number of enterprises and people. Competitive advantage, through concurrent engineering, comes from the cooperative, coevolving relationships with principal stakeholders. Out of such cooperation come improved products, services, and organizational processes, resulting in earlier commercialization with products having lower development and production costs and higher quality. The basic idea is simple—by cooperating more fully with key stakeholders, understanding the potential contributions of key stakeholders, and marshaling their efforts as members of the project team, major opportunities exist for improving organizational competitiveness through new or improved products, services, and organizational processes.

Two simple models, Figs. 21.1 and 21.2, depict the differences between the concurrent and the serial design process.

Business Process Reengineering

The focus of reengineering is to set aside the current ways of working and painstakingly examine the processes involved in doing the work, to discover new, innovative, and breakthrough ways of improving both operational and strategic work in the enterprise. Reengineering is a clean, fresh start—no preconceived ideas, no limiting assumptions, no preexisting conditions or restrictions, no limiting factors, and no shibboleths to limit the imagination and creativity of how the members of the reengineering team work. There are benefits—and limitations—to what reengineering can do for the enterprise. For example, during one of the largest process reengineering projects ever undertaken, GTE telephone operations management was stunned to find out that the administrative bureaucracy of the company was reducing productivity by as much as 50 percent. As part of its reengineering effort, GTE examined its own processes and benchmarked 80 companies in a wide variety of industries. Reengineering teams then created new concepts, approaches, policies, and procedures for the new processes. To provide incentive

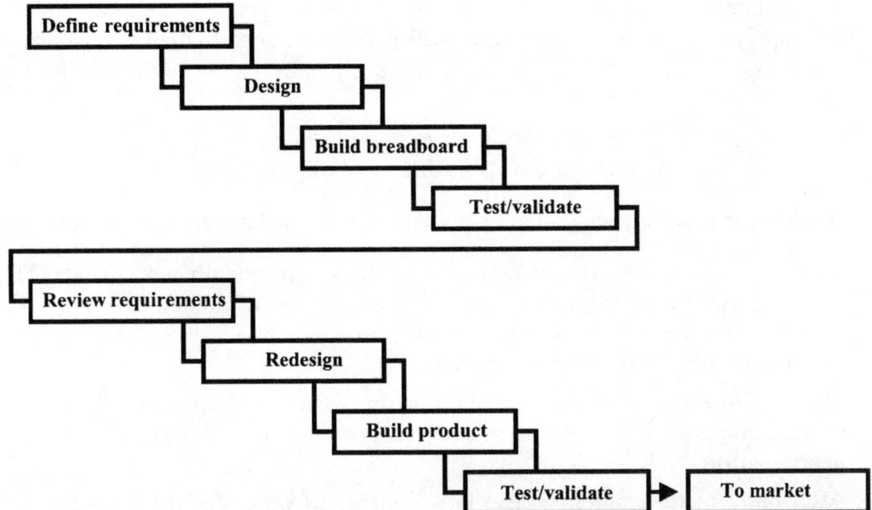

FIGURE 21.1 The concurrent design process. [*Source: in chap. 3, "System Engineering Process Overview,"* Systems Engineering Fundamentals (*Fort Belvoir, Va.: Defense Systems Management College, 2001), pp. 31–32.*]

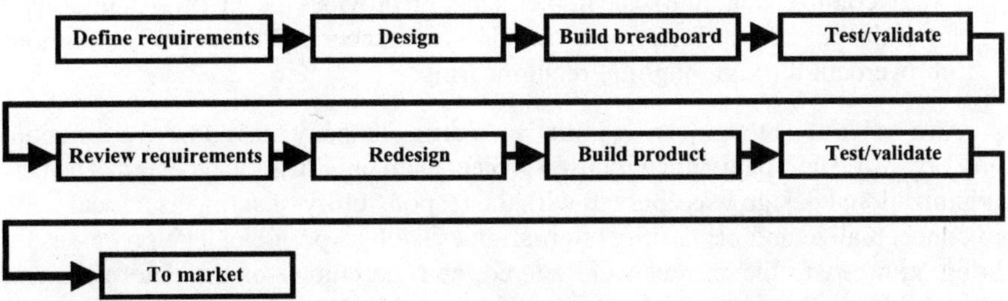

FIGURE 21.2 The serial design process. [*Source: in chap. 3, "System Engineering Process Overview,"* Systems Engineering Fundamentals (*Fort Belvoir, Va.: Defense Systems Management College, 2001), pp. 32–33.*]

to the benchmarking teams, specific goals were set: (1) double revenues while cutting costs in half, (2) cut cycle time in half, (3) cut product rollout time by three-quarters, and (4) cut systems development time in half. The company's reengineering efforts helped integrate everything it learned into a customer value–added path.

One key result of the reengineering effort at GTE was the promotion of a cultural change—a change that promoted a sharing among employees so they would be open to any and all possibilities for improving the way they work.

At GTE, four critical components of reengineering were used:

- Executive support for reengineering through setting the example, promoting reengineering, and providing the required resources.
- Creation of a process-reengineering department, which functioned as the owner of the methodology, assigned support work flow coordinators, and functioned as the corporate "cheerleading squad."

- Appointment and oversight of the process reengineering teams, including membership from front-line hourly employees and supervisors. In the first year, over 100 people were assigned to teams, with each team having responsibility for each process being reengineered.
- Establishment of teams of consultants to provide technical assistance and facilitation of the reengineering efforts through working with the teams.

During the process reengineering at GTE, the following were carried out:

- There was a complete examination of how things were actually working at GTE. It was found that after investing hundreds of millions of dollars in computers, what GTE had put in place was a maze of functionally based systems that created rather than solved downstream problems.
- The reengineering effort was carried out so that whatever savings and efficiencies existed could be harvested as quickly as possible from the process reengineering effort.
- After best practices among select U.S. companies were identified, conceptual platforms for the new processes were created, followed by the integration of everything that had been learned during the benchmarking work.
- The findings from the reengineering were put into practice.
- It was recognized that full-scale implementation provides the payoff. An important part of the payoff was fostering a "mindset reengineering" to support continuous improvement through ongoing reengineering.

As a result of a reengineering initiative, a drug company moved from a functionally organized company into a focused project team organizational design. The new organizational design was charged with the responsibility of acting as a focal point to conceptualize and bring drugs to market as soon as possible. The processes for bringing drugs to the market were altered, and the culture of the enterprise was changed from a "command-and-control" hierarchical top-down bureaucracy to a cross-functional matrix organization.

But reengineering projects have a mixed record of producing meaningful results. Studies indicate that more than half of all business process reengineering (BPR) projects have failed to deliver on objectives. Some of the causes were:

- Flawed study of the current situation
- Incorrect process selected for assessment and a poorly organized effort
- Failure to follow through on initiatives selected
- Insufficient buy-in from senior management
- Evangelism which overshadowed action planned and taken

The use of traditional project management tools and methodologies in BPR initiatives has not met expectations. These projects tend to overrun budget and schedule, and resources are incorrectly allocated, resulting in a sense of frustration on the part of the project team. An added problem has been the failure to take on BPR initiatives as a distinct departure from normal operations—rather than getting the process mixed up in the daily operational problems of the enterprise.

Yet properly planned and executed project management techniques and processes can facilitate effectively the BPR process. As a focal point in the enterprise, a duly appointed and charged project team can deal with the change inherent in a BPR process. Project management can be the organizational "glue" that binds the BPR effort together.[9]

Three years after he launched this decade's hottest management fad with the best-selling book *Reengineering the Corporation: A Manifesto for Business Revolution,* Michael Hammer pointed out a flaw: He and others in the $4.7 billion reengineering industry forgot about people. Companies that have experimented with reengineering initiatives have learned that simply cutting staff, rather than reorganizing the way people in different functions work, won't yield the "quantum" leaps heralded in Hammer and Champy's book.[10]

Reengineering projects create both *direct* and *secondary* results. These secondary results impact people. Sometimes the companies that engage in reengineering lay off people—in particular middle managers because of new technologies and strategies and because growing use of project managers to do many of the things that middle managers formerly did makes it possible to do without these middle managers. Traditional hierarchies are broken down along with the old functional fiefdoms. Old work titles, old organizational designs, and the more cumbersome work-cycle mentality are replaced.

The likely direct results of reengineering are found to be:

- Earlier commercialization
- Reduced costs and improved schedules
- Enhanced profitability
- Greater empowerment of people
- Greater use of process-driven strategies
- Higher quality
- Improved customer and other stakeholder relationships
- Enhanced work-cycle time
- Enhanced competitiveness

Along with these direct results of reengineering efforts there are bold secondary results that impact people. These are:

- Need for enhanced interpersonal skills by members of the organization because of the increase in working across organizational boundaries
- Increased requirements for additional knowledge and skills

[9]Ira King, "The Road to Continuous Improvement: BPR and Project Management," *IIE Solutions,* October 1996, pp. 23–27.

[10]Joseph B. White, "Next Big Thing: Reengineering Gurus Take Steps to Remodel Their Stalling Vehicles," *The Wall Street Journal,* November 26, 1996, p. 1.

- Changes in attitudes caused by the need to work with many different stakeholders, many of whom are outside the existing organizational framework
- Changes in job and work performance
- Improved operational efficiency
- Improved strategic effectiveness
- Modified management philosophies and processes
- Impact on the individual to learn new skills and roles, and even to seek other employment if downsizing or restructuring of the enterprise results in reduction of the work force
- Increase in adversary roles in the enterprise as people feel more freedom to criticize existing policies, procedures, techniques, or processes

When a reengineering project is carried out, there will be both direct and secondary results that impact the organization. With this realization, those people who initiate and carry out reengineering projects can better plan for and execute a project that will have a higher likelihood of being successful.

Crises Management

Teams are being used as a focal point for damage control and to develop remedial strategies in order to correct the situation that caused the crises to occur. Aircraft crashes, oil spills, fires, tornadoes, hostage-taking situations, and earthquakes are just a few of the emergencies that can happen to an enterprise.

Recent history has shown that the costs to an enterprise of a crisis can be staggering. Government policy requires that the owners of plants and facilities that use hazardous material have an emergency plan in place that includes how a "damage control" team is organized and trained in advance to respond to a crisis. The outside forces, such as the media and others that appear when a crisis occurs, dictate that the organization be prepared to respond. A timely and calculated response has the real promise of limiting the range of legal and stakeholder relationships and liabilities—and consequently minimizing the damage done by an emergency. A crisis, such as an oil spill, will have legal, media, and political stakeholders involved in a matter of hours. The enterprise has to be prepared to respond to environmental, legal, media, and political questions in a minimum of time. It is absolutely necessary to be prepared in advance as much as possible for such responses.

Self-Managed Production Initiatives

A self-managed production team (SMPT) is generally a small, independent, self-organized, and self-controlling group of people in which the members carry out the management functions of planning, organizing, motivating, leading, and controlling themselves. The team usually works without immediate supervision and often has the authority to choose, hire, promote, and discharge its members. The term "production" in this definition is used in the sense of creating value in the delivery of products and services to customers. These teams are used in many different contexts and in many different industries. The importance of such teams can be noted in reviewing the vision statement of a company that has been successful in the use of such teams:

> Self-directed teams will be responsible for "whole" processes or pieces of processes as defined by the company. They will be the focal point of the organization, with remaining functional departments acting as suppliers to the work teams. We will try to keep teams from 5 to 15 people, although more or less could be necessary depending on the nature of the respective team. Teams will be independent and rely on internal resources assigned to each team. They will work with pre-established boundaries set jointly with their sponsors. They will be empowered to make all decisions within their boundaries. They will set and execute goals based on corporate objectives and needs of their customers. Individuals will be committed to the team and the company. They will be properly trained in all team-related activities and will have the willingness to make decisions. They must also be willing to promote the team over self and trust each other to do what is necessary for team success. Communication lines between teammates and others will be open and frank. Managers, supervisors, and group leaders become sponsors, advisers, and coordinators. They support the team, break down barriers, and focus on employee development. They exist to make the team successful.[11]

SMPTs perform a wide variety of management and administrative duties in their area of work:

- Design jobs and work methods.
- Plan the work to be done and make job assignments.
- Control material and inventory.
- Procure supplies.
- Determine the personnel required.
- Schedule team member vacations.
- Provide backup for absentees.
- Set goals and priorities.
- Deal with customers and suppliers.
- Develop budgets.
- Participate in fund planning.
- Keep team records.
- Measure individual and team performance.
- Maintain health and safety requirements.
- Establish and monitor quality standards and measures.

[11] Process Manual–Self-Directed Work Teams, Company Document, MEDRAD Technology for People.

- Improve communications.
- Select, train, evaluate, and release team members.[12]

Some other examples of SMPTs follow:

1. In one factory the manufacturing workers manage themselves. There is a deep belief by the workers at this factory that constant change is the only constant. At this factory the work is technical and teachable. What isn't teachable is initiative, curiosity, and collegiality. Accordingly, during the hiring process every attempt is made to weed out loners and curmudgeons. The team selects its own leaders, who maintain oversight of the team's activities to include quality, training, scheduling, and communication with other teams. Management establishes the mission for the plant, but the workers are expected to design and implement strategies for fulfilling that mission. The professionals have cubicles next to the assembly cells. Every procedure is written down—but workers can recommend changes to procedures. Care is taken to display the plant's operating data so that everyone knows how the plant is doing. Employees work with suppliers and customers and have the opportunity to participate in trade shows and visiting installation sites. A yearly bonus is based on both individual achievement and team performance.[13]
2. At a Lucent Technologies plant self-managed work teams are used effectively. In the brutal global market for digital cellular stations, high speed and low cost are everything. At this plant in 2 years there hasn't been a single delivery deadline missed. Labor costs are an exceedingly low 3 percent of product cost. Some of the key strategies followed by the managers of this plant utilizing self-managed work teams include:
    - Hiring of new employees is based on attitude over aptitude—the work is technical and teachable, but what is not teachable is curiosity and collegiality. New employees come in as contractors and are hired as employees only after proving that they are self-starters and team players.
    - Teams elect their own leaders to oversee quality, training, scheduling, and communications with other teams.
    - The mission for the plant is created from above, the methods from below. The plant follows a one-page list of "working principles," which is a contract signed by every employee committing each one to speed, innovation, candor, deep respect for colleagues, and other plainly stated goals.
    - Teams continually alter the manufacturing process and design. Engineers and other professionals are located next to the assembly cells—with problems solved in the hallways rather than in the conference rooms.
    - All procedures are written down—but workers can change these procedures if needed.
    - Operating information and statistics are displayed everywhere—to include both the good and the bad news.

[12]David I. Cleland, *Strategic Management of Teams* (New York: Wiley, 1996), p. 170.

[13]Thomas Petzinger, Jr., "How Lynn Mercer Manages a Factory That Manages Itself," *The Wall Street Journal,* March 7, 1997, p. B8.

- Employees work closely with customers—and the plant manager encourages attendance at trade shows and installation sites. Workers conduct the plant tours for new workers.
- The yearly bonus, equivalent of 15 percent of regular pay in 1996, is based equally on individual and team performance.[14]

3. Sun Life Assurance Society Plc., an insurer, has eliminated most middle management and reorganized once-isolated customer service representatives, each of whom was in charge of a small part of processing a customer's files. Teams now handle jobs from start to end—with the result that turnaround time to settle claims was reduced by half, and new business grew 45 percent.[15]

Resolution of Short-Term Initiatives

Sometimes operational or short-term initiatives come up which require an interdisciplinary approach in their resolution. The appointment of an ad hoc team of people to study, analyze, and make recommendations concerning these initiatives becomes necessary.

Some examples of such initiatives where interdisciplinary teams were used include the following:

- Evaluation of a company's procurement policy, which resulted in a centralization of the procurement function for common items of equipment and supplies.
- Development of a "continuous performance improvement process" by a team at an electrical utility. The process was developed by a joint union-management team, which talked and worked with hundreds of employees before making recommendations on how organizational processes could be improved in the company. After the process was launched, another joint union-management team was charged with the responsibility to oversee the evolution and maturation of the process in the company.
- An ad hoc team was appointed by an electrical product manufacturer to study and make recommendations for an improved merit and promotion evaluation program. The team did benchmarking with other companies, studied the literature on the subject, interviewed company employees, and worked with a couple of consulting companies in reaching their decisions regarding the changes that should be instituted in the current evaluation program.
- A food processor used small groups of managers, supervisors, and workers to conduct ad hoc studies and recommend strategies to senior management for improvement strategies in the areas of:
    - Purchasing practices
    - Overhead costs
    - Advertising of company products and services
    - Selection and purchase of automated factory equipment
    - Collective bargaining strategies

[14]Ibid.

[15]Special Report, "Rethinking Work," *Business Week,* October 17, 1994, pp. 75–117.

As companies become more team-driven, more austere in the use of resources, and more competitive in the global marketplace, and develop cultures in which people are anxious and willing to participate in the design and execution of strategies that are likely to impact them, the use of small short-term ad hoc teams will likely increase.

Quality Improvement

The use of teams in total quality management (TQM) has enjoyed considerable acceptance in contemporary organizations. These teams can facilitate quality management and productivity improvements, improve labor-management communication, and improve job satisfaction and the quality of work life for employees. Some of the companies that have been notably successful in setting up superior TQM programs include L. L. Bean, Caterpillar, General Electric, the Boeing Company, and the Exxon Company, to name a few. Another example: At the Chevron Company, a major oil refiner, a "best-practices" discovery team was formed in 1994. It consisted of 10 quality improvement managers and computer experts from different functions of the enterprise to include oil production, chemicals, and refining. The team uncovered numerous examples of people sharing best practices. After a year of operation the company published a "best-practice resource map" to facilitate the sharing of knowledge across the company. The map contains brief descriptions of the various official and grass-roots teams along with directions on how to contact them. The map and its information help connect people working on diverse things in a diverse company.[16]

Audit Processes

Audit teams, which perform independent performance audits on some aspect of an organization, can provide valuable insight into successes and "failures" in the enterprise's operations. Audit of a project is best done at critical points in the project's life cycle. The value of using an interdisciplinary team to conduct an organizational or a project audit is that the organization of the audit can be viewed from the perspective of the disciplines or functions that are involved.

Senior-Level Decision Making

The management of large, complex organizations presents major forces that need to be evaluated when key decisions are made concerning the enterprise. Product and service development, market strategies, spending in R&D, global business strategies, and the impact of competitive and stakeholder initiatives require that decision making be made in a timely manner, fully cognizant of the potential issues, and done in a manner that best preserves the operational and strategic competency of the enterprise.

---

[16]Justin Martin, "Are You as Good as You Think You Are?" *Fortune,* September 30, 1996, pp. 150–152.

Probably the first senior-level decision-making team, or plural executive, was organized by the E. I. du Pont de Nemours & Company in 1921, in the form of a 10-person committee-line team devoted full time to company affairs.

A plural executive, acting as a body, is usually a senior team in the enterprise. A board of directors might be considered a plural executive, but usually the plural executive is one level below the board of directors. Many large companies today have some form of team design that functions as a plural executive, such as General Electric, General Motors Company, and the Boeing Company, to name a few. Two other examples are the following:

- At the Merck Corporation a plural executive in the form of a management committee composed of 10 experienced, highly talented executives was formed to lead the company into the future. One of the first works undertaken by this plural executive team was a thorough review of Merck's objectives and operations and consideration of the trends in the pharmaceutical industry, new directions of major markets, and emerging needs of customers.[17]
- Sometimes senior management plural executive teams are appointed to manage the integration of two or more enterprises that are merging. For example, Burlington Northern, Inc., and Santa Fe Pacific Corporation announced a senior management team that drew membership from the two railroads during the process of their merger. By appointing a joint team to oversee the development and execution of strategies for the pending merger, the infighting and rivalries that have plagued some rail mergers were avoided.

One of the principal responsibilities of a plural executive team is to maintain oversight over the contribution that projects are making in the enterprise in furthering the operational and strategic performance of the enterprise. As additional projects are initiated in the enterprise, the plural executive team's responsibilities grow in terms of:

- Assuring the "strategic fit" of the stream of projects that are under way in the enterprise.
- Determining through regular review the progress that the major projects are making in meeting their objectives—and ensuring that for smaller projects other managers in the enterprise are tracking these projects.
- Establishing general priorities for the use of resources on the projects, including removal of obstacles in the availability and use of such resources.
- Ensuring that contemporaneous state-of-the-art project management systems are used in the management of enterprise projects, which includes appropriate training of project teams.
- Ensuring that appropriate organizational design initiatives have been established that provide the project managers and the project team members with the empowerment needed to manage the enterprise's projects successfully.

[17] Raymond V. Gilmartin, "Letter to Our Stockholders," Merck 1994 Annual Report, p. 2.

- Maintaining contact with the principal stakeholders of the project in order to assure their continued commitment to the projects.
- Finally, by word and deed, communicating the important message to the enterprise employees that projects are vital building blocks in the design and execution of enterprise strategies and should be supported accordingly.

New Business Development Initiatives

The use of interdisciplinary teams to provide a focus for product development, production, and thorough launch is growing in popularity. These teams become involved in marketing and sales promotion strategies, selection of distribution channels, inventory levels, and customer training, and an ongoing measurement of the firm's ability to meet customers' needs on a timely and quality basis. These teams can have responsibility for the development of financial strategies, including estimates and tracking of revenues, costs, and likely profit contributions of the product(s).

At the Gillette Company, currently more than 40 percent of sales have come from new products over the past 5 years. This remarkable track record is done through a growing cadre of people who know how to manage product development projects from ideas through successful product launch. The company's new products are typically those that represent significant improvements. This incessant attention to innovation has been the primary mover of the company's innovative product lines, beyond just razors and blades, such as the Duracel battery acquisition. The company is cannibalizing its current products—assisted by the innovative and effective use of project management techniques and processes to create new products and services.[18]

21.6 THE PERSONAL IMPACT OF TEAMS

Careers are being impacted by the growing use of project and alternative teams in the design and execution of organizational initiatives. The promotion of people will be shaped less by tenure in a given company's hierarchy and more by what the individual has done in his or her career. Adaptive, rapidly changing organizational designs using alternative teams will be used more frequently where the individual's credentials will be determined by how well he or she works with teams of diverse individuals from within the organizational hierarchy and from outside stakeholders. Project managers, and the project teams, will be used to acquire resources from diverse sources and put these resources to work developing new products, services, and organizational processes. Project management has become the principal cauldron in which careers are formed, inasmuch as the people on these teams have usually been placed there because of a special talent and capability that they bring to the project. Those team members that have made a notable

[18]Linda Grant, "Gillette Knows Shaving—and How to Turn Out Hot New Products," *Fortune,* October 14, 1996, pp. 207–210.

contribution in creating something new for the enterprise that did not previously exist, such as a new product, service, organizational process or capability, or a new factory, will be the elite from which new managers are selected. What will be the special capability of these project managers? They will:

- Have demonstrated competence in working with diverse groups of people in the enterprise and with diverse demanding stakeholders in the company's environment.
- Have sufficient technical skills—such as engineering, procurement, manufacturing, and so on—to be noticed as people who produce quality results in their professional lives.
- Be able to understand how the enterprise "makes money" and be able to use the enterprise's resources to achieve revenue-producing results.
- Have people and communicative skills that are absolutely necessary—how to communicate, to network, to build and maintain alliances, to build the team, and to use empowerment as a means of exercising authority in the enterprise.
- Have the motivation to take the initiative and seek careers in the project management arena where new initiatives to better the enterprise are being forged.
- Recognize and accept that what facilitates a successful career is the impact that the individual has had on the organization—not that person's title.[19]

21.7 ROLE CHANGE

One of the concerns that a traditional supervisor has when asked to become a team leader is how the new position will impact his or her supervisory salary. Supervisors who become team leaders are usually expected to take a cut in pay, usually between 5 and 10 percent. Another concern is how the new role will impact their future careers. The so-called soft skills of communication, conflict resolution, mentoring, and teaching might not have much impact on their résumés. Other skills such as the patience to share information, the ability to trust members of the team, and the willingness to let go of authority and power and empower the members of the team are important and subtle skills.

New team leaders should recognize that they are in a role that is different from what they have seen or have performed as a traditional supervisor. In addition to emphasis on the soft skills mentioned above, the new team leader has to be willing to admit ignorance about the new role—and how he or she is expected to work with the team members—and not try to figure out each team member's job. He or she must also be willing to be an active agent for the team in getting the resources that the team needs to do its job. Working with the team in the spirit of team ownership is important as well, so the team members feel that there is a direct connection between their individual work and the work and objectives of the team.

---

[19]Paraphrased from Thomas A. Stewart, "Planning a Career in a World without Managers," *Fortune*, March 20, 1995, pp. 72–80.

Besides working with the team in planning, organizing, and controlling the use of resources on the team, the new leader should have a clear understanding of the correlation between the work and objectives of the team as well as the higher-level organizational objectives, goals, and strategies which provide support to the team—and is expected to be supported by the team in creating value for the customer and the parent organization itself. Yet work on the team seldom goes without any problems or frustrations. A common mistake that team leaders make is that leadership of the team is a "hands-off" strategy—leaving the team members to solve their problems. However, if the team is bogged down on a problem and the clock is running on a schedule requirement, the involvement of the team leader is essential. The secret is knowing when to intervene and how to intervene so that the members of the team are not insulted or alienated. The training and orientation given to new team leaders—and to existing team members as well—should acquaint the new people with what is considered typical and accepted team and leader behavior. When team leaders—and team members—do not buy into the team's work and its culture, chances are high that the lack of buy-in is because of an inadequate understanding of how the team is going to operate and what the behavioral patterns will be of people who serve on it. An individual who is confused or doesn't understand some of the things that are going on with respect to the team should ask questions of the team members, other team leaders, and the team facilitator who works out of the human resources unit of the enterprise.

The typical traditional supervisor, particularly one who has been successful in that role for many years, will likely have difficulty giving up the "perks" of the former role. Having absolute power over the subordinates, telling people what to do, having a title, and the unilateral responsibility in doing a merit evaluation and salary assessment on subordinates are "perks" that can build up one's ego—and when lost can be regretted, and the supervisor can easily slip into a mental state of longing for the "good old days." Another requirement that new team leaders face is the requirement for continuous learning on the technology the team is using, the managerial and leadership challenges, the new culture, and the requirement of working with diverse stakeholders—a few of the new things that have to be learned and can be learned only by getting immersed in what the team is doing and the means that the team uses to meet its goals and objectives.

Team leadership is about keeping the team members focused as well as understood and accepted as old organizational structures and hierarchies crumble. The new team leader recognizes that the team members cannot be expected to solve all the problems, but that the multidisciplinary activities going on in the team's work have to be dealt with by the team members who are experts in the activities under way. The team leader must have an understanding of what the team's mission is and how the team's objectives and goals support that mission—and work continually at communicating that understanding to the team members. The team leader must work proactively at creating a culture within the team that supports the team's work and is rewarding to the team members in terms of their social, psychological, and economic interests. Leadership of a team cannot be accomplished by barking orders.

21.8 TO SUMMARIZE

The major points that have been described in this chapter include:

- Alternative teams are becoming key elements in the modern enterprise.
- The use of teams has modified the theory and practice of management and has broadened the opportunities for young professionals to wend their way into management and leadership positions.
- Although alternative teams are much like a traditional project team, there are differences that are described in the chapter.
- The nature of traditional project teams and alternative teams was described in the chapter.
- The nature and the typical work of the nontraditional teams were described in the chapter and included examples of the use of such teams.
- Alternative teams play key roles in the design of operational and strategic initiatives in the enterprise. Such teams have become part of the permanent organizational design of many contemporary enterprises.
- Business process reengineering teams have turned in mixed results—and have not been the panacea for improved business processes that was originally believed.
- Many of the alternative teams described in this chapter have been ad hoc in nature. The self-managed production team is a permanent organizational design for the management of manufacturing or production operations. These teams are found wherever there is a distinct "production function" in the enterprise, which includes manufacturing enterprises, insurance companies, banks, and any enterprise that "produces" a product or service for a user.
- A company's vision statement for its use of self-directed work teams gives a clear indication of the importance that such teams have in the design and execution of organizational purposes.
- Teams have a major impact on the career plans for an enterprise's organizational members. For the most part, more career opportunities, particularly in providing opportunities to move into management and leadership positions in the enterprise, are enhanced.
- The use of alternative teams in the enterprise is likely to have a major impact on the roles and careers of traditional supervisors.
- Alternative teams are clearly an idea whose time has come.

21.9 ADDITIONAL SOURCES OF INFORMATION

The following additional sources of project management information may be used to complement this chapter's topic material. This material complements and expands on various concepts, practices, and the theory of project management as it relates to areas covered here.

- Karen M. Bursic, "Self-Managed Production Teams"; Gwenn C. Carr, Gary L. Englehardt, and John Tuman, Jr., "Reengineering Teams"; and Preston G. Smith, "Concurrent Engineering Teams," chaps. 33, 30, and 32 in David I. Cleland (ed.), *Field Guide to Project Management* (New York: Van Nostrand Reinhold, 1997).
- Lisa W. Churitch, Denis P. Couture, and Clement L. Valot, "Saturn's Vision for Program Management: A Different Kind of Approach"; Stephen W. T. O'Keeffe, "Chrysler and Artemis: Striking Back with the Viper"; and Bruce Watson, "A Town Makes History by Rising to New Heights," in David I. Cleland, Karen M. Bursic, Richard J. Puerzer, and Alberto Y. Vlasak, *Project Management Casebook,* Project Management Institute (PMI). (Originally published in *Proceedings,* PMI Seminar/Symposium, Pittsburgh, 1992, pp. 74–80; *Industrial Engineering,* December 1994, pp. 15–17; and *Smithsonian,* June 1996, pp. 110–120.)
- David I. Cleland, *Strategic Management of Teams* (New York: Wiley, 1996). This book reveals the enormous potential of alternative, cross-functional teams as forces for change with an organization. Such teams are perceived also as a focus for the management of the building blocks in the design and execution of competitive strategies. This pioneering work gives managers the tools, techniques, and information they need to integrate teams into the overall strategic management of the enterprise.
- Nicky Hayes and Clive Fletcher (ed.), *Successful Team Management,* Thomson Learning, 1996. This book provides a clear and readable text about the organizational psychology of teams and team management. The authors show how and why teams can have positive forces and results in contemporary organizations. The readers will gain information to improve their awareness of how to make the most of teams and enhance their understanding of what makes the team tick.
- William R. Bigler, "The New Science of Strategy Execution: How Incumbents Become Fast, Sleek Wealth Creators," *Strategy and Leadership,* May/June 2001, p. 29. The author advises managers that if they want to develop world-class strategy-execution skills they must accept a new strategic paradigm. Although the implementation of strategy was thought to be simply good project management, today's fast market rhythms require the application of more science and art in the use of project management to implement strategies.
- M. L. Swink, J. C. Sandvig, and V. A. Mabert, "Customizing Concurrent Engineering Processes: Five Cast Studies," *The Journal of Product Innovation Management,* vol. 13, issue 3, 1996. The authors present the strategies of concurrent engineering in several case studies: (1) the Boeing 777 aircraft, (2) the heavy-duty diesel engine of Cummins Engine Co., (3) the thermoplastic olefin automotive coating of Red Spot Paint and Varnish Co., (4) the airborne forward-looking infrared night vision system of Texas Instruments, and (5) the digital satellite system of Thomson Consumer Electronics. The authors found that three

types of teams appeared frequently in these projects—a program management team, a technical team, and a design-build team.

- J. E. Prescott and D. C. Smith, "A Project-Based Approach to Competitive Analysis," *Strategic Management Journal,* vol. 8, issue 5, 1987. The authors describe the use of a project-based approach to competitive analysis. They describe the implementation of the analysis in four steps. The first task is to define the project clearly, and obtain an understanding of the constraints involved. Then the steps of the project are defined, such as success expectations, identification of relevant competitors, information sources, and collection of information, and finally the team integrates and disseminates the information along with appropriate recommendations.

21.10 DISCUSSION QUESTIONS

1. What are some of the principal reasons that alternative teams are becoming key elements in the design of contemporary organizational forms?
2. What are some of the major ways that the use of teams has modified the theory and practice of management?
3. What are the principal differences between "traditional" and "nontraditional" project teams?
4. What is the principal work that is carried out by alternative teams?
5. Identify and present the primary objectives of the alternative teams that are found in use today. What are the similarities and differences of these teams?
6. In what ways can the use of alternative teams enhance the competitiveness of contemporary enterprises?
7. What is the principal reason for using competitive assessment teams? How should the results of the use of these teams meld into the strategic planning process of the enterprise?
8. Benchmarking teams are a relatively new phenomenon in the business enterprise. What are the principal advantages to using a team to carry out a competitive assessment?
9. What are some of the principal competitive reasons for using concurrent engineering teams? What are the potential linkages between such teams and the use of teams to design and carry out the strategic planning process in the enterprise?
10. What are some of the principal reasons that business-reengineering teams have not been able to deliver useful results, as had been anticipated when such teams were launched in the enterprise?
11. What are some of the major differences between a self-directed work team and the other alternative teams?
12. Is it possible in the modern industrial enterprise to deal with operational and strategic change without using alternative teams? Why or why not?

21.11 USER CHECKLIST

1. To what extent have alternative teams been used in your organization for the management of change?
2. Do the principal decision makers recognize the differences between traditional project teams and alternative teams as described in this chapter?
3. Is any of the work that is carried out by alternative teams being done in your enterprise? Why or why not?
4. If alternative teams are not being used in your enterprise, what is the reason?
5. Have the potential benefits of the use of concurrent engineering teams been explored in your company? If not, how is the product/service and organizational process work being carried out in the enterprise?
6. Has your enterprise tried using business-reengineering teams to examine and improve business processes in the enterprise? If used, how successful has such use been? Are there any opportunities to improve the use of such teams in the enterprise?
7. What key strategies have been developed in your enterprise that facilitate the better utilization of alternative teams in the management of operational and strategic change? How successful have these strategies been?
8. Have crises management teams been appointed in your enterprise to deal with emergencies that might arise? If not, why not?
9. Has the senior management of your enterprise considered the appointment and development of a plural executive team to deal with key operational and strategic issues in the enterprise?
10. If alternative teams are used in the enterprise, what are the principal means or philosophies by which change is managed in the enterprise?
11. How has the role of the traditional supervisors changed in your enterprise? Has any training been initiated to deal with these role changes?
12. Can any of the successes of your enterprise be related to the use of alternative teams? Or is the enterprise's business such that alternative teams are not likely to play a meaningful role?

21.12 PRINCIPLES OF PROJECT MANAGEMENT

1. The use of teams has modified the theory and practice of management.
2. Teams can be categorized as "traditional" and "nontraditional."
3. Nontraditional teams have as their primary focus the changing of organizational processes.
4. Teams, like any organizational unit, must be designed for the purpose they will serve in the enterprise.

5. Teams are key initiatives in the modern enterprise.
6. Teams can be used to manage operational and strategic initiatives.

21.13 PROJECT MANAGEMENT SITUATION—POSITIVE AND NEGATIVE RESULTS

Teams are not a panacea for the management of operational and strategic change in the enterprise. Teams can make a significant change in an organization when they are employed in areas that typically do not receive adequate attention by any element within the organization.

One has to identify the potential areas in which teams can most effectively perform and make a difference for the organization. These "opportunities" can be used to develop teams with the expertise to address the specific problems.

Teams do not perform just because groups of people are designated to work as a team. It takes time to develop their interactions and a camaraderie among team members. "Shake and bake" teams do not exist. It takes a central purpose and time for the team to gel and work smoothly together. Thus it is not just the composite skills and knowledge of the team, it is the ability to function as a single unit prosecuting a purpose or mission.

Practitioners that have used teams have found both positive and negative results. Some of the positive results of teams include:

- Productivity increases
- Quality improvements
- Cost reductions
- Earlier commercialization
- Enhanced profitability

Some of the negative results of teams include:

- Cloudy delineation of authority
- Adversarial relationships with subordinates
- The manager/supervisory role undercut
- Interpersonal difficulties created
- No fixation of responsibility and accountability

21.14 STUDENT/READER ASSIGNMENT

Students/readers are requested to examine the positive and negative summary results of teams as expressed above. In such an examination answers to the following questions should be provided and the additional assignments should be carried out:

- Are there any other positive or negative results that could be added to the summary list indicated above?
- How might these results be utilized in an enterprise considering the use of teams in its operational and strategic initiatives?
- If possible, the students/readers should examine an organization with which they are familiar to determine if the use of teams has produced any of the results indicated above.
- What could be done by way of policies and procedures to enhance the positive results of the use of teams in the organization examined by the students/readers?
- What could be done to reduce the opportunity for negative results in the organization?

CHAPTER 22
THE FUTURE OF PROJECT MANAGEMENT[1]

"The future ain't what it used to be."
YOGI BERRA

22.1 INTRODUCTION

Josh Billings once said: "Don't ever prophesy: for if you prophesy wrong, nobody will forget it; and if you prophesy right, nobody will remember it." With these words in mind, it takes a certain amount of intellectual courage and emotional recklessness to make predictions about the future of project management.

As the authors were finishing writing this chapter, the horrific events of September 11, 2001, had happened—terrorists struck a devastating blow to America by crashing two airplanes into the World Trade Center towers in New York City, crashing an airplane into the Pentagon in Washington, D.C., and causing a fourth airplane to crash in southwestern Pennsylvania. This unprecedented attack on the American people changes the future of U.S. citizens and their guests in America as perhaps no other event could.

22.2 RECENT DISASTER SHAPES THE FUTURE

The authors envision that the events of September 11, 2001, will bring out the latent talents of Americans and recovery will be accomplished, for the most part, through projects. Projects can be used to address the challenges facing people everywhere. Such areas as (1) removal of debris at the crash sites, (2) rebuilding of damaged or destroyed facilities, (2) relocation of services from the crash sites, (3) investigations of terrorist activities, (4) changes in laws regarding civil liberties, (5) changes to airport security, and (6) changes to emergency response practices are candidates for the use of project management practices.

[1]This chapter is an extension of David I. Cleland, "The Strategic Pathway of Project Management," *Proceedings*, Project Management Institute, 28th Annual Seminar/Symposium, Chicago, September 27–October 2, 1996.

Today, many people are making predictions about how the world has been changed by these events. Some of these predictions may be correct—but in all probability a limited number of these predictions will not come about. Given the intellectual liability that authors assume in writing this chapter, we choose to press on. Perhaps some of our predictions will be correct!

This chapter starts with a review of some of the past influences and some of the probable general future trends likely to impact project management. The authors attempt to answer the question: What does it mean? Then the chapter makes some predictions about the future of project management, and includes how management philosophies might change in the future. The growth of the Project Management Institute in recent years, the likely continued increase in megaprojects, and some additional project management changes are noted. As readers peruse this chapter, the authors would like everyone to think about the question: What is beyond project management?

In considering this question it would be useful to remember that project management evolved from the theory and practice of general management. As the discipline of project management continues to grow toward maturity, it is quite probable that new meanings, new ideas, changed concepts, enhanced philosophies, and improved management processes will develop. All of this could cause, at some future time, a new focus in management that just might replace project management as we know it today.

22.3 THE PRACTICE OF PROJECT MANAGEMENT

The practice of project management has been with us for a long time. This discipline and its evolution as a profession are described elsewhere in this book. Today, project management has reached a maturity that entitles it to a rightful place in the practice and the literature of the management field. Its application has spread to many "nontraditional" uses as it continues to be one of the principal means by which operational and strategic changes are managed in the enterprise. Pinto noted that "the importance of project management for organizational success will expand rather than wane in the years to come."[2]

In this chapter some of the recent major changes in the management field are presented, along with a summary of the key contributions that have been made by project and team management in recent times. From that summary, a few predictions will be made concerning the likely nature of project and team management in the future.

22.4 PAST AND CURRENT INFLUENCES

In considering the future of project management, an examination of some of the major forces in the field of the management discipline in the past should be considered.

[2]Jeffrey K. Pinto, (ed.), "The Future of Project Management," chap. 25 in *The Project Management Institute Project Management Handbook* (San Francisco: Jossey-Bass, 1998).

In the last 10 to 20 years, there have been major forces that have impacted the theory and practice of management. A summary of these forces follows:

- Product and service life cycles are becoming shorter, impacting how companies compete in the global marketplace.
- A growing realization that as new or improved technology is integrated into products and services, the organizational processes to support such technology have to be improved as well.
- An increasing trend in the use of alternative team organizational designs to cope with the need for the integration of interfunctional and interorganizational activities to support product, service, and process development.
- Downsizing and restructuring of organizations to improve efficiency and effectiveness, which has resulted in the elimination and shifting of managerial and professional positions.
- Growing realization that the persons doing the work know the most about how that work should be done, and through enhanced participation of these individuals on alternative teams, improved organizational performance has been sustained.
- Computer and telecommunications technology has made remarkable progress in the use and management of information and has helped increase enterprise productivity.
- The relative roles of "managers" and "leaders" have come under scrutiny and redefinition, with the use of alternative teams enhancing the opportunities for more people to move into managerial and leadership positions than previously could.
- More emphasis is being placed on the interpersonal capabilities of executives, with the role of traditional first-level supervisors changing from a traditional "boss" to a facilitator, coach, mentor, counselor, coordinator, and oversight person in obtaining and using resources in the enterprise.
- Global competition has become the "name of the game" for the survivability and growth of industrial enterprises. The surge in demand for products and services in developing countries, as well as the need for infrastructure improvement throughout the world, has influenced the demand for project management services.
- Customers and suppliers are taking more active roles in the design and execution of enterprise strategies, even to the point of serving on new and improved product and service design and development teams.
- Collective bargaining and unionism are becoming more sophisticated to include the presence of union leaders and members on the alternative teams of the enterprise.
- The maturation of a philosophy of strategic management of the enterprise—the management of the enterprise as if its future mattered—is reflected in more proactive strategic planning and execution strategies in contemporary enterprises.
- Engineering and other technologies appear to be doubling every few years, with innovators in such technologies arising from many places in the global environment.
- The growing success of the application of project management in the managing of operational and strategic change.

- A broader application of project management processes and techniques beyond the traditional construction and defense industries—expressed in reengineering, benchmarking, concurrent engineering, and self-managed production initiatives, to name a few.
- Growing influences of enterprise and project stakeholders as legitimate claimants of those things of value being created by the enterprise, including an increased membership of such stakeholders on project and alternative teams in the enterprise.
- Traditional jobs are being modified and lost in today's organizations, and similar losses will likely continue in the future. Bank tellers, secretaries, administrative typists, factory workers who do rote assembly work, and service station attendants, to name a few, are those whose jobs are being changed or lost.

Robert Reich, secretary of labor under President Clinton, argues that businesses must stay competitive in an unprecedented manner—and must constantly innovate, compete on razor-thin margins, and operate around-the-clock to stem competition from the global marketplace.[3]

In the material that follows, a few general predictions of what the future holds will be given. These predictions will, in general, be presented within the context of a systems framework: political systems, economic systems, social systems, legal systems, and technological systems.

22.5 SOME GENERAL FUTURE TRENDS

Before becoming too enamored with trying to predict the future, we must revisit Peter Drucker's caution about trying to do so. He cautions about organizations becoming too fascinated with the future, and makes the argument that organizations that forecast and make strategic decisions about the future are unlikely to succeed in the long term.[4]

Those "futurists" and other people who spend most of their time making predictions about likely trends and events of the future have missed some of the major changes that impact societies of the world. Who was able to predict the outbreak of AIDS, the rapid collapse of world communism, and the strong movement of drastic corporate restructuring in the United States? How many forecasters were able to see the rapid technological change and the growing experimentation in the way enterprises are organized and managed? How many stock market analysts envisioned the perceptible decline of high-tech stocks in 2001?

Not many were able to anticipate the growing influence of cooperative and long-term relationships between companies and their various stakeholders. And, in the field of project management, only a few visionaries sensed the change that the use of project teams would have on the structure and management style used

[3]Robert B. Reich, *The Future of Success* (Random Audio Books, 2001).

[4]Peter Drucker, "Planning for Uncertainty," *The Wall Street Journal,* July 22, 1999, p. A12.

by contemporary organizations, and the growing use of alternative teams in the management of operational and strategic change in the enterprise. But, given our modest capabilities in predicting the future, we keep trying.

22.6 CHANGING MANAGEMENT PHILOSOPHIES

Firms today, throughout the world, are attempting major shifts in their management philosophies through reengineering and restructuring of their organizations. Firms have restructured their organizational design and have introduced the use of alternative teams to focus more on core operations by eliminating unrelated operations, improving organizational processes, downsizing, flattening their organizational hierarchies, and developing strategies to commercialize their products sooner and reduce their manufacturing cycle time. Continuous improvement of products, services, and organizational processes dominated the strategy of many of the premier firms of the 1980s and 1990s. Major changes in manufacturing include a departure from mass production strategies to lean manufacturing where flexibility, rapid responses, and improved human resources practices, product development strategies, manufacturing methods, supplier and customer relations, and management methods are practiced.

22.7 DISTRIBUTION CHANGES

Channels of distribution of products and services are likely to change significantly. Witness the past emergence of large discount chains, the evolution of telemarketing, and the changes already under way in the manner in which automobiles are purchased. Retailers are continuing their trend to fewer frills, more warehouse stores, and fewer face-to-face interactions with salespersons. Smaller retailers will continue to be threatened and are being replaced by large discounters, where personal selling has declined and purchasers find what they want from large displays of products. The need for better product distribution has helped motivate the rise of integrated logistic support initiatives in the business community.

22.8 INFRASTRUCTURE NEEDS

Rich and emerging nations are becoming more aware of the need to revalue and update their infrastructures. In Asia, the flow of people into the cities continues, with predictions that approximately 1.5 billion people will be added to the urban centers of Asia in the next 10 years. Radical solutions and innovations are needed in a focus of project management to improve the infrastructure of these areas. China alone is expected to spend something over $200 billion in the next few years on infrastructure development. Other emerging nations are not far behind this rate of spending.

Transportation

The shifting focus of economic opportunities in the Far East will make air transportation and air cargo product distribution key considerations in the development of infrastructure to support these initiatives.

Strategic Partnering

Strategic alliances and joint ventures will continue to become more common as companies learn about the advantages of cooperation. Established western companies will form alliances with competitive Asian firms, whose home marketing clout cannot be ignored.

Population Growth

Of all the challenges facing global society today, the growing population of the world is sowing seeds for future opportunities and problems. Demand increases for supporting infrastructures, more food, more medical systems, more of everything.

Today, it is not uncommon to hear of famine and deprivation in Africa, Latin America, and Asia. The predictions of Thomas Malthus of the excess of population over available resources may become a reality. Thomas Malthus (1760–1834), an English economist and demographer, is probably best known for his concept that the population will increase in geometrical progression, whereas the means of subsistence will increase at only an arithmetical progression.

He said that population will always expand to the limit of subsistence and be held there by famine, war, and ill health. He may yet be recognized for his clairvoyance—many places in the world today cannot produce enough food for their citizenry—and there is not enough transportation capability to get the food to the people that need it. Millions of human beings are likely to have a subsistence existence.

Technology Changes

In the twentieth century, awesome changes have occurred in technology. There is little reason to expect that rate of change to slow down in the next century. Indeed, it may very well accelerate. What will be the outcome of the miniaturization of computers, the outcome of telecommunications via satellites, and what about the advances in optical transmission technologies? Will biotechnology fulfill its promises? What will be the impact of ceramics on the design of engines of the future? What additional changes will come about in manufacturing, where today there is a strong movement from mass production to lean manufacturing strategies? The linking of computers from the design laboratory through to the production floor of the factories has—and will likely continue to have—enormous implications for global manufacturing competitiveness, the work force, and flexibility in producing goods to order, reducing the need for warehouses. The practical promise of solar energy and electric batteries continues to evade us—but what will the future bring in these technologies? The information highway, and its promises and threats, has caught the attention of "futurists" around the globe.

Diverse Work Force

A growing, diverse work force greater than we have today, with more women and so-called minorities, will increasingly represent the work force of the future. Training and retraining initiatives will become a way of life in the enterprises of the future. Diversity in the work force will parallel the changes impacting the manner and style in which organizations are managed today, with additional changes coming in the future as we grapple with the means of improving productivity through people, as well as through capital equipment such as robotic systems, information systems, and alternative organizational designs. Our ability to develop and implement new managerial paradigms will be a principal test of whether or not the enterprise survives. Strategic management—the management of the enterprise as if its future mattered—will become the principal game to play in managing the organizations of the future. Changing organizational cultures to provide for more employee participation, empowerment, and higher motivation through transitory membership on alternative teams is a major challenge facing today's managers—a challenge that will certainly continue in the future. The key question facing all managers throughout the globe is clear: Will the company have the management systems and culture to move forward against the unforgiving global competition in the marketplace today?

People Skills

As we move away from the traditional "command-and-control" paradigm of management suggested by Peter Drucker, interpersonal skills are becoming more important. Seasoned managers realize that many of the skills required for success revolve around the ability to communicate, the ability to work with people, the competency to negotiate, and the patience to listen. As the use of teams grows in the management of change in contemporary and future organizations, people skills will continue to be one of the most important assets in the management style of managers and professionals at all levels of the enterprise.

Political Discontinuities

Without question, the world is in a sea of constant discontinuities. Countries like China, Russia, and others are departing from their traditional roles. The restructuring of their economies, political changes, and the struggle to maintain social equality in their citizenry are posing challenges of unprecedented balances in the utilization of resources. Over the next couple of decades the countries that best resolve their political uncertainties and the optimum use of their resources will be the winners. Others may well lose their existing power bases in the world.

The cyber revolution will continue to change human life. We are likely to become richer and better educated as we expand the use of computers in our work and private lives. Yet there is the probability that we will become lonelier and less healthy as we work more in isolation and neglect physical activity. People may socialize less in order to spend more time in the "make believe world" of entertainment.

22.9 WHAT DOES IT MEAN?

Throughout the changes impacting modern organizations one constant has remained: Resistance to change in organizations most often comes from individuals who fear that change may bring losses to the organization and to themselves. Change usually creates winners and losers, although enlightened management can deal with the change in such a way that a loss to individuals is reduced through retraining, early retirement, placement assistance, and other strategies. Those organizations that have undergone significant changes have gradually come to recognize that successful change comes only with the commitment and cooperation of individuals in the working levels of the enterprise.

Some, if not all, of the trends that have been put forth in this chapter may never come about. This is unlikely. What is likely is that the political, social, economic, technological, and competitive changes in the future will be somewhat like what has happened in the past, but some of those changes will be extraordinary and will pose challenges of major proportions, particularly in how the societies deal with these changes. Part of successfully dealing with these changes will come from a judicious use of a prescription of alternative teams in the management philosophy and style of those entities—industrial, military, governmental, social, or whatever—that have to deal with the changes as interested stakeholders.

Knowing of a pending change and having no management philosophy with which to deal with that change can be terrifying. Muddling through will be an invitation to failure. But an organized team of resources managed through a project management systems approach shows considerable promise as a way to cope with, and in the process influence, the effects of a change.

22.10 THE FUTURE OF PROJECT MANAGEMENT

One of the fascinating predictions about the future of project management was set forth by Stewart, who wrote, "Project management is the wave of the future."[5]

In considering the future of project management, first it will be necessary to review some of the expected changes in our society that will, in some manner, influence the theory and practice of project management. As changes in political, social, economic, competitive, legal, and technological systems occur, project

[5]Tom Stewart, "The Corporate Jungle Spawns a New Species: The Project Manager," *Fortune,* July 10, 1995, pp. 179–180.

management will likely be impacted, because project management provides the principal means by which such change is managed in today's and tomorrow's organizations. For example, the promise of an improved or new technology in a planned product or service requires the use of concurrent engineering teams to simultaneously develop the new product, service, and supporting organizational processes. A strategic plan that calls for the expansion of an enterprise's manufacturing capacity will require the appointment of a project team to conceptualize, design, construct, and arrange for the start of the plan. An enterprise that faces a downturn in the demand for its products and services, and must consider cutting back its expenses, will require a reengineering team to examine its operation with an assessment of how well its organizational processes are being carried out, the basic design of the organizational structure, and other means for improving its overall efficiency and effectiveness. Some of the project management–related changes follow.

Political Changes

Political changes foster the development and application of project management strategies. The sweeping political changes in the former Eastern European bloc fostered the abandonment of many of the autocratic social and management philosophies of the communist-dominated countries in this bloc. Major changes in the infrastructure of the former communist countries and the need to design and develop competitive business strategies, update plant and equipment, and improve the managerial and professional skills of employees became critical. Project management, sometimes used in an informal process, helped these former communist countries move toward capitalistic and democratic societies.

Social Considerations

Bringing about major social changes in a country calls for the use of project management. The push for affirmative action in the United States motivated executives to develop proactive programs in their organizations to support equality in the management of personnel. Equality standards in U.S. colleges and universities prompted the need to develop projects and programs to ensure that such standards were carried out in the hiring of faculty and in undergraduate and graduate study programs. Lawyers had to develop initiatives in their work to support the objectives of affirmative action programs, as well as other state, local, and national legislation in the environmental area.

Competitive Alterations

Competitive changes have fostered increased interest on the part of companies in the use of teams. In some cases, the benchmarking of their competitors has disclosed the use of teams in the product and service strategies. Sometimes they have benchmarked the "best-in-the-industry" performance and found that teams have been used as key elements in the strategy of these best performers. Nothing will catch the attention of senior executives more quickly than to find out that they are being defeated in the marketplace by a competitor. Finding out that a competitor is outperforming a company in the marketplace will motivate the executives to find out why the competitors have been able to do so much better. For example, when Xerox found out that Japanese companies were selling their copiers for what it cost Xerox to manufacture its copiers, an immediate investigation was conducted by Xerox to determine the reason. Xerox found that the use of concurrent engineering processes and techniques accounted for the difference. Xerox then initiated concurrent engineering in a forthright manner to reduce the time it took for them to develop their copiers—subsequently becoming able to compete with the Japanese competitors.

Modification of Project Management Practices

Project management, as we know it today, will likely change in the future. Advances in communication through computer technology, the integrated voice data and imaging techniques, and the Internet will provide higher levels of ability to communicate. The growing ability to exchange information on a global basis will help foster a different world of project management. The ability to use technology for gathering, analyzing, and interpreting data should provide more opportunity for the improvement of the processes and techniques involved in managing projects. With the help of such technologies the typical project manager should have more time to deal with the human element in the management of projects—and this asset will become more valuable in the future world. Continued refinement and improvement of the tools of project management, principally by way of schedule and cost management, will reduce the amount of time that the typical project manager will spend on these things.

Future project managers will focus more on the people who are responsible for executing the project, including greater time with all of the stakeholders as their influence in the project affairs is likely to increase, facilitated by the increasing availability of information about the project.

Project Management Research

A few years ago the Project Management Institute launched an initiative to determine the need for project-related research in the field. Through a team of PMI notables a development effort was organized and developed to study and make recommendations concerning the need for research in the project management field. As a result of the deliberations of this team, researchers were funded to pursue a variety of research topics in order to advance the state-of-the-art of the discipline.

The first major contribution of this research initiative was a conference held during June 2000, in Paris, France. This conference, titled "Project Management Research at the Turn of the Millennium" (*Proceedings,* PMI Research 2000, Paris, France, June 21–24, 2000), attracted over 350 people from 27 countries. Fifteen invited and 30 contributed papers reported on a wide variety of different project management issues. Another conference to review research in the project management field will be held July 2002 in Seattle, Washington. The reader who wishes to learn more about PMI's research strategy should look at the PMI website at www.pmi.org.

In 1999, the Project Management Institute commissioned a study to evaluate the future of project management. This study was sponsored by the 1998 PMI Research Program team and included garnered input from PMI members, reviewed PMI reports, and publications, and through thoughtful consideration of the forces at work involving project management came up with a "first of its kind" report. A paraphrasing of the major conclusions of this study follows:

The nature of project teams is changing dramatically, and the work of such teams is becoming more complex.[6]

- Project management is evolving into a profession.
- There is growing evidence of rapid participation in the project management profession.
- Attention to the people and interpersonal aspects of the practice of project management is increasing.
- The complexity of the project management profession is putting increased demands on those people who work on project teams.

Nontraditional Projects[7]

In the last 10 years project management has moved rapidly out of its traditional redoubts and is moving from a specialty strategy into a central task of management for the process of managing product, service, and organizational process change. Project management's original application was principally in construction and Department of Defense projects. As it demonstrated its ability to provide a process and techniques for pulling together cross-functional and cross-organizational activities, organizations began experimenting with the use of teams to deal with other applications. As these applications grew in importance and use, project management and its success in the use of teams extended the use of project management techniques to other areas of the enterprise, such as:

- Reengineering applications teams used to bring about a fundamental rethinking and radical redesign of business processes to achieve extraordinary improvements in organizational efficiency.
- Concurrent engineering applications—concurrent product, service, and organizational process development teams to develop, produce, and market products and services earlier, of a higher quality, and at a lower cost.

---

[6] *The Future of Project Management,* published by the Project Management Institute, Newtown Square, Pa., 1999.

[7] The reader will doubtless note that the terms in this subsection are discussed more fully in Chapter 21. The summary given here is meant to place them in the context of current and future changes in project management.

- Benchmarking initiatives—using teams to measure organizational products, services, and processes against the most formidable competitors and industry leaders to use as performance standards for the enterprise.
- Development of new business opportunities using teams to explore, design, develop, and execute new ventures for the enterprise.
- Total quality management initiatives accomplished through the use of teams that utilize cross-functional organizational designs to integrate enterprise quality improvement strategies.
- Improved quality and productivity in manufacturing and production operations through the use of self-managed production teams.

22.11 MANAGEMENT PHILOSOPHY MODIFICATIONS

One of the major contributions of project management, and alternative team management during the 1980s and 1990s, has been a modification to the management discipline. These modifications have changed management philosophies, processes, and techniques and the performance standards by which an enterprise's efficacy can be judged. These modifications and their likely continuation include:

- Acceptance and virtual institutionalization of the "matrix" organizational design. In such acceptance, the use of project teams overlaid on the traditional organizational structure has become simply "the way we do things around here" in the project- and team-driven enterprise entity. This acceptance will continue in the future, with the potential of leading to the gradual disappearance of the "matrix" organizational design as a distinct entity as it is further amalgamated into the culture of enterprises.
- Acceptance of the singularity and importance of project planning as a means for determining the resources required—and how these resources will be used—during the life cycle of the project. In the future, project planning will gain in importance as the means for identifying and committing resources to deal with operational and strategic change in the enterprise.
- New organizational design initiatives will be assessed as a means for enhancing the use of a focal point for product, service, and organizational process change. The virtual organization, further extension of the "matrix" design to deal with a more active participation of project stakeholders, and the means for the extemporaneous emergence of formal and informal teams in the organization will deal with the pressures of change coming from competitive and environmental sources. Ad hoc teams, both formal and informal, will become more of a way of life in the enterprises of the future.
- Organizational members have found that the opportunities for their growing participation in both the operational and strategic initiatives of the enterprise have heightened their role in decision making in the enterprise. This has led to

an enhancement of their feelings of belonging and contributing to the organization. Through having growing opportunities for participation in the affairs of the enterprise, people are motivated to share in the results and rewards of the enterprise.

- In the late 1980s and the 1990s, the growing use of teams has opened broadened opportunities for workers to perform managerial and leadership functions as they work on operational and strategic teams in the enterprise. As workers have participated on these teams, their appreciation of the challenges facing organizational managers and leaders has increased, leading to greater support of such managers and leaders—and a growing desire of workers to become more proactive in the opportunities to carry out managerial and leadership activities in the spheres of work. In the future these trends will continue, with a growing base of people in the enterprise who can perform managerial and leadership roles.
- Managerial control systems in the 1980s to the present time have grown in importance and in sophistication. Yet the role of individuals to "monitor, evaluate, and control" their personal work in the enterprise has never been more important and pronounced. As individuals have served on the various teams available to the enterprise, there has been a growing recognition on the part of these individuals that "self-control" is important and is a major and effective way to ensure that the planned use of resources is consistent with organizational goals and objectives.
- The growing use of teams in the 1980s to the present time has brought the management of stakeholders into greater play than in the past. The growing importance of stakeholders to the destiny of the project and the enterprise has sharpened the awareness of the "systems approach" to the management of the project and the enterprise. A philosophy has developed in contemporary managers and leaders that "everything is related to everything else." Accordingly, the making and execution of decisions are being done more and more with an awareness of the likely system reverberations of the use of resources in both the operational and the strategic sense, to include an assessment of the growing influence of stakeholders. This trend will continue in the future and will likely stimulate the emergence of models and paradigms on how such "systems" considerations can be melded into the design and execution of decisions by future project managers.
- In recent years there have been new applications of project management, reflecting its use in a wide variety of different industries and organizations. Today, industrial, educational, military, social, governmental, and ecclesiastical organizations use project management to varying degrees. This trend will continue in the future, as the recognition that project management is an effective means of dealing with change grows among more managers, leaders, commanders, administrators, and ministry and lay people who see project management for what it is: a means for dealing with operational and strategic change in the enterprise.
- The concept and process of strategic management of the enterprise gained acceptance in the late 1980s and 1990s. An important part of strategic management is the process of strategic planning, whose objective is to develop a sense of direction and acceleration for the enterprise in its future. The growing use of

a philosophy of strategic management of the enterprise has been accompanied by growing use of project management and alternative team management as means for dealing with the environmental and competitive changes that face all organizational entities today. Strategic planning teams have been used successfully by modern enterprises to facilitate the strategic planning process. This trend will likely continue, stimulated in part because the growing competitiveness in product and service change is becoming more the performance standard in the global marketplace.

22.12 INCREASE IN MEGAPROJECTS

A megaproject is defined as one that has the following characteristics: (1) extraordinary financing from both private and public sources; (2) major political, economic, and environmental considerations—both during its design and engineering and construction phases as well as when the project results become operational; (3) ongoing involvement by knowledgeable stakeholders from political, intervenor, financial, economic, and user communities; (4) major "systems" planning and control challenges; (5) likely intense scrutiny by stakeholders to include media and local community interests; and (6) projects that deal with the likely shortages of water, energy, and transportation resources.

An example of a current megaproject keynotes some of the major "systems" challenges that these projects face. Bechtel was responsible for planning, designing, engineering, procurement, right-of-way acquisition, and construction of a second gas pipeline which extends 875 miles from Canada into central California, with a cost of approximately $1.6 billion, including a new compressor station and retrofitting of 17 other compressor stations and 3 major meter stations. The new pipeline parallels the first line that was built in the early 1960s. Concern about a wide range of environmental factors was paramount throughout the pipeline expansion. Careful planning by Bechtel resulted in the development of extensive safeguarding of environmental factors on the pipeline. Certain measures dealt with the control of erosion, noxious weeds, hazardous material, and construction noise, as well as extensive training for all personnel on environmental awareness and work practices.[8]

Concerning the growing increase in megaprojects:

- The demand for megaprojects for the improvement of infrastructures in the world will continue to accelerate. This demand is particularly strong in the emerging nations in the world. Much of this demand will center around major power, water, and transportation initiatives and health care, as well as social "reengineering" projects.
- Key stakeholders will continue to gain in their influence on the conceptualization, design, construction, and operation of megaprojects.

[8]Gary Walker and John Myrick, "Doubling a Pipeline," *Civil Engineering,* January 1994, pp. 50–52.

- The financing of megaprojects will continue to become more critical, because public funds are increasingly used to develop economic and social infrastructures to help meet the growing needs of population increases, resource shortages, resource exploration and use, and increases in global standards of living.

22.13 OTHER PROJECT MANAGEMENT CHANGES

- Continuing the influence that it has had in the past, project management will continue to contribute to the further modification of traditional management approaches and vertical hierarchies in the organizations of the future.
- Experience and competency in team management and leadership will become major considerations in the selection and promotion of senior managers in the future. The opportunity to gain experience in the management and leadership of teams will continue to broaden as teams grow in use in the organizations of the future.
- The strategic and operational management of technology through projects will become a key pacing factor in the enterprise's ability to offer new and improved products and services supported by innovative organizational processes.
- New products and services will be created at unprecedented rates in the future through project management, with the ability to develop such products and services in effective and efficient ways as a key competitive factor for existing enterprises and for new enterprises that will come forth as the relentless changes foster the need for new products and services for the global marketplace.
- As the use of project management and alternative team management grows, the need for training in the concepts and processes of team management will also accelerate. Universities and colleges will continue to recognize the need for undergraduate and graduate courses and research to advance the state of the art in the theory and practice of project and team management. The developing nations of the world will be particularly anxious to learn more about project management as a way to deal with the awesome changes that are facing them.

22.14 ADDITIONAL CHANGES

Some additional likely changes to project management in the future include:

- Continued growth in the sophistication of software to be used in cost and schedule tracking, and further development of group software, which can be used to conduct project planning and review meetings even though the people are geographically separated.

- Professional organizations representing the project management community will likely continue to grow in membership. Alternative professional organizations in the field of project management will likely emerge.
- The growing extension of project management and alternative team management will continue to change the role of traditional managers—moving away from the traditional "command-and-control" mode to a role that functions as counselors, consultants, coaches, teachers, and trainers.
- Less attention will be given to the "matrix organization" in the literature as future organizations institutionalize the use of the matrix design in the management of projects and other alternative team initiatives.
- There will be a growing, closer link of project planning and strategic planning in the enterprise, as it becomes increasingly clear that project results are the pathway to the organization's future.
- There will be limits to innovative changes in project cost and schedule techniques. The greatest opportunities for the improvement of project management will deal with human and organizational issues. Project monitoring, evaluation, and control will be assumed more by the members of the project team than by formal review and reporting procedures.
- International issues over scarce resources having geopolitical origins will likely not be reduced. The risk of armed conflict over the use of scarce resources such as petroleum, food, water, and critical minerals will be present. Military projects will be undertaken by the stronger nations of the world to contain these conflicts and work toward settlements that will have some compromise in the allocation of the world's scarce resources.
- Project managers—and other managers—will have to become conversant with the means for international competition and the marketing and development of products and services that will be appropriate for diverse markets and varying customers throughout the world.
- The technological advancements in computer technology and information sources such as the Internet as applied to project management will continue to grow.

22.15 A TURNING POINT

We have just turned the corner on a new century, which invites speculation about what our future will be like. Will the theory and practice of project management undergo as much change as we have seen in the last 50 years in this discipline? Will the future of project management be another time of unprecedented fulfillment, or will it be "business as usual" in this remarkable discipline that is still undergoing change in its theoretical foundations, in its practice, and in its application? A more important question is: How do we look at the future of project management? Will project management be the same game with little change in the way it is played? Or will project management become a new game, with new concepts, processes, techniques, and applications?

Project management—and its likely changes—has to be considered from the perspective of the major changes that have impacted our society in the past and in the present. We are in the middle of major transformations cutting across our society, and the outcomes of these transformations are far from certain. Political, economic, social, and technological changes are causing dislocations, and even today, Heraclitus's thought that "all is flux, nothing stays still, nothing endures but change" has relevancy.

Project management—and the use of alternative teams—has influenced and has been influenced by the forces of change in today's world of management. The interdependency of these forces and the use of teams as elements of operational and strategic change in today's organization will likely continue into the future. Traditional organizational structures are likely to undergo continued modification as the use of teams and higher degrees of employee participation are likely to be found in the organizations of the future. Those authors who write the history of management for the late twentieth century and the first couple of decades after the turn of the century are likely to see that project management was one of the principal forces that influenced change in the theory and practice of management. What will be the likely continued influence of project management and alternative teams in the change of management theory and practice in the future?

One prediction about the future of project management can be made with little risk as to its likely outcome: That prediction is that the future promises many changes cutting across all of society—in the entire world. Such political, social, economic, technological, and competitive changes will stimulate an already strong demand for project management. Consequently, the demand for people who are competent to serve on project teams and who are able to manage the enterprise using project management as a key strategy in the management of operational and strategic change will likely continue.

But throughout all of our thinking about the future of project management, we are reminded: "Predicting the future is easy. Getting it right is the hard part."[9]

22.16 TO SUMMARIZE

Some of the major points that have been expressed in this chapter include:

- The future is important to all of us because we will be spending the rest of our lives there.
- The management discipline has experienced considerable change in the last 50 years.
- The project management discipline has emerged with considerable rapidity in the last several decades. During this evolution it has changed in many ways. Changes to this discipline are likely to continue in the future.

[9]Howard Frank, Director, Information Technology, Office of the Defense Advanced Research Projects Agency, Arlington, Va., quoted in Garry H. Anthes, "Predicting the Future," *Computer World,* vol. 30, no. 23, June 3, 1996, p. 70.

- Project management and alternative team management have had major influences in how contemporary organizations are able to deal with change in their environment.
- Although "futurists" and other prognosticators have been able to some degree to predict future trends and events, some of the major changes in the political, social, economic, technological, and competitive areas have come as a surprise to most of us.
- Project management has an interlink with the future through its ability to provide a management system and a philosophy for the management of operational and strategic change in contemporary organizations.
- Many of the changes likely to occur in the future will probably be a simple extrapolation of what has happened in the past. Other changes could be of major significance for which there is no past or present precedent.
- Many changes to be expected in the future will have their roots in the changing world's population, the diversity of the work force, and the demands of people for a greater share of the economic benefits of their societies.
- There will likely be increased emphasis on the interpersonal skills of managers and professionals. This need will be particularly important as project stakeholders become more influential in the practice of project management in their domain.
- Much of the future application of project management will occur in the emerging nations of the world.
- There will be a growing interest in how best to manage megaprojects as emerging nations see the need to change their infrastructures.
- A leading U.S. business magazine, *Fortune,* predicted, "Project management is the wave of the future."
- The increasing interest in project management has been caused in part by more frequent use of nontraditional teams in the management of contemporary organizations.
- Membership in project management professional associations is increasing, particularly in the Project Management Institute, the "flagship" of such professional organizations. This membership increase is likely to continue into the future.
- Attention to the development of the theory and practice of megaprojects will likely gain renewed interest in the future, as emerging nations see the need for such projects in order to provide higher standards of living to their citizenry.
- The ability to lead projects and other team efforts will likely be a principal consideration in the promotability of managers in the future.
- Project management will become a distinct career path in organizations of the future. A demonstrated track record in management of successful projects and nontraditional projects will be a prerequisite for promotion to a senior management position in the enterprise of the future.
- The unanticipated changes of the future in the social, political, economic, technological, and competitive systems will likely bring about some changes in the theory and practice of project management.

But we should always be reminded that the prediction of the future is easy. Getting that prediction right is the hard part. It is even harder to provide the leadership of the enterprise to capitalize on that future.

22.17 ADDITIONAL SOURCES OF INFORMATION

The following additional sources of project management information may be used to complement this chapter's topic material. This material complements and expands on various concepts, practices, and the theory of project management as it relates to areas covered here.

- Elvin Isgrig, "Developing Project Management Skills for the Future," chap. 22 in David I. Cleland (ed.), *Field Guide to Project Management* (New York: Van Nostrand Reinhold, 1997).
- Seth Lubove, "Destroying the Old Hierarchies," and Larry Martin and Paula Green, "Gaining Project Acceptance," in David I. Cleland, Karen M. Bursic, Richard Puerzer, and Alberto Y. Vlasak, *Project Management Casebook,* Project Management Institute (PMI). (First published in *Forbes,* June 3, 1996, pp. 62–70 and *Civil Engineering,* August 1995, pp. 51–53.)
- "Project Management Research at the Turn of the Millennium," *Proceedings,* PMI Research Conference 2000, Paris, France, June 21–24, 2000. This conference was a "first of its kind" in the field of project management. Fifteen invited papers and 30 contributed papers were presented dealing with research initiatives of the past, present, and future. A reading of these proceedings, along with some reflective thinking, should provide valuable insight into what likely changes will occur in the theory and practice of project management in the future.
- Virginia Postrel, *The Future and Its Enemies: The Growing Conflict over Creativity, Enterprise, and Progress* (New York: Simon & Schuster Trade Paperbacks, 1999). The author talks about the future of economic prosperity, technological progress, and cultural innovation—and says that understanding these forces depends upon embracing principles of choice and competition. Postrel readily notes that we cannot manage tomorrow by acting today. This book is intellectually sweeping and reader-friendly. She argues that if we do not meet the future in the proper spirit, we will miss its benefits, or be run over by it. This book will also provoke one into thinking about the future—and how we can better prepare to meet that future. As members of the project community read this book, they should ask themselves the question, What is in this book that can help me to better understand the future of project management, and how to prepare for that future?
- William J. Swanston and William C. Carney, "Institutionalizing Project Management: A Necessity for Project Management to Provide Value and Thrive in the Coming Century," *Proceedings,* 28th Annual Project Management Institute, September 29–October 1, 1997 Seminars/Symposium, Chicago, p. 530. This brief article deals with the subject of the "institutionalization" of project

management—a subject that has received considerable attention in recent times when the future of project management is presented. The authors state that such institutionalization requires a dedicated long-term commitment by both the organization and its project management community. When a project management system is so converted, its value to the organization can have far-reaching implications.

- Jeffrey K. Pinto (ed.), "The Future of Project Management," in *The Project Management Institute Project Management Handbook* (San Francisco: Jossey-Bass, 1998). The author asks the question, "What is the future of project management?" He then admits that we have no more prescient crystal ball than do others. Pinto offers some suppositions based on our experience and the state of the world in which project management is found. He sees a bright future for the expanding role of project management on a worldwide basis.
- N. M. Barnes and S. H. Wearne, "The Future for Project Management," *International Journal of Project Management,* vol. II, issue 3, 1993. The book in which this article appears presents a view of what major project management may be like in the future. An examination of 4th edition chaps. 21–27 presents the demands communities around the world may put upon groups of people who are capable of mounting and completing major projects. Factors likely to be the basis of successful and unsuccessful projects are listed. It is suggested that experience and expertise in project risks will become recognized at higher and higher levels in business.

22.18 DISCUSSION QUESTIONS

1. On the basis of some of the general predictions of changes in the future, what is your prediction of the future of project management? Of team management?
2. Select an industry and a few of the major products and services of that industry. What are some of the likely changes in those products and services caused by genetic changes in the marketplace?
3. What have been some of the major changes in the theory and practice of the management discipline in the past 30 years? How has the project management discipline been impacted by those changes?
4. What has been the cause of the role change of the traditional first-level "boss" or supervisor? What further changes might be anticipated in this role in the future?
5. Discuss the implications of the statement: "The future comes like a funeral gone by." (Edmund Gosse, 1849–1928, *The White Throat.*)
6. What is the important message in the quote: "You can never plan the future by the past?" (Edmund Burke, 1729–1797, letter to a member of the National Assembly, 1791.)
7. From the personal perspective of the reader, what are some of the likely generic trends that may impact your future career and retirement?

8. In this chapter, it was suggested that the root cause of major changes likely to happen in the future could be traced to the probable increases in the world's population. Discuss the rationale behind this statement.
9. What are some of the leading changes likely to impact project management in the future? What forces and factors will cause these changes?
10. What is meant by the statement, "In the last 10 years project management has moved out of its traditional redoubts?"
11. What are some of the likely forces that will increase or decrease membership in the existing project management professional organizations? What is the likelihood of new associations appearing in the field of project/team management?
12. Select a geographic area, such as the Far East, and identify some of the megaprojects that are likely to come forth in that area. Why are these megaprojects coming forth? Be specific!

22.19 USER CHECKLIST

1. Is any attempt made by your organization to develop and implement a rigorous process to identify and predict some of the major future trends likely to impact your company?
2. What are some of the likely specific future market changes that could impact the existing products and services of your organization?
3. Has anyone been designated a focal point in your organization to identify and track the relevant environmental changes that could impact the well-being of your enterprise? If such an individual has been appointed, how well has that person discharged his or her responsibility in this regard?
4. What is the means of translating expected future environmental trends into the strategic planning process of your organization? If such a process exists, how effective is this process?
5. What future trends in global population will likely impact the demand for your company's products and services?
6. The demand for the products and services of some companies is directly impacted by changes in the development of technology by competitors. Is this the case regarding your company? Why or why not?
7. What are some of the likely changes in your enterprise that will affect the manner in which project management is applied to the management of product, service, and organizational process change in your enterprise?
8. Do the "general managers" of your company understand the importance of having an appreciation of project management as a means for managing changes in the enterprise? Why or why not?
9. Are there any areas in your company where the use of project management or alternative team management would likely pay off? Why or why not?

10. What are some of the summary strengths and weaknesses of the practice of project management in your enterprise? If there are such weaknesses, are strategies being developed to rectify those weaknesses?
11. Does a process exist in your company to determine the strengths and weaknesses of competitors and the prediction of the likely trends in competitive performance in the future?
12. What are the major likely trends in project management presented in this chapter that will have the greatest impact on your enterprise?

22.20 PRINCIPLES OF PROJECT MANAGEMENT

1. There is unprecedented change facing all organizations today.
2. Project management has impacted the evolution and practice of the management discipline.
3. The use of project teams, as well as nontraditional teams, has provided a major means for contemporary organizations to deal with change in their competitive environments.
4. Project management will grow in importance throughout the world as a means for dealing with change.
5. The strategic and operational management of technology through projects will be a key pacing factor in the enterprise's ability to offer new products and services.

22.21 PROJECT MANAGEMENT SITUATION— WHAT DOES THE FUTURE HOLD?

Throughout history noted people have thought about the future, and have written or spoken some elegant words about the hereafter. A few of these thoughts are provided below:

"You can never plan the future by the past." (Edmund Burke)

"The best of prophets of the future is the past." (Sir William Henry Maule)

"The Future comes like an unwelcome guest." (Edmund Gosse)

"I believe the future is only the past again, entered through another gate." (Sir Arthur Wing Pinero)

"If we open a quarrel between the past and the present, we shall find that we have lost the future." (Winston Spencer Churchill)

"The enemies of the Future are always the very nicest people." (Christopher Morley)

"The future is something which everyone reaches at the rate of sixty minutes an hour, whatever he does, whoever he is." (CLIVE STAPLES LEWIS)

"The wave of the future is coming and there is no fighting it." (ANNE MORROW [MRS. CHARLES] LINDBERGH)

22.22 STUDENT/READER ASSIGNMENT

Given this sample of words about the future and the predictions about the future of project management offered in this chapter, the students/readers are requested to identify a potential project. This project should have a future life cycle of at least 10 years, and the students/readers should have personal knowledge of the project or be able to find out about the project by perusing the literature. Identify and describe the likely future changes that could impact this project.

INDEX

The *n* after a page number refers to a note.

ABOUT THE AUTHORS

David I. Cleland is a Fellow of the Project Management Institute and has received the Institute's Distinguished Contribution to Project Management award three times. Dr. Cleland has been described as the "Father of Project Management" and has been honored through the establishment of the annual David I. Cleland Excellence in Project Management Literature Award sponsored by the Project Management Institute. He is currently Professor Emeritus in the School of Engineering, University of Pittsburgh, and is the author/editor of 34 books in the fields of project management and engineering management. He resides in Pittsburgh, Pennsylvania.

Lewis R. Ireland is a fellow of the Project Management Institute and has received the Institute's Distinguished Contribution Award, Person of the Year Award, and many certificates of appreciation. He has more than three decades of management experience and has focused intensely on managing projects and supporting organizational change to project-based enterprise for the past 22 years. Dr. Ireland is active in advancing the project management discipline through participation in professional associations. He is a 20-year veteran of the Project Management Institute and served as its President and Chair in 1998. Dr.Ireland currently serves as Director of Research for the American Society for the Advancement of Project Management. He resides in Monument, Colorado.